Radiopharmaceuticals and Brain Pathology Studied with PET and SPECT

Editors

Mirko Diksic, Ph.D.
Associate Professor of Neurosciences
Department of Neurology and Neurosurgery
Director of the Cyclotron-Radiochemistry Unit
Montreal Neurological Institute
McGill University
Montreal, Canada

Richard C. Reba, M.D.
Professor of Radiology and Medicine
Director of the Division of Nuclear Medicine
Department of Radiology
George Washington University Medical Center
Washington, D.C.

CRC Press
Taylor & Francis Group
Boca Raton London New York

CRC Press is an imprint of the
Taylor & Francis Group, an **informa** business

CRC Press
Taylor & Francis Group
6000 Broken Sound Parkway NW, Suite 300
Boca Raton, FL 33487-2742

Reissued 2019 by CRC Press

A Library of Congress record exists under LC control number:

Publisher's Note
The publisher has gone to great lengths to ensure the quality of this reprint but points out that some imperfections in the original copies may be apparent.

Disclaimer
The publisher has made every effort to trace copyright holders and welcomes correspondence from those they have been unable to contact.

ISBN 13: 978-0-367-23253-5 (hbk)
ISBN 13: 978-0-367-23258-0 (pbk)
ISBN 13: 978-0-429-27898-3 (ebk)

Visit the Taylor & Francis Web site at http://www.taylorandfrancis.com and the
CRC Press Web site at http://www.crcpress.com

PREFACE

*"Not only our pleasure, our joy, and our laughter
but also our sorrow, grief and tears
arise from the brain, and the brain alone."*

Hippocrates (c. 460 to c. 370 B.C.)

In the past decade, there has been an almost exponential increase in technological advances in PET and SPECT. These advances, in association with the development of new radiopharmaceuticals, have given us extremely powerful tools for studying the physiology of normal and diseased brain. These methodologies also allow us to study the influence of drugs on metabolism, receptor densities, and functional organization of the brain.

PET, SPECT, the synthesis of radiopharmaceuticals, and the application of these methodologies to normal and diseased brain are reviewed by experts in the field. We hope this volume will assist scientists in selecting and developing new methods to study *in vivo* brain biochemistry/physiology in normal and diseased states. The overview of recent technological advances and their application to the study of *in vivo* brain functional biochemistry with PET and SPECT should give readers a good grasp of both the potentials of the techniques and the real limitations imposed by radioactive decay, specific radioactivity of the radiopharmaceuticals, scanner design, and heterogeneity of human brain diseases.

We believe this book will be of value to basic neuroscientists, neurologists, neurosurgeons, psychiatrists, radiologists, psychologists, neuro-oncologists, and individuals at various stages of training in these disciplines. It should help them define the relationship between biochemical data and finding/discovering the pathophysiological meaning of these measurements. The book will be especially valuable to those not yet familiar with setting up *in vivo* research projects using these technologies as it discusses the development of radiopharmaceuticals, scanner requirements, biological modeling, the conversion of brain radioactivity distribution into biological parameters (e.g., glucose utilization, receptor densities), and the interpretation of brain pathology as measured with these methods. Readers already familiar with these methodologies will, we hope, be interested in discussions of the usefulness of PET and SPECT in studying pathology in living brain by relatively noninvasive means.

We express our sincere thanks to all contributing authors. The extremely valuable help of Ms. Carolyn Elliot, who performed the secretarial work associated with this book for one of us (M.D.) on her own time, is greatly appreciated. Our thanks are also extended to everyone at CRC Press, especially Ms. Carolyn C. Lea.

Mirko Diksic
Richard C. Reba

THE EDITORS

Mirko Diksic, Ph.D., is Associate Professor of Neurosciences, Department of Neurology and Neurosurgery, and Director, Cyclotron-Radiochemistry Unit, Montreal Neurological Institute, both of McGill University, Montreal, Quebec, Canada.

Dr. Diksic received his B.Sc. in Chemical Engineering from the University of Zagreb, Zagreb, Croatia, Yugoslavia, in 1966. He obtained his M.Sc. and Ph.D. degrees in 1968 and 1970 from the same university while working at the Rudjer Boskovic Institute in Zagreb. After post-doctoral work in the chemistry departments of McGill University and the University of Florida, Gainesville, he was appointed Research Associate in the Department of Chemistry at McGill. Dr. Diksic joined the Montreal Neurological Institute in 1979 as Assistant Professor and became Director of the Cyclotron-Radiochemistry Unit in 1984 and Associate Professor in 1986.

Dr. Diksic is a member of the Croatian Chemical Society, the American Chemical Society, the Society of Nuclear Medicine, the Society for Neuroscience, the Radiopharmaceutical Science Council, the Brain Imaging Science Council, the Canadian Association of Radiopharmaceutical Scientists, the International Society of Cerebral Blood Flow and Metabolism, the New York Academy of Sciences, the Canadian Association for Neuroscience, and the International Isotope Society. In 1988 he was a member of the Forbeck Forum on brain tumor treatment.

Dr. Diksic has received research grants from the Medical Research Council of Canada, the National Institutes of Health, and the American Health Assistance Foundation. He has published more than 100 papers in peer reviewed journals and has given approximately 30 invited lectures. His current major research interests include studies of the brain's *in vivo* metabolism by PET and autoradiography and the development of new tracers for these studies.

Richard C. Reba, M.D., is Professor of Radiology and Medicine and Director of the Division of Nuclear Medicine in the Department of Radiology at the George Washington University Medical Center, Washington, D.C.

Dr. Reba attended Loyola College of Baltimore and obtained his M.D. degree from the University of Maryland, College of Physicians and Surgeons, in 1957. After a clinical residency in Internal Medicine at the University of Maryland in Baltimore, Dr. Reba was a Fellow in Nuclear Medicine at the Johns Hopkins Medical Institutions in Baltimore in 1961 and 1962. He was appointed Senior Investigator and subsequently Chief, Department of Isotope Metabolism, in the Division of Nuclear Medicine at the Walter Reed Army Institute of Research and from 1966 to 1970 was on the faculty and staff of the Johns Hopkins University and Hospital, where he was Associate Professor of Radiology, Radiologic Science and Internal Medicine, and Chief, Clinical Section, Division of Nuclear Medicine. From 1970 to 1976 he was Chairman, Department of Nuclear Medicine, and Senior Attending Physician in the Departments of Nuclear Medicine and Internal Medicine at the Washington Hospital Center where Dr. Reba was in private practice from 1970 to 1976.

In 1976 Dr. Reba returned to academic medicine when he accepted an appointment at the George Washington University Medical Center.

Dr. Reba has served as a consultant to numerous organizations, including the Veterans Administration, the Federal Aviation Agency, the Department of Defense, the International Atomic Energy Agency, the Department of Energy, and the National Institutes of Health. He is a member of the Society of Nuclear Medicine, the American Association for the Advancement of Science, and the American Medical Association. He has been elected a Counselor of the Eastern Section, American Federation of Clinical Research and a Fellow in the American College of Physicians and the American College of Nuclear Physicians, where he has also served as Speaker of the House of Delegates and President.

Dr. Reba has lectured extensively throughout the U.S. and abroad. He is the author or coauthor of more than 200 papers and is an editor of 3 books. He has been the recipient of many research grants from the National Institutes of Health and the Department of Energy. His current major research interests relate to the development of radioligands for quantitative *in vivo* neuroreceptor imaging and their application to human disease processes and to developing improved techniques for radiolabeling of monoclonal antibodies.

CONTRIBUTORS

Pascale Abadie, M.D.
INSERM U. 320
Cyceron
Caen, France

Abass Alavi, M.D.
Department of Radiology
Hospital of the University of
  Pennsylvania
Philadelphia, Pennsylvania

Nathaniel M. Alpert, Ph.D.
Department of Radiology
Massachusetts General Hospital
Boston, Massachusetts

Nancy C. Andreasen, M.D., Ph.D.
Department of Psychiatry
The University of Iowa
Iowa City, Iowa

J.-C. Baron, M.D.
INSERM U. 320
Cyceron
Caen, France

Samuel F. Berkovic, M.D., F.R.A.C.P.
Department of Neurology
Austin Hospital
Melbourne, Australia

Karen Faith Berman, M.D.
Clinical Brain Disorders Branch
Intramural Research Program
National Institute of Mental Health
Neuroscience Center at St. Elizabeth's
Washington, D.C.

J. A. Correia, Ph.D.
Department of Radiology
Massachusetts General Hospital
Boston, Massachusetts

Mirko Diksic, Ph.D.
Department of Neurology and
  Neurosurgery
and Montreal Neurological Institute
McGill University
Montreal, Canada

W. C. Eckelman, Ph.D.
Diagnostic Drug Discovery
Bristol Myers Squibb Pharmaceutical
  Research Institute
New Brunswick, New Jersey

**William Feindel, O. C., M.D.C.M.,
  F.R.C.S.(C.)**
Department of Neurology and
  Neurosurgery
and Montreal Neurological Institute
McGill University
Montreal, Canada

Ludwig E. Feinendegen, M.D.
Institute of Medicine
Jülich Research Center
Jülich, FRG

Joanna S. Fowler, Ph.D.
Department of Chemistry
Brookhaven National Laboratory
Upton, New York

Albert Gjedde, M.D., Ph.D.
Department of Neurology and
  Neurosurgery
and Montreal Neurological Institute
McGill University
Montreal, Canada

Mark Guttman, M.D., F.R.C.P.C.
Department of Neurology and
  Neurosurgery
and Montreal Neurological Institute
McGill University
Montreal, Canada

W.-D. Heiss, M.D.
Max Planck Institute for Neurological
  Research and Clinic for Neurology
University of Köln
Köln, FRG

Karl Herholz, M.D.
Max Planck Institute for Neurological
  Research and Clinic for Neurology
University of Köln
Köln, FRG

Hans Herzog, D.Sc.
Institute of Medicine
Jülich Research Center
Jülich, FRG

B. Leonard Holman, M.D.
Department of Radiology
Brigham and Women's Hospital
Harvard Medical School
Boston, Massachusetts

Ronald J. Jaszczak, Ph.D.
Department of Radiology
Duke University Medical Center
Durham, North Carolina

Iwao Kanno, Ph.D.
Department of Radiology and Nuclear
  Medicine
Research Institute of Brain and Blood
  Vessels
Akita City, Japan

Hiroto Kuwabara, M.D., Ph.D.
Department of Neurology and
  Neurosurgery
and Montreal Neurological Institute
McGill University
Montreal, Canada

Gabriel Léger
Department of Neurology and
  Neurosurgery
and Montreal Neurological Institute
McGill University
Montreal, Canada

Ernst Meyer, Ph.D.
Department of Neurology and
  Neurosurgery
and Montreal Neurological Institute
McGill University
Montreal, Canada

Geerd-J. Meyer, Ph.D.
Institut für Nuklearmedizin und spezielle
  Biophysik
Medizinische Hochschule Hannover
Hannover, Germany

Daniel S. O'Leary, Ph.D.
Department of Psychiatry
The University of Iowa
Iowa City, Iowa

Jörg J. Pahl, M.D., F.C.P.(S.A.)
Department of Psychiatry
The University of Iowa
Iowa City, Iowa

William J. Powers, M.D.
Departments of Neurology and Radiology
Washington University School of
  Medicine
and Department of Neurology
Jewish Hospital at Washington University
  Medical Center
St. Louis, Missouri

Richard C. Reba, M.D.
Department of Radiology
George Washington University Medical
  Center
Washington, D.C.

Karim Rezai, M.D.
Department of Radiology
The University of Iowa
Iowa City, Iowa

Christopher C. Rowe, M.D.,
  F.R.A.C.P.
Departments of Neurology and Nuclear
  Medicine
Austin Hospital
Melbourne, Australia

Otmar Schober, M.D., Ph.D.
Nuclear Medicine Clinic
Westfälischen Wilhelms University
Münster, Germany

Michio Senda, M.D.
Department of Radiology
Massachusetts General Hospital
Boston, Massachusetts

Elaine Souder, Ph.D.
Department of Radiology
Hospital of the University of
  Pennsylvania
Philadelphia, Pennsylvania

Henry N. Wagner, Jr., M.D.
Departments of Medicine, Radiology, and
 Radiation Health Sciences
The Johns Hopkins Medical Institutions
Baltimore, Maryland

Daniel R. Weinberger, M.D.
Clinical Brain Disorders Branch
Intramural Research Program
National Institute of Mental Health
Neuroscience Center at St. Elizabeth's
Washington, D.C.

David W. Weiss, M.D.
Department of Radiology
Hospital of the University of
 Pennsylvania
Philadelphia, Pennsylvania

Lennart Widén, M.D., Ph.D.
Department of Clinical Neurophysiology
 of the Karolinska Institute
Karolinska Hospital
Stockholm, Sweden

Alfred P. Wolf, Ph.D.
Department of Chemistry
Brookhaven National Laboratory
Upton, New York

TABLE OF CONTENTS

Chapter 1

HISTORICAL BACKGROUND OF PET

William Feindel

TABLE OF CONTENTS

I. ORIGIN OF THE POSITRON

At the turn of this century, Ernest Rutherford and Frederick Soddy (who introduced the term "isotope") reported from McGill University their studies on the nature and cause of radioactivity.[1] Their work on exponential decay and radioactive half-life were fundamental for the eventual medical applications of radioisotopes, including the exciting developments since 1951 in positron emission tomography.

The origin of the idea of the positron and the connection between the positron of physics and the application to positron emission tomography (PET) imaging were reviewed in 1982 by Bell,[2] then Rutherford Professor of Physics, McGill University. Carl Anderson[3] in 1932, announced the observation of "a positively charged particle having a mass comparable with that of an electron." The first use of the word "positron" was in his reports in 1933 and 1934.[4-6]

The origin of the idea of the positron, and hence of antimatter in general, according to Bell's account, can be attributed to Dirac, who rectified some of the flaws in the new theories of quantum mechanics in the mid-1920s. However, his speculations were incomplete; the topic was argued by some of the biggest names in physics research, including Oppenheimer, Fermi, Jeans, and Millikan. Studies of scattering of high energy gamma rays by heavy elements led Anderson, using a cloud chamber operated in a magnetic field, to become convinced that he was demonstrating positive electrons. The findings were confirmed by Blackett and Occhialini, who verified that gamma rays could produce positive/negative electron pairs. Irène Curie and Frédéric Joliot, also working with cloud chambers, further confirmed these results; by bombarding aluminum foil with alpha-particles, they produced the first artificial radioactive substance, ^{30}P, with a half-life of 2.5 min. This was the beginning of the production of a wide range of positron emitting atoms some of which we now use in PET.

The origin of the term positron was also noted by Bell:[2]

> *Originally "electron" (Greek; amber) was proposed as a name for the amount of electric charge carried by the particle, as opposed to a name for the particle itself (the present usage). This subtle change of meaning seems to have happened spontaneously. In the word electron, the stem is "electr-" and the ending is "-on". If we want separate names for the positive and negative electrons, we arrive at "negaton" and "positon". It is very easy, though, to assume that the word electron consists of "elec-" and "-tron", thus producing "negatron" and "positron". (In fact the ending "-tron" implies machine or device, e.g., cyclotron.) As expected, the Europeans, especially the British, favored negaton and positon. Nowadays, the American usage is much the more common, although negaton and positon are still seen. Often we simply use electron (implying negatron) and positron.*

II. THE BASIS OF PET

When an electron and a positron annihilate each other, it turns out that the overwhelmingly probable outcome is for the energy to be carried off by two oppositely directed electromagnetic quanta, indistinguishable from hard X-rays or gamma rays. They are exactly equal in energy and exactly opposite in direction. Each one has precisely the energy equivalent to the mass of the electron, $m_0c^2 = 511$ keV. This characteristic "annihilation radiation" is highly correlated near 180°.

Swift positrons from a radioactive source shot into a sample of ordinary matter will usually slow down to a very low energy without annihilating. In so doing, each positron travels perhaps a millimeter or so and then finds itself wandering slowly in a sea of electrons, mostly the outer electrons of the atoms of the sample. (It is repelled if it tries to enter an atom deeply.) Typically within less than a billionth of a second, this electron hole (the positron) is filled (annihilated) by an atomic electron, and the two "annihilation quanta" proceed simultaneously outward at 180°.

If two radiation detectors are situated on opposite sides of the above sample along a straight line drawn through the sample, some of the pairs of annihilation quanta will be registered as what the physicists call coincidences (simultaneous events). Conversely, if the two radiation detectors observe such a coincidence, it proves that a positron has been annihilated somewhere along the line joining them and that the atom which emitted that positron was located somewhere within, say, 1 mm of that line. This is the basis of PET; the 180° correlation of the annihilation quanta establishes a straight-line path in space, permitting a technique of tomography mathematically similar to the earlier CAT that was based on narrow beams of X-rays.

Bell[2] concludes:

> The number of scientific laws and concepts involved in the successful use of PET is remarkable. The positron arose as the first known instance of antimatter from a highly abstract development in quantum mechanics. Within this theory, the positron was explicitly a creature of the theory of relativity. The energy of the annihilation quanta is a direct demonstration of Einstein's $E = mc^2$. The angular correlation near 180° of the annihilation quanta is a direct proof of the conservation of energy and momentum, both basic to physics. The detailed angular distributions near 180° and the lifetimes of the positrons in the sample both involve the detailed picture of the solid or liquid state.

III. EARLY PET STUDIES

The use of positron emitting radioisotopes in medicine was first proposed in 1951 for the localization of brain tumors by Wrenn et al.[7] Their brief paper outlined the main principles of positron scanning. This included the use of sodium iodide crystal detectors in place of Geiger counters, collimation to sharpen the field of view of the detectors, and coincidence counting to detect simultaneously the two photons emitted back to back and thus to relocate the point of positron-electron annihilation in the brain. Shortly after that, Sweet and Brownell[8] applied these principles with the use of ^{74}As to rectilinear scanning in patients with brain tumors. Their method indicated lateralization and localization of meningiomas, malignant gliomas, and metastatic tumors because of a preferential uptake of the radioactive tracers in tumors as compared to brain. Low-grade gliomas, however, were not detected because of insufficient entry of the tracer into the tumor.

Shy[9] and associates later devised an ingenious device utilizing gold collimators for external detection of tumors by positron emission.

Partly because of difficulties in the availability of ^{74}As and its associated radiation dosage and partly because of improved developments in the use of gamma emitters for brain scanning, these early efforts at positron emission scanning were superseded by techniques depending upon gamma-emitting tracers such as albumin labeled with ^{131}I, ^{203}Hg and ^{197}Hg, and ^{99m}Tc. These were utilized in many cases by manual serial scanning of multiple areas of the head and plotting of the radioactive count-rates (e.g., see Chou and French[10] and Planiol[11]). Many of the factors affecting entry of radioactive tracers into tumors in relation to the blood-brain and blood-tumor barriers that were defined in regard to a gamma-emitting isotopes (see Bakay[12] and Moore[13]) are also applicable to the use of positron emission indicators.

A system of contour scanning which was sufficiently quantitative to show the higher differential uptake between the tumor and brain of mercury-compounds as compared to radioactive serum albumin was designed and constructed by a team of physicists headed by Johns (Reid and Johns[14]) to neurosurgical specifications (Feindel[15]). This allowed for a higher percentage of detection of supratentorial tumors other than low-grade gliomas. The design of the scanning patterns avoided the foreshortening of the curved surface of the head that was unavoidable with the rectilinear scanning. An automatic printout displayed the differential uptake between the two sides of the head in a reproducible pattern.[15-18] In the pre-computer era, this instrument was of historical interest as one of the devices applied over two decades for diagnostic isotope scanning, before the more versatile and highly computerized PET systems were developed.

IV. ADVANCES IN TECHNOLOGY

A. POSITRON CAMERAS

In 1966 Yamamoto and Robertson[19] published a Brookhaven Laboratory report on the first physiological application of positron emitting ^{79}Kr as a means of measuring cross-sectional cerebral blood flow. From 1970 onward, several centers focused on the design and development of positron cameras. Thus, Brownell et al.[20] in 1973 introduced the multicrystal camera. In 1975, Ter-Pogossian et al.[21] developed a device based on a hexagonal array of sodium iodide crystals.

Positron tomography imaging in patients was achieved at the Montreal Neurological Institute in 1975 with the revision of a positron camera based on a circular array of 32 sodium iodide detectors, on loan from the Brookhaven National Laboratory. This updated Brookhaven instrument, using ^{68}Ga derived from a germanium generator, yielded the first PET images of brain tumors and infarcts (Yamamoto et al.[22]). After Cho and associates[23] pointed out theoretical advantages in detection of gamma emission by bismuth germanate (BGO), Thompson et al.[24] developed an imaging system using BGO crystals. Their results were reported at the 1st International Symposium of PET[25] held in Montreal in June 1978. Summarized also at this symposium were significant new developments in positron cameras, positron radiopharmacology, the mathematics of three-dimensional display, the potential application of the multiwire proportional chamber camera, and early results on measuring glucose utilization in the brain by fluorine-labeled deoxyglucose.

The characteristics of the MNI BGO camera detailed by Thompson et al.[26,27] at this same symposium included higher resolution and more efficient gamma capture than previous cameras. The full circle design of the camera using 64 BGO-crystal detectors together with the computer program was adapted by Atomic Energy of Canada to an upgraded copy (THERASCAN). The MNI camera was the prototype for PET cameras based on this full-circle array of BGO detectors.

The relative quantitative uptake of ^{68}Ga in brain tumors, in strokes, and in other organic brain lesions gave images that provided early kinetic PET studies of tracer entry through the blood-brain barrier. ^{77}Kr produced uniquely from the McGill Synchrocyclotron by bombardment of NaBr at 50 MeV and 1.5 μA using techniques developed by Diksic and Yaffe[28] and detailed for positron application by Mark[29] was applied to a positron method for the measurement of perfusion blood flow in the human brain based upon the original Fick principle.[30,31] These two approaches for identifying dynamic changes in the blood brain barrier and for measuring focal blood flow were invaluable in establishing with PET a data bank on neurological disorders and for developing further technology and methodology. Since then, miniaturization of electronic components has provided an increase in the number of detectors per ring and in the number of rings to provide a simultaneous anatomical survey of the whole brain.

B. MINI-CYCLOTRON

The rate of progress in PET over the following 5 years was reviewed in 1983.[40] Several significant advances were noted at that time. One of these was the innovative design of a mini-cyclotron, initiated by the Japanese Steel Works in 1980. This compact unit, with automated controls and the capability of producing positron emission nuclides of the "metabolic" elements ^{15}O, ^{13}N, and ^{11}C facilitated a wide range of studies into the biochemical processes in living tissues and organs and especially the brain. In addition, production of the halide ^{18}F enabled the radiochemical syntheses and labeling of many natural substances such as carbohydrates, amino acids, and lipids as well as a large number of drugs.

V. PET STUDIES ON BRAIN TUMORS

A. RECURRENT GLIOMAS

Many of these points relating to the application of PET may be exemplified by PET studies from various centers of brain tumors. Following the early work with ^{68}Ga from many centers to apply positron tomography in brain tumors, it became possible to study the kinetics of uptake in various compartments of gliomas using the better resolution of the BGO camera.[32-34] The radioactive labeling of the commonly used nitrosourea antitumor drug BCNU, by both ^{11}C and ^{13}N, opened up a new avenue for the evaluation by PET of the entry, concentration, distribution, duration, and breakdown of this antitumor drug in gliomas compared to normal brain.[35-39]

The improved methodology for measuring glucose metabolism, blood flow, and oxygen extraction by PET initially offered the promise of identifying the histological types and grades of malignancy by these metabolic profiles in human gliomas.[41,42] Studies by DiChiro and associates[43,44] suggested a correlation of high glucose turnover with malignancy in recurrent gliomas of different types. However, others[45-47] published reports which demonstrated the complexity of the pathophysiology of these tumors in relation to the metabolism of the surrounding white matter and cortex, the difficulty of obtaining quantitative results in malignant gliomas with an inherent heterogenicity not only of tumor cell types, but as well, variations in vascularity, and the inclusion of necrosis, thrombosis, and hemorrhage of variable degrees from one glioma to another.[48] Thus, the earlier suggestion by DiChiro[49,50] of the matching of the deoxyglucose activity and the grade of malignancy of cerebral gliomas became less clear as studies were reported from other centers. The intricate biochemical changes in gliomas as well as those instituted by the presence of gliomas in surrounding brain were evidently extremely complex as more parameters, such as oxygen extraction, volume, blood flow, and acidity, were used to define a broader metabolic profile.[51-55]

B. PRETREATED GLIOMAS

To reduce somewhat these heterogenous factors that were compounded in recurrent gliomas already treated surgically and by radiation and/or chemotherapy, analysis by PET was carried out in a series of patients with untreated cerebral gliomas which were later proven by biopsy. Measures of cerebral glucose and oxygen metabolism, oxygen extraction, blood flow, and blood volume as well as pH values provided a broad indication of metabolism. Rates of glucose utilization were variable and did not correlate with tumor size or grade. Tumor pH, however, differed significantly from that of contralateral brain in the same patient, with consistent alkalotic values being obtained. The evidence for relating degrees of malignancy with (specific) metabolic changes was critically reviewed by Tyler et al.[56] The limits of PET and matching changes in the different vascular and chemical parameters were pointed out both in regards to methodology and to the heterogeneous nature of gliomas. Both the average and the maximum glucose utilization values in untreated gliomas were lower than those reported from a combination of patients with both newly diagnosed and recurrent gliomas. The depression of metabolism in the brain distant from the tumor has been demonstrated by several studies[57,58] and may be related in part to transneuronal interaction as suggested by Timperley.[59] The alkalinity of tumor tissue is in keeping with the more exact demonstration of this by magnetic resonance spectroscopy as reported by Arnold et al.[60] The common characteristics of untreated gliomas were lower oxygen extraction and oxygen utilization and increased blood volume and pH, whereas blood flow and glucose utilization were variable within these tumors. These findings suggest intrinsic metabolic differences between untreated tumors and those occurring after therapy in which high glucose utilization values have been reported. Studies matching the more detailed tissue imaging and chemical profiles obtained by PET can be expected to resolve some of the present discrepancies in the interpretation of the metabolic results obtained by different PET groups in human gliomas. In particular, analysis by magnetic resonance spectroscopy with ^{31}P and proton spectra[61,62] can be expected to elucidate the chemical shift in gliomas and in surrounding normal brain and to supplement the quantitation of chemical measurements obtained by PET. This avenue of investigation by identifying the metabolic profiles of gliomas will undoubtedly lead to further understanding of the growth characteristics of gliomas.

VI. SCOPE AND LIMITATIONS OF PET

Continuing improvements in methodology, camera design, spatial resolution, and imaging quality coupled with innovative radiochemical techniques have led to greatly improved results in PET research. Most of all, more sophisticated quantitation and mathematical formulation for exact kinetic studies have been developed from a number of PET centers.[63,64]

The value of PET imaging has undergone a significant advance by the advent of magnetic resonance imaging that provides matching topographic templates to improve the anatomical localization of the PET studies. Together with higher resolution cameras, this has made possible more precise investigation into localization of metabolic changes and particularly into focal cerebral distribution relating to receptor pharmacology. In a critique of the scope and limitations of PET, Budinger[65] concluded five major aspects of the problem of quantitation demanded consideration.

1. Quantitation of the input function (tracer concentration in the blood as a function of time)
2. Resolution of the PET sampling system
3. Models for metabolic rates and rate constants
4. Nonspecific uptake of tracers
5. The chemical form of the label when imaged

As we noted in 1978 at the 1st International Symposium on PET,[25] "The great attraction of positron imaging is that it can show us not only how the brain looks but how it works. We can reasonably expect that this technique will yield information to develop new methods of treatment for some of the many unsolved neurological disorders that affect the human brain and mind." The wide ranging investigations now under way worldwide in some 70 PET units attest to the exciting application of this dynamic method.

ACKNOWLEDGMENTS

I am grateful to Dr. Robert E. Bell, Rutherford Professor of Physics, McGill University, for providing me with notes on the origin of the positron. This review was done as part of the Neuro-History Project, Montreal Neurological Institute and Hospital, and McGill University; the manuscript was processed by Mrs. Irene Elce, and the work was supported by a grant from the Donner Canadian Foundation.

REFERENCES

1. **Rutherford, E. and Soddy, F.,** The cause and nature of radioactivity, *Philos. Mag.,* 6(iv), 370, 1902.
2. **Bell, R. E.,** The positron and positron emission tomography, personal communication, 1982.
3. **Anderson, C.,** The apparent existence of easily deflective positives, *Science,* 76, 238, 1932.
4. **Anderson, C.,** Free positive electrons resulting from the impact upon atomic nuclei of the photons from ThC, *Science,* 77, 432, 1977.
5. **Anderson, C.,** The positive electron, *Phys. Rev.,* 43, 491, 1933.
6. **Anderson, C. and Neddermeyer, S. H.,** Positrons from gamma rays, *Phys. Rev.,* 43, 1034, 1934.
7. **Wrenn, F. R., Jr., Good, M. L., and Handler, P.,** Use of positron-emitting radioisotopes for localization of brain tumors, *Science,* 113, 525, 1951.
8. **Sweet, W. H. and Brownell, G. L.,** Localization of intracranial lesions by scanning with positron-emitting arsenic, *J. Am. Med. Assoc.,* 157, 1183, 1955.
9. **Shy, G. M., Bradley, R. B., and Matthews, W. B.,** *External Collimation Detection of Intracranial Neoplasia with Unstable Nuclides,* E. & S. Livingston, Edinburgh, 1958.
10. **Chou, S. N. and French, L. A.,** Graphic interpretation of isotope localization of intracranial lesions, *J. Neurosurg.,* 14, 421, 1957.
11. **Planiol, T.,** *Diagnostic des Lésions Intra-Craniennes par les Radio-Isotopes (Gammaencéphalographie),* Masson & Cie, Paris, 1959.
12. **Bakay, L.,** *The Blood-Brain Barrier with Special Regard to the Use of Radioactive Isotopes,* Charles C Thomas, Springfield, IL, 1956.
13. **Moore, G. E.,** *Diagnosis and Localization of Brain Tumors. A Clinical and Experimental Study Employing Fluorescent and Radioactive Tracer Methods,* Charles C Thomas, Springfield, IL, 1953.
14. **Reid, W. B. and Johns, H. E.,** An automatic brain scanner, *Int. J. Appl. Radiat. Isot.,* 3, 1, 1958.
15. **Feindel, W., Stratford, J., Cowan, G. A. B., and Fedoruk, S.,** Radioactive encephalography: automatic brain scanning using radioactive iodinated albumin, *J. Neurol. Neurosurg. Psychiatry,* 22, 342, 1959.
16. **Cowan, G. A. B., Fedoruk, S., Feindel, W., and Stratford, J. G.,** Localization of intracranial lesions using radioactive iodinated human serum albumin and an automatic brain scanner, *Can. J. Radiol.,* 11, 15, 1960.
17. **Feindel, W., Rovit, R. L., and Stephens-Newsham, L.,** Localization of intracranial vascular lesions by radioactive isotopes and automatic contour brain scanner, *J. Neurosurg.,* 18, 811, 1961.
18. **Feindel, W.,** Detection of intracranial lesions by contour brain scanning with radioisotopes, *Postgrad. Med.,* 31, 15, 1962.
19. **Yamamoto, Y. L. and Robertson, J. S.,** Study of Quantitative Assessment of Section Micro-Regional Cerebral Blood Flow in Man by Multiple Positron Detecting System Using Krypton-79, BNL Med. Dept. Circ. No. 28, Brookhaven National Laboratory, Islip, NY, 1966.
20. **Brownell, G. L. and Burnham, C. A.,** MGH positron camera, in *Tomographic Imaging in Nuclear Medicine,* Freedman, G. S., Ed., Society of Nuclear Medicine, New York, 1975, 154.
21. **Ter-Pogossian, M. M., Phelps, M. E., and Hoffman, E. J.,** A positron-emission transaxial tomograph for nuclear imaging (PETT), *Radiology,* 114, 89, 1975.
22. **Yamamoto, Y. L., Thompson, C. J., Meyer, E., Robertson, J. S., and Feindel, W.,** Dynamic positron emission tomography for study of cerebral hemodynamics in a cross section of the head using positron-emitting ^{68}Ga-EDTA and ^{77}Kr, *J. Comput. Assist. Tomogr.,* 1, 43, 1977.
23. **Cho, Z. H. and Farukhi, M. R.,** New bismuth germanate scintillation crystal — a potential detector for the positron camera applications, *J. Nucl. Med.,* 18, 840, 1977.
24. **Thompson, C. J., Yamamoto, Y. L., and Meyer, E.,** A positron-imaging system for measurement of regional cerebral blood flow, *Proc. Soc. Photo Opt. Inst. Eng.,* 96, 263, 1976.
25. **Feindel, W. and Yamamoto, Y. L.,** Physiological tomography by positrons: introduction and historical note, *J. Comput. Assist. Tomogr.,* 2, 637, 1978.
26. **Thompson, C. J., Yamamoto, Y. L., and Meyer, E.,** Positome II: a high efficiency PET device for dynamic studies, *J. Comput. Assist. Tomogr.,* 2, 637, 1978.
27. **Meyer, E., Yamamoto, Y. L., and Thompson, C. J.,** Confidence limits for topographical cerebral blood flow values obtained by ^{77}Kr positron emission tomography, *J. Comput-Assist. Tomogr.,* 2, 662, 1978.

28. **Diksic, M. and Yaffe, L.,** Production of carrier-free [77]Kr and [123]Xe by proton irradiation, *J. Comput. Assist. Tomogr.,* 2, 640, 1978.

29. **Mark, S. K.,** Cyclotron production of positron emitting nuclides, *J. Comput. Assist. Tomogr.,* 2, 638, 1978.

30. **Yamamoto, Y. L., Little, J., Meyer, E., Thompson, C., Ethier, R., and Feindel, W.,** Evaluation of Krypton[77] positron emission tomography studies in strokes, *J. Comput. Assist. Tomogr.,* 2, 663, 1978.

31. **Yamamoto, Y. L., Thompson, C. J., Meyer, E., Little, J., and Feindel, W.,** Positron emission tomography: a new method for examinations of the circulation and metabolism of the brain in man, in *Advances in Neurosurgery,* Vol. 7, Marguth, F., Brock, M., Kazner, E., Klinger, M., and Schmiedek, P., Eds., Springer-Verlag, Berlin, 1979, 3.

32. **Nukui, H., Yamamoto, Y. L., Thompson, C. J., and Feindel, W.,** Positron emission tomography with positome II: with special reference to ^{68}Ga-EDTA positron emission tomography in cases with brain tumor, *Neurol. Med. Chir.,* 19, 941, 1979.

33. **Feindel, W., Yamamoto, L., Arita, N., and Villemure, J.-G.,** Positron emission tomography for detection and dynamic analysis of brain tumors, *Proc. Am. Acad. Neurol. Surg.,* November 1981.

34. **Yamamoto, Y. L., Diksic, M., Meyer, E., Thompson, C., and Feindel, W.,** Positron emission tomography (PET) for regional cerebral blood flow (rCBF) and brain tumor kinetic studies, in *MARIA Design Symposium IV: Positron Emission Tomography,* Filipow and Menon, Eds., University of Alberta, Edmonton, 1982, 132.

35. **Diksic, M., Yamamoto, Y. L., and Feindel, W.,** Recent developments of tracers for BCNU pharmacokinetics and regional cerebral blood flow (rCBF) measurements, in *MARIA Design Symposium IV: Positron Emission Tomography,* Filipow and Menon, Eds., University of Alberta, Edmonton, 1982, 163.

36. **Diksic, M., Farrokhzad, S., Yamamoto, L., and Feindel, W.,** Synthesis of ''no-carrier-added'' 1,3-bis-(2-chloroethyl) nitrosourea (BCNU), *J. Nucl. Med.,* 23, 895, 1982.

37. **Diksic, M., Sako, K., Feindel, W., Kato, A., and Yamamoto, Y. L.,** Pharmacokinetics of positron-labeled 1,3-bis(2-chloroethyl) nitrosourea in human brain tumors using positron emission tomography, *Cancer Res.,* 44, 3120, 1984.

38. **Diksic, M., Yamamoto, Y. L., and Feindel, W.,** An on-line synthesis of ^{15}O-N$_2$O: new blood flow tracer for PET imaging, *J. Nucl. Med.,* 24, 603, 1983.

39. **Feindel, W., Yamamoto, L., Gotman, J., Gloor, P., Meyer, E., and Thompson, C.,** Correlation of focal cerebral blood flow and EEG abnormalities in epilepsy using positron emission tomography, 12th Int. Epilepsy Congr. September 17-21, 1981, Kyoto, Japan, *Ann. RCPSC,* 15, 321, 1982.

40. **Feindel, W.,** Progress review of cyclotron medicine in neurology, in *Proc. 16th Japan Conf. on Radiation and Radioisotopes,* 1983, 173.

41. **Yamamoto, Y. L., Diksic, M., Sako, K., Arita, N., Feindel, W., and Thompson, C. J.,** Pharmacokinetic and metabolic studies in the human malignant glioma, in *Radionuclide Imaging of the Brain,* Vol. 5, Magistretti, P. L., Ed., Raven Press, New York, 1983, 327.

42. **Yamamoto, Y. L., Meyer, E., Menon, D., Diksic, M., Thompson, C. Matsunaga, M., Shibasaki, T., Sako, K., and Feindel, W.,** Dynamic positron emission tomography for measurement of regional cerebral blood flow using [77]Kr and ^{15}O-labeled tracers, in *Functional Radionuclide Imaging of the Brain,* Vol. 5, Magistretti, P. L., Ed., Raven Press, New York, 1983, 269.

43. **DiChiro, G., DeLa Paz, R. L., Brooks, R. A. et al.,** Glucose utilization of cerebral gliomas measured by ^{18}F fluorodeoxyglucose and positron emission tomography, *Neurology,* 32, 1232, 1982.

44. **Patronas, N. J., DiChiro, G., Kafta, C. et al.,** Prediction of survival in gliomas patients by means of positron emission tomography, *J. Neurosurg.,* 63, 816, 1985.

45. **Ito, M., Lammertsma, A. A., Wise, R. S. J. et al.,** Measurement of regional cerebral blood flow and oxygen utilization in patients with cerebral tumors using [15]O and positron emission tomography: analytical techiques and preliminary results, *Neuroradiology,* 23, 63, 1982.

46. **Lammertsma, A. A., Wise, R. S. J., Gibb, J. et al.,** The pathophysiology of human cerebral tumors and surrounding white matter and remote cortex, *J. Cereb. Blood Flow Metab.,* 3, S9, 1983.

47. **Wise, R. J. S., Thomas, D. G. T., Lammertsma, A. A. et al.,** PET scanning of human brain tumors, *Prog. Exp. Tumor. Res.,* 27, 154, 1984.

48. **Kato, A., Sako, K., Diksic, M. et al.,** Regional glucose utilization and blood flow in experimental brain tumors studied by double tracer autoradiography, *J. Neuro Oncol.,* 3, 271, 1985.

49. **DiChiro, G., Brooks, R. A., Patronas, N., Jr. et al.,** Issues in the *in vivo* measurement of glucose metabolism of human central nervous system tumors, *Ann. Neurol.,* 15, S138, 1984.

50. **DiChiro, G., Brooks, R. A., Sokoloff, L. et al.,** Glycolytic rate and histologic grade of human cerebral gliomas: a study with ^{18}F fluorodeoxyglucose and positron emission tomography, in *Positron Emission Tomography of the Brain,* Phelps, M. E. and Heiss, W.-D., Eds., Springer-Verlag, Berlin, 1983, 181.

51. **Kato, A., Diksic, M., Yamamoto, Y. L., Strother, S. C., and Feindel, W.,** An improved approach for measurement of regional cerebral rate constants in the deoxyglucose method with positron emission tomography, *J. Cereb. Blood Flow Metab.,* 4, 555, 1984.

52. **Yamamoto, Y. L., Thompson, C. J., Diksic, M., Meyer, E., and Feindel, W. H.,** Positron emission tomography, *Neurosurg. Rev.*, 7, 233, 1984.

53. **Kato, A., Diksic, M., Yamamoto, Y. L., and Feindel, W.,** Quantification of glucose utilization in an experimental brain tumor model by the deoxyglucose method, *J. Cereb. Flow Metab.*, 5, 1908, 1985.

54. **Tyler, J. L., Villemure, J.-G., Diksic, M., Evans, A. C., Yamamoto, Y. L., and Feindel, W.,** Metabolic and hemodynamic studies of gliomas using positron emission tomography, *Can. J. Neurol. Sci.*, 13, 280, 1986.

55. **Rottenberg, D. A., Ginos, J. Z., Kearfott, K. J. et al.,** Local interrelationships of cerebral oxygen consumption and glucose utilization in normal subjects and in ischemic stroke patients: a positron tomography study, *J. Cereb. Blood Flow Metab.*, 4, 140, 1984.

56. **Tyler, J. L., Diksic, M., Villemure, J.-G., Evans, A. C., Meyer, E., Yamamoto, Y. L., and Feindel, W.,** Metabolic and hemodynamic studies of gliomas using positron emission tomography, *J. Nucl. Med.*, 28, 1123, 1987.

57. **Kessler, R. M., Ellis, J. R., and Eden, M.,** Analysis of emission tomographic scan data: limitations imposed by resolution and background, *J. Comput. Assist. Tomogr.*, 8, 514, 1984.

58. **DeLa Paz, R. L., Patronas, N. J., and Brooks, R. A.,** Positron emission tomography of gray matter glucose utilization by brain tumors, *AJNR*, 4, 826, 1983.

59. **Timperley, W. F.,** Glycolysis in neuroectodermal tumors, in *Brain Tumors*, Thomas, D. G. T. and Graham, D. I., Eds., Butterworths, London, 1980, 145.

60. **Arnold, D. L., Shoubridge, E. A., Feindel, W., and Villemure, J.-G.,** Metabolic changes in cerebral gliomas within hours of treatment with intra-arterial BCNU demonstrated by phosphorous magnetic resonance spectroscopy, *Can. J. Neurol. Sci.*, 14, 570, 1987.

61. **Shoubridge, E. A., Arnold, D. L., Villemure, J.-G., Emrich, J. F., and Feindel, W.,** Phosphorus magnetic resonance spectroscopy and characterization of astrocytomas, meningiomas, and pituitary adenomas, Symp. on Positron Emission Tomography and Magnetic Resonance Spectroscopy in Oncology, Heidelberg, FRG, May 12 to 14, 1988

62. **Feindel, W., Robitaille, Y., Arnold, D., Shoubridge, E., Emrich, J., and Villemure, J.-G.,** Diagnostic value of magnetic resonance imaging and spectroscopy in brain tumors, in *Cerebral Gliomas*, Broggi, G. and Gerosa, M. A., Eds., Elsevier Science, Amsterdam, 1989, 109.

63. **Mazziotta, J. C.,** *PET Scanning Principles and Applications. Discussions in Neurosciences,* Vol. 2, Fondation pour l'Etude du Système Nerveux Central et Périphérique, Geneva, Switzerland, 1985.

64. **Feindel, W., Frackowiak, R. S. J., Gadian, D., Magistretti, P. L., and Zalutsky, M. R., Eds.,** *Brain Metabolism and Imaging,* Vol. 2, Fondation pour l'Etude du Système Nerveux Central et Périphérique, Geneva, Switzerland, 1985.

65. **Budinger, T. F.,** Scope and limitations of external imaging for studying the brain, in *Brain Metabolism and Imaging,* Vol. 2, Feindel, W., Frackowiak, R. S. J., Gadian, D., Magistretti, P. L., and Zalutsky, M. R., Eds., Fondation pour l'Etude du Système Nerveux Central et Périphérique, Geneva, Switzerland, 1985, 111.

Chapter 2

RECENT ADVANCES IN RADIOTRACERS FOR PET STUDIES OF THE BRAIN

Joanna S. Fowler and Alfred P. Wolf

TABLE OF CONTENTS

I. INTRODUCTION

The development of radiotracers for studying the brain remains one of the major thrusts in positron emission tomography (PET) research today. This focus on the brain was stimulated, in large, by the early development of quantitative PET methods for measuring regional brain glucose metabolism and blood flow using [18]FDG[1,2] and [15]O tracers.[3,4] It is significant that these PET methods were developed well before the advent of commercial PET instrumentation and thus could be relatively rapidly applied to medical problems at new PET centers. Thus, the early demonstrations of the feasibility of quantitating glucose metabolism and blood flow in human brain *in vivo*, and the potential of expanding this approach to the measurement of other biochemical transformations, such as neurotransmitter properties, were in large responsible for the current worldwide growth of PET and the establishment of new PET centers. Today a broad spectrum of radiotracers has been developed and applied to the study of the brain with a special focus on problems in neurology, psychiatry, and cerebral malignancy.[5,6] More recently, PET and appropriate radiotracers have been used to study the action of therapeutic drugs and substances of abuse.[7]

This chaper will begin with a summary of the technology required to apply PET. This will be followed by sections describing a number of recent developments in radiotracers for probing neurotransmitter properties, protein synthesis, enzyme activity, cerebral malignancy, and drug binding properties. A discussion of the problems in radiotracer validation and mechanisms will also be included where appropriate. This chapter will conclude with a section on the challenges facing the PET field as more institutions acquire the instrumentation to apply this modality in basic and clinical research.

II. THE TECHNOLOGY

The technology required to apply PET to problems in biology and medicine varies, depending on the particular applications and interests of the program. The most common configuration of a cyclotron PET program consists of a cyclotron, a positron emission tomograph, and a radiotracer chemistry laboratory. However, clinical PET studies can be carried out with a PET alone with the radiotracer being supplied either from a generator or from a regional center.

A. CYCLOTRONS

While the cyclotron field has not remained static over the past several years, the nature of innovation is primarily directed toward a more convenient and automated operation. The basic machine has varied little, and radionuclide production from any of the current machines is adequate, and peripherals supplied with the machine are generally adequate.[8] One does have a choice between positive ion and negative ion machines and single or multiple particle acceleration. However, the advantages and disadvantages of each machine cannot be covered here. Suffice it to say, they all work reasonably well. What one can look forward to is the possibility of alternate modes of particle acceleration such as linacs being used at high beam currents and low accelerating voltages competing with the more traditional cyclotron. These devices will become available in the near future assuming that the design and fabrication of targetry which will perform adequately beam currents can be accomplished. The advantages of these projected devices are smaller size, decreased shielding requirements, simplified operation and maintenance, and lower cost. It remains to be seen if some or all of these benefits will be realized.

Finally, a point needs to be made about the future of PET and the need for an in-house accelerator. While the in-house machine is necessary for the isotopes of carbon, oxygen, and nitrogen, this need is not determining for compounds labeled with [18]F. The 110-min half-life of this isotope and the lowest energy positron (0.635 MeV max; radial FWHM range, 1.0 mm) of all the available positron emitters (thus making the resolution obtainable the least perturbed by the positron range) make it near ideal as a label for biomolecules of all types. Furthermore, ^{18}F can be produced on a daily basis in a centralized facility for routine shipments to PET centers as much as 300 miles distant. The advantages of a regional PET center are clear since "in house" isotope production and compound preparation are not required. However, at this writing, no regional centers exist. This chapter has not addressed the other positron emitters such as ^{68}Ga, ^{82}Rb, ^{62}Cu, etc. available from generators and also deliverable to sites distant from the source of production.[9]

B. PET INSTRUMENTATION

The future of PET instrumentation is more complex. Here, the design and operation of the currently available machines are quite different. Furthermore, changes occurring almost yearly make current models obsolete. Reduced to the most elementary terms, resolution and sensitivity are improved with every new announcement. While cyclotron energies from 10 to 17 MeV are standard, the 5- to 6-mm resolution PET machine quoted today is already being superceded by the 4- to 5-mm machine and at the time of publication of this book, it may be the 3- to 4-mm machine. At present, resolution is determined by the particular device used and is poorer in even the best machine available today than the ultimate resolution defined by the positron emitter utilized. There is, however, little question that PET devices will become available in the next decade which approach the "theoretical" limit of spatial resolution and which will overcome other problems inherent in present day devices (sensitivity, speed, uniformity, etc.). Physical aspects of PET including resolution, sensitivity, speed, quantitation, image definition, partial volume effects, noise, etc. have been dealt with in detail elsewhere.[5]

Much room for improvement is necessary in the software and hardware of the PET machine and a particular problem, the high cost of these machines, needs to be addressed by the manufacturers. Whereas the cost of accelerators has remained reasonably stable, the same cannot be said for PET devices. Nevertheless, this is a time of ferment and innovation and one can expect the PET of 1995 to be very different from the PET of 1990.

III. PROGRESS IN RADIOTRACER DEVELOPMENT

The history of PET research clearly demonstrates that it is advances in chemistry and PET instrumentation coupled with a detailed examination of the biochemistry of new radiotracers which have allowed PET to be applied to new areas of biology and medicine. The radiotracer remains as the driving force for new avenues of PET research. New compounds being reported are showing increasing sophistication in synthesis and a greater diversity of specialized applications. Here PET has a tremendous advantage over single photon emission computed tomography (SPECT) in that the range and type of tracer that can be developed is almost infinitely larger. One should also remember that PET radiopharmaceuticals are frequently identical to the endogenous compounds whose actions they probe and identical to many drugs whose action is of interest. Thus, one is probing biochemistry and physiology directly rather than with a foreign substance whose basic biochemistry is poorly understood or not understood at all.

A. RADIOTRACER SYNTHESIS

Radiotracers whose regional distribution reflects glucose metabolism, neurotransmitter activity, and enzyme activity have all required the development of rapid synthetic methods for the radiotracers themselves as well as the characterization of their biochemical behavior *in vivo*.[10,11] The continued vitality of the field is thus inextricably related to innovation in chemistry and to advances in basic neurochemistry and neuropsychopharmacology which can be applied to the development of the radiotracers for the PET methods of the future.

Although several hundred different organic molecules have been labeled with ^{11}C and with ^{18}F, only a dozen or so are being applied in PET research today.[12] [11]C- and ^{18}F-labeled organic molecules constitute the majority of the radiotracers either in use or under development. Challenges in synthetic chemistry center about the development of rapid methods for introducing a limited spectrum of precursor molecules into relatively complex organic molecules and developing rapid, efficient purification methods. No-carrier-added syntheses are absolutely necessary in applications where the saturation of a biological substrate such as a receptor is possible or when the use of pharmacologically active tracers is contemplated. Thus, reactions and purifications must be carried out on the microscale.

Because of the relatively short half-life of ^{11}C, its widespread use outside of the research environment and especially in clinical PET requires rapid, simple, and dependable methods. Alkylation with [^{11}C]methyl iodide still represents the most widely used method for labeling with ^{11}C and provides access to a large number of neurotransmitter receptor active compounds and a number of amino acids having the N, O, or S-[11]CH$_3$ functionality. Labeled methyl iodide can be obtained simply, in high yield and in high specific activity, and alkylations are generally straightforward.[13,14] New developments include resin-supported methylation[15] and direct carboxylation of nitrogen.[16] Problems are generally reduced to those of rigorously excluding unlabeled carbon dioxide in order to maximize the specific activity of the product and developing chromatographic methods which separate the labeled product from starting material. New developments in the synthesis of ^{11}C-labeled compounds include the synthesis of other labeled alkyl iodides[17] and the assymetric synthesis of amino acids which holds the promise of higher radiochemical yields by reducing the production of the unwanted enantiomer.[18]

Perhaps the greatest challenge in organic synthesis related to development of the short-lived positron emitters lies in the synthesis of high specific activity ^{18}F-labeled molecules. With the 110-min half-life of ^{18}F, the distribution of ^{18}F-labeled radiotracers from a regional center is certainly feasible when clinically useful radiotracers can be produced dependably and in high yield. While this is certainly the case with [18]FDG where hundreds of millicuries can be produced using automated systems, other ^{18}F-labeled tracers such as the ^{18}F-labeled butyrophenones are more problematic in terms of yields and reproducibility. Progress continues to be made in the efficient use of NCA [^{18}F] fluoride, although problems with competing nucleophiles and metal ions from target materials and other factors which limit yields and affect reproducibility continue to arise.[19] An understanding of the factors influencing the reactivity of NCA [^{18}F] fluoride is important as is the development of new methods which will allow the synthesis of a wider spectrum of different structures. Rapid methods for elaborating simply prepared ^{18}F-labeled molecules into useful radiotracers are also needed. Since virtually all ^{18}F-labeled molecules for human use will have no endogenous naturally occurring counterparts, it is of great advantage to have them available at no carrier added levels in the interest of eliminating potential pharmacological effects and in facilitating approval for human use.

The nucleophilic aromatic substitution reaction with [^{18}F] fluoride which was reported several years ago continues to be examined with a view to increasing its synthetic utility.[20,21] For example, new leaving groups have been identified.[22,23] Progress has been made in the synthesis of labeled aryl fluorides on aromatic rings having electron rich substituents in addition to the required electron withdrawing and leaving groups.[24,25] This observation has opened up the possibility of synthesizing NCA ^{18}F-labeled molecules having hydroxyl substituents and increases the spectrum of ^{18}F-labeled molecules for PET. Using these strategies, it should be possible to prepare physiologically potent radiotracers such as ^{18}F-labeled biogenic amines in sufficiently high specific activity for human PET studies.

The use of ^{18}F-labeled alkyl halides as alkylating agents to produce *N*-[^{18}F]fluoroalkylamines, amides, and other compounds has also been explored as a more general method for labeling organic molecules with [18]F. Thus far, it has been applied to many different compounds and has been most successfully used in the synthesis of *N*-[^{18}F]fluoroalkylspiroperidol. These compounds all appear to bind to dopamine D_2 receptors *in vivo* although there are differences in metabolism for different chain lengths.[26-30]

B. ANALYTICAL AND QUALITY CONTROL

Quality control, an issue which most researchers working in the PET field have been aware of from the beginning, is now gaining greater prominence in PET research.[31] This is clearly driven by the increasing numbers of clinical PET centers where routine methods for testing compounds for human use are necessary and by the diversity of new compounds appearing yearly where accurate methods of quality control are becoming an increasingly important area of investigation. The use of automated and robotic procedures will play an important role in displacing what are labor intensive methods involving purely routine operations.

From a radiopharmaceutical perspective, the first issue would seem to be purity (chemical, radiochemical, and radionuclidic) along with sterility and apyrogenicity. Radionuclidic purity is rarely a problem. Emphasis is usually given to radiochemical purity. The question of chemical purity is, however, not always emphasized, yet can be a much greater problem than radiochemical purity. For example, in the preparation of a neuroreceptor ligand, the radiochemical purity may be excellent, but the chemical purity (e.g., a precursor molecule which does not get labeled, but which carries along in the synthesis and mimics the labeled compound and is not separate from it) can create havoc if its concentration is such as to compete with the labeled compound and thus yield false information as to the specificity of the labeled compound. Another issue is specific activity. It is necessary to know how high is "high" and how high is necessary when one uses a neuroreceptor probe as a tracer. Basically "high" is defined as the maximum specific activity reached in a particular lab by the methods used. Routine and reliable methods need to be developed to determine specific activity. While some centers routinely determine specific activity on compounds used in human research where "high" specific activity is necessary, e.g., probing receptor density, it is as yet not generally done. Of course many compounds in use are applied to studies where "high" specific activity is not needed, e.g., many [18]FDG studies; yet it must always be kept in mind that we are dealing with a tracer method and the application of a dose eliciting a physiological response may not be in the interest of the experiment. New sensitive analytical methods continue to be developed and many of these are applicable to the analysis of the chemical purity and the specific activity of radiopharmaceuticals at no carrier added levels. For example, ion chromatography combined with pulsed amperometric detection has proved to be a very sensitive method for the analysis of sugars such as [18]FDG at no carrier added levels.[32] With the increasing use of high specific activity radiotracers, there is the added possibility that radiochemical purity can be degraded by radiolysis.[33]

C. THE BIOCHEMICAL BASIS OF THE PET IMAGE

In going from the labeled compound to the biochemical basis of the image, one enters an area which is the heart of the PET method and one which this condensed outlook can hardly begin to discuss and define. Mechanistic information can be obtained from *in vivo* PET studies in animals where the stereoselectivity of the localization process, pharmacological intervention with unlabeled compounds of known pharmacological specificity, and kinetic isotope effects and pharmacokinetics can be determined.

In probing biochemistry *in vivo*, PET can lead to definitive answers where other methods may provide data which can be misinterpreted. A simple example emerged in PET studies

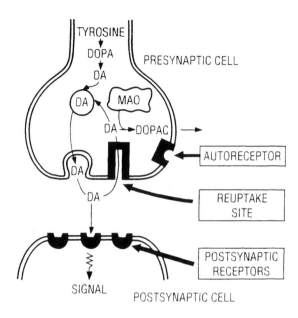

FIGURE 1.   Simplified sketch of the dopaminergic synapse.

of the stereoselectivity of cocaine binding using [^{11}C]-(+)- and (−)-cocaine.[34] (−)-Cocaine is the behaviorally active natural product. Behavioral research with (+)-cocaine showed low or no biological activity in monkeys.[35] PET provided a very straightforward explanation of the behavioral impotency of the (+)-enantiomer which had not been previously considered. After intravenous injection, ^{11}C-labeled (+)-cocaine cannot be detected in the baboon brain. In a subsequent investigation of this effect, it emerged that the D-isomer is metabolized in plasma so rapidly that there is essentially no intact tracer available to cross the blood-brain barrier. Thus, biological activity of the D-isomer can only be assessed if the compound could be directly administered into the brain and then only would be available if brain enzymatic activity relative to metabolism of the D-isomer is negligible.

Many more examples of the biochemical basis of the PET image where PET provides unique and quantitative answers could be given. The current controversy over what constitutes an accurate measure of B_{max} for the D_2 dopamine receptor, and how best to probe this, is an example of the complexities which can arise in PET work, but also represents an example of where solutions can and will be found (see discussion below). The prospects for PET in this area are great. The point of view that the modeling and mathematics are too complex for routine use are without merit. All of these methods, when they become definitive, can be reduced to canned programs where the clinical question is answered by automatic manipulation of the PET data.

IV. NEUROTRANSMITTER STUDIES

A. THE DOPAMINE SYSTEM

The observation of disease-associated abnormalities in neurotransmitter properties in post-mortem human brain tissue has provided an impetus for the development of many radiotracers for examining the properties of neurotransmitters directly in the human brain. By far, the most effort has been focused on the study of the dopaminergic system. This interest stems from observations of abnormalities in dopamine metabolism in Parkinson's disease and in schizophrenia.[36,37]

Radiotracers for probing the dopaminergic synapse from different neurochemical perspectives have been developed as shown in Figure 1. Thus dopamine metabolism, postsyn-

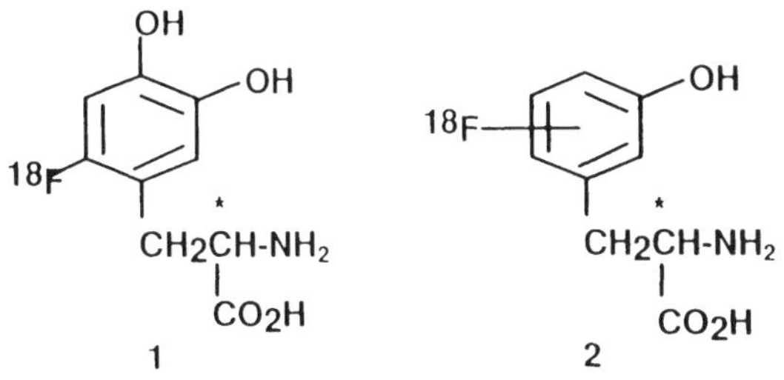

FIGURE 2.   Structures of radiotracers for dopamine metabolism.

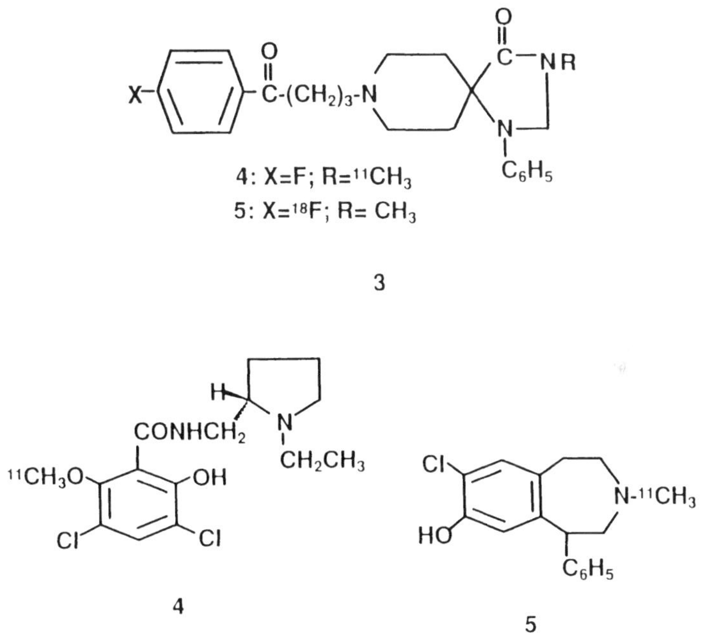

FIGURE 3.   Structures of radiotracers for dopamine D_1 and D_2 receptors.

aptic D_1 and D_2 receptors, and dopamine reuptake sites have all been targets of radiotracer
research and development.

Dopamine does not cross the blood-brain barrier and therefore the investigation of brain
dopamine metabolism with PET uses a ^{18}F-labeled derivative of DOPA, 6-[^{18}F]fluoro-DOPA
(1, Figure 2), which crosses the blood-brain barrier and is metabolized to 6-[^{18}F]fluorodopamine
by aromatic amino acid decarboxylase.[38] One complicating feature of quantitative PET studies
with 6-[^{18}F]fluoroDOPA is its metabolism by catechol O-methyl transferase (COMT) to 3-
O-methyl-6-[^{18}F]fluoroDOPA. This compound also crosses into the brain thus requiring its
input to the brain be accounted for in quantitation.[39-42] A new approach to the study of
dopamine metabolism has been taken in the development of ^{18}F-labeled derivatives of
m-tyrosine (2, Figure 2).[43-45] m-Tyrosine is not a substrate for COMT. However, it is a
substrate for aromatic amino acid decarboxylase, the enzyme responsible for the conversion
of DOPA to dopamine. Thus, its use may increase uptake and therefore simplify the signal
in brain and eliminate the need to correct for input of labeled metabolites.

Dopamine (D_2) receptor activity has also been examined with PET using positron emitter
labeled antagonists of the D_2 receptor (Figure 3). A key development in the application of

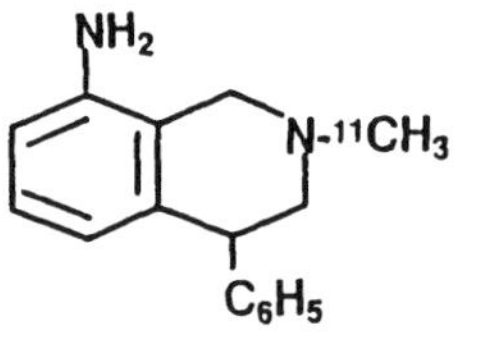

FIGURE 4.   Structures for radiotracers for dopamine reuptake sites.

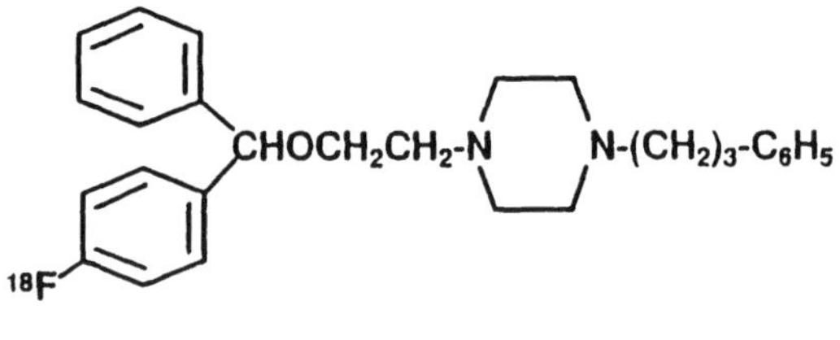

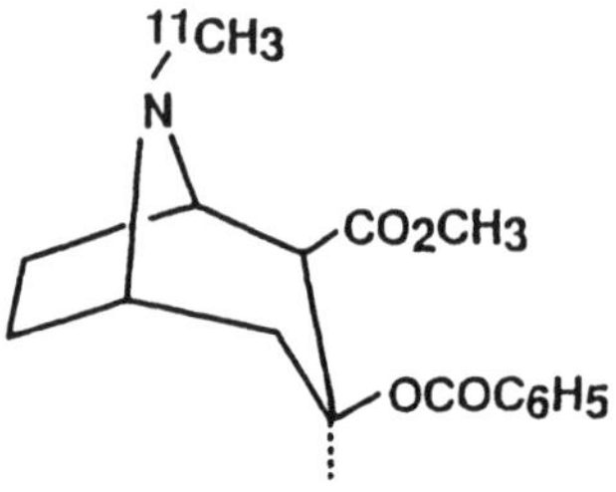

PET to the study of neurotransmitter receptors was the discovery of synthetic methods such as the nucleophilic aromatic substitution reaction with [^{18}F]fluoride for producing high specific activity tracers which also have high affinity and selectivity for the receptor.[46] High specific activity, where the dilution of radioisotope by the naturally occurring element can be in the range of factors of 100 to 10,000, is required in order to avoid saturation and measurable physiological reactions to the labeled compound. For example, *N*-methylspiroperidol has been labeled with ^{11}C and ^{18}F (3, Figure 3) and has been used to study dopamine receptors in normal and diseased brain and to probe dopamine receptor occupancy by antipsychotic drugs.[47-50] The benzamide, [^{11}C]raclopride (4, Figure 3), has also been used in human studies.[51,52] One of the most important issues in PET research related to schizophrenia are two recent reports of the use of PET to measure dopamine D_2 receptor cencentration in schizophrenics where one group has reported a significantly elevated B_{max} in schizophrenics using [^{11}C]*N*-methylspiroperidol[53] and another group reports no deviation from normal values using [^{11}C]raclopride.[54] The resolution of this apparent discrepancy is now requiring a detailed examination of the behavior of the tracers and the models used in the quantitation of receptor density as well as other factors such as the patient population.[55] One possible explanation for the results of the measurements with [^{11}C]raclopride is that endogenous dopamine may compete with labeled raclopride lowering the measured B_{max}.[56,57]

The D_1 receptor antagonist, SCH-23390 (5, Figure 3), has been successfully labeled with ^{11}C.[58-60]

The dopamine reuptake system has also been the subject of radiotracer development (Figure 4). Dopamine reuptake sites appear to be related to neuronal mass and thus these tracers may be of particular value in monitoring the progress of neurodegenerative diseases such as Parkinson's disease. For example, studies of post-mortem brain tissue with [^{3}H]cocaine and [^{3}H]GBR-12935, tracers which bind to the dopamine reuptake system, show a significant decrease in binding in Parkinson's patients relative to normal controls.[61,62] [^{11}C]Nomifensine (6, Figure 4), and [^{18}F]GBR 13119 (7, Figure 4) have been prepared and used to probe the dopamine reuptake system.[63,64] [^{11}C]Cocaine (8, Figure 4), which has been developed to

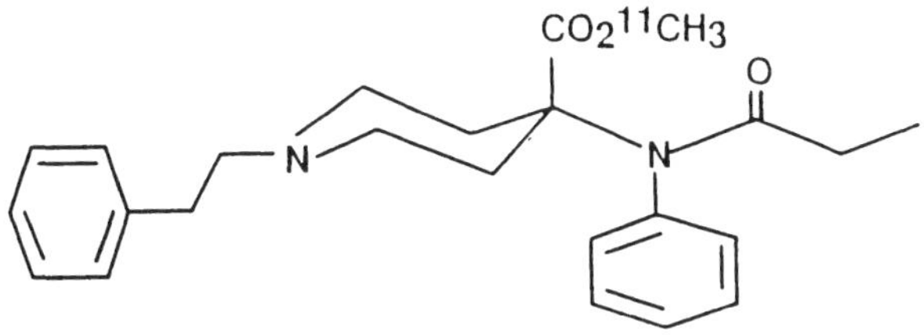

9

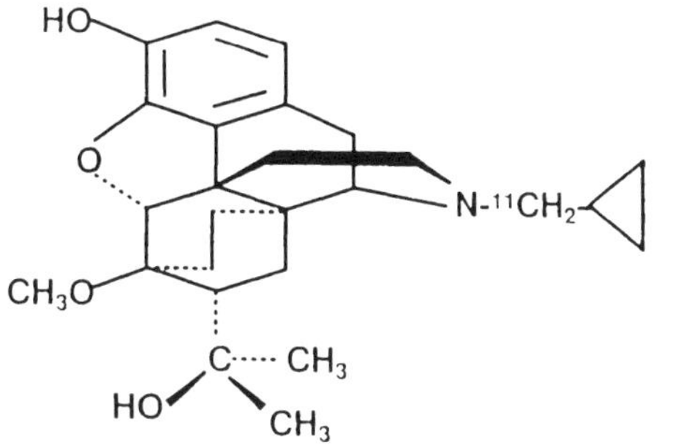

10

FIGURE 5. Structures of radiotracers for opiate receptors.

characterize cocaine binding sites in normal human subjects and in chronic cocaine abusers, may also prove useful in longitudinal studies of neurodegenerative disorders.[65]

B. THE OPIATE SYSTEM

Opioid peptides appear to be involved in seizure mechanisms. ^{11}C-labeled carfentanil (9, Figure 5), a tracer which binds selectively to the μ-opiate system, has been shown to concentrate in regions of the brain such as the basal ganglia and the thalamus which contain high levels of μ-opiate receptors.[66] Its uptake can be reduced by pretreatment with nalaxone, an opiate antagonist. This tracer has been applied to the study of seizure mechanisms where it has recently been demonstrated that μ-opiate receptors are increased in temporal lobe epilepsy.[67] When [^{11}C]carfentanil is used in conjunction with [18]FDG, the accuracy in the location of the epileptic focus with PET is increased. More recent studies have demonstrated that quantitative estimates of *in vivo* rate constants for μ-opiate receptor binding can be determined with PET and [^{11}C]carfentanil. The ratio of a region of interest to the occipital cortex can be used for quantification.[68] [^{11}C]Diprenorphine (10, Figure 5), a ligand which is not specific for any particular opiate receptor subtype, has also been synthesized and used in human studies.[69,70]

C. THE BENZODIAZEPINE RECEPTOR SYSTEM

Benzodiazepines are a group of chemical compounds which are potent anxiolytics, anticonvulsants, and hypnotics. Their mechanism of action is through the facilitation of inhibitory gamma-aminobutyric acid (GABA) neurotransmission in the CNS.[71] Radiotracers have been developed for studying both central and peripheral benzodiazepine receptors in the brain with PET (Figure 6). One of these, [^{11}C]RO-15-1788 (11, Figure 6), is a selective antagonist of central benzodiazepine receptors.[72] With this tracer and PET, a clear dose dependent inhibition of *in vivo* binding of trace doses of [^{11}C]RO-15-1788 was demonstrated

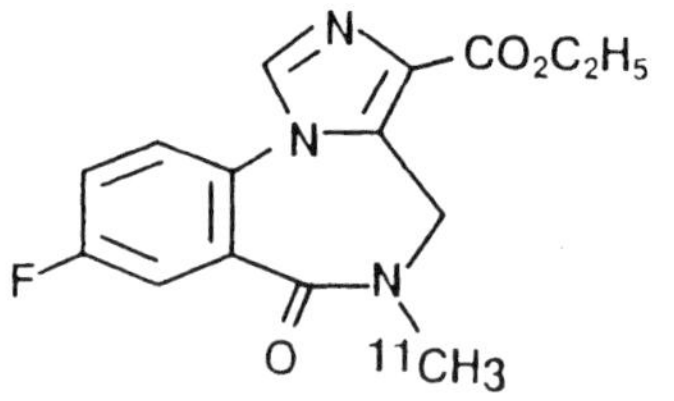

11

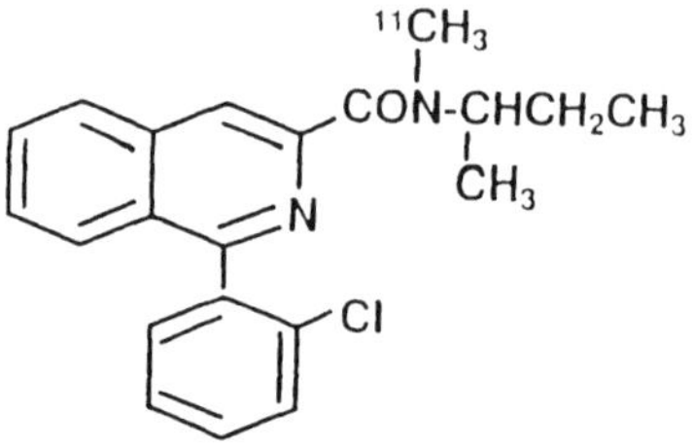

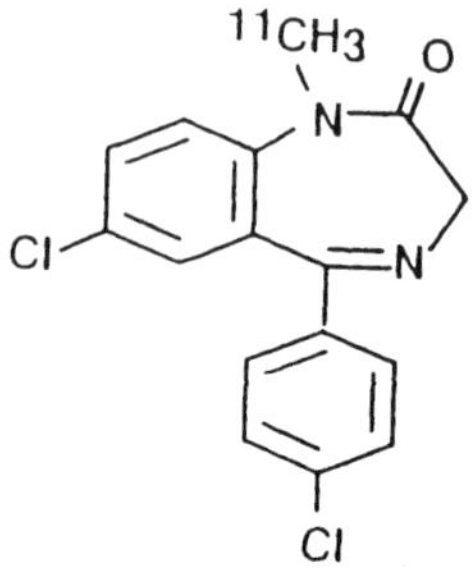

12   **13**

FIGURE 6.   Structures of radiotracers for central and peripheral benzodiazepine receptors.

by co-injecting increasing amounts of unlabeled RO15-1788.[73] Moreover, the distribution
of this radiotracer parallels regional concentrations of benzodiazepine receptors in human
brain tissue. Labeled benzodiazepine receptor ligands are of interest in the study of anxiety
and epilepsy. For example, in a recent study, benzodiazepine receptor binding was shown
to be significantly lower in an epileptic focus than in the contralateral reference region.[74]

Peripheral benzodiazepine receptors differ from central benzodiazepine receptors in their
pharmacological specificity and by the fact that they are not coupled to GABA. These
receptors are present in low concentrations in normal brain tissue but are present in high
concentrations in malignant gliomas.[75] The potential for imaging peripheral benzodiazepine
receptors with PET has been examined using two different peripheral benzodiazepine receptor
ligands, [11C]PK-11195 (12, Figure 6) and [11C]RO-5-4864 (13, Figure 6), with different
results. Whereas [11C]RO-5-4864 showed a lower uptake in tumor than in normal tissue,[76]
[11C]PK-11195 showed a high uptake in glioma relative to normal tissue.[77] Moreover, tumor
uptake was saturable indicating receptor-mediated uptake and a high signal to noise was
obtained in noncontrast enhancing gliomas demonstrated that uptake was not related to blood
brain barrier breakdown.

D. THE MUSCARINIC-CHOLINERGIC SYSTEM

One of the primary reasons for developing PET radiotracers for probing the cholinergic
system is its association with the pathophysiology of Alzheimer's disease.[78] [11C]Scopolamine
(14, Figure 7) has been used to examine the muscarine-cholinergic system, but factors in
addition to the concentration of muscarinic receptors appear to contribute to its distribution
in vivo.[79] Other 11C-labeled ligands such as tropanyl benzilate (15, Figure 7) have been
developed more recently and show more promise since their extraction does not appear to
be flow-limited as is the case with labeled scopolamine.[80] The anticholinergic drug cogentin
(benztropine mesylate, 16 Figure 7) has been labeled with 11C and its regional distribution

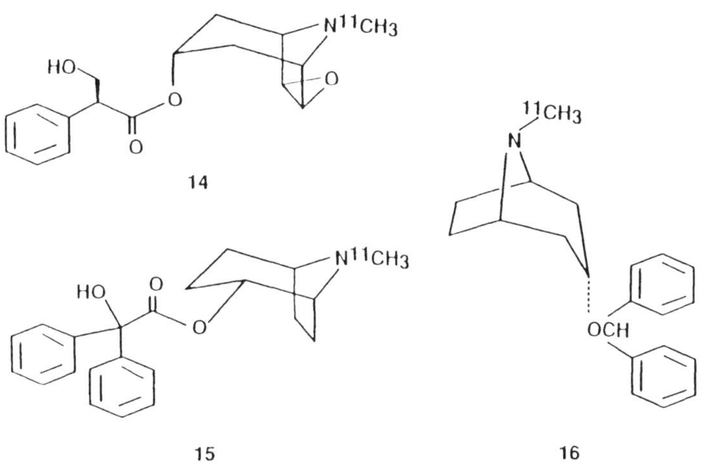

FIGURE 7. Structures of radiotracers for the muscarinic-cholinergic system.

in human brain parallels cholinergic receptors. Its uptake in baboon brain can be reduced by blockade with both cholinergic antagonists such as scopolamine and agonists such as pilocarpine.[81] ^{11}C-labeled dexetimide, a potent muscarinic-cholinergic antagonist, and levetimide, its inactive enantiomer, have also been prepared for studies of this receptor system.[82]

E. OTHER NEUROTRANSMITTER SYSTEMS

Labeled PET ligands for the serotonergic system include both ^{11}C and ^{18}F containing compounds such as [N-^{11}C-methyl]bromoLSD[83] and ^{18}F-labeled setoperone,[84] ritanserin,[85] and altanserine.[86] A method for quantitating the rate of serotonin synthesis using ^{11}C-labeled α-methyl tryptophan has also been recently developed.[87] Radiotracers for probing the NMDA receptor are under development.[88-91] Imaging agents for brain glucocorticoid receptors are being studied to investigate the role of glucocorticoids in hippocampal degeneration.[92,93] Nicotine binding in human brain has also been examined with PET.[94]

V. BRAIN PROTEIN SYNTHESIS

Protein synthesis and amino acid turnover represent important biochemical reactions in the human brain, and thus their quantitative assessment in normal human brain and in disease states is of interest. Positron emitter-labeled amino acids have been examined as tracers for the quantitative measurement of the regional incorporation of amino acids into protein. One strategy uses a carboxyl-labeled amino acid, [^{11}C-carboxyl]L-leucine, where the products of amino acid metabolism (loss of [^{11}C]O$_2$) are not labeled, but proteins of interest are labeled.[95,96] The recent detailed examination of the use of carboxyl-labeled L-leucine as a tracer for protein synthesis has revealed considerable recycling of unlabeled amino acids derived from steady state protein degradation which must be taken into account in the estimation of the precursor pool specific activity in tissue from measurements in plasma.[97] [^{11}C-S-methyl]L-Methionine and [1-^{11}C]L-methionine have also been examined as substrates for protein synthesis.[98,99] [S-^{11}C-methyl]Methionine has been investigated most extensively, although studies are complicated by the fact that it can undergo both protein synthesis and transmethylation. Recently, L-[1-^{14}C]tyrosine and L-[2-^{18}F]fluorotyrosine have been synthesized and shown to be incorporated into protein almost completely to the exclusion of other metabolic pathways.[100,101] The incorporation of L-[2-^{18}F]fluorotyrosine into tissue proteins was proven by sodium dodecyl sulfate (SDS) gel electrophoresis and a three compartment model for the quantitation of protein synthesis has been proposed.

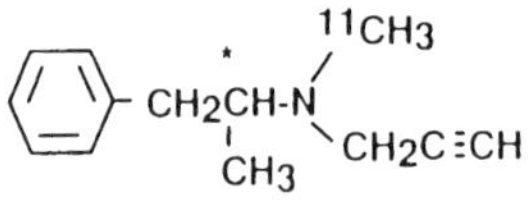

FIGURE 8.   Structures of radiotracers for probing MAO activity.

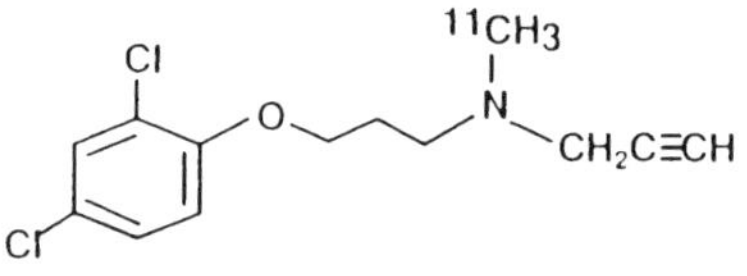

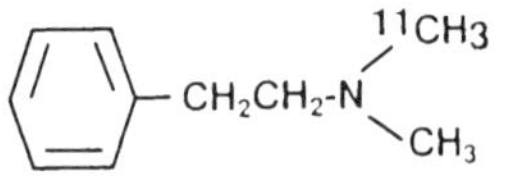

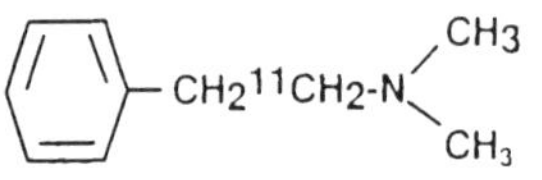

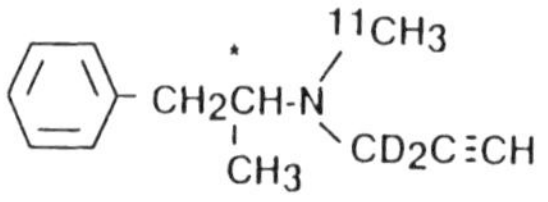

VI. FUNCTIONAL ENZYME ACTIVITY (MONOAMINE OXIDASE)

The enzyme monoamine oxidase is responsible for the oxidative deamination of biogenic amines and has been widely studied in neurological and psychotic illness.[102] It is typically measured in platelets or in post-mortem human brain tissue. Interest in monoamine oxidase (MAO) has intensified recently with the observation that the inhibition of MAO B by the suicide inactivator (−)-deprenyl prevents the development of 1-methyl-4-phenyl-1,2,3,6-tetrahydropyridine (MPTP)-induced Parkinson's disease.[103] Moreover, a recent prospective study in patients with early Parkinson's disease showed that (−)-deprenyl significantly retarded the rate of progression of the disease.[104]

Radiotracers for measuring functional MAO activity directly in human brain with PET have been developed (Figure 8). One approach uses a labeled substrate for MAO, [^{11}C]N,N-dimethylphenethylamine and measures MAO activity through the MAO catalyzed oxidation of this tracer to labeled dimethylamine which is intracellularly trapped in tissue.[105] Another uses labeled suicide inactivators [^{11}C]-(−)-deprenyl (17, Figure 8) or [^{11}C]clorgyline (18, Figure 8) which become covalently attached to the flavin cofactor of the enzyme as a result of MAO catalyzed activation to produce a highly reactive species which covalently binds to the enzyme within the enzyme-substrate complex.[106] Both of these tracers have a regional distribution in human brain which parallels the distribution of MAO B in human brain tissue as measured *in vitro*. In addition, mechanistic studies have been carried out to demonstrate that the accumulation of labeled products in tissue reflects functional MAO activity. For example, N,N-dimethylphenethylamine has been labeled both in the N-methyl group (19, Figure 8) and in the methylene group α to the nitrogen atom (20, Figure 8). The time course of the ^{11}C in the brain has been compared.[107] As would be predicted if MAO catalyzed deamination were responsible for the accumulation of ^{11}C in tissue, ^{11}C is retained with the N-methyl labeled tracer. This has been attributed to the intracellular production of labeled dimethylamine which is charged at physiological pH and cannot diffuse from the cell. In contrast, ^{11}C is rapidly cleared from tissue for the methylene-labeled compound because the MAO produced dimethylamine is not labeled. In this case, the metabolite carrying the label is phenylacetaldehyde which is not charged and therefore not trapped.

For labeled (−)-deprenyl, uptake has been demonstrated to be stereoselective with the (−) enantiomer being retained, whereas the (+) enantiomer (a weaker MAO inhibitor) is cleared from the brain.[106] In addition, [^{11}C](−)-deprenyl has been synthesized with deuterium atoms replacing hydrogen atoms on the methylene carbon of the propargyl group (21, Figure 8).[108] These hydrogen (or deuterium) atoms are at the reactive center for MAO and the initial step of catalysis by MAO is cleavage of one of these carbon-hydrogen (or deuterium) bonds. The demonstration of a deuterium isotope effect *in vivo* with [^{11}C](−)-deprenyl-α,α-D$_2$ demonstrated that MAO catalyzed oxidation is the rate limiting step in the trapping of radioactivity in brain.

VII. CEREBRAL MALIGNANCY

The development of radiotracers for examining brain tumors has focused on the following areas: the detection of malignant tissue and measurement of the degree of malignancy; the characterization of tumor tissue from a biochemical viewpoint; the differentiation of tumor tissue from normal tissue or from other pathologies such as radiation necrosis; the detection of receptors, tumor antigens or growth factors; the measurement of the delivery and efficacy of chemotherapeutic drugs.

A. PROBES FOR METABOLISM

A predominant biochemical feature of rapidly growing tumor cells is an ability to sustain high rates of glycolysis under anaerobic conditions. Although glucose metabolism in human gliomas has been examined with ^{18}FDG and used to grade tumors[109,110] and to differentiate recurrent tumor from treatment-related necrosis,[111-113] questions have been raised regarding the ability to relate ^{18}FDG uptake or metabolism to tumor grade.[114,115] Moreover, recent estimation of the lumped constant in tumor vs. normal brain in transplanted rat glioma has demonstrated that tumor differs significantly from that in normal brain tissue, suggesting the need for measuring the lumped constant in human gliomas for the accurate quantitation of tumor glucose metabolism.[116] Clearly, this complex issue warrants a careful examination of the many factors which can impact on the measurement of glucose metabolic rate in tumors to optimize the use of PET in the clinical management of cerebral malignancy. In addition to glucose metabolism, ^{15}O-labeled tracers ([^{15}O]O$_2$ and [^{15}O]H$_2$O) have been used to assess the relationship between the oxygen supply and demand of tumors.[117]

Enhanced protein synthesis in tumors may also be a useful biochemical measure of tumor grade. Labeled methionine has been extensively utilized to delineate tumor tissue although the mechanism for sequestration has not been fully characterized. [^{11}C]D- and L-methionine have been used to show that methionine transport into tumor is nonselective.[118] However, only [^{11}C]L-methionine showed continuous irreversible trapping in some tumors.[119] The examination of protein synthesis in tumors using 2-[^{18}F]fluorotyrosine may prove to provide very useful information since it appears to be used almost exclusively for protein synthesis.[120]

B. OTHER RADIOTRACERS FOR TUMORS

The incorporation of labeled thymidine (tritium or ^{14}C) into cells has long been the gold standard for measuring tissue proliferation and growth kinetics.[121] The extension of this approach to *in vivo* studies using [^{11}C]thymidine was explored in the early 1970s prior to modern PET instrumentation.[122] Currently, efforts are focused on the biochemistry and kinetics of [^{11}C]thymidine in tumors[123,124] as well as imaging with PET.[125] ^{18}F-Labeled 5-fluorouridine and 5-fluoro-2-deoxyuridine have also been prepared and studied in small animals.[126] Polyamine biosynthesis, which accompanies rapid tissue growth, has been studied in gliomas with [1-^{11}C]putrescine.[127] Although PET studies of labeled putrescine have shown that this substance is taken up in high-grade gliomas, the uptake is not specific for tumor

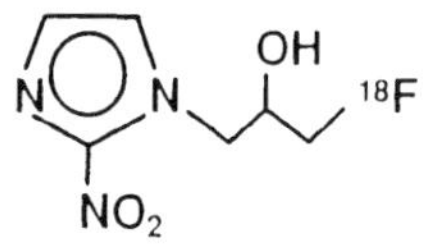

22

FIGURE 9.   Structure of [^{18}F]fluoromisonidazole.

tissue.[128] Hypoxic cell sensitizers such as [^{18}F]fluoromisonidazole (22, Figure 9) are under investigation for the delineation of hypoxic regions of tumor tissue.[129,130]

The use of PET to examine the presence of peripheral benzodiazepine receptors in gliomas was described in a previous section. The presence of dopamine receptors which are expressed on some pituitary adenomas and which may be directly related to the cellular origins of the neoplastic cells has been detected by [*N*-^{11}C-methyl]spiroperidol, a dopamine D$_2$ receptor antagonist.[131]

PET has been used to probe the delivery of chemotherapeutic agents administered via different routes to tumors through the use of positron emitter labeled agents.[132,133]

VIII. RADIOTRACERS IN DRUG RESEARCH AND SUBSTANCE ABUSE

The use of PET to examine the behavior of therapeutic drugs as well as drugs of abuse holds tremendous promise. Among the parameters which can be measured by PET are the following: the effect of a drug on metabolism and neurotransmitter properties: binding sites and target organs for the drug; absolute brain uptake of the drug; pharmacokinetics; binding mechanisms; drug interactions; duration of action; effect of route of administration on target uptake; relationship between drug binding to target receptors and plasma drug concentration.

The effect of drugs on regional glucose metabolism has been probed with ^{18}FDG or [^{11}C]2-deoxy-D-glucose for a number of drugs such as neuroleptics,[134] amphetamine,[135] alcohol,[136-139] azidothymidine,[140] and cocaine.[141-143] Similarly, measurements have been made of dopamine D$_2$ receptor availability as it is influenced by antipsychotic drugs,[49,50,151] and cocaine.[144,145] The effects of both cocaine and alcohol on brain blood flow as measured by ^{15}O-labeled water have also been reported.[146,147]

The regional distribution and pharmacokinetics as well as the binding sites for cocaine have been determined using PET and ^{11}C-labeled cocaine in human and baboon brain.[65] Cocaine is rapidly taken up and cleared from the striatum, and its time profile parallels the mean subjective high, experienced by human volunteers after the intravenous administration of cocaine.[148] Moreover, by selectively blocking the dopamine reuptake site with nomifensine, cocaine binding in the striatum was significantly reduced, demonstrating that cocaine binding is associated with the dopamine reuptake site. This study demonstrates the usefulness of PET in examining the relationship between the binding of a drug or substance of abuse to a target site in the brain and its behavioral effects and holds the potential of allowing the design of effective therapeutic interventions in chronic cocaine abusers through a knowledge of the neurochemical basis for its binding.

The regional distribution of binding sites for the most widely used antipsychotic drug, haloperidol, has also been determined using [^{18}F]haloperidol (23, Figure 10) and PET.[149] It is known that haloperidol binds to dopamine D$_2$ receptors. Although the structure of haloperidol is very similar to *N*-methylspiroperidol which is selectively taken up in the striatum, the brain uptake of haloperidol is widespread. This indicates that other binding sites such as sigma, in addition to D$_2$ receptors, are operative and may be partially responsible for the regional distribution and the therapeutic characteristics of the drug.

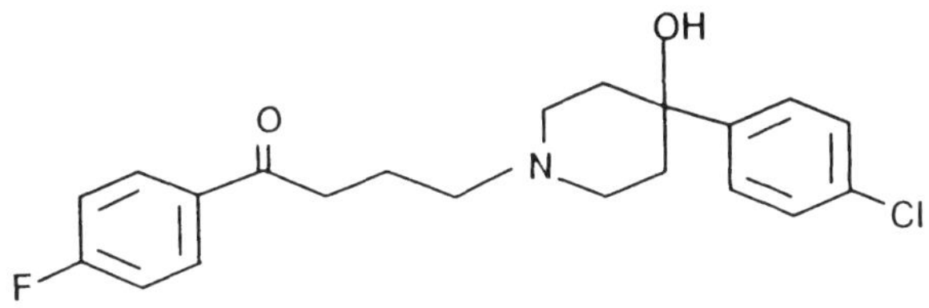

23

FIGURE 10. Structure of [¹⁸F]haloperiodol.

The question of whether receptor occupancy by haloperidol parallels the concentration of drug in the plasma was also addressed with PET in order to determine whether plasma drug levels are a good measure of receptor occupancy by antipsychotic drugs.[49,50] The study was carried out using serial PET studies with [¹⁸F]N-methylspiroperidol to probe receptor availability (using the ratio index as a measure of availability). Plasma drug levels were measured at the time of each PET study. It was found that receptor occupancy by haloperidol increased with increasing plasma concentrations of the drug, leveling off at 5 to 15 ng/ml. Interestingly, clinical studies show this to be the initial phase of therapeutic levels for haloperidol. However, at the higher plasma drug concentrations achieved by high haloperidol doses, there was no increase in receptor occupancy. This supports a growing clinical consensus that there is little added benefit in increasing plasma haloperidol levels above 20 ng/cc. [¹⁸F]N-Methylspiroperidol was also used to address the question of whether schizophrenic subjects who respond positively to antipsychotic drug therapy have a different receptor occupancy by the drug than those who do not respond.[150] The conclusion of this study was that receptor blockade by haloperidol occurs to the same extent in responders and nonresponders and that the treatment of nonresponders with high neuroleptic doses in order to increase receptor occupancy is not warranted. In another PET study with ⁷⁶Br-bromospiperone, receptor occupancy by neuroleptic drugs showed a clear cut saturation curve with increasing oral daily doses.

The duration of action of drugs directly at their target sites has been measured with PET. One study examined the duration of action of a single dose of (−)-deprenyl, the suicide enzyme inactivator of MAO B which is currently being used in the therapy of Parkinson's disease.[152] Tracer doses of [¹¹C](−)-deprenyl were used to probe MAO activity in baboon brain before and 34 min after a single dose of 1 mg/kg of (−)-deprenyl and then at 1, 5, 22, 34, 61 and 111 days following the dose. Since (−)-deprenyl acts by covalent attachment to MAO which results in the destruction of the enzyme, the recovery of MAO activity represents the synthesis of enzyme. Here it was found that the recovery of the enzyme activity is surprisingly slow with a half-life of 30 days. This type of information is of importance in designing a drug dosage regimen especially in the case of drugs which act by irreversibly inhibiting a biological substrate such as an enzyme.

Another report addressed the duration of occupancy of μ-opiate receptors by naltrexone, an orally administered opiate antagonist.[153] Occupancy was probed by the μ-specific ligand [¹¹C]carfentanil and an external detector before naltrexone and at 1, 48, 72, 120, and 168 h after the oral administration of 50 mg of naltrexone. The half-time of blockage by naltrexone in the brain ranged from 72 to 108 h which corresponds to the half-time of the terminal plasma phase of naltrexone clearance. Both of these studies illustrate the power of *in vivo* imaging and detection not only to probe the target sites for a drug, but also the duration of its pharmacological action and provide a direct assessment of the relevance of measurements of drug concentration in plasma to receptor occupancy by the drug.

IX. OUTLOOK FOR PET

Positron emission tomography (PET) as a tool for investigating *in vivo* biochemistry and physiology in normal and pathological states has made great strides since the early 1970s and particularly so in the last 10 years due mainly to the increased sophistication in radiopharmaceutical probes and rapid advances in PET instrumentation. The coming years will see increasing clinical application, establishment of PET in drug research, and a widening of our understanding of the biochemistry and functioning of the human brain.

Issues remain which need to be addressed before full confidence in the biochemical basis of any PET image can be achieved. Unfortunately, the advantages and limitations of PET are rarely considered in an objective manner. The basis for the PET method is the use of tracer kinetics in the study of human biochemistry and physiology *in vivo*. Thus, any PET study involves a multiplicity of processes which together result in the PET image and the pharmacokinetics that define that image.

One aspect of PET that must be kept in mind is that PET cannot be used to define a biochemical pathway in a single cell and it does not compete with the elegant basic work on the structure and function of single cells or small groups of cells. By the same token, and this is one of the great powers of the PET method, the pharmacokinetics of a large aggregate of cells *in vivo*, especially the human brain involving systems with numerous complex interactions in quantitative terms, is unique to PET at present.

An underdeveloped area in PET research is the question of reproducibility, i.e., "test-retest" data.[154-156] Will the same radiopharmaceutical give the same answer in two sequential studies on the same individual on the same day, on sequential days, or months or years apart? Reproducibility will vary with the functionality of the compound and the variability of the biochemical aspect of the individual's brain. Too often generalizations are based on small sample differences where no evidence is presented that in a normative data base experiment no changes are observed. However, here again PET can provide unique new answers as to the variability and constancy of particular biochemical aspects of the human brain. The measurement of a plasma input function where the amount of unchanged radiotracer is known and where the nature of the labeled metabolites and their properties vis-a-vis their entry into the brain is thus of critical importance in this regard. The value of the plasma input function is manyfold, including its use in tracer kinetic model development, in normalizing tissue uptake for changes in peripheral metabolism and radiotracer delivery, and in inter- and intrasubject comparisons.

One should perhaps separate research with PET and what it can tell us about the brain and the use of PET as a clinical tool. Clinical PET, to some the end result of PET research, is no more than the routine application of PET methods which prove clinically useful and can be converted to routine procedures whether this be the reading of a PET scan generated by a specific labeled probe or a quantitative number reflecting a particular condition generated by mathematical manipulation of the PET data. There is little argument that simplification of mathematical models in interpreting PET is necessary for a clinical environment. The PET image itself, produced by a particular radiopharmaceutical, e.g., an [18]FDG image, is already being used today in a purely clinical context. The outlook is particularly bright for methods of assessing metabolic dysfunction, tumor localization and tumor viability, patients at risk in the case of diseases which do not yet present clinical symptoms, and the quantification of psychotic disorders.

It is safe to say that basic research in labeling biomolecules with positron emitters has shaped the PET field as we know it today. The growth of the field can be appreciated when one considers that in 1976 there were only 4 cyclotron-PET facilities, whereas there are currently more than 60 centers worldwide. It is important to point out that PET is by no means a mature field. For example, new developments in rapid organic synthesis have a far-reaching impact, especially when they can be applied to radiotracers which are in demand for basic and clinical research. Basic research is still required to better understand and expand the physiological information that these tracer experiments can provide and accelerate the clinical availability of new tracers.

The outlook for PET is indeed bright.

ACKNOWLEDGMENT

This chapter was prepared at Brookhaven National Laboratory under Contract DE-AC02-76CH0016 with the U.S. Department of Energy and supported by its Office of Health and Environmental Research.

REFERENCES

1. **Reivich, M., Kuhl, D., Wolf, A. P., Greenberg, J., Phelps, M., Ido, T., Casella, V., Hoffman, E., Alavi, A., and Sokoloff, L.,** The [^{18}F]fluorodeoxyglucose method for the measurement of local cerebral glucose utilization in man, *Circ. Res.*, 44, 127, 1979.
2. **Fowler, J. S. and Wolf, A. P.,** 2-Deoxy-2-[^{18}F]fluoro-D-glucose for metabolic studies: current status, *Int. J. Appl. Radiat. Isot.*, 37, 663, 1986.
3. **Ter-Pogossian, M. M., Eichling, J. O., Davis, D. O., Welch, M. J., and Metzger, J. M.,** The determination of regional cerebral blood flow by means of water labeled with radioactive oxygen 15, *Radiology*, 93, 31, 1969.
4. **Ter-Pogossian, M. M. and Herscovitch, P.,** Radioactive oxygen-15 in the study of cerebral blood flow, blood volume, and oxygen metabolism, *Sem. Nucl. Med.*, XV, 377, 1985.
5. **Phelps, M. E., Mazziotta, J. C. and Schelbert, H. R., Eds.,** *Positron Emission Tomography and Autoradiography: Principles and Applications for the Brain and Heart,* Raven Press, New York, 1986.
6. **Lucignani, G., Moresco, R., and Fazio, F.,** PET based neuropharmacology: state of the art, *Cereb. Vascular Brain Metab. Rev.*, 1, 281, 1989.
7. **Fowler, J. S., Wolf, A. P., and Volkow, N. D.,** New directions in positron emission tomography, Part II, in *Annual Reports in Medicinal Chemistry,* Vol. 25, Allen, R. C., Ed., Academic Press, Orlando, 1990, 261.
8. **Wolf, A. P.,** Cyclotrons for clinical and biomedical research with PET, in *PET/SPECT '87,* American College of Nuclear Physicans, Washington, D.C., 1988, 109.
9. **Guillaume, M. and Brihaye, C.,** Generators for short-lived gamma and positron emitting radionuclides: current status and prospects, *Nucl. Med. Biol.*, 13, 89, 1986.
10. **Jacobson, H. G., Ed.,** Cyclotrons and radiopharmaceuticals in positron emission tomography, *JAMA*, 259, 1854, 1988.
11. **Feliu, A. L.,** The role of chemistry in positron emission tomography, *J. Chem. Eng. Ed.*, 65, 654, 1988.
12. **Fowler, J. S. and Wolf, A. P.,** Positron emitter labeled compounds—priorities and problems, in *Positron Computed Tomography,* Phelps, M. E., Mazziotta, J. C., and Schelbert, H., Eds., Raven Press, New York, 1986, 391.
13. **Langstrom, B. and Lundqvist, H.,** The preparation of ^{11}C-methyl iodide and its use in the synthesis of ^{11}C-methyl-L-methionine, *Int. J. Appl. Radiat. Isot.*, 27, 357, 1976.
14. **Langstrom, B., Antoni, G., Gullberg, P., Halldin, C., Malmborg, P., Nagren, K., Rimland, A., and Svard, H.,** Synthesis of L- and D-[methyl-^{11}C]methionine, *J. Nucl. Med.*, 28, 1037, 1987.
15. **Mulholland, G. K., Watkins, G. L., Jewett, D. M., Sugimoto, H., and Kilbourn, M., R.,** Resin supported C-11-methylation approach for phenols amides and acidic precursors which can avoid need for HPLC purification, *J. Nucl. Med.*, 30, 822, 1989.
16. **Ram, R. and Ehrenkaufer, R. L.,** Synthesis of ^{11}C-radiopharmaceuticals using direct fixation of [^{11}C]carbon dioxide and [^{11}C] carbon monoxide, *Int. J. Radiat. Appl. Instrum. Part B*, 15, 345, 1988.
17. **Langstrom, B., Antoni, G., Gullberg, P., Halldin, C., Nagren, K., Rimland, A., and Svard, H.,** The synthesis of 1-11C-labeled ethyl, propyl, butyl and isobutyl iodides and examples of alkylation reactions, *Int. J. Radiat. Appl. Instrum. Part A*, 37, 1141, 1986.
18. **Fasth, K.-J., Antoni, G., and Langstrom, B.,** Asymmetric synthesis of L-[3-^{11}C]alanine and L-[3-^{11}C]phenylalanine by a phase-transfer alkylation reaction, *J. Chem. Soc. Perkin Trans.*, 1, 3081, 1988.
19. **Tewson, T. J., Berridge, M. S., Bolomey, L., and Gould, K. L.,** Routine production of reactive fluorine-18 fluoride salts from an oxygen-18 water target, *Nucl. Med. Biol.*, 15, 499, 1988.
20. **Attina, M., Cacace, F., and Wolf, A. P.,** Displacement of a nitro group by [^{18}F]fluoride ion. A new route to aryl fluorides of high specific activity, *J. Chem. Soc. D.*, 108, 1983.

21. **Attina, M., Cacace, F., and Wolf, A. P.,** Labeled aryl fluorides from the nucleophilic displacement of activated nitro groups by [18]F-, *J. Labelled Compd. Radiopharm.*, XX, 501, 1983.

22. **Angelini, G., Speranza, M., Wolf, A. P., and Shiue, C.-Y.,** Nucleophilic aromatic substitution of activated cationic groups by ^{18}F-labeled fluoride. A useful route to no carrier added ^{18}F-labeled aryl fluorides, *J. Fluorine Chem.*, 27, 177, 1985.

23. **Haka, M. S., Kilbourn, M. R., Watkins, G. L. and Toorongian, S. A.,** Aryltrimethylammonium trifluoromethanesulfonates as precursors to ary [^{18}F]fluorides: improved synthesis of [^{18}F]GBR-13119, *J. Labelled Compd. Radiopharm.*, XXVII, 823, 1989.

24. **Ding, Y.-S., Shiue, C.-Y., Fowler, J. S., Wolf, A. P., and Plenevaux, A.,** No-carrier-added (NCA) aryl [^{18}F]fluorides via the nucleophilic aromatic substitution of electron rich aromatic rings, *J. Fluorine Chem.*, 48, 189, 1990.

25. **Lemaire, C., Guillaume, M., and Christiaens, L.,** Asymmetric synthesis of 6-[F-18]fluoro-L-dopa via NCA nucleophilic fluorination, *J. Nucl. Med.*, 30, 752, 1989.

26. **Satyamurthy, N., Bida, G. T., Barrio, J. R., Luxen, A., Mazziotta, J. C., Huang, S.-C., and Phelps, M. E.,** No-carrier-added 3-(2'-^{18}F]fluoroethyl)spiperone, a new dopamine receptor-binding tracer for positron emission tomography, *Int. J. Radiat. Appl. Instrum. Part B*, 13, 617, 1986.

27. **Chi, D. Y., Kilbourn, M. R., Katzenellenbogen, J. A., Brodack, J. W., and Welch, M. J.,** Synthesis of no-carrier-added N-(^{18}F]fluoroalkyl)spiperone derivatives, *Int. J. Radiat. Appl. Instrum. Part A*, 37, 1173, 1986.

28. **Shiue, C.-Y., Bai, L.-Q., Teng, R.-R., Arnett, C. D. and Wolf, A. P.,** No-carrier-added N-(3-^{18}F]fluoropropyl)spiroperidol: biodistribution in mice and tomographic studies in a baboon, *J. Nucl. Med.*, 28, 1164, 1987.

29. **Welch, M. J., Katzenellenbogen, J. A., Mathias, C. J., Brodack, J. W., Carlson, K. E., Chi, D. Y., Dence, C. S., Kilbourn, M. R., Perlmutter, J. S., Raichle, M. E., and Ter-Pogossian, M. M.,** N-(3-^{18}F]Fluoropropyl)-spiperone: the preferred ^{18}F labeled spiperone analog for positron emission tomographc studies of the dopamine receptor, *Int. J. Radiat. Appl. Instrum. Part B*, 15, 83, 1988.

30. **Coenen, H. H., Wienhard, K., Stocklin, G., Laufer, P., Hebold, I., Pawlik, G., and Heiss, W.-D.,** PET measurement of D_2 and S_2 receptor binding of 3-N-([2'-18]F]fluoroethyl)spiperone in baboon brain, *Eur. J. Nucl. Med.*, 14, 80, 1988.

31. **Vera-Ruiz, H., Marcus, C. S., Pike, V. W., Coenen, H. H., Fowler, J. S., Meyer, G. J., Cox, P. H., Vaalburg, W., Cantineau, R., Helus, F., and Lambrecht, R. M.,** Special Report: Report of an International Atomic Energy Agency's Advisory Group Meeting on ''Quality Control of Cyclotron-Produced Radiopharmaceuticals,'' *Nucl. Med. Biol.*, 17, 445, 1990.

32. **Alexoff, D. L., Casati, R., Fowler, J., Wolf, A. P., Shea, C., Schlyer, D., and Shiue, C.-Y.,** Ion chromatographic analysis of [18]FDG produced by [^{18}F]fluoride displacement: production of 2-chloro-2-deoxy-D glucose as an impurity in the presence of chloride ion, in 8th Int. Symp. on radiopharmaceutical Chemistry, Princeton, NJ, 1990, 183.

33. **MacGregor, R. R., Schlyer, D. J., Fowler, J. S., and Wolf, A. P.,** Fluroine-18-N-methylspiroperidol: radiolytic decomposition as a consequence of high specific activity and high dose levels, *J. Nucl. Med.*, 28, 60, 1987.

34. **Gatley, S. J., MacGregor, R. R., Fowler, J. S., Wolf, A. P., Dewey, S. L., and Schlyer, D. J.,** Rapid stereoselective hydrolysis of (+) cocaine in baboon plasma revents its uptake in the brain: implications for behavioral studies, *J. Neurochem.*, 54, 720, 1990.

35. **Spealman, R. D., Kelleher, R. T., and Goldberg, S. R.,** Stereoselective effects of cocaine and a phenyltropane analog, *J. Pharm. Exp. Ther.*, 225, 509, 1983.

36. **Hornykiewicz, O.,** Dopamine (3-hydroxytyramine) and brain function, *Pharmacol. Rev.*, 18, 925, 1966.

37. **Seeman, P.,** Dopamine receptors and the dopamine hypothesis of schizophrenia, *Synapse*, 1, 133, 1987.

38. **Firnau, G., Garnett, E. S., Chirakal, R., Sood, S., Nahmias, C., and Schrobilgen, G.,** [18]Fluoro-L-dopa for the *in vivo* study of intracerebral dopamine, *Int. J. Radiat. Appl. Instrum. Part A*, 37, 669, 1986.

39. **Boyes, B. E., Cumming, P., Martin, W. R. W., and McGeer, E. G.,** Determination of plasma ^{18}F]-6-fluorodopa during positron emission tomography: elimination and metabolism in carbidopa treated subjects, *Life Sci.*, 39, 2243, 1986.

40. **Firnau, G., Sood, S., Chirakal, R., Nahmias, C., and Garnett, E. S.,** Metabolites of 6-[^{18}F]fluoro-L-dopa in human blood, *J. Nucl. Med.*, 29, 363, 1988.

41. **Hoffman, J. M., Melega, W. P., Grafton, S. T., Mahoney, D. K., Hawk, T. C., Barrio, J. R., Huang, S. C., Mazziotta, J. C., Perlmutter, M., and Phelps, M. E.,** The importance of carbidopa in 6-[^{18}F]fluoro-L-dopa (FD) PET studies, *J. Nucl. Med.*, 30, 761, 1989.

42. **Huang, S. C., Barrio, J. R., Hoffman, J. M., Mahoney, D. K., Hawk, T. C., Melega, W. P., Luxen, A., Grafton, S., Mazziotta, J. C., and Phelps, M. E.,** A compartmental model for 6-[F-18]fluoro-L-dopa kinetics in cerebral tissues, *J. Nucl. Med.*, 30, 735, 1989.

43. DeJesus, O. T., Sunderland, J. J., Chen. C.-A., Nickles, R. J., Mukherjee, J., and Appelman, E. H., Synthesis of radiofluorinated M-tyrosine analogs as potential L-DOPA PET tracers, J. Nucl. Med., 30, 930, 1989.

44. Gildersleeve, D. L., Van Dort, M. E., Rosenspire, K. C., Toorongian, S., Sherman, P. S., and Wieland, D. M., Synthesis of [H-3]-L-m-tyrosine, [H-3]-U-methyl-L-m-tyrosine, and 4-[F-18]fluoro-D.L m-tyrosine as potential protracers for dopamine neuron mapping, J. Nucl. Med., 30, 752, 1989.

45. Melega, W. P., Perimutter, M. M., Luxen, A., Nissenson, H. K., Grafton, S. T., Huang, S.-C., Phelps, M. E., and Barrio, J. R., 4-[^{18}F]Fluoro-L-m-tyrosine: an L-3,4-dihydroxyphenylalanine analog for probing presynaptic dopaminergic function with positron emission tomography, J. Neurochem., 53, 311, 1989.

46. Shiue, C.-Y., Fowler, J. S., Wolf, A. P., McPherson, D. W., Arnett, C. D., and Zecca, L., No-carrier-added fluorine-18-labeled N-methylspiroperidol: synthesis and biodistribution in mice, J. Nucl. Med., 27, 226, 1986.

47. Wagner, H. N., Jr., Burns, H. D., Dannals, R. F., Wong, D. F., Langstrom, B., Duelfer, T., Frost, J. J., Ravert, H. T., Links, J. M., Rosenbloom, S. B., Lukas, S. E., Kramer, A. V., and Kuhar, M. J., Imaging dopamine receptors in the human brain by positron tomography, Science, 221, 1264, 1983.

48. Arnett, C. D., Wolf, A. P., Shiue, C.-Y., Fowler, J. S., MacGregor, R. R., Christman, D. R., and Smith, M., Improved delineation of human dopamine receptors using [^{18}F]-N-methylspiroperidol in PET, J. Nucl. Med., 27, 1878, 1986.

49. Wolkin, A., Brodie, J. D., Barouche, F., Rotrosen, J., Wolf, A. P., Smith, M., Fowler, J., and Cooper, T. B., Dopamine receptor occupancy and plasma haloperidol levels, Arch. Gen. Psychiatry, 46, 482, 1989.

50. Smith, M., Wolf, A. P., Brodie, J. D., Arnett, C. D., Barouche, F., Shiue, C.-Y., Fowler, J. S., Russell, J. A. G., MacGregor, R. R., Wolkin, A., Angrist, B., Rotrosen, J. and Peselow, E., Serial [^{18}F]-N-methylsproiperidol PET studies to measure changes in antipsychotic drug D$_2$ receptor occupancy in schizophrenic patients, Biol. Psychiatry, 23, 653, 1988.

51. Farde, L., Ehrin, E., Eriksson, L., Greitz, T. Y., Hall, H., Hedstrom, C. G., Litton, J. E., and Sefvall, G., Substituted benzamides as ligands for visualization of dopamine receptor binding in the human brain by positron emission tomography, Proc. Natl. Acad. Sci. U.S.A., 82, 3863, 1985.

52. Farde, L., Wiesel, F.-A., Halldin, C., and Sedvall, G., Central D$_2$-dopamine receptor occupancy in schizophrenic patients treated with antipsychotic drugs, Arch. Gen. Psychiatry, 45, 71, 1988.

53. Wong, D. F., Wagner, H. N., Jr., Tune, L. E., Dannals, R. F., Pearlson, G. D., Links, J. M., Tamminga, C. A., Broussolle, E. P., Ravert, H. T., Wilson, A. A., Toung, J. K. T., Malat, J., Williams, J. A., O'Tuama, L. A., Snyder, S. H., Kuhar, M. J., and Gjedde, A., Positron emission tomography reveals elevated D$_2$ dopamine receptors in drug-naive schizophrenics, Science, 234, 1558, 1986.

54. Farde, L., Wiesel, F.-A., Hall, H., Halldin, C., Stone-Elander, S., and Sefvall, G., No D$_2$ receptor increase in PET study of schizophrenia, Arch. Gen. Psychiatry, 44, 672, 1987.

55. Andreasen, N. C., Carson, R., Diksic, M., Evans, A., Farde, L., Gjedde, A., Hakim, A., Lal, S., Nair, N., Sedvall, G., Tune, L., and Wong, D., Workshop on schizophrenia, PET, and dopamine D$_2$ receptors in the human neostriatum, Schiz. Bull., 14, 471, 1988.

56. Seeman, P., Guan, H.-C., and Niznik, H. B., Endogenous dopamine lowers the dopamine D$_2$ receptor density as measured by [^{3}H]raclopride: implications for positron emission tomography of the human brain, Synapse, 3, 96, 1989.

57. Ross, S.B. and Jackson, D. M., Kinetic properties of the accumulation of ^{3}H-raclopride in the mouse brain in vivo, Naunyn Schmiedebergs Arch. Pharmakol., 340, 6, 1989.

58. DeJesus, O. T., Van Moffaert, G. J. C., and Friedman, A. M., Synthesis of [^{11}C]SCH 23390 for dopamine D$_1$ receptor studies, Int. J. Radiat. Appl. Instrum. Part A, 38, 345, 1987.

59. Halldin, C., Stone, Elander, S., Farde, L., Ehrin, E., Fasth, K.-J., Langstrom, B., and Sedvall, G., Preparation of ^{11}C-labelled SCH 23390 for the in vivo study of dopamine D-1 receptors using positron emission tomography, Int. J. Radiat. Appl. Instrum. Part A, 37, 1039, 1986.

60. Ravert, H. T., Wilson, A. A., Dannals, R. F., Wong, D. F., and Wagner, H. N., Jr., Radiosynthesis of a selective dopamine D-1 receptor antagonist: R(+)-7-chloro-8-hydroxy-3-[^{11}C]methyl-1-phenyl-2,3,4,5-tetrahydro-^{1}H-3-benzazepine ([^{11}C]SCH 23390), Int. J. Radiat. Appl. Instrum. Part A, 38, 305, 1987.

61. Schoemaker, H., Pimoule, C., Arbilla, S., Scatton, B., Javoy-Agid, F., and Langer, S. Z., Sodium dependent [^{3}H]cocaine binding associated with dopamine uptake sites in the rat striatum and human putamen decrease after dopaminergic denervation and in Parkinson's disease, Naunyn Schmiedebergs Arch. Pharmakol., 329, 227, 1985.

62. Maloteaux, J.-M., Vanisberg, M.-A., Laterre, C., Javoy-Agid, F., Agid, Y., and Laduron, P. M., [^{3}H]GBR 12935 binding to dopamine uptake sites: subcellular localization and reduction in Parkinson's disease and progressive supranuclear palsy, Eur. J. Pharmacol., 156, 331, 1988.

63. **Aquilonius, S.-M., Bergstrom, K., Eckernas, S.-A., Hartvig, P., Leenders, K. L., Lundquist, H., Antoni, G., Gee, A., Rimland, A., Uhlin, J., and Langstrom, B.,** *In vivo* evaluation of striatal dopamine reuptake sites using ^{11}C-nomifensine and positron emission tomography, *Acta Neurol. Scand.,* 76, 283, 1987.

64. **Kilbourn, M. R.,** *In vivo* binding of [18F]GBR 13119 to the brain dopamine uptake system, *Life Sci.,* 42, 1347, 1988.

65. **Fowler, J. S., Volkow, N. D., Wolf, A. P., Dewey, S. L., Schlyer, D. J., MacGregor, R. R., Hitzeman, R., Logan, J., Bendriem, B., and Christman, D.,** Mapping cocaine binding sites in human and baboon brain, *in vivo, Synapse,* 4, 371, 1989.

66. **Frost, J. J., Wagner, H. N., Jr., Dannals, R. F., Ravert, H. T., Links, J. M., Wilson, A. A., Burns, H. D., Wong, D. F., McPherson, R. W., Rosenbaum, A. E., Kuhar, M. J., and Snyder, S. H.,** Imaging opiate receptors in the human brain by positron tomography, *J. Comp. Assist. Tomog.,* 9, 231, 1985.

67. **Frost, J. J., Mayberg, H. S., Fisher, R. S., Douglass, K. H., Dannals, R. F., Links, J. M., Wilson, A. A., Ravert, H. T., Rosenbaum, A. E., Snyder, S. H., and Wagner, H. N., Jr.,** Mu-opiate receptors measured by positron emission tomography are increased in temporal lobe epilepsy, *Ann. Neurol.,* 23, 231, 1988.

68. **Frost, J. J., Douglass, K. H., Mayberg, H. S., Dannals, R. F., Links, J. M., Wilson, A. A., Ravert, H. T., Crozier, W. C., and Wagner, H. N., Jr.,** Multicompartmental analysis of [11C]-carfentanil binding to opiate receptors in humans measured by positron emission tomography, *J. Cereb. Blood Flow Metab.,* 9, 398, 1989.

69. **Luthra, S. K., Pike, V. W., and Brady, F.,** The preparation of carbon-11 labelled diprenorphine: a new radioligand for the study of the opiate receptor system *in vivo, J. Chem. Soc. D,* 1423, 1985.

70. **Lever, J. R., Dannals, R. F., Wilson, A. A., Ravert, H. T., and Wagner, H. N., Jr.,** Synthesis of carbon-11 labeled diprenorphine: a radioligand for positron emission tomographic studies of opiate receptors, *Tetrahedron Lett.,* 28, 4015, 1987.

71. **Nutt, D. J.,** Benzodiazepine receptor ligands, *Neurotransmissions,* IV, 1, 1988.

72. **Persson, A., Ehrin, E., Eriksson, L., Farde, L., Hedstrom, C.-G., Litton, J. E., Mindus, P., and Sedvall, G.,** Imaging of [11C]-labelled RO 15-1788 binding to benzodiazepine receptors in the human brain by positron emission tomography, *J. Psychiatr. Res.,* 19, 609, 1985.

73. **Pappata, S., Samson, Y., Chavoix, C., Prenant, C., Maziere, M., and Baron, J. C.,** Regional specific binding of [11C]RO 15 1788 to central type benzodiazepine receptors in human brain: quantitative evaluation by PET, *J. Cereb. Blood Flow Metab.,* 8, 304, 1988.

74. **Savic, I., Roland, P., Sedvall, G., Persson, A., Pauli, S., and Widen, L.,** *In vivo* demonstration of reduced benzodiazepine receptor binding in human epileptic foci, *Lancet,* Oct. 15, 863, 1988.

75. **Olson, J. M. M., Junck, L., Young, A. B., Penney, J. B., and Mancini, W. R.,** Isoquinoline and peripheral-type benzodiazepine binding in gliomas: implications for diagnostic imaging, *Cancer Res.,* 48, 5837, 1988.

76. **Bergstrom, M., Mosskin, M., Ericson, K., Ehrin, E., Thorell, J.-O., von Holst, H., Noren, G., Persson, A., Halldin, C., Stone-Elander, S., and Collins, V. P.,** Peripheral benzodiazepine binding sites in human gliomas evaluated with positron emission tomography, in *13th Symp. Neuroradiologicum,* Greitz, T., Ed., Stockholm, June 1986, in press.

77. **Junck, L., Olson, J. M. M., Ciliax, B. J., Koeppe, R. A., Watkins, G. L., Jewett, D. M., McKeever, P. E., Wieland, D. M., Kilbourn, M. R., Starosta-Rubinstein, S., Mancini, W. R., Kuhl, D. E., Greenberg, H. S., and Young, A. B.,** PET imaging of human gliomas with ligands for the peripheral benzodiazepine binding site, *Ann. Neurol.,* 26, 752, 1989.

78. **McGeer, P. L. and McGeer, E. G.,** Cholinergic systems and cholinergic pathology, in *Handbook of Neurochemistry,* Lajtha, A., Ed., Plenum Press, New York, 1984, 379.

79. **Frey, K. A., Koeppe, R. A., Jewett, D. M. et al.,** The *in vivo* distribution of [11C]scopolamine in human brain determined by positron emission tomography, *Soc. Neurosci. Abstr.,* 13, 1658, 1987.

80. **Mulholland, G. K., Otto, C. A., Jewett, D. M., Kilbourn, M. R., Sherman, P. S., Koeppe, R. A., Frey, K. A., and Kuhl, D. E.,** Radiosynthesis and comparisons in the biodistribution of carbon-11 labeled muscarinic antagonists: (+)2U-tropanyl benzilate and *N*-methyl-4-piperidyl benzilate, *J. Labelled Compd. Radiopharm.,* 26, 202, 1989.

81. **Dewey, S. L., MacGregor, R. R., Bendriem, B., King, P. T., Volkow, N. D., Schlyer, D. J., Brodie, J. D., Fowler, J. S., Wolf, A. P., Gatley, S. J., and Hitzeman, R.,** Mapping muscarine receptors in human and baboon brain using [N-11C-methyl]benztropine, *Synapse,* 5, 213, 1990.

82. **Dannals, R. F., Langstrom, B., Ravert, H. T., Wilson, A. A., and Wagner, H. N., Jr.,** Synthesis of radiotracers for studying muscarinic cholinergic receptors in the living human brain using positron emission tomography: [11C]dexetimide and [11C]levetimide, *Int. J. Radiat. Appl. Instrum. Part A,* 39, 291, 1988.

83. **Wong, D. F., Lever, J. R., Hartig, P. R., Dannals, R. F., Villemagne, V., Hoffman, B. J., Wilson, A. A., Ravert, H. T., Links, J. M., Scheffel, U., and Wagner, H. N., Jr.,** Localization of serotonin 5-HT$_2$ receptors in living human brain by positron emission tomography using N1-([^{11}C]-methyl)-2-Br-LSD, *Synapse*, 1, 393, 1987.

84. **Blin, J., Pappata, S., Kiyosawa, M., Crouzel, C., and Baron, J. C.,** [^{18}F]Setoperone: a new high-affinity ligand for positron emission tomography study of the serotonin-2 receptors in baboon brain *in vivo*, *Eur. J. Pharmacol.*, 147, 73, 1988.

85. **Crouzel, C., Venet, M., Sanz, G., and Denis, A.,** Labelling of a new serotoninergic ligand: [^{18}F]ritanserin, *J. Labelled Compd. Radiopharm.*, XXV, 827, 1988.

86. **Lemaire, C., Cantineau, R., Christiaens, L., and Guillaume, M.,** N.C.A. radiofluorination of altanserine a potential serotonin receptor-binding radiopharmaceutical for positron emisson tomography, *J. Labelled Compd. Radiopharm.*, 26, 336, 1989.

87. **Diksic, M., Nagahiro, S., Sourkes, T. L., and Yamamoto, Y. L.,** A new method to measure brain serotonin synthesis *in vivo*. Theory and basic data for a biological model, *J. Cereb. Blood Flow Metab.*, 10, 1, 1990.

88. **Brady, F., Luthra, S. K., Pike, V. W., and Zecca, L.,** Nucleophilic substitution at cyclic sulpha-mates — routes to [^{18}F]fluoroanalogues of MK 801 as potential radioligands for the NMDA receptor, *J. Labelled Compd. Radiopharm.*, 26, 381, 1989.

89. **Wieland, D. M., Kilbourn, M. R., Yang, D. J., Laborde, E., Gildersleeve, D. L., Van Dort, M. E., Pirat, J.-L., Ciliax, B. J., and Young, A. B.,** NMDA receptor channels: labeling of MK-801 with iodine-125 and fluorine-18, *Int. J. Radiat. Appl. Instrum. Part A*, 39, 1219, 1988.

90. **Burns, H. D., Eng, W.-S., Dannals, R. F., Wong, D. F., Guilarte, T. R., Wilson, A. A., Ravert, H. T., Gibson, R. E., Britcher, S. F., Frost, J. J., Wagner, H. N., Jr., and Solomon, H. F.,** Design, synthesis and preliminary evaluation of (+)-[C-11]-8-methoxy-MK-801: a potential radiotracer for NMDA receptor imaging via PET, *J. Nucl. Med.*, 30, 930, 1989.

91. **Monn, J. A. and Rice, K. C.,** A bridgehead a-amino carbanion: facile preparation of C5(bridgehead)-substituted analogues of ()-5H-dibenzo[a,d]cyclohepten-5,10-imine including a stable α-iodo secondary amine, *Tetrahedron Lett.*, 30, 911, 1989.

92. **Feliu, A. L. and Rottenberg, D. A.,** Synthesis and evaluation of fluorine-18 21-fluoroprednisone as a potential ligand for neuro-PET studies, *J. Nucl. Med.*, 28, 998, 1986.

93. **Pomper, M. G., Kochanny, M. J., Thieme, A. M., Carlson, K. E., VanBrocklin, H. F., Mathias, C. J., Welch, M. J., and Katzenellenbogen, J. A.,** Imaging agents for brain corticosteroid receptors: synthesis and tissue distribution of fluorine-18 substituted corticosteroids, *J. Nucl. Med.*, 30, 821, 1989.

94. **Nyback, H., Nordberg, A., Langstrom, B., Halldin, C., and Sedvall, G.,** Nicotine binding in the living human brain as studied by positron emission tomography, *J. Nucl. Med.*, 30, 742, 1989.

95. **Phelps, M. E., Barrio, J. R., Huang, S.-C., Keen, R. E., Chugani, H., and Mazziotta, J. C.,** Criteria for the tracer kinetic measurement of cerebral protein synthesis in humans with positron emission tomography, *Ann. Neurol.*, Suppl. 15, S192, 1984.

96. **Smith, C. B., Davidsen, L., Deibler, G., Patlak, C., Pettigrew, K., and Sokoloff, L.,** A method for the determination of local rates of protein synthesis in man, *Trans. Am. Soc. Neurochem.*, 11, 94, 1980.

97. **Smith, C. B., Deibler, G. E., Eng, N., Schmidt, K., and Sokoloff, L.,** Measurement of local cerebral protein synthesis *in vivo*: influence of recycling of amino acids derived from protein degradation, *Proc. Natl. Acad. Sci. U.S.A.*, 85, 9341, 1988.

98. **Bustany, P., Henry, J. F., Cabanis, E., Soussaline, F., Crouzel, M., and Comar, D.,** Incorporation of L-(^{11}C)-methionine in brain proteins studied by PET in dementia, *Proc. 3rd World Congr. Nucl. Med. Biol.*, 2, 2216, 1982.

99. **Ishiwata, K., Vaalburg, W., Elsinga, P. H., Paans, A. M. J., and Woldring, M. G.,** Comparison of L-[1-^{11}C]methionine and L-methyl-[^{11}C]methionine for measuring *in vivo* protein synthesis rates with PET, *J. Nucl. Med.*, 29, 1419, 1988.

100. **Ishiwata, K., Vaalburg, W., Elsinga, P. H., Paans, A. M. J., and Woldring, M. G.,** Metabolic studies with L-[1-^{14}C]tyrosine for the investigation of a kinetic model to measure protein synthesis rates with PET, *J. Nucl. Med.*, 29, 524, 1988.

101. **Coenen, H. H., Kling, P., and Stocklin, G.,** Cerebral metabolism of L-[2-^{18}F]fluorotyrosine, a new PET tracer of protein synthesis, *J. Nucl. Med.*, 30, 1367, 1989.

102. **Tipton, K. F., Dostert, P., and Benedetti, M. S., Eds.,** *Monoamine Oxidase and Disease*, Academic Press, New York, 1984.

103. **Markey, S. P., Johannessen, J. N., Chiueh, C. C., Burns, R. S., and Herkenham, M. A.,** Intraneuronal generation of a pyridinium metabolite may cause drug-induced parkinsonism, *Nature*, 311, 464, 1984.

104. **Tetrud, J. W. and Langston, J. W.,** The effect of deprenyl (selegiline) on the natural history of Parkinson's Disease, *Science*, 245, 519, 1989.

105. **Shinotoh, H., Inoue, O., Suzuki, K., Yamasaki, T., Iyo, M., Hashimoto, K., Tominaga, T., Itoh, T., Tateno, Y., and Ikehira, H.,** Kinetics of [^{11}C]N,N-dimethylphenyethylamine in mice and humans: potential for measurement of brain MAO-B activity, *J. Nucl. Med.*, 28, 1006, 1987.

106. **Fowler, J. S., MacGregor, R. R., Wolf, A. P., Arnett, C. D., Dewey, S. L., Schlyer, D., Christman, D., Logan, J., Smith, M., Sachs, H., Aquilonius, S. M., Bjurling, P., Halldin, C., Hartvig, P., Leenders, K. L., Lundqvist, H., Oreland, L., Stalnacke, C.-G., and Langstrom, B.,** Mapping human brain monoamine oxidase A and B with ^{11}C-labeled suicide inactivators and PET, *Science*, 235, 481, 1987.

107. **Halldin, C., Bjurling, P., Stalnacke, C.-G., Jossan, S. S., Oreland, L., and Langstrom, B.,** ^{11}C-labelling of dimethylphenethylamine in two different positions and biodistribution studies, *Int. J. Radiat. Appl. Instrum. Part A*, 40, 557, 1989.

108. **Fowler, J. S., Wolf, A. P., MacGregor, R. R., Dewey, S. L., Logan, J., Schlyer, D. J., and Langstrom, B.,** Mechanistic positron emission tomography studies: demonstration of a deuterium isotope effect in the monoamine oxidase-catalyzed binding of [11C]L-deprenyl in living baboon brain, *J. Neurochem.*, 51, 1524, 1988.

109. **Di Chiro, G., Brooks, R. A., Patronas, N. J., Bairamian, D., Kornblith, P. L., Smith, B. H., Mansi, L., and Barker, J.,** Issues in the *in vivo* measurement of glucose metabolism of human central nervous system tumors, *Ann. Neurol.*, Suppl. 15, S138, 1984.

110. **Alavi, J. B., Alavi, A., Chawluk, J., Kushner, M., Powe, J., Hickey, W., and Reivich, M.,** Positron emission tomography in patients with glioma, *Cancer*, 62, 1074, 1988.

111. **Patronas, N. J., Di Chiro, G., Brooks, R. A., DeLaPaz, R. L., Kornblith, P. L., Smith, B. H., Rizzoli, H. V., Kessler, R. M., Manning, R. G., Channing, M., Wolf, A. P., and O'Connor, C. M.,** Work in progress: [18F]fluorodeoxyglucose and positron emission tomography in the evaluation of radiation necrosis of the brain, *Radiology*, 144, 885, 1982.

112. **Di Chiro, G., Oldfield, E., Wright, D. C., De Michele, D., Katz, D. A., Patronas, N. J., Doppman, J. L., Larson, S. M., Ito, M., and Kufta, C. V.,** Cerebral necrosis after radiotherapy and/or intraarterial chemotherapy for brain tumors: PET and neuropathologic studies, *Am. J. Roentgenol.*, 150, 189, 1988.

113. **Doyle, W. K., Budinger, T. F., Valk, P. E., Levin, V. A., and Gutin, P. H.,** Differentiation of cerebral radiation necrosis from tumor recurrence by [18F]FDG and [82]Rb positron emission tomography, *J. Comp. Assist. Tomog.*, 11, 563, 1987.

114. **Tyler, J. L., Diksic, M., Villemure, J.-G., Evans, A. C., Meyer, E., Yamamoto, Y. L., and Feindel, W.,** Metabolic and hemodynamic evaluation of gliomas using positron emission tomography, *J. Nucl. Med.*, 28, 1123, 1987.

115. **Di Chiro, G. and Brooks, R. A.,** PET-FDG of untreated and treated cerebral gliomas, *J. Nucl. Med.*, 29, 421, 1988; **Tyler, J. L., Diksic, M., Villemure, J.-G., Evans, A. C., Meyers, E., and Yamamoto, Y. L.,** Reply, *J. Nucl. Med.*, 29, 422, 1988.

116. **Spence, A. M., Graham, M. M., Muzi, M., Abbott, G. L., Krohn, K. A., Kapoor, R., and Woods., S. D.,** Deoxyglucose lumped constant estimated in a transplanted rat astrocytic glioma by hexose utilization index, *J. Cereb. Blood Flow. Metab.*, 10, 190, 1990.

117. **Jacobson, H. G., Ed.,** Positron emission tomography in oncology, *JAMA*, 259, 2126, 1988.

118. **Schober, O., Duden, C., Meyer, G.-J., Muller, J. A., and Hundeshagen, H.,** Nonselective transport of [11C-methyl]-L-and D-methionine into a malignant glioma, *Eur. J. Nucl. Med.*, 13, 103, 1987.

119. **Bergstrom, M., Muhr, C., Lundberg, P. O., Bergstrom, K., Lundqvist, H., Antoni, G., Fasth, K.-J., and Langstrom, B.,** Amino acid distribution and metabolism in pituitary adenomas using positron emission tomography with D-[11C]methionine and L-[11C]methionine, *J. Comp. Assist. Tomog.*, 11, 384, 1987.

120. **Wagner, H. N., Jr.,** Scientific Highlights 1988: the future is now, *J. Nucl. Med.*, 29, 1329, 1988.

121. **Cronkite, E. P., Fliedner, T. M., Bond, V. P., Rubini, J. R., Brecher, G., and Quastler, H.,** Dynamics of hemopoietic proliferation in man and mice studied by H^3-thymidine incorporation into DNA, *Ann. N.Y. Acad. Sci.*, 77, 803, 1959.

122. **Christman, D. R., Crawford, E. J., Friedkin, M., and Wolf, A. P.,** Detection of DNA synthesis in intact organisms with positron-emitting methyl ^{11}C-thymidine, *Proc. Natl. Acad. Sci. U.S.A.*, 69, 988, 1971.

123. **Shields, A. F., Coonrod, D. V., Quackenbush, R. C., and Crowley, J. J.,** Cellular sources of thymidine nucleotides: studies for PET, *J. Nucl. Med.*, 28, 1435, 1987.

124. **Shields, A. F., Larson, S. M., Grunbaum, Z., and Graham, M. M.,** Short-term thymidine uptake in normal and neoplastic tissues: studies for PET, *J. Nucl. Med.*, 25, 759, 1984.

125. **Martiat, Ph., Ferrant, A., Labar, D., Cogneau, M., Bol, A., Michel, C., Michaux, J. L., and Sokal, G.,** *In vivo* measurement of carbon-11 thymidine uptake in non-Hodgkin's lumphoma using positron emission tomography, *J. Nucl. Med.*, 29, 1633, 1988.

126. **Fowler, J. S. and Wolf, A. P.,** Positron emitter labeled compounds — priorities and problems, in *Positron Computed Tomography*, Phelps, M. E., Mazziotta, J. C., and Schelbert, H., Eds., Raven Press, New York, 1986, 391 and references therein.

127. **McPherson, D. W., Wolf, A. P., Fowler, J. S., Arnett, C. D., Brodie, J. D., and Volkow, N.,** Synthesis and biodistribution of no-carrier-added [1-11C]putrescine, *J. Nucl. Med.*, 26, 1186, 1985.

128. **Hiesiger, E., Fowler, J. S., Wolf, A. P., Logan, J., Brodie, J. D., McPherson, D., MacGregor, R. R., Christman, D. R., Volkow, N. D., and Flamm, E.,** Serial PET studies of human cerebral malignancy with [1-¹¹C]putrescine and [1-¹¹C]2-deoxy-D-glucose, *J. Nucl. Med.,* 28, 1251, 1987.

129. **Grierson, J. R., Link, J. M., Mathis, C. A., Rasey, J. S., and Krohn, K. A.,** A radiosynthesis of fluorine-18 fluoromisonidazole, *J. Nucl. Med.,* 30, 343, 1989.

130. **Hwang, D.-R., Dence, C. S., Bonasera, T. A., and Welch, M. J.,** No-carrier-added synthesis of 3-[¹⁸F]fluoro-1-(2-nitro-1-imidazolyl)-2-propanol. A potential PET agent for detecting hypoxic but viable tissues, *Int. J. Radiat. Appl. Instrum. Part A,* 40, 117, 1989.

131. **Muhr, C., Bergstrom, M., Lundberg, P. O., Bergstrom, K., Hartvig, P., Lundqvist, H., Antoni, G., and Langstrom, B.,** Dopamine receptors in pituitary adenomas: PET visualization with ¹¹C-N-methylspiperone, *J. Comp. Assist. Tomog.,* 10, 175, 1986.

132. **Ginos, J. Z., Cooper, A. J. L., Dhawan, V., Lai, J. C. K., Strother, S. C., Alcock, N., and Rottenberg, D. A.,** [¹³N]Cisplatin PET to assess pharmacokinetics of intra-arterial versus intravenous chemotherapy for malignant brain tumors, *J. Nucl. Med.,* 28, 1844, 1987.

133. **Tyler, J. L., Yamamoto, Y. L., Diksic, M., Theron, J., Villemure, J. G., Worthington, C., Evans, A. C., and Feindel, W.,** Pharmacokinetics of superselective intra-raterial and intravenous [¹¹C]BCNU evaluated by PET, *J. Nucl. Med.,* 27, 755, 1986.

134. **Volkow, N. D., Brodie, J. D., Wolf, A. P., Angrist, B., Russell, J., and Cancro, R.,** Brain metabolism in patients with schizophrenia before and after acute neuroleptic administration, *J. Neurol. Neurosurg. Psychiatry,* 49, 1199, 1986.

135. **Wolkin, A., Angrist, B., Wolf, A., Brodie, J., Wolkin, B., Jaeger, J., Cancro, R., and Rotrosen, J.,** Effects of amphetamine on local cerebral metabolism in normal and schizophrenic subjects as determined by positron emission tomography, *Psychopharmacology,* 92, 241, 1987.

136. **Volkow, N. D., Hitzemann, R., Wolf, A. P., Logan, J., Fowler, J., Dewey, S., Schlyer, D., Christman, D., and Bendriem, B.,** Decreased brain metabolism during alcohol intoxication, *J. Nucl. Med.,* 30, 801, 1989.

137. **Kessler, R. M., Parker E. S., Clark, C. M., Martin, P. R., George, D. T., Weingarten, H., Sokoloff, L., Ebert, M. H., and Mishing, M.,** Regional cerebral glucose metabolism in patients with alcoholic Korsakoff's syndrome, *Soc. Neursci.,* 10 (Abstr.), 541, 1985.

138. **deWit, H., Metz, J. T., Gatley, S. J., Brunner, J. B., and Cooper, M. D.,** Relationship between mood and regional cerebral metabolism after a moderate dose of alcohol, *J. Cereb. Blood Flow Metab.,* 9, 5325, 1989.

139. **Sachs, H., Russell, J. A. G., Christman, D. R., and Cook, B.,** Alteration of regional cerebral glucose metabolic rate in non-Korsakoff chronic alcoholism, *Arch. Neurol.,* 44, 1242, 1987.

140. **Yarchoan, R., Brouwers, P., Spitzer, A. R., Grafman, J., Safai, B., Perno, C. F., Schmidt, P. J., Larson, S. M., Berg, G., Fischl, M. A., Wichman, A., Thomas, R. V., Brunetti, A., Myers, C. E., and Broder, S.,** Response of human-immunodeficiency-virus-associated neurological disease to 3′-azido-3′-deoxythymidine, *Lancet,* January 17, 132, 1987.

141. **Volkow, N. D.,** Brain metabolism in chronic cocaine users, in *Proc. American Psychiatric Association Annual Meeting,* Puerto Rico, 1988, 117.

142. **London, E. D., Cascella, N. G., Wong, D. F., Sano, M., Dannals, R. F., Links, J., Herning, R. I., Toung, J. K. T., Wagner, H. N., Jr., and Jaffe, J. H.,** Acute Cocaine Decreases Regional Cerebral Glucose Utilization in Human Substance Abusers, *Society of Neuroscience,* 1988, 919.

143. **Baxter, L. R., Jr., Schwartz, J. M., Phelps, M. E., Mazziotta, J. C., Barrio, J., Rawson, R. A., Engel, J., Guze, B. H., Selin, C., and Sumida, R.,** Localization of neurochemical effects of cocaine and other stimulants in the human brain, *J. Clin. Psych.,* 49(Suppl. 2), 23, 1988.

144. **Wong, D. F., Ross, C., Wagner, H. N., Jr., Pearlson, G., Links, J. M., Broussolle, E., Fanaras, G., Fischman, M., Danashvar, D., Wilson, A., Ravert, H., and Dannals, R. F.,** The effect of IV cocaine on the kinetics of C11-3-N-methylspiperone and binding in the human caudate, *J. Nucl. Med.,* 27(Abstr.), 853, 1986.

145. **Volkow, N. D., Fowler, J. S., Wolf, A. P., Schlyer, D., Shiue, C.-Y., Alpert, R., Dewey, S. L., Logan, J., Bendriam, B., Christman, D., Hitzemann, R., and Henn, F.,** Effects of chronic cocaine abuse on postsynaptic dopamine receptors, *Am. J. Psychiatry,* 147, 719, 1990.

146. **Volkow, N. D., Mullani, N., Gould, K. L., Adler, S., and Krajewski, K.,** Cerebral blood flow in chronic cocaine users: a study with positron emission tomography, *Br. J. Psychiatry,* 152, 641, 1988.

147. **Volkow, N. D., Mullani, N., Gould, L., Adler, S. S., Guynn, R. W., Overall, J. E., and Dewey, S.,** Effects of acute alcohol intoxication on cerebral blood flow measured with PET, *Psychiatr. Res.,* 24, 201, 1988.

148. **Cook, C. E.,** Pharmacokinetic studies of cocaine and phencyclidine in man, in *Pharmacokinetics and Pharmacodynamics of Psychoactive Drugs,* Biomedical Publications, Foster City, CA, 1984, 48.

149. **Wolf, A. P., Shiue, C.-Y., Dewey, S. L., Schlyer, D. J., MacGregor, R. R., Logan, J., Volkow, N., and Fowler, J. S.,** Regional and temporal distribution of the antipsychotic drug haloperidol in human brain. A PET study with [¹⁸F]haloperidol, *J. Nucl. Med.,* 29, 767, 1988.

150. **Wolkin, A., Barouche, F., Wolf, A. P., Rotrosen, J., Fowler, J. S., Shiue, C.-Y., Cooper, T. B., and Brodie, J. D.,** Dopamine blockade and clinical response: evidence for two biological subgroups of schizophrenia, *Am. J. Psychiatry,* 146, 905, 1989.
151. **Cambon, H., Baron, J. C., Boulenger, J. P., Loc'h, C., Zarifian, E., and Maziere, B.,** *In vivo* assay for neuroleptic receptor binding in the striatum. Positron tomography in humans, *Br. J. Psychiatry,* 151, 824, 1987.
152. **Arnett, C. D., Fowler, J. S., MacGregor, R. R., Schlyer, D. J., Wolf, A. P., Langstrom, B., and Halldin, C.,** Turnover of brain monoamine oxidase measured *in vivo* by positron emission tomography using L-[^{11}C]deprenyl, *J. Neurochem.,* 49, 522, 1987.
153. **Lee, M. C., Wagner, H. N., Jr., Tanada, S., Frost, J. J., Bice, A. N., and Dannals, R. F.,** Duration of occupancy of opiate receptors by naltrexone, *J. Nucl. Med.,* 29, 1207, 1988.
154. **Bartlett, E. J., Brodie, J. D., Wolf, A. P., Christman, D. R., Laska, E., and Meissner, M.,** Reproducibility of cerebral glucose metabolic measurements in resting human subjects, *J. Cereb. Blood Flow Metab.,* 8, 502, 1988.
155. **Tyler, J. L., Strother, S. C., Zatorre, R. J., Alivisatos, B., Worsley, K. J., Diksic, M., and Yamamoto, Y. L.,** Stability of regional cerebral glucose metabolism in the normal brain measured by positron emission tomography, *J. Nucl. Med.,* 29, 631, 1988.
156. **Grady, C. L., Berg, G., Carson, R. E., Daube-Witherspoon, M. E., Friedland, R. P., and Rapoport, S. I.,** Quantitative comparison of cerebral glucose metabolic rates from two positron emission tomographies, *J. Nucl. Med.,* 30, 1386, 1989.

Chapter 3

BRAIN PERFUSION RADIOTRACERS

Richard C. Reba and B. Leonard Holman

TABLE OF CONTENTS

I. INTRODUCTION

Radiotracers must meet three physiological requirements to be useful for brain perfusion SPECT: (1) they must cross the blood brain barrier, (2) their extraction must approximate unity and be independent of flow so that their initial distribution will be proportional to regional cerebral blood flow, and (3) they must be retained within the brain in their initial distribution long enough for statistically valid tomographic imaging to be accomplished. It is also useful if the radiopharmaceutical can be formulated as a simple kit so that the preparation and quality control are suitable for routine clinical use. The compounds that have been developed are lipophilic, moving across the blood-brain barrier with nearly complete extraction during a single passage through the cerebral circulation. Once inside the brain, they are either bound to nonspecific receptors or are metabolized to nonlipophilic compounds. As a result, they maintain their distribution within the brain for some time after injection.

II. ^{123}I IMP

^{123}I *N*-isopropyl-*p*-iodoamphetamine IMP is a highly lipophilic compound, moving across the blood-brain barrier with almost complete extraction during a single passage through the cerebral circulation.[1,2] Under normal physiologic conditions, the initial distribution of ^{123}I IMP is proportional to regional cerebral blood flow over a wide range of flow rates, but may be decreased when plasma pH is low, as in cerebral ischemia or any acidotic state.[3]

IMP uptake within the brain is rapid, reaching 45% of maximum radioactivity by 2 min, about 75% of the peak activity within 15 to 20 min, and 6 to 9% of the injected dose by 30 min.[4] The clearance of IAMP from the brain is balanced by continued brain uptake from plasma concentration maintained by slow release of the tracer from the lungs.[5] As a result, brain activity remains constant from 20 min to at least 60 min after injection. The gray/white matter activity ratio remains constant during this time, resulting in high spatial resolution images capable of providing good contrast between gray and white matter. By 24 h, the gray/white matter activity ratio is reversed and activity is higher in the white matter. By 4 h after administration, it is not possible to distinguish very well between gray and white matter structures. Thus, except for specific circumstances, such as the luxury perfusion state or a brain tumor (vide infra), it is the SPECT images of ^{123}IMP distribution obtained early after injection that will reflect most accurately the expected correlation with regional cerebral blood flow. Subsequent to the early times, not only does continuous uptake influence IMP distribution, but intracerebral distribution changes with time not only in abnormal brain tissues but in normal brain structures as well.[6,7] Therefore delayed images are not well suited to assess regional cerebral blood flow.

Specific transport systems are characterized by saturability, competitive inhibition, and stereospecificity. Several studies indicate that uptake and retention of IMP are not blocked by IMP carrier doses nor by sympathetic amines[8,9] nor are they affected by the spatial orientation of the optically active carbon atom.[10] Although there is poor correlation between lipid solubility (pKa) and brain uptake of various amines,[11] IMP uptake appears to be most consistent with lipophilicity alone, i.e., simple diffusion through the blood-brain barrier. Almost all of the IMP retained in the brain is bound, probably to high-density, relatively low-affinity binding sites and not to any of the well-known amine receptors.[11a] These features confound the assessment of regional cerebral blood flow using IMP. However, regional cerebral IMP uptake and oxygen metabolism in ischemic lesions are highly correlated.[12]

The extraction fraction of IMP is approximately 1.0 at reduced cerebral blood flows, but deteriorates rapidly with increasingly high cerebral blood flows. However, with corrections for the extraction fraction deviations from unity and the precise determination of the input function, full kinetic modeling may result in relatively reliable absolute quantitation of cerebral blood flow. The kinetic modeling approach, with appropriate corrections, will also allow determination of cerebral blood flow using HM-PAO.

Most brain tumors, even those that are highly vascular, do not extract IMP probably because of a deficiency in the lipophilic uptake system. Cerebral metastases from malignant melanoma and oat cell tumors do take up the tracer, particularly if they are actively metabolizing amphetamine precursors.

III. ^{99m}Tc-HM-PAO

^{99m}Tc-HM-PAO (hexamethyl propyleneamine oxime), an optically active lipid soluble diastereoisomer, a racemic mixture of the D and L isomers, was developed by modifying a neutral and stable [^{99m}Tc]PnAO.[13-16] It is the first FDA approved ^{99m}Tc brain perfusion imaging agent and is now widely available for routine human use.

The compound is a macrocyclic amine, like IMP, able to penetrate the intact blood-brain barrier because of its lipophilicity, resulting in a fairly high, but significantly less than complete, first pass extraction. Brain uptake is therefore rapid and within 10 min after injection reaches its maximum. Its initial cerebral distribution remains constant for several hours and is proportional to regional cerebral blood flow.[16] ^{99m}Tc-HM-PAO reacts in seconds after crossing the cell membrane, probably with intracellular glutathione, and is converted from its highly lipophilic state to a hydrophilic species and thus is trapped intracellularly.[17] This conversion occurs at a slower rate in the extracellular fluids. Unlike IMP, after the initial 6 to 10 min and after a plateau has been established, ^{99m}Tc-HM-PAO does not redistribute within the brain for at least 8 h after injection and brain washout is insignificant, 0.4% h.[18] Because blood clearance is slow and its brain uptake is approximately 75% that of IMP, degradation of image quality results. It is apparent that delayed ^{99m}Tc-HM-PAO images will demonstrate areas of reduced cerebral blood flow relatively accurately. However, the incomplete first-pass extraction results in underestimation of CBF in high flow areas. These may be important features resulting in the lowered sensitivity for detecting ischemia when HM-PAO is used.

^{99m}Tc-HM-PAO has a number of advantages relative to that of ^{123}I IMP, including on-site preparation, lower radiation dose, lower cost, ideal nuclear characteristics, and a higher available photon flux. However, the radiopharmaceutical is chemically unstable *in vitro* by 30 min after preparation. Because of slow blood clearance, there is high blood activity of ^{99m}Tc-HM-PAO and defects due to stroke or other focal cerebral pathology are not seen as sharply as with ^{123}I IMP. On the other hand, ^{99m}Tc-HM-PAO may mask the ischemic lesion because of hyperemia and normal or increased tracer uptake.[19] These regions may appear abnormal on early ^{123}I IMP imaging but normal on delayed imaging, reflecting the dependence of IMP distribution on receptor interaction and pH and, therefore, on metabolism, a phenomenon apparent when blood flow and metabolism are uncoupled.

IV. ^{99m}Tc ECD (ETHYLENE CYSTEINE DIMER)

Another class of compounds is based on the diamine dithiol (DADT) backbone of one of the N_2S_2 ligands with free SH groups.[20,21] Several members of this class of compounds form a neutral complex with ^{99m}Tc, are lipid soluble, have a Tc^{V}O center core structure, and are characterized by high initial cerebral extraction. One of these compounds, the L,L-isomer of ^{99m}Tc ethyl cysteine dimer (ECD) couples the high initial cerebral extraction with very slow clearance from the brain.[22,23] Its initial distribution is proportional to regional cerebral blood flow. The proposed mechanism for retention within the brain is a specific enzymatic rapid hydrolysis of one of the ester groups to an acid (a polar metabolite) that is trapped within the brain cells and does not recross the blood-brain barrier. The enzymatic process is stereospecific and only occurs in the brains of primates, i.e, monkeys and humans, but not in rats or rabbits. Brain uptake is rapid and peak brain activity compares favorably with other brain perfusion agents, reaching 6% of the injected dose by 5 min after intravenous injection.

The blood clearance is rapid, resulting in high brain-to-soft tissue activity ratios early after injection and improving with time for at least several hours.[24,25] Rapid lung clearance reduces the problem of adjacent background activity during imaging. The rapid washout from facial muscles and from the salivary glands and the more rapid blood clearance of ECD results in higher brain-to-soft tissue ratios than have been reported for ^{99m}Tc-HM-PAO at comparable times after injection. The specific mechanism of cerebral localization (vide supra) is worth noting because the perfusion images obtained by this agent are probably a combination of perfusion and the specific enzyme reaction. When the ubiquitous enzymatic reaction is tightly coupled to regional perfusion, the regional distribution of the ^{99m}Tc radiopharmaceutical reflects blood flow.

Unlike ^{99m}Tc-HM-PAO, ECD clears slowly from the brain at the rate of 12% during the first hour and 5 to 6%/h thereafter.[26] There does not appear to be intracerebral redistribution of the tracer as seen with ^{123}I IMP. The ^{99m}Tc ECD gray/white matter activity ratio remains high and constant over time when measured from sequential SPECT studies.[27] [99m]Tc ECD appears to have the same advantages relative to ^{123}I IMP as does ^{99m}Tc-HM-PAO, but, in addition, is stable and has a rapid blood clearance, resulting in high-quality SPECT images with high target-to-background ratios.

ACKNOWLEDGMENTS

We thank Dr. Hank Kung for his critical review of the manuscript and his suggestions. The preparation of this chapter was supported by DHHS/NIH Grant NS 22215 and Department of Energy Grant ER 60649.

REFERENCES

1. **Winchell, H. S., Baldwin, R. M., and Lin, T. H.,** Development of I-123-labeled amines for brain studies: localization of I-123 iodophenylalkyl amines in rat brain, *J. Nucl. Med.,* 21, 940, 1980.
2. **Winchell, H. S., Horst, W. A., Braum, L., Oldendorf, W. H., Hattner, R., and Parker, H.,** *N*-isopropyl (I-123) *p*-iodoamphetamine: single-pass brain uptake and washout, binding to brain synaptosomes and localization in dog and monkey brain, *J. Nucl. Med.* 21, 947, 1980.
3. **Kuhl, D. E., Barrio, J. R., Huang, S.-C., Selin, C., Ackermann, R. F., Lear, J. L., Wu, J. L., Lin, T. H., and Phelps, M.,** Quantifying local cerebral blood flow by *N*-isopropyl-*p*-123-I-iodoamphetamine (IMP) tomography, *J. Nucl. Med.,* 23 196, 1982.
4. **Holman, B. L., Lee, R. G. L., Hill, T. C. et al.,** A comparison of two cerebral perfusion tracers - isopropyl I-123 *p*-iodoamphetamine and I-123 HIPDM in the human, *J. Nucl. Med.,* 25, 25, 1984.
5. **Yonekura, Y., Fujita, T., Nishizawa, S., Iwasaki, Y., Mukai, T., and Konishi, J.,** Temporal changes in accumulation of *N*-isopropyl-*p*-iodoamphetamine in human brain: relation to lung clearance, *J. Nucl. Med* 30, 1977, 1989.
6. **Lassen, N. A., Henriksen, L., Holm, S. et. al.,** Cerebral blood flow tomography: xenon-133 compared with isopropyl-amphetamine-iodine-123; concise communication, *J. Nucl. Med.,* 24, 17, 1983.
7. **Nishizawa, S., Tanada, S., Yonekura, Y., Fujita, T., Mukai, T. et. al.,** Regional dynamics of *N*-isopropyl-(^{123}I) p-iodoamphetamine in human brain, *J. Nucl. Med.* 30, 150, 1989.
8. **Wu, J. L., Baldwin, R. M., and Lin, T. H.,** Preparation and tissue distribution no-carrier added I-125 *N*-isopropyl-*p*-iodoamphetamine, *J. Nucl. Med.,* 27, 940, 1986.

9. **Moretti, J. L. Holman, B. L., Delmon, L., Carmel, A., Johnson, D., Moingeon, P., and Blau, M.,** Effect of antidepressant and narcoleptic drugs on N-isopropyl-p-iodoamphetamine biodistribution in animals, *J. Nucl. Med.,* 28, 354, 1987.

10. **Baldwin, R. M., Lin, T. H., and Wu, J. L.,** Synthesis and brain uptake of isomeric I-123 iodoamphetamine derivatives, *J. Labeled Comp. Radiopharm.,* 19, 1305, 1985.

11. **Pardridge, W. M. and Connor, J. D.,** Saturable transport of amphetamine across the blood-brain barrier, *Experientia,* 29, 302, 1973.

11a. **Mori, H., Shiba, K., Matsuda, H., Tusji, S., and Hisada, K.,** Characteristics of the binding of N-isopropyl-p-iodoamphetamine [^{125}I] in rat brain synaptosomal membranes, *Nucl. Med. Commun.,* 11, 327, 1990.

12. **Raynaud, C., Rancurel, G., and Samson, Y.,** Pathophysiologic study of chronic infarcts with I-123-isopropyl-iodoamphetamine (IMP). The importance of the periinfarct area, *Stroke,* 18, 21, 1987.

13. **Volkert, W. A., Hoffman, T. J., Seger, R. M., Troutner, D. E., and Holmes, R. A.,** ^{99m}Tc propylene amine oxime (^{99m}Tc-PnAO); a potential brain radiopharmaceutical, *Eur. J. Nucl. Med.,* 9, 511, 1984.

14. **Nowotnik, D. P., Canning, L. R., and Cummings, S. A.,** Development of a 99mTc-labelled radiopharmaceutical for cerebral blood flow imaging, *Nucl. Med. Commun.,* 6, 499, 1985.

15. **Troutner, D. E., Simon, J., Ketring, A. R. et. al.,** Complexing of Tc99m with cyclam; concise communication, *J. Nucl. Med. Suppl.,* 21, 443, 1980.

16. **Neirinckx, R. D., Canning, L. A., Piper, J. M., Nowotnik, D. P., Pickett, R. D., Holmes, R. A., Volkert, W. A., Forster, A. M., Weisner, P. S., Marriott, J. A., and Chaplin, S. B.,** Technetium 99 d,l HM-PAO: a new radiopharmaceutical for SPECT imaging or regional cerebral blood perfusion, *J. Nucl. Med.,* 28, 191, 1987.

17. **Neirinckx, R. D., Burke, J. F., Harrison, R. C., Forster, A. M., Anderson, A. R., and Lassen, N. A.,** The retention mechanism of technetium-99m-HM-PAO: intracellular reaction with glutathione, *J. Cereb. Blood Flow Metab. Suppl.,* 8, S4, 1988.

18. **Anderson, A. R., Friberg, H. H., Schmidt, J. F., and Hesselbach, S. G.,** Quantitative measurements of cerebral blood flow using SPECT and [^{99m}Tc]-d,l-HM-PAO compared to xenon-133, *J. Cereb. Blood Flow Metab.,* 8(Suppl. 1), S69, 1988.

19. **Moretti, J.-L., Defer, G., Cinotti, L., Cesaro, P., Degos, J.-D., Vigneron, N., Ducassou, D., and Holman, B. L.,** ''Luxury perfusion'' with ^{99m}Tc-HM-PAO and ^{123}I-IMP SPECT imaging during the subacute phase of stroke, *Eur. J. Nucl. Med.,* 16, 17, 1990.

20. **Lever, S. Z., Burns, H. D., Kervitsy, T. M., et. al.,** The design, preparation and biodistribution of a technetium-99m triaminodithiol complex to assess regional cerebral blood flow, *J. Nucl. Med.* 26, 1287, 1985.

21. **Kung, H. F., Molnar, M., and Billings, J.,** Synthesis and biodistribution of neutral lipid-soluble Tc-99m complexes that cross the blood-brain barrier, *J. Nucl. Med.,* 25, 326, 1984.

22. **Cheesman, E. H., Blanchette, M. A., Ganey, M. V., Maheu, L. J., Miller, S. J., and Watson, A. D.,** Technetium-99m ECD: ester-derivatized diamine-dithiol Tc complexes for imaging brain perfusion, *J. Nucl. Med.,* 29 (Abstr.), 788, 1988.

23. **Walovitch, R. C., Williams, S. J., Morgan, R. A., Garrity, S. T., and Chessman, E. H.,** Pharmacology characterization of Tc-99m ECD in non-human primates as a new agent for brain perfusion imaging, *J. Nucl. Med.,* 29 (Abstr.), 788, 1988.

24. **Vallabhajosula, S., Zimmerman, R. E., Picard, M. et. al.,** Technetium-99m EDC: a new brain imaging agent: in vivo kinetics and biodistribution studies in normal human subjects, *J. Nucl. Med.,* 30, 599, 1989.

25. **Holman, B., L., Hellman, R. S., Goldsmith, S. J., Mena, I. G., Levelle, J. et. al.,** Biodistribution, dosimetry, and clinical evaluation of technetium-99m ethyl cysteine dimer in normal subjects and in patients with chronic cerebral infraction, *J. Nucl. Med.,* 30, 1018, 1989.

26. **Franceschi, M., Picard, M., Zimmerman, R. E., Kronauge, J. F., Jones, A. G., and Holman, B. L.,** Brain washout of Tc99m-L, L-ethyl cysteinate dimer (ECD) in normal volunteers, *Eur. J. Nucl. Med.,* 14, 227, 1988.

27. **Leveille, J., Demonceau, G., Rigo, P., De Roo, M., Taillefer, R., Burgess, B. A., Morgan, R. A., and Wallovitch, R. C.,** Brain tomographic imaging with Tc-99m-ethyl cysteinate dimer (Tc-ECD): a new stable brain perfusion agent, *J. Nucl. Med.,* 29 (Abstr.), 758, 1988.

Chapter 4

THE TESTING OF PUTATIVE RECEPTOR BINDING RADIOTRACERS *IN VIVO*

W. C. Eckelman

TABLE OF CONTENTS

I. INTRODUCTION

The testing of putative receptor-binding radiotracers *in vivo* is one of the most complicated tasks in radiopharmaceutical chemistry. Although the synthetic challenges are also formidable for this class of compounds, the analysis of a newly synthesized, receptor-binding radiotracer can be made in terms of physicochemical properties and the specific activity can be determined by well-known analytical techniques. On the other hand, the testing of putative receptor-binding radiotracers *in vivo* is based entirely on circumstantial evidence. No one piece of evidence proves that the receptor-binding radiotracer is localizing by a receptor-mediated mechanism much less that the sensitivity of the concentration of a receptor-binding radiotracer to changes in receptor concentration is appropriate. In this chapter, a series of tests are proposed that define a radiolabeled ligand as both a receptor-binding ligand and a sensitive measure of receptor change. These criteria are summarized and then applied to a series of putative receptor-binding radiotracers.

II. SCOPE OF THE PROBLEM

The scope of the problem falls into two distinct categories: the development of a radiotracer that binds preferentially to a chosen receptor after intravenous injection and the development of a pharmacokinetic scheme that allows the change in receptor concentration to be detected with accuracy and sensitivity. The first implies that nonreceptor binding, whether to plasma proteins or to binding sites in the target tissue, is minimal and that transport to the target tissue is uninhibited. Transport is a problem especially for neuroreceptor ligands when the blood brain barrier is normal and cellular transport is a necessity. Having achieved the efficient delivery of the radiotracer to the receptor, the second challenge is to develop a scheme whereby the distribution of the radioactivity is an accurate and sensitive measure of the receptor concentration. The two parts of the problem are contradictory in a sense that a radioligand which binds preferentially to one receptor and little else must by its nature bind with a high affinity constant. However, if the trapping mechanism is so avid, having more or fewer receptors will not change the amount in the target tissue greatly and therefore the agent will not be a sensitive measure of the change in receptor concentration, i.e., the quantity of radioactivity detected in the target tissue will not change drastically for a change in receptor concentration. Both of these properties are related to a third tacit assumption which will not be addressed in this chapter, namely, that the receptor system has been identified as one connected to a certain disease state. Many receptor-binding ligands have been developed before the question of their clinical usefulness is postulated. Likewise, many receptor-binding radiotracers have been developed and studied in humans without validating the sensitivity to change in receptor concentration.

A. DEVELOPING A RADIOTRACER THAT BINDS PREFERENTIALLY TO A CHOSEN RECEPTOR

The greatest strength of nuclear medicine lies in its ability to detect and quantitate biologic function rather than fixed anatomical properties. These functional studies can be divided into two categories: the determination of flow and the determination of a biochemical interaction, such as receptor binding. These are not mutually exclusive phenomena; certainly, the material must be delivered to the target organ so flow is a necessary part of the distribution. Rather, the distinction is based on the relative kinetics of the two processes. Whether the distribution of a particular agent is a reflection of flow differences or of receptor concentration differences is kinetically determined. One example of a predominately flow-dominated distribution is radiolabeled microspheres where the trapping of the particles in the capillaries is much faster than the flow in comparable units, e.g., time^{-1}, so that flow is the rate determining step and hence the distribution is primarily a function of flow. In most cases, the relative kinetics are such that one process does not dominate the other kinetically over a wide range of flows.

Many of the radiopharmaceuticals used in routine clinical diagnosis measure changes in flow to an organ as a function of disease. The gold standard for flow measurements in experimental animals is usually radiolabeled microspheres that are extracted by the capillary vessels with an extraction fraction of one. This extraction fraction is independent of flow. Many radiopharmaceuticals are compared to microspheres over a flow range expected in the clinic. For example, monocationic salts of the group I alkali metals, K, Rb, and Cs, and the group III element, Tl, are indicators of flow and correlate with microspheres in a linear fashion over a limited range of flows.[1] This is in spite of the fact that these cations are not strictly flow-limited tracers, i.e., their extraction fraction is not one at physiologic flow rates. Although the uptake of these cations is determined by capillary surface area and membrane permeability as well as flow, in practice the uptake is directly proportional to flow. However, this does not mean that the correlation between microspheres and a cation such as Rb will have a slope of one and an intercept of zero.[2] The relative rate constants are such that the transport phenomena do not become kinetically dominant until the flow has reached multiples of the normal flow, but they do decrease the sensitivity to flow changes at lower flows.[3] Other agents also measure flow. Certainly ^{99m}TcMAA, having an identical mechanism of localization as microspheres, is a measure of flow in the lung.[4] Likewise, the extraction of colloid by the phagocytic cells in liver is flow limited.[5] In the case of reduced flow as a result of hepatitis, the uptake of ^{99m}Tc sulfur colloid is decreased proportionately to flow.[6] ^{131}I o-iodohippurate (Hippuran) is also used to measure flow, in this case, renal plasma flow because of its high extraction by the kidneys.[7] McAfee et al. have shown that radiochemically pure Hippuran is extracted with high efficiency from plasma.[8] Likewise, iodoamphetamine,[9] and the proposed pH shift agents of Kung et al.[10] are a measure of cerebral blood flow.[11] Both seem to correlate well with microspheres immediately after injection in normal tissue. However, in abnormal tissue, LaFrance at al. showed that the extraction decreased although the flow stayed the same.[12] On the other hand, Szasz et al. showed that increased concentration of radioactivity could be obtained in brain metastases with normal flow presumably due to increased amine uptake.[13] Lucignani et al. showed that a related compound, HIPDM, was not a chemical microsphere, but rather a fortuitous balance of input and efflux that led to constant levels of radioactivity.[14] The extraction fraction for radiopharmaceuticals usually applies only to normal organs; no systematic study has identified the relative effect of decreased flow or changes in the physiological chemistry on decreased concentration of radioactivity in the diseased organ. Clinically useful flow tracers that are insensitive to biochemical changes, especially in disease states, have not been validated to date.

In recent years, the research emphasis has been toward not only those tracers that measure flow, but also those radiotracers that measure predominantly a biochemical process, such as receptor binding. This emphasis has come about for a number of reasons: the competition of other imaging modalities, the success in measuring biochemical pathways by using cyclotron-produced radionuclides by isotopic substitution, and the improvements in both positron emission tomography and single-photon computed tomography. Many new imaging modalities, such as CAT, nuclear magnetic resonance imaging, and ultrasound, are better suited to record anatomical changes. On the other hand, the success of isotopic substitution with such compounds as (^{11}C)-palmitic acid,[15] (^{11}C)-glucose,[16] and (^{18}F)-2-fluoro-2-deoxyglucose[17] has demonstrated that in vivo biochemical tracers are possible. Finally, the demonstration that external imaging can be used to trace physiological chemistry in quantitative terms has had a major impact on the direction of research. The ability to measure glucose metabolism using (^{18}F)-2-fluoro-2-deoxyglucose has been especially impressive. In a similar manner, the ability to measure receptor concentration by external detection would greatly broaden the field of biochemical radiotracers.

Receptor binding radiotracers are especially interesting in that changes in receptor concentration are thought to be related to certain disease states. The receptor concept was developed to explain the specific effect of a minute amount of unchanged substrate on a target organ. Throughout the history of science, investigators realized that there must be a substance that imparts specificity to a particular cell type. Ehrlich realized this in his study of the interaction of antigens with specific, complementary, preformed receptors.[18] In 1906, Langley postulated the existence of "receptive substances" on cell surfaces.[19] Few receptors have been isolated for structural determination. As a result, the definition of a receptor is operational.[20] It is defined by certain properties observed *in vitro*. The usual criteria are high affinity, specificity, saturability, and distribution in relation to physiologic response. These criteria separate the receptor from a binding site in that the latter is not involved in a physiologic action. An example of such a binding site is thyroxine-binding globulin which binds thyroxine in plasma with high affinity and can be saturated, but causes no physiologic action by virtue of the binding. The receptor can be differentiated from an enzyme in that the ligand causes a physiologic effect without molecular changes. The most important property of an enzyme is its catalytic activity which is influenced by the nature of the substrate, temperature, and pH. The receptors have no catalytic activity, but rather produce a physiologic change indirectly. In addition, enzyme systems are often much higher in capacity than receptor systems.

The determination of the properties of receptors has been made possible by the development of high specific activity radiotracers. Jensen and Jacobson used radiolabeled estradiol to identify the cytosolic estradiol receptor in the early 1960s.[21] In general, receptor protein is present in limited concentration, about 10^{-7} to $10^{-10} M$ in homogenized tissue. Therefore, the receptor is easily saturated by the appropriate ligand. Specificity and high ligand affinity are closely related because, by the nature of the high affinity between the receptor and the ligand, specificity results. Often ligands at high concentration can cause a physiologic effect at numerous receptors, but are specific for only one receptor at low concentration. In the central nervous system, specificity is imparted to relatively low affinity agonists by virtue of the synaptic connections which are insulated both morphologically and biochemically by diffusional barriers, high affinity neurotransmitter uptake systems, and strategically located catabolic enzymes.[22] One of the most important properties of receptors is stereoselectivity. In the case when a pair of stereoisomers exist, one is usually far more potent than the other in binding affinity and physiologic response. For *in vivo* studies, this property will allow proof of receptor binding because it can be tested using high specific activity physiologically active and physiologically inactive radiotracers. *In vivo* tests for saturability would require the injection of a physiologically active amount of the biochemical or drug. This would probably be unacceptable as a routine diagnostic procedure if the ligand produces a pharmacologic effect. Many of the antagonists will produce adverse pharmacologic effects and some toxic reactions.

The final criterion is that the binding of the ligand to the sites can be related to the biological effect of the ligand. Experimentally, this has often been shown by comparing the affinity of various ligands with their *in vivo* biological effect. If a correlation is obtained, then the receptor is defined. In the radiotracer context, the distribution of the radioligand is determined *in vivo* and this is correlated with the known distribution of receptors from biological studies *in vivo* or from *in vitro* assays of various organs and tissues.

B. DEVELOPING A PHARMACOKINETIC SCHEME THAT ALLOWS THE CHANGE IN RECEPTOR CONCENTRATION TO BE DETECTED WITH ACCURACY AND SENSITIVITY

Wong et al. have first applied the use of tissue ratios as an indication of neuroreceptor concentration change to imaging *in vivo*.[23] This was based on the work Kuhar et al., who observed that the ratio between a tissue containing receptors and one that did not was an indication of the receptor concentration.[24] This has been derived in mathematical terms by Patlak for the general system and for the particular receptor system and has been proposed by Gjedde for the FDG studies.[25-27] In general, steady state assumptions are needed and the binding process is irreversible during the time of measurement.

Recently, Farde et al.[28] compared four approaches to analyzing pharmacokinetic data in an attempt to understand the different results obtained by Wong et al.[29] and his group[30] in the study of drug-naive schizophrenics. The approaches were (1) an equilibrium approach (more appropriately a steady state approach) previously published by Farde et al.,[31] (2) a steady state approach using the cerebellum as the reference organ, (3) a steady state approach using the pharmacologically inactive enantiomer to determine the nonreceptor localization, and (4) a modeling approach solving for five parameters including the receptor concentration. All the approaches gave similar values for the receptor concentration. In the experiments using the enantiomer, receptor concentrations could not be determined for two of the three patients indicating that both the pharmacologically active and pharmacologially inactive compounds bound equally in the target organ. Binding to other sites than the D2 dopamine receptor could be the cause of the cited discrepancy. This approach to determining the sensitivity of the distribution of the radioligand to changes in receptor concentration is far from satisfactory because it assumes that one of the approaches is indeed sensitive to changes in receptor concentration. Another similar approach to determining sensitivity is to compare the receptor concentration obtained by *in vivo* analysis to that obtained at autopsy using *in vitro* methods. Because of the many problems with comparing *in vivo* data with *in vitro* data, this method also cannot be considered a rigorous validation.

The method of choice is to alter systematically the various parameters that could control distribution of the radioligand *in vivo* and determine the sensitivity. The most obvious example of this approach is the determination of the kinetic sensitivity of ^{99m}Tc galactosyl-neoglycoalbumin.[32] The flow, ligand-receptor affinity, and total receptor concentration were determined under various conditions and the value and hence the sensitivity were confirmed by the simultaneous measurement of flow and the measurement of receptor concentration and affinity constant by *in vitro* techniques.

III. SOLUTION TO THE PROBLEM

A. SYNTHESIS OF THE RADIOLABELED RECEPTOR ANTAGONIST

In most cases, an antagonist has been chosen as the potential receptor binding radiotracer because the affinity of the antagonist is more likely to be in the range to give an appropriate target to nontarget ratio than the affinity of an agonist. Great strides have been made in choosing receptor-binding radiotracers by using the affinity constant alone, determined *in vitro*, to predict the target to nontarget ratio using the equilibrium expression developed by Scatchard:[33]

$$\text{Bound/free} = \text{target to nontarget} = R_oK - BK$$

where R_o is the total receptor concentration, K is the affinity constant, and B is the bound concentration. At small B/R_o ratios, the target to nontarget ratio can be predicted by the product of the receptor concentration and the affinity constant, R_oK.

The choice of radionuclide for single photon emitting radiotracers is limited to ^{99m}Tc and ^{123}I. ^{99m}Tc is by far the best radionuclide from an imaging point of view because of the ideal nuclear properties for the standard gamma camera and the ready availability afforded by the ^{99}Mo/^{99m}Tc generator system. However, the incorporation of ^{99m}Tc into a receptor antagonist is a formidable synthetic task which in most cases will perturb the interaction of the receptor ligand with the receptor. One important consideration is the preparation of a neutral chelate to better match the lipophilicity of most receptor binding ligands and to cross cell membranes in the case of cerebral receptors and peripheral intracellular receptors. Neutral ^{99m}Tc chelates have been prepared by Yokoyama et al. with the KTS chelate,[34] Dannals with the diaminodithio ligands,[35] Troutner et al. with the amineoxime ligands,[36] and Nunn et al. with the seven coordinate tris-dioximes.[37] Although the ^{99m}Tc chelates have been bound to various biochemicals, especially the fatty acids, none have been bound to receptor-binding ligands in such a way as to maintain high receptor affinity. One receptor ligand that is labeled with ^{99m}Tc is neogalactosealbumin (NGA).

Radiohalogens seem to be a more likely prospect for initial studies. The radionuclides with the best nuclear properties for imaging with the gamma camera are ^{123}I and ^{131}I. ^{125}I has also been used but is most effective in *in vitro* tests. Both ^{125}I and ^{131}I are reactor produced and therefore less expensive and more readily available than the cyclotron product ^{123}I. However, the equilibrium absorbed dose to the patient from both reactor products is high because of their long half lives. ^{123}I is the ideal radioisotope for imaging with a 159 KeV gamma ray and low radiation absorbed dose. This gamma ray has a half thickness in water of 4.7 cm and therefore has satisfactory tissue penetration, yet the energy is low enough to be easily collimated.[38] However, this radionuclide is only produced in a cyclotron and cannot easily be made free of ^{124}I and/or ^{125}I with a low-energy cyclotron. High-energy protons, ^{127}I(p,5n), or high-energy helium particles, ^{123}Te(α,3n), produce the purest ^{123}I, but are not readily available.[39] High purity ^{123}I is produced by those nuclear transformations that go through a ^{123}Xe intermediate. The most often used reaction to produce ^{123}I of high purity is the ^{127}I(p,5n)^{123}I reaction. The indirect method of production via intermediates does not result in ^{124}I because ^{124}Xe is stable. ^{124}I is an undesirable radionuclidic impurity because of its long half-life. (4.2 d) and its high-energy photons (511, 603, 723 KeV). A 1% level of ^{124}I at the time of production will usually result in a 5% contamination at the imaging time of 24 h. The ^{124}I degrades the resolution because a significant fraction of the gamma rays within the 159 KeV energy window will be from scattered radiation. With a low energy collimator, the scatter contribution will be 28%. With a high resolution, medium-energy collimator, the contribution is 10%. With a high-energy collimator only 3% is scattered radiation, but the sensitivity is greatly reduced.[40] Since both low resolution and high sensitivity are necessary for the external detection of receptor-binding radioligands, the need for high radionuclide purity is obvious.

^{123}I appears to be the best radiohalogen based on nuclear properties, absorbed radiation dose to the patient, and availability. However, the chemical properties of iodine are less than ideal because of the low bond energy of the carbon iodine bond and the subsequent ease of *in vivo* deiodination. Bromine, on the other hand, has superior chemical properties to iodine in most instances, but ^{77}Br has a complicated gamma ray decay scheme which makes image resolution difficult and the radiation dose to the patient higher per millicurie than that delivered by ^{123}I. A relative Figure of Merit for the halogens shows that ^{123}I $>$ ^{18}F $>>$ ^{77}Br.[41] Presently, hundred-millicurie quantities of ^{77}Br are made only on large accelerators.[42] The radioiodination of receptor binding radiotracers has been reviewed by Eckelman[43] and by Reba and Holman[44] in this volume.

B. SIXTEEN STEPS TO ACHIEVE THE GOAL

There have been many suggestions on how to systematically evaluate a putative receptor-binding radiotracer. One such approach[45] outlines 16 categories in developing and testing a receptor binding radiotracer. They are as follows:

1. The choice of a receptor system
2. Use of a mathematical model to choose potential receptor-specific radiopharmaceuticals

3. The determination of the K_A (app) for the parent compound
4. Use of ^{3}H-labeled parent compound to determine distribution *in vivo*
5. The preparation of the nonradioactive derivative
6. The determination of *in vitro* stability of the nonradioactive derivative
7. The determination of the K_A (app) for the nonradioactive derivative
8. The evaluation of various physical parameters of the nonradioactive derivative (structure-distribution relationship, SDR)
9. *In vivo* displacement of the ^{3}H compound with the nonradioactive derivative
10. Preparation of the radioactive derivative
11. Chromatographic separation
12. The determination of the K_A for the radioactive derivative
13. Correlation of distribution of the radioactive derivative in animals with receptor location
14. Use of preinjection, coinjection, or postinjection to decrease effective specific activity of the radioactive derivative
15. Use of active and inactive stereoisomers of the radioactive derivative
16. Confirmation of animal distribution of the radioactive derivative in humans

The choice of a receptor system is difficult because it is often based on *in vitro* assay of autopsy samples from patients who have been extensively treated. There have been a number of reviews of the change in receptor concentration as a function of disease as obtained from autopsy samples.[46,47] It is outside the scope of this chapter to review those data. Once having chosen a receptor system, the use of a mathematical model to choose potential receptor-specific radiopharmaceuticals has been put forth by Katzenellenbogen[48] and Eckelman[49] as the first screen. Certainly distributional factors, protein binding, and metabolism will tend to decrease the maximal B/F ratio. As a result, this criterion is necessary, but not sufficient. One problem that results from using the *in vitro* equation for *in vivo* studies is the nonequal volume of distribution for the radioligand and the receptor. This can be accounted for by altering the classical quadratic equation for calculating B/F.[50] The determination of the K_A (app) for the parent compound by *in vitro* tests is a necessary part of the determination of the maximum B/F ratio if a reliable literature value is not available.

The use of ^{3}H-labeled compounds is one of the most efficient methods to determine if particular antagonist is suitable for radiolabeling with a gamma-emitting radionuclide. At the maximum specific activity of 30 Ci/mmol for one tritium per molecule, the specific activity is sufficient to produce the maximum B/F ratio in *in vivo* experiments. These experiments are one level of sophistication above the use of the mathematical model in that the actual distribution of the ligand, with all the complicating factors assumed to be negligible in the modeling, is present. Many of the tritium ligands are commercially available, thereby eliminating costly synthetic effort to prove that a particular ligand has a high probability of success as a receptor binding radiotracer. After the particular receptor ligand has been deemed satisfactory either by *in vitro* receptor studies or preferably by *in vivo* distribution studies using the ^{3}H-labeled compound, the preparation of the nonradioactive substituted ligand is often the next step. This avoids the problems with no-carrier-added synthesis until the substituted ligand is shown to be a true tracer for the unsubstituted receptor binding ligand. This also produces the necessary reference compounds for the radiolabeled compound. The preparation of the nonradioactive compound better defines the radioactive compound because classical analytical methods can be used, whereas with the high specific activity compound, chromatography is the only analytical tool usually available.

Many halogenated compounds, especially iodinated compounds, are not stable *in vivo*. Therefore the stability of the new derivative should be determined in plasma and in the presence of target tissue. One difficulty with using the nonradioactive compound for stability testing is caused by the large concentrations used which will most likely result in a second order reaction whereas the high specific activity ligand will most likely be in a pseudo first order reaction.

The preparation of the nonradioactive form of the putative receptor binding radiotracer leads to the determination of the K_A (app) for the nonradioactive derivatives. In comparing the apparent affinities of a series of halogenated compounds, this approach is especially advantageous because of the similar chemistry that can be used for fluorine, bromine, and iodine on the carrier level. On the other hand, the need for different chemistry for the various halogens when radiolabeling on the no-carrier-added level makes comparisons of the halogens more difficult.

Just as structure-activity relationships have been the foundation of classic drug design, structure-distribution studies are the backbone of radiopharmaceutical development. The use of both theoretical and experimental parameters can direct the choice of radioligand. Most radiopharmaceuticals to date are water-soluble, polar compounds that are excreted by the kidneys. As radioligands were developed that cross cell membranes, the properties of these compounds were often correlated with a particular physicochemical property. Most often the property is lipophilicity determined either by measuring the organic-aqueous partition coefficient or by using a related technique such as reversed-phase high pressure liquid chromatography (HPLC). Various organic solvents have been used to determine the partition coefficient. The goal is to choose an organic solvent that most closely models the cell membrane. Franks and Lieb published the classic example of the relationship between the activity of general anesthetics and the partition coefficient.[51] Various solvents were used including vegetable oil, *n*-hexadecane and *n*-octanol. The vegetable oil system did not give a good correlation when hydroxy-containing anesthetics were included, and the hexadecane system was poor for most polar anesthetics. Octanol-water on the other hand, produced a good correlation for a broad range of anesthetics. The authors concluded that octanol best represents the cell membrane involved in anesthesia.

Oftentimes, a poor correlation is not due to the incorrect choice of solvent, but rather to the dependence on a diffusion phenomena.[52] The net flux across a cell membrane is directly proportional to the diffusion coefficient and the partition coefficient of the ligand and indirectly proportional to the thickness of the membrane. The combination of these three parameters is the permeability coefficient. If diffusion is an important parameter in the cell membrane transport, then it must be taken into account. Derivations of the size-corrected permeability coefficient then become an important part of the evaluation. These can be calculated by measuring the permeability coefficient-partition coefficient ratio as a function of the molecular volume. Since the molecular volume is difficult to obtain for anything but the simplest organic compounds, a function of the molecular weight (MW), usually the square root, is often used.[53]

In the radiopharmaceutical context the percent dose per gram of tissue gives a value proportional to the permeability coefficient. The dependence on MW can be determined and then the MW independent percent dose per gram can be plotted against the partition coefficient to determine the best linear model. This is more sophisticated than the model by Fenstermacher et al.[54] because it determines the molecular volume dependence for each membrane. It is more difficult because the molecule volume must be estimated. Recently, Wilson and Pinkerton[55] used various calibrated chromatographic techniques to estimate the volume of ^{99m}Tc phosphonates.

SDRs for various radiopharmaceuticals have been published over the years. Burns et al. correlated percent biliary excretion of various HIDA analogs with the ln MW/Z where Z is the charge on the ^{99m}Tc radiopharmaceutical.[56] Nunn et al. related the calculated lipophilicity and the reversed-phase HPLC retention time to the percent urinary excretion of some 33 HIDA derivatives labeled with ^{99m}Tc.[57] The measured and calculated lipophilicities were correlated, but the ortho substituents and those without ortho substituents gave different slopes and intercepts. However, the measured lipophilicity of ^{99m}Tc HIDAs from both groups correlated well with the percent renal excretion.

Other SDRs have been published, but the emphasis has been on structural changes in the molecule without concomitant measurements of physical chemical properties. Wieland et al. have described various substituted aralkylguanidines and the effect of these substituents on the adrenal medulla uptake of radioactivity.[58] Likewise, Counsell et al. studied the effect of substituents on the distribution of radiolabeled androgens.[59] Spitznagle et al. followed a similar approach for derivatives of cortisol.[60] A number of these and other SDRs have been published as a symposium proceedings.[61]

For receptor-binding radiotracers, especially those for cerebral receptors, the SDR can be subdivided into two aspects: transport and receptor binding. For the cerebral agents, there have been three SDR studies to date. The percent dose per gram in the rat brain was correlated with the octanol-water partition coefficient for a series of dopamine receptor-binding ligands by Moerlein et al.[62] For the various spiperone analogs, the molecular weight varied by only about 30%, but the octanol-water partition coefficient varied by 2.5 log units from 2.7 to 5.2. The striatum to cerebellum ratio is an indication of the receptor binding since the striatum contains D2 dopamine receptors and the cerebellum does not. This ratio is at a maximum for N-methylspiperone at log P = 3.2, but the striatum concentration peaks for bromospiperone with a log P of 3.6. In this case, the log P value is important for both delivery to the brain and to obtaining a specific receptor binding ratio. Fortunately, the *in vitro* binding constant for the series of 6 pharmaceuticals varied from only 2.6 to 10.3 nM so that these were essentially one variable experiments.

Welch et al.[63] looked at similar spiperone derivatives that were labeled with ^{18}F rather than ^{77}Br as in the case of Moerlein et al.[64] Welch et al. did not find the same dependence in the percent dose/brain or in the striatum to cerebellum ratio although the striatum to cerebellum ratio is highest for the N-methyl derivative.

Finally, Rzeszotarski et al. applied the NIH-EPA Chemical Information System to determine various physical-chemical parameters of derivatives of 3-quinuclidinyl benzilate, a strong antagonist of the muscarinic cholinergic receptor.[65]

From the studies of Moerlein et al. and Welch et al., it is clear that transport into the brain is a function of the lipophilicity of the compound. For compounds of near equal affinity, this is the important factor in cerebral concentration. If, however, the affinity constants vary, then the amount in the cerebrum can be correlated with the affinity constant as shown for the QNB series. Katzenellenbogen has shown a similar dependence on affinity constant for a series of estrogens.[66] Using the affinity constant alone was not, however, as highly correlated with the uterine concentration as when the affinity constant was adjusted by the lipophilicity of the estrogen. This approach combines the permeability and receptor binding in a single index.

In vivo displacement with the cold compound is another level of sophistication that avoids the complications of using the radiolabeled ligand. We have reviewed a number of methods to increase the probability that the time spent on preparing the no-carrier-added radioligand will be well spent. Each must be assessed independently and balanced against the time expended to obtain the information. The theoretical model is perhaps the easiest to use but gives only an ideal maximum value. The use of the nonradioactive compound *in vivo* is yet another escalation of the effort. If a tritium-labeled compound is available, the ability of the nonradioactive derivative to compete for the receptor should indicate the distribution of the radioactive form of the same compound. For example a series of halogenated derivatives of 3-quinuclidinyl benzilate have been synthesized and tested for their ability to compete with ^{3}H-QNB for the muscarinic receptor. The logit plot of the displacement of ^{3}H-QNB by each of the analogs was constructed.[67] The Y axis is the percent dose per gram bound in the brain in the presence of 50 nmol of nonradioactive analog per animal to the percent dose per gram bound using ^{3}H-QNB alone. The X axis is the relative binding as measured in the *in vitro* radioreceptor assay. The single compound that is less potent than predicted is 3-quinuclidinyl-4-iodobenzilate.

Animal distribution studies using the radiolabeled ligand are perhaps the most telling experiments. Besides the requirement that the radioactivity be present in the target organ with a target to nontarget ratio of at least 10, receptor-binding radiotracers must also show receptor specific binding. In general this is carried out by using either pre-, co- or post-injection of a known receptor binding biochemical or drug. Since the definition of a receptor is operational and likewise so is the definition of a receptor binding radiotracer, the distribution of the radioligand must fit the criteria of high affinity, specificity including stereo-specificity, saturability, and correlation with biological activity. Localization in the target organ alone does not guarantee a receptor process. [^{125}I]-iodinated tyramine practolol gives high heart to blood ratios in rats.[68] However, in guinea pig, lower ratios were obtained. In addition when propranolol was preinjected, the same heart to blood ratio was obtained.[69]

In general, the animal distribution is verified in human volunteers before investigational studies begin. Because of the pharmacological effect, few tests of saturability are carried out in humans. Because of the synthetic difficulties, few studies with active and inactive stereoisomers have been undertaken.

C.   ANALYSIS OF DATA

Although many ligands have been labeled with radioiodine and neogalactosealbumin has been labeled with Tc, the *in vivo* data to date have been used mostly to support the contention that the radioactivity is receptor bound in the target organ. As we stated earlier, the development of receptor binding radiotracer is a two-part problem: (1) the development of a radioligand that has a high receptor to nonreceptor binding and fulfills the operational definition and (2) the development of an analytical scheme that shows a high sensitivity between the radioactivity in the target organ and the receptor concentration. Many experiments have been put forth to support the former, but few have been put forth to support the latter. The kinetic analysis has involved two approaches: (1) the use of a high specific activity radioligand and (2) the use of a low specific activity ligand or the ligand at varying specific activities.

1. Low Specific Activity Ligands

In a conference on receptor-binding radiotracers in 1981, Krohn et al. put forth the hypothesis that the maximum sensitivity to receptor concentration change will come at ligand to receptor ratios of between 0.2 to 0.8.[70] The ability to separate the total receptor concentration from the rate of ligand-receptor binding was also related to the receptor saturation level, but in most cases, investigators have treated the product of the receptor concentration and the binding rate constant as a single variable. The assumption that the binding rate constant has not changed in disease has avoided the latter issue, but the former issue is still a major point of discussion. The sensitivity of determining receptor concentration change is best studied using simulations since there are few animal models where the receptor concentration can be systematically changed. An elegant analysis has been carried out by Vera et al. that shows the coefficient of variation in the measurement of receptor concentration as a function of the binding affinity and the fractional receptor saturation. The receptor concentration is more precisely determined at higher fractional receptor saturation.[71] The precision of determining receptor concentration has a parabolic dependence on the binding rate constant. At low ligand-receptor binding rate constant, the fraction of receptor binding to nonreceptor binding is low. At a very high rate constant, the radioactivity determined in the target organ is independent of receptor concentration and the rate determining step becomes the flow or membrane transport rather than the ligand receptor binding process.

Vera et al. have recently completed the analysis of the kinetic sensitivity experimentally in a pig model and found that the pharmacokinetics are altered significantly by changes in receptor concentration.

Friedman et al. used thermodynamic equilibrium to describe the binding process.[72] This approach assumes that steady state is reached and that the equilibrium equations apply. Based on these equations, a number of simulations were run showing the effect of injected dose (specific activity) and endogenous ligand. These show the same results put forth by Vera et al., namely, that the sensitivity to receptor change is maximal at about 50% saturation.

Farde et al. have used a similar approach with ^{11}C raclopride.[73] Farde et al. are using the law of mass action in the form of the Hill equation. This is the same equilibrium expression put forth by Katzenellenbogen et al.[74] and Eckelman et al.[75] to predict *in vivo* behavior using *in vitro* data. In general, the equation could predict trends, but did not give the quantitative target to nontarget ratio. Therefore, the assumptions listed by Farde et al. to obtain an accurate receptor concentration *in vivo* are crucial. Three of their assumptions involve the use of the caudate putamen to represent receptor binding and the use of the cerebellum to represent free ligand available for uptake. Then using four different chemical doses of ^{11}C raclopride, a Hill plot can be constructed. Another important assumption is the requirement for equilibrium (or more properly steady state). If a process such as blood clearance or lung clearance is fast compared to the dissociation rate of the ligand-receptor complex, then the target to nontarget ratio will be decreased as shown in the equation derived by John Wagner:

$$\frac{B}{F} = \frac{k_{12}}{k_{21} - \beta}$$

where k_{12} is the rate of ligand-receptor association, k_{21} is the rate of ligand-receptor dissociation, and β is the rate of disappearance of the input function.[76]

These approximations may result in a relative receptor value, but exact correlation with *in vitro* values are most likely coincidental. Also the use of low specific activity ligand may cause physiologic effects that will alter the pharmacodynamics of the ligand. Welch et al. showed some years ago that haloperidol membrane transport was dependent on the haloperidol concentration.

2. High Specific Activity Ligands

The use of high specific activity ligands to estimate the receptor concentrations was first developed on an empirical basis using ^{3}H ligands.[77] Wagner et al. used the same approach and found that the slope of the tissue ratio vs. time plot changes as a function of age. They used the D_2 receptor-binding ligand ^{11}C-N-methylspiperone (NMS) and thus postulated that the D_2 receptor is decreasing as a function of age.[78] The analysis can be derived from basic principles as has been done by Patlak for an irreversible binding ligand[79] and then applied to the three (3) compartment model.[80] In this situation, the slope of the tissue ratio vs. time plot represents a combination of rate constants:

$$Slope = \frac{k_2 k_3}{k_2 + k_3}$$

where k_2 is the efflux from the extracellular fluid to the plasma and k_3 is the first-order process describing the ligand-receptor interaction. It is a pseudo-first-order rate constant derived from the product of the rate constant ($M^{-1}t^{-1}$) and the receptor concentration (M^{+1}). For high specific activity ligands, the free receptor concentration equals the total receptor concentration if the low affinity endogenous ligands are ignored. If k_2 is greater than k_3, then the slope will be a function of k_3 and hence the receptor concentration. However, if

$k_3 > k_2$, then the slope will be a function of membrane transport or blood flow. These constants have not been determined for ^{11}C-NMS.

Sawada et al. have recently analyzed 3-quinuclidinyl 4-iodobenzilate (IQNB) using numerous kinetic models.[81] They used a combination of low error and accuracy (using *in vitro* data as the gold standard) to choose the best model. They also applied the Patlak equation to IQNB for the three compartment model.[82] In the rat, the k_3/k_2 ratio was 1.4, showing that the slope of the tissue ratio vs. time plot will be sensitive to changes in receptor concentration. From these values, one could expect that the tissue ratio (in the case of IQNB the caudate-putamen to cerebellum ratio) vs. time plot would yield an apparent receptor concentration decrease to 70% when the receptor concentration actually decreased to 50% of the normal value in the rat.

Mintun et al.[83] have argued against the tissue ratio approach put forth by Wagner et al.[84] because it lumps various physiologic process by using a single three (3) compartment model. Much controversy has resulted, but few definitive experiments have been carried out to show the responsiveness of any parameters to a change in receptor concentration. The lack of a reliable animal model where well-validated receptor concentration changes take place has been a major impediment.

IV. RECEPTOR SYSTEMS PROPOSED FOR SPECT RECEPTOR BINDING RADIOPHARMACEUTICALS

A review of the literature shows that quite a few receptor binding radiotracers have been proposed for various receptor types. These will be analyzed for completeness of the study based on those criteria presented earlier that deal directly with the evaluation of putative receptor binding radiotracers *in vivo*. The criteria and the appropriate abbreviations used in the following tables (Tables 1 to 8) are

1.   Use of a mathematical model to choose potential receptor binding radiopharmaceuticals (MATH MODEL)
2.   The determination of the K_A (app) for the parent compound (K_A PARENT).
3.   The use of ^3H-labeled parent compounds *in vivo* (^3H-PARENT)
4.   The synthesis of the nonradioactive substituted (halogenated) compound (SYN OF COLD)
5.   The determination of *in vitro* stability (IN VITRO STAB)
6.   The determination of the K_A (app) for the nonradioactive derivative (K_A-DERIV)
7.   The evaluation of various physicochemical parameters of the nonradioactive compound (SDR)
8.   *In vivo* displacement of ^3H-labeled compound with nonradioactive substituted compound (DIS BY COLD)
9.   Correlation of the distribution of the radioactive derivative in animals with the known receptor location (CORRELATE)
10. The determination of the K_A (app) for the radioactive derivative (K_A-HOT)
11. Preinjection, coinjection, or postinjection to decrease the effective specific activity of the radioactive derivative (SATURATE)
12. Use of active and inactive stereoisomers of the radioactive derivative (STEREOMER)
13. Confirmation of the animal distribution in volunteers (HUMAN)
14. Attempt to validate sensitivity of radioligand distribution on receptor concentration change (KIN MODEL)

A number of radioligands for the D1 and D2 dopamine receptors have been synthesized based on reports of their pharmacologic activity (see Table 4). De Paulis et al.[104] have reviewed the pertinent attempts. For the D1 receptor, the iodinated ligand SCH 23982 (also

TABLE 1
Evaluation of Putative Adenosine Receptor
Binding Radiotracers *In Vivo*

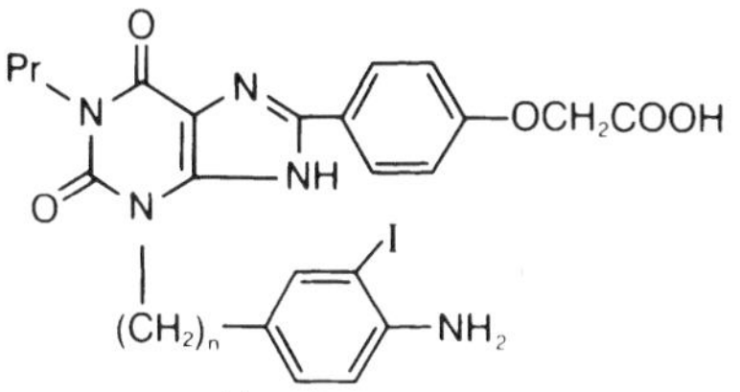

I-8-PX
A-1

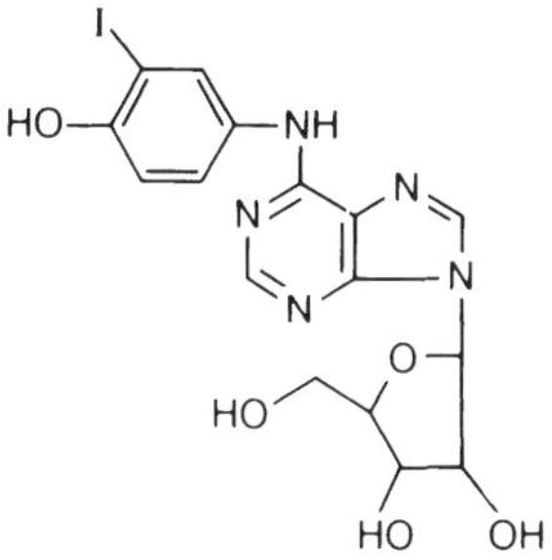

IHPIA
A-2

	I-8-PX[85]	IHPIA[86,87]
Structure #	A-1	A-2
MATH MODEL	No	No
$1/K_A$ PARENT	32/60 nM	$EC_{50} = 24$ nM
^{3}H-PARENT	No	No
SYN OF COLD	No	Yes
IN VITRO STAB	No	No
$1/K_A$-DERIV	37/29 nM*	$ED_{50} = 28$ nM
SDR	Yes	No
DIS BY COLD	No	No
$1/K_A$-HOT	1 nM	1.94 nM
CORRELATE	Yes	Yes
SATURATE	Yes	Yes
STEREOMER	No	No
HUMAN	No	No
KIN MODEL	No	No

Note: A-1 (n = 2), 2-[4-[3-(4-amino, 3-iodophenylethyl)-1,2,3,6-
Tetrahydro-2,6-dioxo-1-propyl-9H-purin-8-yl]phenoxyl]-
acetic acid; A-2, (−)-N⁶-(R-3-iodo-4-hydroxyphenyliso-
propyl) adenosine; asterisk indicates n = 1/n = 2; terms
in column 1 are defined in Section IV.

called IBZP by Kung et al.[98]) appears to be the choice, although deiodination is high (0.45%
at 24 h vs. 0.1% for FISCH). Of the butyrophenones, 2′- and 4-iodospiperone have the
highest affinities *in vitro*, but the 4-iodospiperone has a higher lipophilicity than spiperone
itself that leads to high nonreceptor binding *in vivo*. The 2′-iodospiperone is the butyro-
phenone of choice. A number of benzamides have been iodinated, but most including
iodoclebopride, the iodinated derivative of amisulpride, and its *N*-cyclopropylmethyl analogue and iodosulpride seem to suffer from low affinity and poor brain uptake. Iodolisuride
has high potency ($K_i = 1.2$ nM), but the brain uptake in rats is only 10% of that found for
IBZM.[105] IBZM developed by Kung et al.[99] and the 5-iodo-2-methoxybenzamide containing
the pyrrolinyl moiety of sulpride developed by de Paulis et al. appear to be the best. Kung
et al. have recently added another benzamide FISCH to the list because of its low deiodination
rate.[106]

The muscarinic acetylcholine receptor (mAChR) system has been studied extensively
(see Table 5). The radioiodinated analog of 3-quinuclidinyl benzilate (QNB) has been shown

TABLE 2
Evaluation of Putative Adrenergic Receptor Binding
Radiotracers *In Vivo*

	Yohimbine[88]	PIC[89]	I-HEAT[90,91]
Structure #	B-1	B-2	B-3
MATH MODEL	No	No	No
$1/K_A$ PARENT	No	IC_{50} = 21 nM	5 nM
^{3}H-PARENT	No	No	No
SYN OF COLD	No	Yes	No
IN VITRO STAB	No	No	No
$1/K_A$-DERIV	11 nM	IC_{50} = 0.5 nM	
SDR	No	No	No
DIS BY COLD	No	No	No
$1/K_A$-HOT	1.8 nM	No	87 pM
CORRELATE	Yes	No	Yes
SATURATE	Yes	No	Pre
STEREOMER	No	No	No
HUMAN	No	No	No
KIN MODEL	No	No	Yes

Note: Yohimbine = 17-hydroxyyohimbine-16-carboxylic acid methyles-
ter; PIC = 2-[(2,6-dichloro-4-iodophenyl)imino]imidazolidine; I-
HEAT = 2[β-(4-hydroxy,3-iodophenyl)-ethyl-aminomethyl]-
tetralone; abbreviations in column 1 are defined in Section IV.

to bind to the mAChR by testing the saturability and the stereoselectivity in the corpus striatum, cerebellum, and the heart of rats. 3-Quinuclidinyl 4-iodobenzilate (4-IQNB) receptor binding can be inhibited by coinjection of small amounts of nonradioactive mAChR ligands as well as displaced by the same ligand after the 4-IQNB has bound to the receptor. These *in vivo* experiments involve a complicated set of variables including the total receptor concentration, the dissociation rate, transport of the displacing ligand, and the input function (4-IQNB still available for uptake from the blood). Nevertheless, the combined evidence along with the regional distribution indicates a receptor-mediated localization.[109-115]

Another important proof is obtained by using two stereoisomers of IQNB differing in the chirality at the quinuclidinyl carbon. The difference in distribution between the pharmacologically active form, the 3-R-quinuclidinyl-4-iodobenzilate and the pharmacologically inactive form, 3-S-quinuclidinyl 4-iodobenzilate in those organs containing mAChR furnishes that proof. 4-IQNB has two chiral centers. The work was carried out using either R-quinuclidinol or S-quinuclidinol. The chiral center on the benzylic acid portion was a stereoisomer mixture (R,RS) in some batches and pure R (R,R) in other batches. The quinuclidinyl center is given first followed by the benzylic center. One of the most interesting

TABLE 3
Evaluation of Putative Benzodiazepine Receptor Binding
Radiotracers *In Vivo*

IFLUNIT	I-PK 1195	I-Ro 54864
C-1	C-2	C-3

	IFLUNIT[92]	IPK 11195[93]	IRo54864[94]
Structure #	C-1	C-2	C-3
MATH MODEL	No	No	No
$1/K_A$ PARENT	No	14 nM	7.3 nM
^{3}H-PARENT	No	Yes	Yes
SYN OF COLD	No	Yes	No
IN VITRO STAB	No	No	No
$1/K_A$-DERIV	IC$_{50}$ = 2.9 nM	No	13 nM
SDR	No	Yes	No
DIS BY COLD	No	No	No
$1/K_A$-HOT	No	8.0 nM	No
CORRELATE	Yes	No	Yes
SATURATE	No	No	Pre
STEREOMER	No	No	No
HUMAN	No	No	No
KIN MODEL	No	No	No

Note: IFLUNIT = 7-iodo-1,3 dihydro-5-(2-fluorophenyl)-1-methyl-2H-1,4 ben-
zodiazepin-2-one; IPK 11195 = [1-(2-chlorophenyl)-*N*-methyl-*N*-(1-meth-
ylpropyl)-3-isoquinoline carboxamide; Ro54864 = 7-iodo-1, 3-dihydro-
1-methyl-5-(4-chlorophenyl)-2H-1,4 benzodiazepin-2-one; abbreviations
in column 1 are defined in Section IV.

experiments with 4-IQNB is the comparison of the high affinity R,R 4-IQNB with the
inactive S,RS 4-IQNB. Assuming a first-order process, these data can be plotted on a semilog
graph to show the relative net efflux from key tissues. For a major source of muscarinic
cholinergic receptor such as the corpus striatum, the percent dose per gram is higher and
the net efflux is slower for the active material (R,R 4-IQNB) than for the inactive material
(S,RS 4-IQNB). For the pure R,S compound, the efflux is faster than that for the R,R
compound but not as fast as found for the weakly bound S,R isomer.[115] This has led Gibson
et al.[116] to conclude that the receptor concentration can be measured by the efflux kinetics
with R,S 4-IQNB. For an organ such as the lung that contains few mAChR but takes up 4-
IQNB by another mechanism, the percent dose per gram and the net efflux are the same
for the active and inactive isomers. For the cerebellum, which contains about 10% of the
mAChRs found in the corpus striatum, rapid efflux is recorded for all three stereoisomers.
The use of an active and inactive form of the same receptor ligand is one of the most powerful
proofs for the presence of receptor binding *in vivo*.

If the first human study with 4-IQNB was completed on May 11, 1983,[117] why are so

TABLE 4
Evaluation of Putative Dopamine Receptor Binding Radiotracers *In Vivo*

Iodopride
D-1

IBZM
D-2

IBZP
D-3

ISPIP
D-4

SCH 23982
D-5

FISCH
D-6

	Iodopride[95]	IBZM[96,97]	IBZP[98,99a]	ISPIP[100,101]	I-SCH 23982[102]	FISCH[103]
Structure#	D-1	D-2	D-3	D-4	D-5	D-6
MATH MODEL	No	No	No	No	No	No
$1/K_A$ PARENT	33 nM[b]	No	197 nM	0.28 nM	0.53 nM	1.7 nM
'H-PARENT	No	No	No	Yes	No	No
SYN OF COLD	Yes	Yes	Yes	Yes	No	Yes
IN VITRO STAB	No	No	No	No	No	No
$1/K_A$-DERIV	1.5 nM	0.7 nM	0.7 nM	1 nM	7.2 nM	13.4 nM
SDR	No	No	No	Yes	Yes	No
DIS BY COLD	No	No	No	No	No	No
$1/K_A$-HOT	1.2 nM	No	No	0.25 nM	1.5 nM	1.4 nM
CORRELATE	Yes	Yes	Yes	Yes	Yes	Yes
SATURATE	No	Yes	Yes	Pre	Co	No
STEREOMER	No	No	Yes	No	No	No
HUMAN	No	Yes	No	No	No	No
KIN MODEL	No	No	No	No	No	No

Note: Iodopride = (S)-N-[(1-ethyl-2-pyrrolidinyl)methyl]-5-iodo-2-methoxybenzamide; IBZM = 3-iodo-N-[(1-ethyl-2-pyrrolidinyl)]methyl-2-hydroxy-6-methoxybenzamide; IBZP = R-(+)-8-iodo-2,3,4,5 tetrahydro-3-methyl-5-phenyl-1H-3-benzazepin-7-ol; ISPIP = 8-[4-(4-fluoro-2-iodophenyl)-4-oxobutyl]-1-phenyl-1,3,8-triazaspiro]4.5[decan-4-one; I-SCH23982 = (R)-(+)-7-iodo-8-hydroxy-3-methyl-1-phenyl-2,3,4,5-tetrahydro-1H-3-benzazepine; FISCH = (+)-7-chloro-8-hydroxy-1-(4'-iodophenyl)-3-methyl-2,3,4,5-tetrahydro-1H-3-benzazepine; abbreviations in column 1 are defined in Section IV.

ᵃ   Also known as SCH23982.
ᵇ   Sulpride.

few studies recorded to date? This question goes to the heart of the matter for radioiodinated receptor binding radiotracers. Receptor binding radiotracers offer great promise in that they will lead to the noninvasive measurement of the change in receptor concentration as a function of disease. However, the single photon emitting radiotracers, labeled with ^{123}I or ^{99m}Tc, are compromised by the instrumentation available, especially for neuroreceptor studies and by the lack of routine availability of [123]I.

TABLE 5
Evaluation of Putative Muscarinic Cholinergic
Receptor Binding Radiotracers *In Vivo*

IQNB
E-1

IDEX
E-2

	IQNB[107]	IDEX[108]
Structure #	E-1	E-2
MATH MODEL	Yes	No
$1/K_A$ PARENT	27 pM	IC_{50} = 15 nM
³H-PARENT	Yes	Yes
SYN OF COLD	Yes	Yes
IN VITRO STAB	Yes	No
$1/K_A$-DERIV	41 pM	IC_{50} = 17 nM
SDR	Yes	Yes
DIS BY COLD	Yes	No
$1/K_A$-HOT	18 pM	5.8 nM
CORRELATE	Yes	Yes
SATURATE	Co, pre, post	Co, pre
STEREOMER	Yes	Yes
HUMAN	Yes	No
KIN MODEL	Yes	No

Note: IQNB = (R)-(−)-1-azabicyclo[2.2.2]oct-3-yl-(R)-(+)-α-hydroxy-α-(4-iodophenyl) α-phenyl acetate; IDEX = (S)-(+)-3-phenyl-3-[(1-phenyl-methyl)-4-piperidinyl]-2,6-piperidinedione; abbreviations in column 1 are defined in Section IV.

A specific reason for the delay in widespread investigational studies with 4-IQNB is the difficult synthesis of the precursor and the radioiodinated product.[118] The synthesis of the triazene precursor is a five step synthesis of some difficulty. Attempts to iodinate QNB directly by electrophilic iodination were unsuccessful,[119] forcing the development of the nucleophilic approach via the triazene. Although the triazene can be made in relatively large amounts and only 1 mg of the triazene precursor is used in each radioiodination, the isolation and purification of the triazene prevented widespread use. In addition, the overall radioiodination yields were low (<20%) necessitating the use of large quantities of radioiodine especially when carrying out clinical studies with ¹²³I. The isolation of high specific activity 4-IQNB involves a separation of the early eluting degradation products from the 4-IQNB. Although the separation of the early eluting impurities from later eluting 4-IQNB is of the order of 80 ml in the best case, the specific activity was still not near the maximum because of tailing. Using ³H-QNB as a tracer for the impurities, 0.3% of the degradation products per 10 ml contaminates the 4-IQNB even in the best case.

TABLE 6
Evaluation of Putative Serotonin Receptor
Binding Radiotracers *In Vivo*

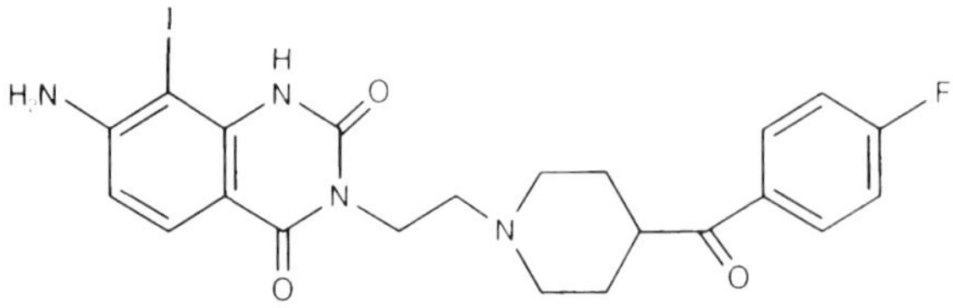

AMIK
F-1

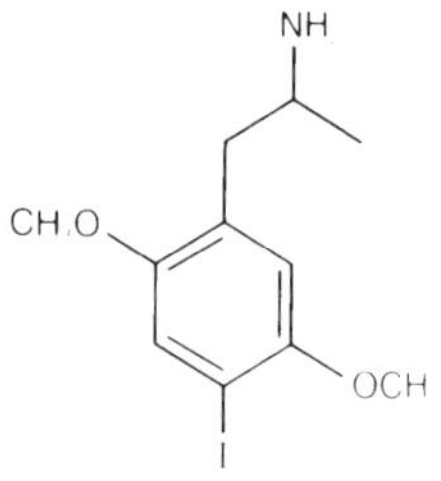

DOI
F-2

	AMIK[133]	DOI[134]
Structure #	F-1	F-2
MATH MODEL	No	No
$1/K_A$ PARENT	0.37 nM	0.4 nM*
^{3}H-PARENT	No	No
SYN OF COLD	No	No
IN VITRO STAB	No	No
$1/K_A$-DERIV	No	0.7 nM
SDR	No	No
DIS BY COLD	No	No
$1/K_A$-HOT	0.14 nM	2.2 nM
CORRELATE	No	No
SATURATE	No	No
STEREOMER	No	No
HUMAN	No	No
KIN MODEL	No	No

DOI = 1-(2,5-dimethoxy-4-iodophenyl)-2-aminopropane;
AMIK = 7-amino-8-iodo-ketanserin; asterisk indicates
DOB; abbreviations in column 1 defined in Section IV.

Recently, Kabalka et al. published an alternate method of synthesis involving electrophilic deboronation of the 4-boronic acid derivative of QNB.[120] This was not carried out to produce no-carrier-added 4-IQNB, but presumably the reaction could be maximized. Kabalka et al. used 95 mg of the QNB-boronic acid with a ten times excess of sodium iodide on a mole basis to achieve a yield of 60%. Recently Nordion Corp. (formerly AECL) has developed an exchange labeling procedure for high specific activity 4-IQNB, based on earlier methods for exchange labeling.[121-123] The difficulty in producing high specific activity 4-IQNB has led to a small number of clinical studies. After the publication of the first study in a normal volunteer, Holman et al. studied the distribution of 4-IQNB in a normal subject and a patient with clinically diagnosed Alzheimer's disease.[124] They found a decrease in the blood flow (as measured using iodoamphetamine) in the posterior temporal and the parietal cortex. Cerebral uptake of 4-IQNB was slow but continued to increase during the 15 h of data acquisition with the efflux of 4-IQNB from the lungs serving as the extended input function. The ratios for temporoparietal cortex to caudate, occipital cortex to caudate, and frontal cortex to caudate were reduced compared to the ratios in a normal subject.

Another study at the National Institutes of Health using planar imaging in Alzheimer's patients and normal subjects showed that the corpus striatum to cerebellum ratio increased

TABLE 7
Evaluation of Putative Steroid Receptor
Binding Radiotracers *In Vivo*

| | MIVE₂ G-1 | IVNT G-2 | MIE G-3 |

MIVE₂[135]   IVNT[136]   MIE[137]

	MIVE₂[135]	IVNT[136]	MIE[137]
Structure #	G-1	G-2	G-3
MATH MODEL	Yes	No	No
$1/K_A$ PARENT	60 pM[a]	0.8 nM[b]	60 pM[a]
³H-PARENT	Yes	No	No
SYN OF COLD	Yes	Yes	No
IN VITRO STAB	No	No	No
$1/K_A$-DERIV	No	No	No
SDR	No	No	No
DIS. BY COLD	No	No	Yes
$1/K_A$-HOT	147 pM	1.0 nM	167 pM
CORRELATE	Yes	No	Yes
SATURATE	Co, post	No	Pre, post
STEREOMER	No	No	No
HUMAN	No	No	No
KIN MODEL	No	No	No

Note: MIVE₂ = 11-β-methoxy 16-α-iodovinylestradiol;
IVNT = 17-α-[2(E)-iodovinyl]-4-estren-17-β-ol-
3-one; MIE = 11-β-methoxy 16-α-iodoestradiol;
abbreviations in column 1 are defined in Section
IV.

[a] Estradiol.
[b] R5020.

with time at the same rate as found in rats when corrected for the body surface area.[125] More recently, Weinberger et al. have shown images in Pick's disease patients using 4-IQNB and ¹⁸FDG.[126] In most cases, the glucose metabolism and the mAChR levels showed similar decreases. In one Pick's disease patient, basal ganglia hypometabolism was detected despite normal IQNB uptake. In another patient, the glucose metabolism was slightly decreased in the frontal cortex, but the 4-IQNB concentration was depressed to a greater extent. Unfortunately, cerebral blood flow changes were not reported. This latter case won the 1989 image of the year at the Society of Nuclear Medicine meeting.[127]

Since 4-IQNB has pharmacologic effects, pharmacokinetic schemes using high specific activity radioligands are necessary. One such approach, the Patlak plot, has been used with animal data to determine the forward flux of 4-IQNB. The corpus striatum to cerebellum ratio is used as an indication of specific to nonspecific binding and the resulting rate is similar to that obtained for ¹¹C-*N*-methyl spiperone. The use of the nonactive form S,SR 4-IQNB as a control for the nonspecific processes has shown similar results.[128]

Sawada et al. have made a complete study of the pharmacokinetics by analyzing the

TABLE 8
Evaluation of Some Other Putative Receptor Binding
Radiotracers *In Vivo*

Receptor system	Ligand
Calcium channel	Iodipine,[145,146] Conotoxin[147]
Ah	2-Iodo-7,8-dibromodibenzo-*p*-dioxin[148]
Substance P	Iodinated substance P[149]
Parathyroid hormone (PTH)	Iodinated PTH[150]
Serotonin	Iodo-lysergic acid diethylamide (LSD)[151]
Benzodiazepine	Iomazenil[152,153]
Opioid	2-Iodomorphine[154]
HCG	Iodinated HCG[155]
Vasopressin	I-d($CH_2)_5$[Tyr(Me)^2,Tyr(NH$_2$)^9]AVP[156]
Opioid	[I-Tyr^1,D-Pro^{10}]dynorphin[157]
Oxytocin	d($CH_2)_5$[Tyr(Me)^2,Thr^4,Tyr-NH$_2^9$]OVT[158]
Interferon	Iodinated interferon[159]
Thromboxane A$_2$/prostaglandin H$_2$	Iodinated thromboxane[160]

distribution of the R and S quinuclidinyl 4-iodobenzilate in rats using various models of increasing sophistication.[129] The radiochemical purity of 4-IQNB is high in the target tissue but the blood radioactivity had to be corrected for metabolism.

The transport parameter k_1 and the receptor off-rate k_4 were similar in each brain structure for four different models. Although k_3, the on rate for receptor binding, and k_2, the efflux from tissue to plasma, were in a constant ratio for each model, each varied over a wide range. In the caudate putamen k_3 varied from 2.10 ± 0.89 to 22.3 ± 9.3 min^{-1}. The extraction of 4-IQNB is between 43 to 50% which agrees with the extraction of 4-IQNB in monkeys of 54% at normal flow and 37% at 100 ml/min/100 g.[130] Since $k_3 \cong k_2$, the steady-state uptake rate is dependent on both transport and receptor binding. Therefore, the distribution of radioactivity in the caudate putamen should be related to changes in receptor concentration. This satisfies the second criterion on a mathematical basis. Recently, Gibson presented data that satisfied the second criterion on an experimental basis. Using autoradiography, he determined the concentration of 4-IQNB in different parts of the brain known from *in vitro* experiments to contain different concentrations of mAChR. The plot of the *in vivo* concentration of 4-IQNB vs. the *in vitro* receptor concentration showed a linear dependence, suggesting a sensitivity of 4-IQNB to a change in receptor concentration.[131]

After many years of research, it appears that 4-IQNB has been validated as a receptor binding radiotracer whose distribution is sensitive to changes in receptor concentration. Another mAChR antagonist, ^{123}I-4-iododexetimide has been studied by Wilson et al.[132] Based on early work by Laduron et al., various nonradioactive halogenated derivatives of benzetimides were prepared. The 4-iododerivative had the same IC_{50} as the parent compound so a radiolabeling technique using the 4-trimethyl silyl derivative with Chloramine-T in trifluoroacetic acid was developed. The radioiodinated derivative had the correct biological distribution in mice and the ratio of receptor binding to nonreceptor binding (i.e., the striatum or cortex to cerebellum ratio) decreased as a function of either increased dose of pre- or co-injected dexetimide or iododexetimide. The use of the radiolabeled inactive isomer further defined the receptor binding properties of this ligand. 4-Iododexetimide clears the target organ with a faster net rate than does 4-IQNB. Whether this is due to a sharper input function, faster metabolism, or a faster receptor dissociation has not been determined.

Katzenellenbogen et al.[138] in their review of the history of steroid receptor binding radiotracers quote the Albert et al. 1949 work[139] as the earliest study in that field. This interest in radiolabeling steroid receptor ligands is reflected in the large number of compounds that soon appeared in the literature.[23,140] Most were evaluated by *in vivo* distribution studies in small animals. The identification of cytosolic estradiol receptors in 1960 led to the systematic study of the requirements for receptor binding by those in the drug industry and those developing radiotracers[141] (see Table 7). One of the first iodinated estrogens with a high affinity constant and high specific activity was 16-α-estradiol reported in 1979.[142] Of the iodinated estrogens, 11-β-methoxy, 17-α-iodovinyl estradiol is the best of the lot based on ease of synthesis, attainable effective and chemical specific activity, affinity constant, and nonspecific binding.[143] The major focus of gamma-emitting receptor-binding radiotracers has definitely been on the steroid hormones.[144]

A number of other putative receptor binding radiotracers have been suggested, but the characterization, especially with respect to neuroreceptors, is not complete enough to warrant analysis of the 16 steps toward validation (see Table 8). The list demonstrates the widespread activity in developing effective radioiodinated receptor binding ligands.

V. CONCLUSIONS

Instrumentation for single photon emitting radiotracers has not progressed to the extent as that for positron emitting radiotracers. Only recently have multihead/ring machines for single photon emitting radiotracers become available: the TRIAD,[161] the PRISM,[162] the Strichman SPET machine,[163] the MUMPI,[164] the ASPECT,[165] and the TOMOMATIC.[166] The lack of readily available ring machines is most detrimental in cerebral receptor studies where the structures are small and the percent dose per gram is also small, but this also applies to cardiac receptors and receptor dependent tumors as well.

The lack of availability of ^{123}I is also a major problem, especially in the U.S. The success of nuclear medicine has depended on the ready availability of ^{99m}Tc through the generator system. The positron emitting radiotracers (^{11}C, ^{18}F) can also be produced by an on-site generator, a cyclotron. Although much more expensive than the ^{99m}Tc generator, it does allow for easy access. However, there is no generator system for ^{123}I and it cannot be made at high purity on the small biomedical cyclotrons. A recent publication indicates that high-purity ^{123}I can be produced on a biomedical cyclotron using an enriched xenon target.[167] The lack of availability is one of the reasons that extensive studies of neuroreceptors, such as the mAChR with 4-IQNB, are few and far between. It appears that a breakthrough in the three following technologies must be made before single photon emitting radiotracers can be used routinely in research technology:

1. Biomedical cyclotrons to prepare pure ^{123}I, i.e., without ^{124}I, either alone or with a mass purification system
2. ^{99m}Tc labeling technnology that allows the preparation of high receptor affinity analogs
3. Instrumentation, especially a ring-type machine, to make maximum use of the emitted radioactivity

There have been encouraging preliminary reports in all three of these areas, especially in instrumentation, but to date the positron-emitting radiotracers still have an advantage. However, it is safe to say that many of the advances in receptor-binding radiotracers for external imaging have come from the pioneering work with ^{131}I and ^{123}I. With further advances in cyclotrons, ^{99m}Tc-labeling technology, and the further proliferation of the new multihead/ring instruments, single photon-emitting, receptor-binding radiotracers will become a routine research tool if not a valuable diagnostic clinical test.

REFERENCES

1. **Budinger, T.**, *Physiology and Physics in Nuclear Cardiology. Cardiovascular Clinic Series,* Vol. 10, No. 2, Brest, A., Ed., F. A. Davis, Philadelphia, PA, 1979, 9.

2. **Mullani, N. A., Goldstein, R. A., Gould, K. L., Marani, S. K., Fisher, D. J., O'Brien, H. A., Jr., and Loberg, M. D.,** Myocardial perfusion with Rubidium-82. I. Measurement of extraction fraction and flow with external detectors, *J. Nucl. Med.,* 24, 898, 1983.

3. **Ziegler, W. H. and Goresky, C. A.,** Kinetics of rubidium uptake in the working dog heart, *Circ. Res.,* 29, 208, 1971.

4. **Tow, D. E., Wagner, H. N., Jr., Lopez-Mejano, V., Smith, E. M., and Migita, T.,** Validity of measuring regional pulmonary arterial blood with macroaggregates of human serum albumin, *Am. J. Roentgenol.,* 96, 664, 1966.

5. **Shaldon, S., Chiandussi, L., Guevara, L., Caesar, J., and Sherlock, S.,** The estimation of hepatic blood flow and intrahepatic shunted blood flow by colloidal heat-denatured human serum albumin labeled with I-131, *J. Clin. Invest.,* 40, 1346, 1961.

6. **Sherlock, S.,** Measurement of hepatic blood flow in dynamic clinical studies with radioisotopes, Kniseley, R. M., Tampe, W. N., and Anderson, E. B., Eds., U.S. AEC Technical Information Center, Oak Ridge, TN, 1964, 359.

7. **Tubis, M., Posnick, E., and Nordyke, R. A.,** Preparation and use of I-131 labeled sodium iodohippurate in kidney function tests, *Proc. Soc. Exp. Biol. Med.,* 103, 497, 1960.

8. **McAfee, J. G., Grossman, Z. D., Gagne, G., Zens, A. L., Subramanian, G., Thomas, E. D., Fernandez, P., and Roskopf, M. L.,** Comparison of renal extraction efficiencies for radioactive agents in the normal dog, *J. Nucl. Med.,* 22, 333, 1981.

9. **Winchell, H. S., Horst, W. D., Braun, L., Oldendorf, W. H., Hattner, R., and Parker, H.,** *N*-isopropyl-[^{123}I]*p*-iodoamphetamine: single pass brain uptake and washout: binding to brain synaptosomes: and localization in dog and monkey brain, *J. Nucl. Med.,* 21, 947 1980.

10. **Kung, H. F., Tramposch, K. M., and Blau, M.,** A new brain perfusion imaging agent [I-123]HIPDM: *N,N,N'*-trimethyl-*N'*-[2-hydroxy-3-methyl-5-iodobenzyl]-1, 3-propanediamine, *J. Nucl. Med.,* 24, 66, 1983.

11. **Kuhl, D. E., Barrio, J. R., Huang, S. C., Selin, C. C., Ackerman, R. D., Lear, J. L., Wu, J. L., Lin, T. H., and Phelps, M. E.,** Quantifying local cerebral blood flow by *N*-isopropyl-*p*-[^{123}I]iodoamphetamine (IMP) tomography, *J. Nucl. Med.,* 23, 196, 1982.

12. **LaFrance, N. D., Wagner, H. N., Whitehouse, P., Corley, E., and Duelfer, T.,** Decreased accumulation of isopropyl-iodoamphetamine (I-123) in brain tumors, *J. Nucl. Med.,* 22, 1081, 1981.

13. **Szasz, I. J., Lyster, D., and Morrison, R. T.,** Iodine-123 IMP uptake in metastasis from lung cancer, *J. Nucl. Med.,* 11, 1342, 1985.

14. **Lucignani, G.,** [^{123}I]HIPDM as a brain perfusion indicator for SPECT, in *New Brain Imaging Techniques in Cerebrovascular Diseases,* Cahn, J. and Lassen, N. A., Eds., John Libbey Eurotext, Paris, 1985, 47.

15. **Schelbert, H. R., Phelps, M. E., and Kuhl, D. E.,** Positron emission tomography of the heart: a new method for the noninvasive assessment of regional myocardial blood flow, function and metabolism, in *Clinical Nuclear Cardiology,* Berman, D. S. and Mason, D. T., Eds., Grune & Stratton, New York, 1981, 167.

16. **Raichle, M., E., Larson, K. B., and Phelps, M. D.,** *In vivo* measurement of brain glucose transport and metabolism employing glucose ^{11}C, *Am. J. Physiol.,* 228, 1936, 1975.

17. **Phelps, M. E., Mazziotta, J. C., and Huang, S. C.,** Study of cerebral function with positron computed tomography, *J. Cereb. Blood Flow Metab.,* 2, 113, 1982.

18. **Himmelweit, F., Ed.,** The collected papers of Paul Ehrlich, in *Immunology and Cancer Research,* Vol. 2, Pergamon Press, Elmsford, N.Y., 1957.

19. **Langley, J. M.,** On nerve endings and special excitable substances in cells, *Proc. R. Soc. B,* 8, 170, 1906.

20. **Kahn, C. R.,** Membrane receptors for hormones and neurotransmitters, *J. Cell Biol.,* 70, 261, 1976.

21. **Jensen, E. V. and Jacobson, H. I.,** Basic guides to the mechanism of estrogen action, *Recent Prog. Horm. Res.,* 18, 387, 1962.

22. **Aronstam, R. S.,** Receptor binding studies: general considerations, in *Receptor-Binding Radiotracers,* Vol. 1, Eckelman, W. C., Ed., CRC Press, Boca Raton, FL, 1982, 5.

23. **Wong, D. F., Wagner, H. N., Jr., Dannals, R. F. et al.,** Effects of age on dopamine and serotonin receptors measured by positron tomography in the living human brain, *Science,* 226, 1393, 1984.

24. **Kuhar, M.,** Localizing drug and neurotransmitter receptor *in vivo* with tritium labeled tracers, in Vol. 1, *Receptor-Binding Radiotracers,* Eckelman, W. E., Ed., CRC Press, Boca Raton, FL, 1982, 37.

25. **Patlak, C. S.,** Derivation of equations for the steady-state reaction velocity of a substance, *J. Cereb. Blood Flow Metab.* 1, 129, 1981.

26. **Patlak, C. S. and Blasberg, R. G.,** Graphical evaluation of blood-to-brain transfer constants from multiple-time uptake data. Generalizations, *J. Cereb. Blood Flow Metab.,* 5, 584, 1985.

27. **Gjedde, A., Wienhard, K., Heiss, W. D., Kloster, G., Diemer, N. H., Herholz, K., and Pawlik, G.,** Comparative regional analysis of 2-fluorodeoxyglucose and methylglucose uptake in brain of four stroke patients. With special reference to the regional estimation of lumped constant, *J. Cereb. Blood Flow Metab.,* 5, 163, 1985.

28. **Farde, L., Eriksson, L., Blomquist, G., and Halldin, C.,** Kinetic analysis of central [¹¹C]raclopride binding to D_2-dopamine receptors studied by PET — a comparison to the equilibrium analysis, *J. Cereb. Blood Flow Metab.,* 9, 696, 1989.

29. **Wong, D. F., Wagner, H. N., Jr., Tune, L. E., Dannals, R. F., Perlsson, G. D., Links, J. M., Tamminga, C. A., Broussolle, E. P., Ravert, H. T., Wilson, A. A., Toung, J. K. T., Malat, J., Williams, F. A., O'Touma, L. A., Synder, S. H., Kuhar, M. J., and Gjedde, A.,** Positron emission tomography reveals elevated D2 dopamine receptors in drug-naive schizophrenics, *Science,* 234, 1558, 1986.

30. **Farde, L., Wiesel, F.-A., Hall, H., Halldin, C., Stone-Elander, S., and Sedvall, G.,** No D2 receptor increase in PET study of schizophrenia, *Arch. Gen. Psychiatry,* 44, 671, 1987.

31. **Farde, L., Hall, H., Ehrin, E., and Sedvall, G.,** Quantitative analysis of dopamine-D2 receptor binding in the living human brain by positron emission tomography, *Science,* 231, 258, 1986.

32. **Vera, D. R., Woodle, E. S., and Stadalnik, R. C.,** Kinetic sensitivity of a receptor-binding radiopharmaceutical: Technetium-99m galactosyl-neoglycoalbumin, *J. Nucl. Med.,* 30, 1519, 1989.

33. **Scatchard, G.,** The attractions of proteins for small molecules and ions, *Ann. N.Y. Acad. Sci.,* 51, 660, 1949.

34. **Yokoyama, A., Terauchi, Y., Horiuchi, K. et al.,** Technetium-99m-kethoxal-bis (thiosemicarbazone), an uncharged complex with a tetravalent ^{99m}Tc state and its excretion into bile, *J. Nucl. Med.,* 17, 816, 1976.

35. **Dannals, R. F.,** The Preparation and Characterization of Nitrogen-Sulfur Donor Ligands and Their Technetium Complexes, Ph.D. thesis, Johns Hopkins University, Baltimore, MD, 1981.

36. **Troutner, D. E., Simon, J., Ketring, A. R., Volkert, W. A., and Holmes, R. A.,** Complexing ^{99m}Tc with cyclam, *J. Nucl. Med.,* 21, 443, 1980.

37. **Nunn, A. D., Feld, T. A., and Treher, E. N.,** Boronic Acid Adducts of Technetium-99m Dioxime Complexes, U.S. Patent 4,705,849, 1987.

38. **Myers, W. G.,** Radioisotopes of iodine, in Radioactive Pharmaceuticals, Andrews, G. A., Kniseley, R. M., and Wagner, H. N., Jr., Eds., U.S. Atomic Energy Commission, Springfield, VA, 1966, 217.

39. **Lambrecht, R. M. and Wolf, A. P.,** Cyclotron and short-lived halogen isotopes for radiopharmaceutical applications, in *Radiopharmaceuticals and Labeled Compounds,* Vol. 1, International Atomic Energy Agency, Vienna, 1973, 275.

40. **Wellman, H. N., Anger, R. T., Jr., Sodd, V., and Paras, P.,** Properties, production, and clinical uses of radioisotopes of iodine, *CRC Crit. Rev. Clin. Radiol. Nucl. Med.,* 81, 1975.

41. **Atkins, H., Fairchild, R. G., Richards, P., and Pate, H.,** *Clinical Needs in Functional Imaging Radiopharmaceuticals II,* Sodd, V. J., Allen, D. R., Hoagland, D. R., and Ice, R. D., Eds., Society of Nuclear Medicine, New York, 1979, 183.

42. **Moody, D.,** Proceedings of the DOE workshop on the role of a high-current accelerator, in The Future of Nuclear Medicine, Conf. LA-11579-C, Los Alamos National Laboratory, Los Alamos, NM, 1984.

43. **Eckelman, W. C.,** The development of single-photon emitting, receptor-binding radiotracers, in *The Chemistry and Pharmacology of Radiopharmaceuticals,* Nunn, A., Ed., Marcel Dekker, New York, 1990.

44. **Reba, R. C. and Holman, B. L.,** Brain perfusion radiotracers, Chapter 3 in this volume.

45. **Eckelman, W. C.,** Receptor-specific radiopharmaceuticals, in *Emission Computed Tomography,* Ell, P. J. and Holman, B. L., Eds., Oxford University Press, Oxford, 1982.

46. **Gibson, R. E.,** Quantitative changes in receptor concentration as a function of disease, in *Receptor Binding Radiotracers,* Vol. 2, Eckelman, W. C., Ed., CRC Press, Boca Raton, FL, 1982.

47. **Wagner, H. N., Jr.,** Introduction: the role of receptors in disease, in *Receptor Binding Radiotracers,* Vol. 2, Eckelman, W. C., Ed., CRC Press, Boca Raton, FL, 1982.

48. **Katzenellenbogen, J. A., Heiman, D. F., Carlson, K. E., and Lloyd, J. E.,** *In vitro* and *in vivo* steroid receptor assays in the design of estrogen radiopharmaceuticals, in *Receptor Binding Radiotracers,* Vol. 1, Eckelman, W. C., Ed., CRC Press, Boca Raton, FL, 1981.

49. **Eckelman, W. C., Reba, R. C., Gibson, R. E., Rzeszotarski, W. J., Vieras, F., Mazaitis, J. K., and Francis, B.,** Receptor binding radiotracers: a class of potential radiopharmaceuticals, *J. Nucl. Med.,* 20, 350, 1979.

50. **Selikson, M., Gibson, R. E., Eckelman, W. C., and Reba, R. C.,** Calculation of binding isotherms when ligand and receptor are in different volumes of distribution, *Anal. Biochem.,* 108, 64, 1980.

51. **Franks, N. B. and Lieb, W. R.,** Where do general anesthetics act?, *Nature (London),* 274, 339, 1978.

52. **Stein, W. D.,** *Transport and Diffusion Across Cell Membrane,* Academic Press, New York, 1986.

53. **Fenstermacher, J. D., Blasberg, R. G., and Patlak, C. S.,** Methods for quantifying the transport of drugs across brain barrier systems, *Pharmacol., Ther.,* 14, 217, 1981.

54. **Fenstermacher, J. D. and Rapoport, S. I.**, Blood-brain barrier, in *Handbook of Physiology, the Cardiovascular System IV*, Renkin, E. M. and Michel, C. C., Eds., American Physiological Society, Bethesda, MD, 1984, 969.

55. **Wilson, G. M. and Pinkerton, T. C.**, Determination of charge and size of technetium diphosphonate complexes by anion-exchange liquid chromatography, *Anal. Chem.*, 57, 246, 1985.

56. **Burns, D. H., Worley, P., Wagner, H. N., Jr., Marzilli, L., and Risch, V.**, Design of technetium radiopharmaceuticals, in *The Chemistry of Radiopharmaceuticals*, Heindel, N., Burns, H., Honda, T., and Brady, L., Eds., Masson Publishing USA, New York, 1978, 269.

57. **Nunn, A. D.**, HPLC as the archetypical animal, *Nucl. Med. Biol.*, 16, 187, 1989.

58. **Wieland, D. and Beierwalters, W.**, A structure-distribution relationship study of adrenomedullary radiopharmaceuticals, in *Radiopharmaceuticals: Structure-Activity Relationships*, Grune & Stratton, New York, 1981, 413.

59. **Counsell, R., Klausmeier, W., Weinhold, P., and Skinner, R. W.**, Radiolabeled androgens and their analogs, in *Radiopharmaceuticals: Structure-Activity Relationships*, Grune & Stratton, New York, 1981, 425.

60. **Spitznagle, L., Eng, R., Marino, C., and Kasina, S.**, Structure-distribution studies with fluorine-18 labeled steroids, in *Radiopharmaceuticals: Structure-Activity Relationships*, Grune & Stratton, New York, 1981, 459.

61. **Spencer, R., Ed.**, in *Radiopharmaceuticals: Structure-Activity Relationships*, Grune & Stratton, New York, 1981.

62. **Moerlein, S. M., Laufer, P., and Stocklin, G.**, Effect of lipophilicity on the *in vivo* localization of radiolabelled spiperone analogues, *Int. J. Nucl. Med. Biol.*, 12, 353 1985.

63. **Welch, M. J., Chi, D. Y., Mathias, C. J., Kilbourn, M. R., Brodack, J. W., and Katzenellenbogen, J. A.**, Biodistribution of *n*-alkyl and *N*-fluoralkyl derivative of spiroperidol: radiopharmaceuticals for PET studies of dopamine receptors, *Nucl. Med. Biol.*, 13, 523, 1986.

64. **Moerlein, S. M., Laufer, P., and Stocklin, G.**, Design, synthesis and evaluation of radiobrominated neuroleptics for *in vivo* mapping of cerebral dopaminergic receptor areas and non-invasive pharmacokinetics studies, *Acta. Pharm. Suec.*, Suppl. 1, 481, 1985.

65. **Rzeszotarski, W. J., Potenzone, R., Jr., Eckelman, W. C. et al.**, Applications of NIH/EPA CIS chemlab programs for the design of radiopharmaceuticals, in *Computer Applications in Chemistry*, Heller, S. R., and Potenzone, R., Jr., Eds., Elsevier, New York, 1983, 341.

66. **Katzenellenbogen, J. A.**, The development of gamma-emitting hormone analogs as imaging agents for receptor-positive tumors, in *The Prostatic Cell: Structure and Function*, Alan R. Liss, New York, 1981, 313.

67. **Eckelman, W. C., Grissom, M., Conklin, J., Rzeszotarski, W. J., Gibson, R. E., Francis, B., Jagoda, E., Eng, R., and Reba, R. C.**, *In vivo* competition studies with analogues of quinuclidinyl benzilate, *J. Pharm. Sci.*, 73, 529, 1984.

68. **Jiang, V., Gibson, R. E., Rzeszotarski, W. J., Eckelman, W. C., and Reba, R. C.**, Radio-iodinated derivatives of beta adrenoceptor blockers for myocardial imaging, *J. Nucl. Med.*, 19, 918, 1978.

69. **Eckelman, W. C., Gibson, R. E., Vieras, F., Rzeszotarski, W. J., Francis, B., and Reba, R. C.**, *In vivo* receptor binding of iodinated beta adrenoceptor blockers, *J. Nucl. Med.*, 21, 436, 1980.

70. **Krohn, K., Vera, D. R., and Stadalnik, R. C.**, A complementary radiopharmaceutical and mathematical model for quantitating hepatic-binding protein receptors, in *Receptor Binding Radiotracers*, Vol. 2, Eckelman, W. C., Ed., CRC Press, Boca Raton, FL, 1982, 41.

71. **Vera, D. R., Krohn, K. A., Scheibe, P. O., and Stadalnik, R. C.**, Identifiability analysis of an *in vivo* receptor-binding radiopharmacokinetic system, *IEEE Trans. Biomed. Eng.*, BME-32, 5, 1985.

72. **Friedman, A. M., DeJesus, O. T., Revenaugh, J., and Dinnerstein, R. J.**, Measurements *in vivo* of parameters of the dopamine system, *Ann. Neurol.*, Suppl. 15, S66, 1984.

73. **Farde, L., Hall, H., Ehrin, E, and Sedvall, G.**, Quantitative analysis of D2 dopamine receptor binding in the living human brain by PET, *Science*, 231, 258, 1986.

74. **Katzenellenbogen, J. A., Heiman, D. F., Carlson, K. E., and Lloyd, J. E.**, *In vitro* and *in vivo* steroid receptor assays in the design of estrogen radiopharmaceuticals, in *Receptor Binding Radiotracers*, Vol. 1, Eckelman, W. C., Ed., CRC Press, Boca Raton, FL, 1981.

75. **Eckelman, W. C., Reba, R. C., Gibson, R. E., Rzeszotarski, W. J., Vieras, F., Mazaitis, J. K., and Francis, B.**, Receptor binding radiotracers: a class of potential radiopharmaceuticals, *J. Nucl. Med.*, 20, 350, 1979.

76. **Wagner, J. G.**, *Fundamentals of Clinical Pharmacokinetics*, Drug Intelligence, Hamilton, IL, 1981.

77. **Kuhar, M.**, Localizing drug and neurotransmitter receptor *in vivo* with tritium labeled tracers, in *Receptor Binding Radiotracers*, Vol. 1, Eckelman, W.C., Ed., CRC Press, Boca Raton, FL, 1982, 37.

78. **Wong, D. F., Wagner, H. N., Jr., Dannals, R. F. et al.**, Effects of age on dopamine and serotonin receptors measured by positron tomography in the living human brain, *Science*, 226, 1393, 1984.

79. **Patlak, C. S.**, Derivation of equations for the steady-state reaction velocity of a substance, *J. Cereb. Blood Flow Metab.*, 1, 129, 1981.

80. **Patlak, C. S. and Balsberg, R. G.**, Graphical evaluation of blood-to-brain transfer constants from multiple-time uptake data. Generalizations, *J. Cereb. Blood Flow Metab.*, 5, 584, 1985.

81. **Sawada, Y., Hiraga, S., Francis, B., Patlak, C., Pettigrew, R., Ito, K., Owens, E, Gibson, R., Reba, R., Eckelman, W., Larson, S., and Blasberg, R.**, Kinetic analysis of 3-quinuclidinyl 4-(^{125}I) Iodobenzilate transport and specific binding to muscarinic acetylcholine receptor in rat brain *in vivo*: implications for human studies, *J. Cereb. Blood Flow Metabol.*, in press.

82. **Patlak, C. S. and Blasberg, R. G.**, Graphical evaluation of blood-to-brain transfer constants from multiple-time uptake data. Generalizations, *J. Cereb. Blood Flow Metab.*, 5, 584, 1985.

83. **Mintun, M. A., Raichle, M. E., Kilbourn, M. R., Wooten, G. F., and Welch, M. J.**, A quantitative model for the *in vivo* assessment of drug-binding sites with positron emission tomography, *Ann. Neurol.* 15, 217, 1984.

84. **Wagner, H. N., Jr., Burns, H. D., Dannals, R. F., Wong, D. F., Langstrom, B., Duelfer, T., Frost, J. J., Ravert, H. T., Links, J. M., Rosenbloom, S., Lukas, S. E., Kramer, A. V., and Kuhar, M. J.**, Assessment of dopamine receptor activity in the human brain with carbon-11 *N*-methyl spiperone. (Subtitle: Dopamine receptors have been imaged in baboon and human brain by positron tomography), *Science*, 221, 1264, 1983.

85. **Linden, J., Patel, A., Earl, C. Q., Craig, R. H., and Daluge, S. M.**, Iodine-125-labeled 8-phenylxanthine derivatives: antagonist radioligands for adenosine A$_1$ receptors, *J. Med. Chem.*, 31, 745, 1988.

86. **Linden, J.**, Purification and characterization of (−)[^{125}I]hydroxyphenylisopropyladenosine, an adenosine R-site agonist radioligand and theoretical analysis of mixed stereoisomer radioligand binding, *Mol. Pharmacol.*, 26, 414, 1984.

87. **Weber, R. G., Jones, C. R., Palacios, J. M., and Lohse, M. J.**, Autoradiographic visualization of A$_1$-adenosine receptors in brain and peripheral tissues of rat and guinea pig using ^{125}I-HPIA, *Neurosci. Lett.*, 87, 215, 1988.

88. **Lanier, S. M., Hess, H. J., Grodski, A., Graham, R. M., and Homcy, C. J.**, Synthesis and characterization of a high affinity radioiodinated probe for the alpha$_2$-adrenergic receptor, *Mol. Pharmacol.*, 29, 219, 1986.

89. **Van Dort, M., Neubig, R., and Counsell, R. E.**, Radioiodinated *p*-iodoclonidine: a high affinity probe for the alpha$_2$ adrenergic receptor, *J. Med. Chem.* 30, 1241, 1987.

90. **Couch, N. W., Greer, D. M., Thonoor, C. M., and Williams, C. M.**, *In vivo* binding in rat brain and radiopharmaceutical preparation of radioiodinated HEAT, an alpha-1 adrenoceptor ligand, *J. Nucl. Med.*, 27, 356, 1988.

91. **Dyne, S., Gjedde, A., Diksic, M., Sherwin, A., and Hakins, A. U.**, *J. Cereb. Blood Flow Metab.*, 7 (Suppl. 1), S347, 1987.

92. **Zecca, L. and Ferrario, P.**, Synthesis and biodistribution of an iodine-123 labeled flunitrazepam derivative: a potential *in vivo* tracer for benzodiazepine receptors, *Appl. Radiat. Isot.*, 39, 353, 1988.

93. **Van Dort, M. E., Ciliax, B. J., Gildersleeve, D. L., Sherman, P. S., Rosenspire, K. C., Young, A. B., Junck, L., and Wieland, D. M.**, Radioiodinated benzodiazepines: agents for mapping glial tumors, *J. Med. Chem.*, 31, 2081, 1988.

94. **Gildersleeve, D. L., Lin, T. Y., Wieland, D. M., Ciliax, B. J., Olson, J. M. M., and Young, A. B.**, Synthesis of a high specific activity ^{125}I-labeled analog of PK 11195, potential agent for SPECT imaging of the peripheral benzodiazepine binding site, *Nucl. Med. Biol.*, in press.

95. **De Paulis, T., Janowsky, A., Kessler, R. M., Clanton, J. A., and Smith, H. E.**, (*S*)-*N*-[(1-Ethyl-2-pyrrolidinyl)-methyl]-5-[^{125}I]iodo-2-methoxybenzamide hydrochloride, a new selective radioligand for dopamine D-2 receptors, *J. Med. Chem.*, 31, 2027, 1988.

96. **Singhaniyom, W., Tsai, Y. F., Bruecke, T., McLellan, C. A., Cohen, R. M., Kung, H. F., and Chiueh, C. C.**, Blockade of *in vivo* binding of iodine-125-labeled 3-iodobenzamide (IBZM) to dopamine receptors by D$_2$ antagonist and agonist, *Brain Res.*, 453, 393, 1988.

97. **Brucke, T., Tsai, Y. F., McLellan, C., Singhaniyom, W., Kung, H. F., Cohen, R. M., and Chiueh, C. C.**, *In vitro* binding properties and autoradiographic imaging of 3-iodobenzamide ([^{125}I]-IBZM): a potential imaging ligand for D-2 dopamine receptors in SPECT, *Life Sci.*, 42, 2097, 1988.

98. **Kung, H. F., Billings, J. J., Guo, Y. Z., Blau, M., and Ackerhalt, R.**, Preparation and biodistribution of [^{125}I]IBZP: a potential CNS D-1 dopamine receptor imaging agent, *Nucl. Med. Biol.*, 15, 187, 1988.

99. **Kung, H. F., Billings, J. J., Guo, Y. Z., and Mach, R. H.**, Comparison of *in vivo* D-2 dopamine receptor binding of IBZM and NMSP in rat brain, *Nucl. Med. Biol.*, 15, 103, 1988.

100. **Nakatsuka, I., Saji, H., Shiba, H., Shimuzu, H., Okuno, M., Yoshitake, A., and Yokoyama, A.**, *In vitro* evaluation of radioiodinated butyrophenones as radiotracers for dopamine receptor study, *Life Sci.*, 41, 1989, 1987.

101. **Saji, H., Nakatsuka, I., Shiba, K., Tokui, T., Horiuchi, K., Yoshitake, A., Torizuka, K., and Yokoyama, A.**, Radioiodinated 2′-iodospiperone: a new radioligand for *in vivo* dopamine receptor study, *Life Sci.*, 41, 1999, 1987.

102. **Felder, R. A. and Jose, P. A.**, Dopamine$_1$ receptors in rat kidneys identified with ^{125}I-SCH 23982, *Am. J. Physiol.*, 255, F970, 1988.

103. **Chumpradit, S., Kung, H. F., Billings, J., Kung, M.-P., and Pan, S.,** (+)-7-Chloro-8-hydroxy-1-(4'-[^{125}I]iodophenyl)-3-methyl-2,3,4,5- tetrahydro-1H-3-benzazepine: a potential CNS D-1 dopamine receptor imaging agent, *J. Med. Chem.,* 32, 1431-1435, 1989.

104. **de Paulis, T., Janowsky, A., Kessler, R. M., Clanton, J. A., and Smith, H. E.,** (*S*)-*N*-[(1-Ethyl-2-pyrrolidinyl)methyl]-5-[^{125}I]iodo-2-methoxybenzamide hydrochloride, a new selective radioligand for dopamine D-2 receptors, *J. Med. Chem.,* 31, 2027, 1988.

105. **Kung, H. F., Kung, M. P., Pan, S., Kasliwal, R., Billings, J., and Guo, Y. Z.,** Radiolabeling and CNS D-2 dopamine receptor specificity of iodobenzamide (IBZM) and iodolisuride (ILIS), *J. Labelled Compd. Radiopharm.,* 27, 98, 1988.

106. **Kung, H. F., Billings, J., Guo, Y. Z., Blau, M., and Ackerhalt, R. A.,** Preparation and biodistribution of [^{125}I]IBZP: a potential CNS D-1 dopamine receptor imaging agent, *Int. J. Nucl. Med. Biol.,* 15, 187, 1988.

107. **Eckelman, W. C.,** Potentials of receptor binding radiotracers, in *New Brain Imaging Techniques in Cerebrovascular Diseases,* Cahn, J. and Lassen, N. A., Eds., J. Libbey Eurotext, Paris, 1985, 113.

108. **Wilson, A. A., Dannals, R. F., Ravert, H. T., Frost, J. J., and Wagner, H. N., Jr.,** Synthesis and biological evaluation of [^{125}I]- and [^{123}I]-4-iododexetimide, a potent muscarinic cholinergic receptor antagonist, *J. Med. Chem.,* 32, 1057, 1989.

109. **Eckelman, W. C., Grissom, M., Conklin, J., Rzeszotarski, W. J., Gibson, R. E., Francis, B. E., Jagoda, E. M., Eng, R., and Reba, R. C.,** *In vivo* competition studies with analogues of quinuclidinyl benzilate, *J. Pharm. Sci.,* 73, 529, 1984.

110. **Gibson, R. E., Weckstein, D. J., Jagoda, E. M., Rzeszotarski, W. J., Reba, R. C., and Eckelman, W. C.,** The characteristics of I-125 QNB and H-3 QNB *in vivo* and *in vitro, J. Nucl. Med.,* 25, 214, 1984.

111. **Gibson, R. E., Rzeszotarski, W. J., Jagoda, E. M., Francis, B. E., Reba, R. C., and Eckelman, W. C.,** (I-125) 3-quinuclidinyl 4-iodobenzilate: a high affinity high specific activity radioligand for the M1 and M2-acetylcholine receptors, *Life Sci.,* 34, 2287, 1984.

112. **Eckelman, W. C., Eng, R., Rzeszotarski, W. J., Gibson, R. E., Francis, B., and Reba, R. C.,** The use of 3-quinuclidinyl 4-iodobenzilate as a receptor binding radiotracer, *J. Nucl. Med.,* 26, 637, 1985.

113. **Eckelman, W. C.,** The design of cholinergic tracers, in *Discussions in Neurosciences,* Vol. 2, Feindel, W., Frackowiak, R. S. J., Gadian, D., Magistretti, P. L., and Zalutsky, M. R., Eds., Foundation for the Study of the Nervous System, Geneva, 1985, 60.

114. **Eckelman, W. C.,** Potentials of receptor binding radiotracers, in *New Brain Imaging Techniques in Cerebrovascular Diseases,* Vol. 2, Cahn, J. and Lassen, N. A., Eds., John Libbey Eurotext, Paris, 1985, 115.

115. **Eckelman, W. C.,** Receptor binding radiotracers, in *Biomedical Imaging,* Hayaishi, O. and Torizuka, K., Eds, Academic Press, Tokyo, 1986, 357.

116. **Gibson, R. E., Schneidau, T. A., Cohen, V. I., Sood, V., Ruch, J., Melograna, J., Eckelman, W. C., and Reba, R. C.,** *In vitro* and *in vivo* characteristics of [Iodine-125] 3-(R)-quinuclidinyl (S)-4-iodobenzilate, *J. Nucl. Med.,* 30, 1079, 1989.

117. **Eckelman, W. C., Reba, R. C., Rzeszotarski, W. J., Gibson, R. E., Hill, T., Holman, B. L., Budinger, T., Conklin, J. J., Eng, R., and Grissom, M. P.,** External imaging of cerebral muscarinic acetylcholine receptors, *Science,* 223, 291, 1984.

118. **Rzeszotarski, W. J., Eckelman, W. C., Francis, B. E., Simms, D. A., Gibson, R. E., Jagoda, E. M., Grissom, M. P., Eng, R. R., Conklin, J. J., and Reba, R. C.,** Synthesis and evaluation of radioiodinated derivatives of 1-azabicyclo (2.2.2) oct-3-yl alpha-hydroxy-alpha-(3-iodophenyl) phenylacetate as potential radiopharmaceuticals, *J. Med. Chem.,* 27, 1156, 1984.

119. **Eckelman, W. C.,** The development of muscarinic cholinergic receptor-binding radiotracers, in *Applications of Nuclear and Radiochemistry,* Marcos, N. and Lambrecht, R., Eds., Pergamon Press, Elmsford, NY, 1982, 287.

120. **Kabalka, G. W., Gai, Y. Z., and Mather, S.,** Synthesis of iodine-125 labeled 3-quinuclidinyl 4-iodobenzilate, *Nucl. Med. Biol.,* 16, 359, 1989.

121. **Mertens, J. J. R., Vanryckeghem, W., and Carlsen, L.,** New fast labeling of iodinated compounds with radioiodine resulting in pure radiopharmaceuticals with high specific activity, *J. Nucl. Med.,* 26, P123, 1985.

122. **Plati, J. T., Strain, W. H., and Warren, S. L.,** Iodinated organic compounds as contrast media for radiographic diagnosis. Ethyl esters of iodinated straight and branched chain phenyl fatty acids, *J. Am. Chem. Soc.,* 65, 1273, 1943.

123. **Verbruggen, R.,** Fast high yield labeling and quality control of (^{123}I)- and (^{131}I)MIBG, *Appl. Radiat. Isot.,* 38, 303, 1987.

124. **Holman, B. L., Gibson, R. E., Hill, T. C., Eckelman, W. C., Albert, M., and Reba, R. C.,** Muscarinic acetylcholine receptors in Alzheimer's disease, *In vivo* imaging with I-123 IQNB and emission tomography, *JAMA,* 254, 3063, 1985.

125. **Eckelman, W. C.,** Receptor binding radiotracers, in *Biomedical Imaging,* Hayaishi, O. and Torizuka, K., Eds., Academic Press, Tokyo, 1986, 364.

126. **Weinberger, D. R., Mann, U., Reba, R. C., Jones, D. W., Coppola, R., Gibson, R. E., and Chase, T. N.,** Comparison of [18]FDG PET and ^{123}I-QNB SPECT in patients, *J. Nucl. Med.,* 30, 896, 1989.

127. **Wagner, H. N., Jr.,** SNM Highlights—1989: "Why not?", *J. Nucl. Med.,* 30, 1283, 1989.

128. **Eckelman, W. C.,** The use of *in vitro* models to predict the distribution of receptor binding radiotracers *in vivo, Nucl. Med. Biol.,* 16, 233, 1989.

129. **Sawada, Y., Hiraga, S., Francis, B., Patlak, C., Pettigrew, R., Ito, K., Owens, E., Gibson, R., Reba, R., Eckelman, W., Larson, S., and Blasberg, R.,** Kinetic analysis of 3-quinuclidinyl 4-([125])iodobenzilate transport and specific binding to muscarinic acetylcholine receptor in rat brain *in vivo:* implications for human studies, *J. Cereb. Blood Flow Metabol., in press.*

130. **Gibson, R. E., Weckstein, D. J., Jagoda, E. M., Rzeszotarski, W. J., Reba, R. C., and Eckelman, W. C.,** The characteristics of I-125 QNB and H-3 QNB *in vivo* and *in vitro, J. Nucl. Med.,* 25, 214, 1984.

131. **Gibson, R. E.,** Imaging and quantification of muscarinic acetylcholine receptors, in *Quantitative Imaging,* Frost, J. J. and Wagner, H. N., Jr., Raven Press, New York, 1990, 129.

132. **Wilson, A. A., Dannals, R. F., Ravert, H. T., Frost, J. J., and Wagner, H. N., Jr.,** Synthesis and biological evaluation of [125]- and [123]-4-iododexetimide, a potent muscarinic cholinergic receptor antagonist, *J. Med. Chem.,* 32, 1057, 1989.

133. **Wouters, W., Van Dun, J., and Laduron, P. M.,** 7-Amino-8-[125]-ketanserin 9[125]AMIK), a highly sensitive, serotonin-S_2 receptor ligand, *Biochem. Pharmacol.,* 35, 3199, 1986.

134. **Glennon, R. A., Seggel, M. R., Soine, W. H., Herrick-Davis, K., Lyon, R. A., and Titeler, M.,** Iodine-125 labeled 1-(2,5-dimethoxy-4-iodophenyl)-2-aminopropane: an iodinated radioligand that specifically labels the agonist high-affinity state of 5-HT-2 serotonin receptors, *J. Med. Chem.* 31, 5, 1988.

135. **Jagoda, E. M., Gibson, R. E., Goodgold, H., Ferreira, N., Francis, B. E., Rzeszotarski, W. J., Reba, R. C., and Eckelman, W. C.,** I-125 17 alpha-iodovinyl II-beta-methoxyestradiol: *in vivo* and *in vitro* properties of a high affinity estrogen-receptor radiopharmaceutical, *J. Nucl. Med.,* 25, 472, 1984.

136. **Hochberg, R. B., Hoyte, R. M., and Rosner, W.,** E-17a-(2-[125]iodovinyl)-19-nortestosterone: the synthesis of a gamma-emitting ligand for the progesterone receptor, *Endocrinology,* 117, 2550, 1985.

137. **Lielinski, J. E., Larner, J. M., Hoffer, P. B., and Hochberg, R. B.,** The synthesis of 11 beta-methoxy-[16 alpha-^{123}I] Iodoestradiol and its interaction with the estrogen receptor *in vivo* and *in vitro, J. Nucl. Med.,* 30, 209, 1989.

138. **Katzenellenbogen, J. A., Heiman, D. F., and Carlson, K. E.,** in *Receptor Binding Radiotracers,* Vol. 1, Eckelman, W. C., Ed., CRC Press, Boca Raton, FL, 1982.

139. **Albert, S., Heard, R. D. H., Leblond, C. P., and Saffran, J.,** Distribution and metabolism of iodo-α-estradiol labeled with radioactive iodine, *J. Biol. Chem.,* 177, 247, 1949.

140. **Counsell, R. E. and Klausmeier, W.,** Radiotracer intraction with sex steroid hormone receptor protein and receptor mapping, in *Principles in Radiopharmacology,* Vol. 2, Colombetti, L., Ed., CRC Press, Boca Raton, FL, 1979, 59.

141. **Jensen, E. V. and Jacobson, H. I.,** Basic guides to the mechanism of estrogen action, *Recent Prog. Horm. Res.,* 18, 387, 1962.

142. **Hochberg, R. D.,** Iodine-125-labeled estradiol: a gamma-emitting analogue of estradiol that binds to the estrogen receptor, *Science,* 205, 1138, 1979.

143. **Jagoda, E. M., Gibson, R. E., Goodgold, H., Ferreira, N., Francis, B. E., Rzeszotarski, W. J., Reba, R. C., and Eckelman, W. C.,** I-125 17 alpha iodovinyl 11 beta methoxyestradiol: *in vivo* and *in vitro* properties of a high affinity estrogen-receptor radiopharmaceutical, *J. Nucl. Med.,* 25, 472, 1984.

144. **Eckelman, W. C.,** Clinical potential of receptor bases radiopharmaceuticals, in *Radiopharmaceuticals: Progress and Clinical Perspectives,* Vol. 2, Fritzberg, A., Ed., 1986, 89.

145. **Ferry, D. R. and Glossmann, H.,** ^{125}I-Iodipine, a new high affinity ligand for the putative calcium channel, *Naunyn Schmiedebergs Arch. Pharmacol.* 325, 186, 1984.

146. **Glossmann, H., Ferry, D. R., Goll, A., and Rombusch, M.,** Molecular pharmacology of the calcium channel: evidence for subtypes, multiple drug-receptor sites, channel subsites, and the development of a radioiodinated 1,4-dihydropyrine calcium channel label, [125]iodipine, *J. Cardiovasc. Pharmacol.,* 5 (Suppl. 4), 608, 1984.

147. **Wagner, N. A.,** Characterization of iodine-125-labeled ω-conotoxin binding to brain N calcium channels and tritium-labeled (−) desmethoxyverapamil binding to novel calcium channels in osteoblast-like osteosarcoma cells, *Diss. Abstr. Int. B.,* 49, 651, 1988.

148. **Bradfield, C. A., Kende, A. S., and Poland, A.,** Kinetic and equilibrium studies of Ah receptor-ligand binding: use of [125]I2-iodo-7, 8-dibromodibenzo-*p*-dioxin, *Mol. Pharmacol.,* 34, 229, 1988.

149. **Li, Z., Mantey, S., Jensen, R. T., and Gardner, J. D.,** An analog of substance P with broad receptor antagonist activity, *Biochim. Biophys. Acta.,* 972, 37, 1988.

150. **Shigeno, C., Yamamoto, I., Kitamura, N., Noda, T., Lee, K., Sone, T., Shiomi, K., Ohtaka, A., and Fujii, N.,** Interaction of human parathyroid hormone-related peptide with parathyroid hormone receptors in clonal rat osteosarcoma cells, *J. Biol. Chem.,* 263, 18, 369, 1988.

151. **Kadan, M. J.,** The development of a high specific activity radioligand, ^{125}I-LSD, and its application to the study of serotonin receptors, *Diss. Abstr. Int. B,* 48, 1258, 1987.
152. **Beer, H. F., Maeder, T., Hasler, P., Blaeuestein, P., and Schubiger, P. A.,** A new iodine-123 labeled benzodiazepine, *Nuklearmedizin,* 24 (Suppl.), 753, 1988.
153. **Holl, K., Deisenhammer, E., Dauth, J., Carmann, H., and Schubiger, P. A.,** Imaging benzodiazepine receptors in the human brain by single photon emission compouted tomography, *Nucl. Med. Biol.,* 16, 759, 1989.
154. **Tafani, M., Escoula, B., Coulais, Y., and Simon, J.,** A simple and original method for the preparation of 2-iodomorphine suitable for *in vivo* and *in vitro* studies, *J. Biophys. Biochem. Cytol.,* 10 (Suppl. 2), 187, 1986.
155. **YoungLai, E. V., Wilkinson, M., Hibbert, P., McMahon, A., and Jarrell, J. F.,** Binding of iodine-125-labeled human chorionic gonadotropin to cell membrane receptors in rabbit ovarian slices, *Can. J. Physiol. Pharmacol.* 66, 337, 1988.
156. **Elands, J., Barberis, C., Jard, S., Lammek, B., Manning, M., Sawyer, W. H., and DeKloet, E.,** I-d(CH$_2$)$_5$[Tyr(Me)^2,Tyr(NH$_2$)^9]AVP: iodination and binding characteristics of a vasopressin receptor ligand, *FEBS Lett.,* 229, 251, 1988.
157. **Jomary, C., Gairin, J.E., Cros, J., and Meunier, J. C.,** Autoradiographic localization of supraspinal k-opioid receptors with [^{125}I-Tyr^1, D-Pro^{10}]dynorphin, *Proc. Natl. Acad. Sci. U.S.A.,* 85, 627, 1988.
158. **Elands, J., Barberis, C., Jard, S., Tribollet, E., Dreifuss, J. J., Bankowski, K., Manning, M., and Sawyer, W. H.,** Iodine-125-labeled d(CH$_2$)$_5$[Tyr(Me)^2,Thr^4,Tyr-NH$_2$^{9}]OVT: a selective oxytocin receptor ligand, *Eur. J. Pharmacol.,* 147, 197, 1988.
159. **Langer, J. A.,** Radiolabeling of the interferon-a receptor, *Biochem. Biophys. Res. Commun.,* 157, 1264, 1988.
160. **Halushka, P. V., MacDermot, J., Knapp, D. R., Eller, T., Saussy, D. L., Jr., Mais, D., Blair, I. A., and Dollery, C. T.,** A novel approach for the study of thromboxane A2 and prostaglandin H2 receptors using an I 125-labeled ligand, *Biochem. Pharmacol.,* 34, 1165, 1985.
161. Trionix Research Laboratory, Twinburg, OH, 44087.
162. Ohio Imaging, Inc., Bedford Heights, OH, 44128.
163. Strichman Medical Equipment, Inc., Medfield, MA.
164. **Logan, K. W., Gu, X, M., Li, J. P., Holmes, R. A., and McFarland, W. D.,** Imaging characteristics of Mumpi: a cylindrical neuro-SPECT system. *J. Nucl. Med.,* 29, 832, 1988.
165. **Smith, A. P. and Genna, S.,** Position analysis and calibration of ASPECT: a totally-digital annular SPECT brain camera, *J. Nucl. Med.,* 29, 786, 1988.
166. Medimatic, Division of M.I.D., Inc., Irvine, CA.
167. **Graham, D., Trevena, I. C., Webster, B., and Williams, D.,** Production of high purity Iodine-123 using Xenon-124, *J. Nucl. Med.,* 26, 105, 1985.

Chapter 5

PET INSTRUMENTATION FOR QUANTITATIVE TRACING OF RADIOPHARMACEUTICALS

Iwao Kanno

TABLE OF CONTENTS

I. INTRODUCTION

This chapter outlines the information necessary to handle data obtained by positron emission tomography (PET), a tool for *in vivo* autoradiography.[1] The principle of this methodology involves measurement of tracer distribution using PET instrumentation.

A. PHYSICAL PRINCIPLES OF PET

A positron (i.e., a positively charged electron) emitted from a radionuclide travels a few millimeters into tissue before it loses kinetic energy and combines with an electron to annihilate. After two particles annihilate, their mass is transformed into two 511 keV photons (gamma rays) emitted in opposite directions (Figure 1). The PET scanner detects this pair of gamma rays by a circular array of detectors. Reconstruction of the data by the dedicated computers produces transaxial images of the spatial and temporal distribution of the positron-emitting tracer in the tissue.

Various factors limit the spatial resolution of a PET scanner. One of the largest limitations is the positron range, the distance traveled by a positron from the point of emission to the point of annihilation. This distance is determined by the maximum energy (E_{max}) of the positron emitted from the radioactive nuclei. E_{max} for ^{11}C, ^{13}N, ^{15}O, ^{18}F, and ^{68}Ga are 0.96, 1.20, 1.73, 0.63, and 1.9 MeV, respectively. The spatial resolution of a PET scanner deteriorates less than 1 mm in FWHM (full-width at half-maximum)[2-4] as a result of this travel. Since paired gamma rays are not emitted at precisely 180 degrees, an angular deterioration of the resolution results. The effect of this angular deviation on the resolution is up to 1 mm.

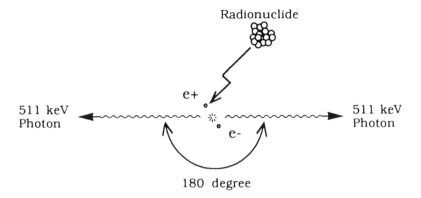

FIGURE 1. A scheme showing positron annihilation and the emission of two gamma rays. The positron and electron annihilate and emit two photons of 511-keV energy in opposite directions. The traveling range of the positron before annihilation and the angular deviation of the two gamma rays from 180 degrees is the theoretical limitation of the spatial resolution of the positron emission tomography.

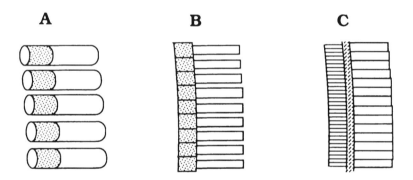

FIGURE 2. A scheme representing the development of detectors for positron emission tomography. (A) The first-generation PET scanners used a large, round scintillator (most commonly, a NaI crystal) and photomultiplier tubes (PMT); (B) the second-generation PET scanners used a square BGO crystal with a one-to-one photocoupled PMT; and (C) the third-generation PET scanners used an assembled block consisting of multiple narrow BGOs, light guide(s), and square PMT(s).

B. HISTORY OF PET DEVELOPMENT

The history of PET can be divided into three periods distinguished by detector development (Figure 2). The first generation of PET scanners used a large sodium iodide (NaI) crystal as a scintillation crystal and had a spatial resolution of about 15 to 30 mm in FWHM.[5,6] The second generation of PET scanners used a relatively large bismuth germanate ($Bi_4Ge_3O_{12}$; BGO) crystal.[7] These crystals were connected one by one to a photomultiplier tube (PMT). This configuration gave a spatial resolution of 7 to 12 mm full-width at half-maximum (FWHM).[8-12] The mid-1980s witnessed the development of the third generation of PET scanners. These scanners use a photo-encoding block detector consisting of multiple narrow BGO crystals connected with single or multiple PMT(s) via sophisticated light guide(s). Their FWHM resolution is 3 to 6 mm.[13-15]

II. PET HARDWARE

A. DETECTOR

A gamma ray detector consists of a scintillation crystal and a PMT. The scintillation

TABLE 1

**Properties of Various Scintillation Crystals Used in Positron
Emission Tomography**

Material	Unit	NaI	BGO	BaF$_2$	CeF$_3$
Density	g/ml	3.67	7.13	4.87	6.16
Decay constant	ns	250	300	0.8	2.0
Relative light yield		100	7	5	3
Time resolution	ps	1000	2500	300	520
Hydroscopic		Yes	No	No	No

crystal emits low-energy photons (visible lights) after interaction of high-energy photons (gamma rays) with the crystal material. The PMT converts the scintillation lights into electric pulses.

1. Scintillation Crystal

The size of a scintillation crystal determines the spatial resolution of the PET scanner. For at least a decade now, researchers have been seeking a narrow crystal with high stopping power. Because of its higher density, BGO is the best material for PET scanner crystals. Since the BGO crystal is nonhydrophilic, it can also be narrowed (3 to 6 mm in width) (Table 1). The small BGO crystals are attached to PMTs larger than the crystal itself (see next section). This disparity in size required development of a new way to join the crystals to the tubes rather than the old one-to-one coupling. Block-encoding detectors have thus been developed for high resolution PET scanners.

The intrinsic resolution of BGO detectors is roughly estimated as follows: with large BGOs more than 10 mm in size, the FWHM is approximately 0.6 times BGO width. However, for small BGOs of 5 mm or less, the FWHM is larger than that estimated from the geometrical rule outlined above. With the block-encoding detector assembly, the FWHM gets smeared even further as a result of the statistical uncertainty associated with decoding (assigning) the detector into which the gamma ray first enters.

The time-of-flight (TOF) PET scanners used barium fluoride (BaF$_2$) crystals.[16] BaF$_2$ emits a fast-decay ultraviolet (UV) light and slow-decay visible lights. Only the UV light is used for TOF detection. The use of this UV light requires that the PMT connected to BaF$_2$ has a quartz window that does not substantially absorb the UV light. Cerium fluoride (CeF$_3$) was also reported as a possible material for TOF scanners.[17] CeF$_3$ crystals have a higher stopping power, but they are slower than BaF$_2$ (see Table 1).

2. Photomultiplier Tube (PMT)

The PMT converts the scintillation lights into an electrical pulse. The PMTs developed for the third-generation PET scanners produce a faster pulse and have a higher linearity than a general purpose PMT. They have square optical windows to couple effectively with the crystal. These features are designed to check timing accuracy, mainly to sharpen the rise point, to extend count-rate linearity, and to improve statistical significance. Recently, multiple modules have been packed onto a single PMT to improve spatial resolution in the block-encoding detector assembly (see Section II.C, Physical Design).

3. Block-Encoding Detector

Several methods have been developed (Figure 3) to identify the detector which first interacts with a gamma ray in the block-encoding detector. One of the early trials of the block-encoding detector was the quad-BGO on a twin-PMT,[18] with four BGOs coupled with two PMTs. A fine block-encoding method in the latest generation PET[19] uses 32 small pieces of BGOs photocoupled with four PMTs via the slit light guide. This configuration

BGO PMT

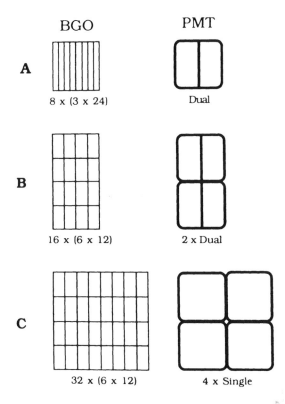

A

8 x (3 x 24)   Dual

B

16 x (6 x 12)   2 x Dual

C

32 x (6 x 12)   4 x Single

FIGURE 3.   Block-encoding detector. Currently, three types of block detectors are used in commercial PET scanners. Type A uses thin 8BGOs (3 × 24 mm) connected to a twin module photomultiplier tube (PMT) via twisted light guide. Type B uses 16 BGO (6 × 12 mm) and 2 twin module PMTs. Type C uses 32 BGOs (6 × 12 mm) and 4 single-module PMTs.

allows two-dimensional determination of the input position. A similar type of design was developed using the twin module PMT.[20] The area of this single unit was one half that of the former, therefore allowing a higher count rate. The other trial was developed using eight small BGOs photocoupled with a single twin module PMT via twisted light guides.[21]

In the block-encoding BGO detector assembly, the location of the crystal impacted by the gamma ray is encoded by dividing the scintillation output into multiple PMTs. The decoding logic is mostly achieved by a simple arithmetic operation of analog outputs of multiple PMTs. Therefore, the key of the block-encoding detector is a good balance of the amplitude gain of multiple PMTs assembled in each single block.

One of the possible deficits of the block-encoding detector is the early decline of count rates due to deadtime, i.e., a limitation in count rates.[22] Because the light from any BGO detector in a block stimulates all PMTs in that block, the duty time of PMT is proportional to the area of the single block. Therefore, the wider block-encoding detector has more deadtime for electronic signal processing.

4. Electronics

The detection electronics have the task of discriminating valid signals from the large number of input signals they receive. A signal from the individual detector consists of two kinds of information, i.e., the timing and the energy of the incident photon. The amount of scintillation light corresponds to the energy of the gamma ray absorbed in the scintillator in the one-by-one detector system and/or provides information on the position in the block-

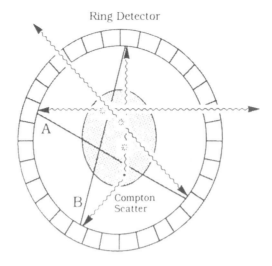

FIGURE 4. Schematic drawing of pseudo-coincidence events. (A) The random events are those resulting from two gamma rays emitted from the independent annihilations and recorded by two detectors within a given time window; (B) the scattered events result from those events where one or both gamma rays were recorded by a detector after being (Compton) scattered. These pseudo-coincidence events are corrected in the hardware and software procedures.

encoding detector system. Because of the relatively low light output (see Table 1), the energy resolution of BGO is normally worse than 20% in FWHM at 511 keV. The energy window is usually set to discriminate between the two thresholds, set at 250 to 350 and 650 to 700 keV. This poor energy resolution of the BGO detector makes it difficult to remove scattered events by energy discrimination.

B. COINCIDENCE

The coincidence circuit will detect not only true coincidence events, but also false events such as accidental coincidences (random events) and scattered coincidences (scatter events). A random event occurs by chance when two or more gamma rays of independent annihilation hit the detector ring. A scatter event occurs when either or both gamma rays have been scattered before reaching a detector (Figure 4). These false events are corrected in hardware and by the software processing as described below.

1. Grouping

Individual timing signals from single detectors of block-encoding detectors are added into a small number of groups (usually 6 to 16 per ring) before being sent into the coincidence logic. Each coincidence logic detects the events occurring simultaneously between the two detector groups. The extent of combinations covered by each group defines the acceptance fan angle. The fan angle determines the effective field of view (FOV) of the PET scanner.

2. Time Window

The time window (2τ) for coincidence is normally set for 10 to 20 ns in a BGO PET system and less than 2 ns in a TOF PET system.[23] The random coincidence rate (Nr) is given by $Nr = 2\tau N_1 N_2$, where N_1 and N_2 are the single count rates of the two detectors.

The random rate rapidly increases when the radioactivity in the FOV increases. Thus, the time window becomes one of the limiting factors of the count-rate linearity, hence a key factor in defining the signal-to-noise ratio and the upper limit of the count rate in a study using a high count technique, e.g., with $H_2^{15}O$ as a tracer (see Section III.B).

3. On-Time and Off-Time Coincidences

Coincidence logic primarily detects events that hit the different detectors simultaneously (on-time coincidences). With the on-time coincidence, the detected events (Non) include the true events (Nt), scattered events (Ns), and random events (Nr). The Nr can be estimated by the hardware off-time coincidences (Noff),[24] which have two inputs, either of which is delayed for a certain time (normally 50 ns). The off-time coincidence technique is commonly used in PET to correct for random rates by subtracting from the on-time events. Measured true rates (Nm = Nt + Ns) are given by Nm = Non − Noff.

C. PHYSICAL DESIGN

In current PET scanners, multiple detector rings are stacked to assess volumetric data of wide axial FOV. The coincidences between adjacent detector rings can independently measure an inter-ring image (cross-plane). At the same time, the coincidences within the same detector ring define an intra-ring image (direct-plane) (Figure 5). These two image planes have different physical characteristics, e.g., axial resolution, sensitivity, and count-rate linearity. These differences are dependent mainly upon the design of the ring diameter and septum.

1. Diameter of Detector Ring

The diameter of the detector ring affects the quantitative performance of a PET scanner. Scanners of smaller diameter suffer from the larger distortion of axial resolution in the cross-plane and from the larger difference in the spatial resolution in the transaxial directions and vice versa.[25] The size of the ring diameter also determines a fraction of scattered coincidence rates. In general, the larger the diameter of the scanner ring, the less the scattered and random coincidences are recorded when the patient aperture is the same. A suitable scanner diameter is 50 to 60 cm for neurological PET and more than 80 cm for whole-body PET.[25]

2. Packing Fraction

A packing fraction is a ratio of the net fraction of scintillators (the number of scintillators times their width) and the circumference of the scanner ring. The sensitivity of a scanner (true event count rate) is proportional to the square of the packing fraction. Therefore, in the most modern PET scanners, a large number of BGOs are housed in a single package to increase the packing fraction up to unity.

3. Septum

A septum is a radiation-opaque (lead) annular plate fixed at the middle of the adjacent detector rings in a multiple-ring PET system. The septum geometrically rejects oblique gamma rays incident from the outside of each slice plane (Figure 6), therefore reducing random and scattered events. The septum also determines the sensitivity and slice thickness difference between direct and cross-planes. Having the septa extended further toward the center of the FOV results in fewer false events being recorded, but this geometrical configuration does increase the differences in the slice thickness between direct and cross planes.[26]

D. GANTRY AND PATIENT COUCH
1. Scan Mechanism

Detectors stacked in a ring are designed to record counts in transaxial and/or axial directions so as to interpolate the sampling gaps. The transaxial scan is needed to meet the

Scintillators

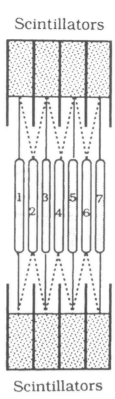

Scintillators

FIGURE 5.   A cross-section of a multi-ring PET scanner. One of the unique features of PET is its capability to image extra planes obtained between the two adjacent detector rings using appropriate coincidence logic. These image planes are called cross-planes (indicated by the even numbers in the figure). The image planes obtained from a single ring are designated as the direct-planes (indicated by the odd numbers in the figure). The physical performance of the two planes differs, and the magnitude of the difference depends on the physical design, e.g., the ring diameter, shape of the septa.

sampling interval required for intrinsic detector resolution.[27] Most PET scanners move transaxially along a small circle parallel to the transaxial plane, the so-called "wobble".[6,28,29] However, the newest PET scanners with BGO detectors only a few millimeters in width tend to remove the transaxial motion because the stationary sampling interval (approximately one half of the detector interval) is smaller than one half or one third of the intrinsic detector resolution. This elimination of mechanical motion will increase the reliability of the PET systems. Another benefit of stationary sampling is a PET system that generates less mechanical noise. Noise is an important factor to be considered, especially for neuropsychological studies.

Axial interpolation between the adjacent slices is also indispensable in obtaining complete axial information.[30,31] Some PET systems with wide axial separation between slices provide the axial scan of the detector ring.[15,23] If the scanner ring does not provide the axial scan motion, then the patient couch should slide axially back and forth.

2. Gantry Posture

On most PET systems, the angle of the gantry can be changed to obtain the transaxial plane as it might be required for a particular study. Tilt, inclining forward or backward, is

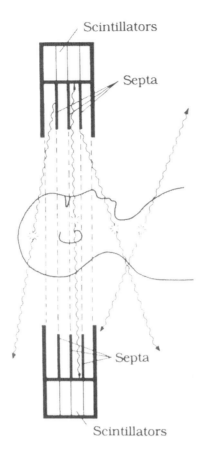

FIGURE 6. A cross-section of the PET gantry showing the shielding effect of the septa between adjacent rings. The septa protect the scintillators from obliquely incident photons emitted from the outside field of view (FOV) of the individual image plane. Because most of these photons from the outside of the whole FOV are the scattered events, rejection of these by the septa reduces the scatter events in the scanner. In addition to this, the septa play a minor role in collimating the photons to sharpen the slice thickness.

the most common feature of a PET scanner and is quite useful in brain studies when an arbitrary slice angle is required. Whole-body PET systems used for heart studies are designed to pivot so as to adjust the section slice to the heart axis.

3. Patient Couch

The patient couch consists of an axially sliding plate and a headrest. The plate can be moved axially by the computer software. Since most PET studies require immobilization of the patient for a couple of hours, patient comfort is the key to obtaining high quality data. Most patient discomfort is attributed to the tight fixation and long immobility. Patient movement during the PET study is the main cause of failure in a routine study.

III. PHYSICAL CHARACTERISTICS OF SCANNERS

The physical characteristics important in assessing PET data are spatial resolution along the transaxial and axial directions, resolution uniformities across the FOV, sensitivity in each slice, and count-rate linearity. In all these areas, brain imagings are superior to whole-body imagings. This improved performance derives mainly from the fact that the smaller

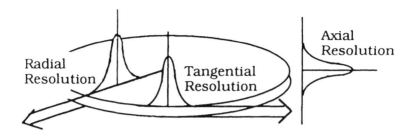

FIGURE 7.   Definition of the three axes of the spatial resolution in PET. The radial and tangential directions are in the transaxial image plane, and the axial direction corresponds to slice thickness. Normally, radial resolution is larger than tangential resolution and worsens along the distance from the center of the field of view. Resolution depends on the physical design of the scanner.

object (the brain) produces less scattered and random events, smaller distortion of resolutions, and less attenuation in the objects. It also derives partially from the fact that the geometry of the head, its relatively uniform, round shape, may diminish additional artifacts introduced by the attenuation correction.

A. SPATIAL RESOLUTION

The quality of the spatial resolution by and large indicates the quality of the PET system. The spatial resolution consists of three-directional components, i.e., radial, tangential, and axial (slice thickness) (Figure 7).

1. Transaxial Resolution

The theoretical limitation of the scanner's spatial resolution resulting from the positron range and the angular deviation of an annhilaton (see Section I.A, Physical Principles of PET) are not major limitations of the current PET systems. Spatial resolution is determined mainly by the intrinsic resolution of the detectors (see Section II.A, Detector). The resolution of a PET system deteriorates additionally with deviation in the individual detector resolution. The PET system resolution expressed as FWHM is normally 1.1 to 1.2 times greater than that of the detector. The resolution in clinical studies is further worsened from this figure as given in the PET system specifications because to reduce statistical noise, the low-pass filter is necessary in clinical image reconstruction (see Section V.A, Image Reconstruction). The clinical FWHM of a PET system is approximately 1.5 times greater than the physical FWHM.

2. Axial Resolution

To permit true volumetric imaging, the axial resolution should ideally be the same as the transaxial resolution. However, this cannot be achieved with the present technology. In most last-generation PET systems, the axial FWHM is 6 to 10 mm, 50 to 100% larger than the transaxial FWHM. This increase in the axial FWHM is required in order to obtain sufficient sensitivity per slice (plane). This derives from the proportionality of the sensitivity with the square of the axial resolution for a diffusely distributed object. The other reason for increasing the axial FWHM is the difficulty in reducing the size of the PMT. There are always serious trade-offs between sensitivity and slice thickness in the design of a PET scanner.

3. Resolution Uniformity

Uniformity is another important characteristic of spatial resolution.[25] The transaxial resolution deteriorates as the distance from the center of the FOV increases because the

FIGURE 8. Radial blurring of point image at the periphery of the FOV. The Derenzo phantom filled with ^{18}F radioactivity was measured by the Headtome III[12] of which tangential and radial resolutions were 6.5 and 8.0 mm in FWHM at the center and at 10 cm off the center, respectively. A total of 60 million counts were acquired. The deterioration of the radial resolution resulted in the radial blurring of the specific pattern. This feature is most often missed in clinical PET images.

incident angles of gamma rays into BGO coming from the center differ from those coming from the edge of a FOV. This deterioration is more significant in the radial than in the tangential direction (radial FWHM > tangential FWHM) (Figure 8). The PET scanner axial resolution is further affected by a nonuniformity between direct and cross-planes which results in a discrepancy in the sensitivity for direct and cross slices (see Section III.C, Sensitivity). The axial FWHM of the cross-plane is smaller than that of a direct plane at the center of the FOV and greater than that of a direct plane in the periphery of the FOV.

B. COUNT-RATE LINEARITY

A PET scanner's count-rate linearity is the most important feature in performing a fast dynamic study using high doses of ^{15}O-labeled tracers, especially for heart studies.[32] The demand for faster PET studies has accelerated improvements of the PET scanner's count-rate capability. Several factors can effectively improve the count rate (i.e., reduce the scanner's deadtime): amplification and decoding of the PMT signal, grouping and coincidence identification of those signals, and storing of coincidence signals into computer memory. In addition, deadtime is corrected by the scanner's software.

The signal processing on the block-encoding detectors carries a rather heavy load because a large BGO area must be covered by a single unit[22] (see Section II.A, Detector). A PET system using a large-area (2 × 2 in.) block-encoding BGO unit is saturated at 12 kcps per slice.[14] However, with a small-area block-encoding BGO unit, saturation is extended to 50 kcps per slice.[15]

C. SENSITIVITY

The theoretical sensitivity of a PET scanner can be approximately estimated, e.g., on the basis of a uniform flood pool of a phantom 20 cm in diameter filled with radioactivity at a concentration of 1 μCi/ml and a slice section 1 cm thick. The total number of annihilation events occurring in this slice can be estimated as $3.7 \times 10^{10} \times 10^{-6} \times 314 \approx 10^7$ events per second. The solid angle of the detector ring with 2-cm axial width and 100-cm diameter is approximately $1/50 = 0.02$. The attenuation coefficient for 511 keV gamma rays through the object is ≈ 0.2. Finally, the total number of events reaching the detector is $10^7 \times 0.02 \times 0.2 \approx 40$ kcps. However, the actual sensitivity of the PET system with this geometry is, at best, approximately 20 to 30 kcps. This discrepancy between estimated and actual sensitivities is due mainly to loss of the efficiency of the BGO crystal and geometrical design factors like the packing fraction (see Section II.C, Physical Design).

We should remember that sensitivity is proportional to the square of the slice thickness for a diffusely distributed object. In most PET scanners, the sensitivities in the cross-plane are 1.3 to 1.5 times higher than in the direct plane. Recent PET systems with the block-encoding detectors normally have 5 to 20 kcps/slice/(μCi/ml) for a 20-cm-diameter pool phantom. This sensitivity is one half or one third that achieved with the second-generation PET scanners.

D. SIGNAL-TO-NOISE (S/N) RATIO

Statistical noise is more of a problem in nuclear medicine imaging than in some other imaging methods. In the field of PET, serious efforts have been made toward studying the relationship between intrinsic resolution, imaging resolution, total counts, and image reconstruction filters. The coefficient of variation (COV) is derived for the cylinder phantom filled with uniformly distributed radioactivity.[33] The COV is inversely proportional to the square root of the total events acquired in an image (Figure 9).

Another important factor in the assessment of noise in PET is the rate of random events recorded. If the PET system uses off-time coincidence (see Section II.B, Coincidence), the number of measured true events (Nm) is given by subtracting the off-time events [Noff = random event (Nr)] from the on-time events [Non = true event (Nt) + scatter event (Ns) + random event (Nr)]. The variance of the Nm is given by the sum of the variances for Non and Noff: var (Nm) = var (Non) + var (Noff) = var (Nt) + var (Ns) + 2 var (Nr). Since Nr is proportional to the square of radioactivity (see Section II.B, Time Window), any study done with a high level of tracer radioactivity decreases the S/N ratio because of the high rate of random coincidences (Nr).

The S/N ratio is enhanced by increasing the intrinsic resolution of the PET systems.[34] Figure 10 shows images taken by two PET systems with different intrinsic resolutions. Even though both of these images have the same total number of events recorded and the same spatial resolution, the higher-resolution PET system has the higher S/N ratio. This is achieved by using of the lower cutoff filter during reconstruction to suppress statistical noise.

IV. PET MEASUREMENTS

A. NORMALIZATION MEASUREMENT

The efficiency of the individual detectors in the PET scanner is normalized by a normalization data. Normalization data are used to correct the emission measurement, pixel by pixel, on raw or sinogram data. A uniform plate source or rotating line source filled with

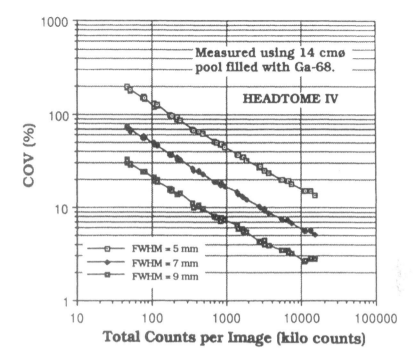

FIGURE 9. Total count vs. coefficient of variation (COV) of the image of cylindrical pool. The images were obtained by the HEADTOME IV[15] continuously during the decay of [15]O radioactivity in a uniform pool phantom of 14-cm diameter. Because of the physical decay, the total number of counts per image is decreased. Three reconstruction filters adjusted to 5-, 7-, and 9-mm resolution in FWHM were used. COV and total counts were plotted in a logarithmic scale.

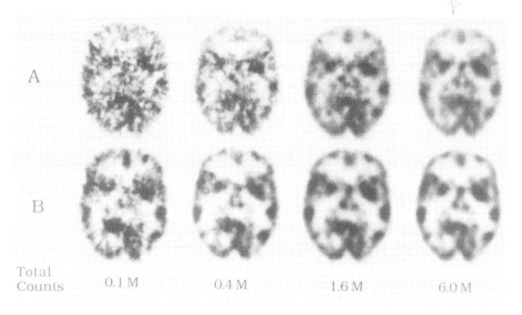

FIGURE 10. Signal-to-noise (S/N) ratio of images measured by the two PET instruments using different intrinsic resolutions and varying the total counts per image. A Hoffman brain phantom filled with ^{68}Ga radioactivity was measured by scanners having a low resolution (A, HEADTOME III,[12] 8 mm in intrinsic FWHM) and a high resolution (B, HEADTOME IV,[15] 4 mm in intrinsic FWHM). Image resolution was adjusted to be the same 9 mm in FWHM by selecting cutoff frequency of the reconstruction filter. Total counts (M, million) per image are shown at the bottom. In spite of the same total counts, the S/N ratio of the images was higher in the high-resolution PET. The difference seems to be distinct in the images with the lower number of total counts.

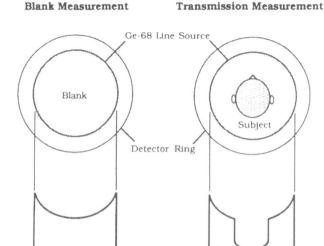

FIGURE 11.   Images of the attenuation factor measurement by transmission and the blank data. The blank measurement (left) provides the projection data of an external ^{68}Ge-^{68}Ga line source without attenuation. The transmission measurement (right) provides the data attenuated by the object to be measured. The ratio of these two data sets reflects the attenuation factor of the object. This is used to correct emission data before the image reconstruction is done.

^{68}Ge-^{68}Ga solution is used once a week or once a month on a routine basis to test the system. The frequency with which normalization is done depends on the stability of the system's electronics and detectors.

B. BLANK MEASUREMENT

The blank measurement provides the reference data needed for the transmission measurement used to correct for the attenuation of 511-keV gamma rays in the tissue. The data are obtained using an annular ring source or rotating line source filled with ^{68}Ge-^{68}Ga radioactivity and with no object present in the FOV.

Newer PET scanners acquire the blank and normalization data simultaneously by using a rotating line source. Normalization data are used to make the additional correction for geometrical factors that is needed to convert a circular orbit to a uniform plate.

C. TRANSMISSION MEASUREMENT

Most PET systems correct for attenuation of 511-keV gamma rays using a transmission measurement. The transmission measurement is done using the same radioactivity source as that used for the blank measurement, but the object is inside the FOV. The ratio between the blank data and the transmission data is equal to the attenuation caused by the medium (object) itself (Figure 11). The transmission data are handled in the same format as the emission data in a sinogram. This data set is multiplied along with the emission data before the image is reconstructed. Since the transmission measurement includes a high level of statistical noise caused by the high random rates, to reduce the influence of this noise, an attenuation correction using the transmission data is done using various noise suppression techniques.[35]

An image of the attenuation coefficient of a subject tissue can be obtained by doing a transmission scan in the same way as X-ray computed tomography. The transmission images can be used as an anatomical reference to define location of the skull or anatomical landmarks (see Section V.B, Anatomical Registration).

D. EMISSION MEASUREMENT

After the calibration measurements outlined above are done on a PET system, an emission measurement to assess the tracer distribution in a particular tissue is performed. There are several modes of emission measurement. Most of them are corrected for the radionuclide decay as well as for the deadtime (see Section III.B, Count Rate Linearity) and interpolate along the Z direction.

1. Dynamic Measurement

Dynamic measurement, which assesses the time course of the tracer distribution in a tissue, is used to study tracer kinetics. Measurement usually begins at the administration of the tracer. Because of the limitation on data storage in a dynamic measurement, the acquisition time is gradually prolonged as the time from an injection increases. This increase in the acquisition time (the frame time) is required in order to obtain comparable temporal information on the tissue tracer distribution. How short the sampling interval can be depends on how fast the scanner can complete the scan sequence. For example, in bolus ^{15}O-tracer studies, the scanner requires several seconds to obtain a scan, whereas [^{18}F]fluorodeoxyglucose studies require about 1 min.

2. Static Measurement

Static measurement of the tissue tracer distribution is used when an analytical model suggests that the tracer distribution is in a steady, near steady, or apparent steady state. In the steady-state measurement, acquisition time is determined by the number of counts per image needed to obtain an acceptable S/N ratio. When the tracer distribution is measured in a transient phase (apparent steady state), the image represents a mean accumulation of the tracer during the same acquisition time. In this case, the acquisition time is determined by the mathematical consideration of a particular model used for analysis.

3. Miscellaneous Emission Measurements

Most PET systems provide several programs for obtaining measurements of physiological variables (e.g., blood flow, glucose utilization). In the cardiac PET scanners, the gated measurements divide the tracer distribution into arbitrary periods dividing the whole heart cycle into 10 to 20 sections triggered by every R wave. This type of measurement requires highly sophisticated software and hardware packages.[15]

Linear scanning is a nontomographic, plane imaging of the tracer distribution performed by sliding the patient couch (Figure 12). This kind of measurement is useful when surveying the distribution in the whole body of newly developed tracers.[5,15]

V. SOFTWARE UTILITIES

Any software used in PET systems must have two basic characteristics. One is the capacity and the speed to permit handling of large data sets obtained by current PET systems. The data set produced in a single PET study often exceeds 20 megabytes of memory and very often serial multiple studies are carried out on the same subject. These large data sets create a bottleneck for transfer and analysis when a small computer is used. Since the present PET scanners are still mainly research rather than clinical tools, PET software should be flexible and reliable in producing quantitative data for the wide varity of parameters studied and the many procedures still being developed.

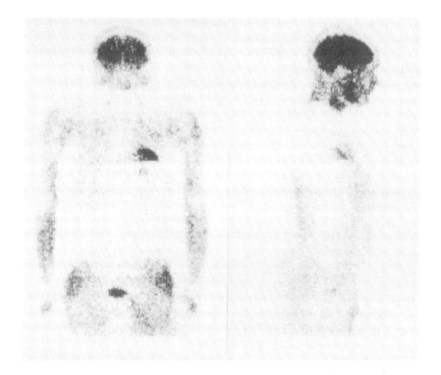

FIGURE 12.   Example of linear scan measurement. Two profiles, the anterior-posterior view (left) and the lateral view (right), are shown here of multiple profiles obtained using HEADTOME IV[15] from a normal subject 60 min after injection of 10 mCi[18]F-fluorodeoxyglucose and an acquisition time of 15 min. High uptakes in the brain, heart, and bladder and diffuse distributions are clearly seen in both profiles. This linear scan is thus useful to survey the whole body distribution of tracers.

A. IMAGE RECONSTRUCTION
1. Convolution Filter

The algorithms used for the reconstruction of images from the projection data are well established. Like conventional nuclear medicine, PET imaging suffers from the problems associated with statistical noise. An optimal convolution filter used in reconstruction is the key to extracting the best information under the limited S/N ratio. Most PET systems are programmed in such a way that the cut-off frequency of the convolution filter is determined on the basis of the total events recorded per image and the intrinsic resolution of the system used. The final image resolution can be expressed as the root-sum-square of different resolutions contributing to the final image resolution. Assuming a Gaussian distribution, the final image resolution is

$$Ri = (Rd^2 + Rf^2 + Ro^2)^{1/2} \qquad (1)$$

where Ri is the resolution of the final image and Rd, Rf, and Ro are the resolution of the detector, reconstruction filter, and object, respectively.

2. Partial-Volume Effect

Any limitation in the scanner's spatial resolution results in a serious deterioraton of the pixel counts. The physical characteristics of the partial volume effect have been extensively discussed in the past and the effect is well documented. Tracer concentration in any object larger than three times the FWHM can be recovered in the reconstructed image.[36]

Furthermore, the partial volume effect influences the functional parameter estimation

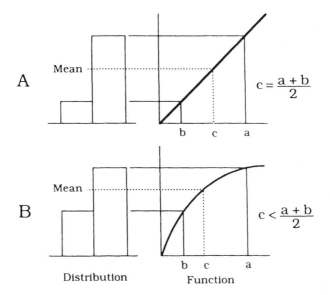

FIGURE 13. Partial volume effect (PVE) on the analysis by a biological model. Tissue mixture will cause an error in the PET measurement, especially when the physiological functions are estimated by means of the look-up table which contains a relationship between tracer distribution and physiological function. If the relationship is linear, the errors are minimum (PVE only) (A), but if this relation is nonlinear, the function estimated from the mean of the distribution becomes smaller than the true functional mean (B). The data shown roughly represent the difference in count rate observed for the gray and white matter (A) in the ^{18}F-fluorodeoxyglucose model; and (B) in the ^{15}O-gas steady state inhalation model. The nonlinear relation between count rate and the physiological variable estimated in the measurement (B; ^{15}O-gas steady state model) produces an underestimation of the true mean value of physiological function.

known as a tissue mixture (heterogeneity) effect. Most physiological models used in PET research are based on the nonlinear relation between the tracer concentration and the physiological function being studied. Usually the tissue heterogeneity causes an underestimation of the physiological variables. Physiological variables estimated by the ^{15}O-labeled gas steady state inhalation method are the most seriously affected by the tissue mixture[37] (Figure 13). However, kinetic analysis used in conjunction with the dynamic PET measurement protocol can correct for the partial volume effect.[32]

3. Attenuation Correction

Transmission data are most often used to correct for gamma ray attenuation. However, when the transmission measurement is not carried out, an arithmetic correction can be performed using an outline determined on the outer rim of the object. This arithmetic correction assumes uniform attenuation coefficient. There are several methods to define the outline, e.g., ellipse shape, automatic rim,[38] or manual rim. No matter what outline is used in conjunction with the arithmetic correction, there are pitfalls, associated mainly with the headrest of the patient couch. The bed pad and headrest used in all PET scanners yield a serious underestimation, in some instances, of more than 30% (Figure 14) when arithmetic correction is applied. In addition, the majority of these arithmetic corrections assume a uniform distribution of the attenuation coefficient. This assumption produces some additional errors.[39]

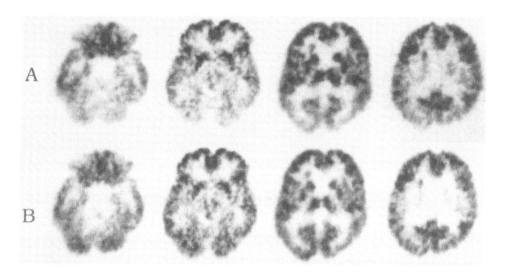

FIGURE 14.   Imperfect correction for the attenuation effect. (A) Using arithmetic correction; (B) using transmission measurement. Since the detection of the outline used in the arithmetic attenuation correction is difficult and includes nontissue-like material such as the headrest or bedpad, the correction causes a nonuniform underestimation. This could result in a misleading image even if the quantitative analysis is not used. Relatively large absorption by the soft material used caused a focal underestimation of more than 30% in the posterior part when compared to the correction done by transmission measurement.

Recently, a technique has been developed that permits the transmission measurement after administration of the tracer.[40,41] This approach is useful because it shortens the time during which a subject must be immobilized.

4. Scattered Event Correction

Because scattered events are estimated to distribute diffusely with the low-frequency component,[12] the PET scans are usually corrected for the scattered events using a filtering technique.[42] The reconstruction filter is designed to suppress low-frequency components. The magnitude of the frequency component is empirically determined from the point spread function in the sinogram. Scattered events in the transmission data are also evaluated, and correction methods continue to be developed.[43-45]

B. ANATOMICAL REGISTRATION

An accurate anatomical registration is the first step for interindividual analyses of any PET data. This registration is achieved by rescaling the PET image along the three axes in order to adjust the image to the standard brain map using approximate landmarks in the brain. Such interindividual standardization will provide new horizons to map higher cortical functions by suppressing differences between individuals or studies created by the misalignment. For brain studies, anatomical landmarks such as the GI (glabella-inion) line[46] or AC-PC (anterior and posterior commissural) line[47] have been proposed to assist neuropsychological functional mapping. Anatomical mapping is even more useful when combined with recent advances in scanner resolution and the utilization of fast computers. A powerful computer permits a real time reconstruction of an arbitrary section from the original stack of transaxial planes and facilitates more relevant anatomical analysis.

C. CROSS-CALIBRATION

Cross-calibration is a most important maintenance procedure for insuring reliability of the PET data. Cross-calibration normalizes the sensitivity of a PET system to that of a well

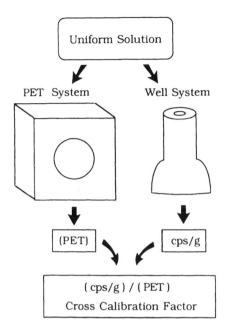

FIGURE 15. Cross-calibration procedures. A radioactive solution of uniform concentration is divided and distributed into a cylindrical pool for the PET system and a pot for the well system. The ratio of the concentration measured in each system is a cross-calibration factor. This ratio is used to convert the PET pixel value (dimensionless) to the tracer concentration (cps/g) in the quantitative *in vivo* study. The cross-calibration factor is vital for maintaining the PET system as a reliable tool.

counter used for the plasma radioactivity measurement. This is done by imaging a cylindrical pool phantom and measuring the corresponding count rate of an aliquote in the well counter (Figure 15). Since cross-calibration is performed under the same physical conditions as a patient study, the procedure cancels out many systematic errors. In addition, by multiplying with this factor, the pixel value is converted from dimensionless numbers to numbers with dimension of the unit of radioactivity per unit mass (volume), the same dimension as the well counter system (cps/g).

D. PHYSIOLOGICAL IMAGING

Many models are used in PET data analysis to derive physiological variables from the tracer kinetics. Some well-established biological models are ^{15}O-labeled gases steady state inhalation,[48] H_2^{15}O autoradiography,[49,50] and ^{18}F-fluorodeoxyglucose methods.[51] The algorithms to calculate physiological parameters from these models are installed in most PET systems. Based on the three-compartment model of ^{14}C-deoxyglucose[52] and ^{18}F-fluorodeoxyglucose,[51] the kinetics of most metabolic tracers can be tested using the standard mathematical and graphic software.[53]

VI. PERIPHERAL EQUIPMENT FOR PET

Quantitation is the key to most PET measurements. Reliable measurements and data analysis are supported by the entire system and procedures, including the peripheral tools used during measurement.

A. WELL COUNTER SYSTEM

The blood radioactivity measurements taken with the well counter and used in determining the radioactivity concentration in the sampled blood are essential if a quantitative

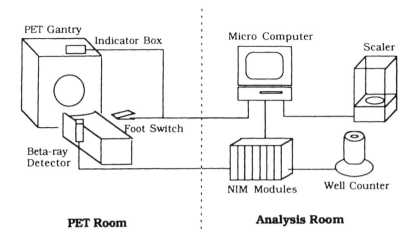

FIGURE 16.   Well-counter system. Blood analysis during the PET study is automated using a microcomputer connected to a well-counter, a scale for weight (the analysis room), a foot switch for the timing of samples, and an indicator box to identify the syringe number (PET room). The system is particularly useful to control the time of blood withdrawal and well-counting in reference to the standard time, which is defined in each study. Proper software will correct for the radioactive decay, well-counter deadtime, and sample weight, giving the output count rates of each sample per unit weight per second at a standard time. The system also supports the beta particle detector.

PET study is contemplated. Because processing of arterial blood is rather time-consuming, many PET centers have automated the well counter system. An automatic system consists of a microcomputer connected to the well counter, a balance for weighing samples, and a foot switch at the patient couch to record sampling time (Figure 16). The system sometimes includes a beta particle detector to measure on-line radioactivity concentraton of continuously withdrawn arterial blood.[54]

B. IMAGE PROCESSOR

An image processor independent of the PET system is an indispensable tool for extracting information from the PET data. Programs are required that can easily accept files containing region-of-interest (ROI) and display data in many different formats. Also needed is the ability to analyze a dynamic data set. Least squares fitting should be available in the various forms of different algorithms.[55] Subtraction and division of the functional images are also required to obtain physiologically meaningful data.

A three-dimensional display able to depict coronal, sagittal, and oblique sections, as well as a surface map of the brain, enables us to extract more information from the PET data. In addition, the capability to superimpose other morphological imaging modalities such as MRI is very important for neuropsychological studies (see Section V.B, Anatomical Registration).

C. TEST PHANTOMS

Phantoms that test the physical property of the PET scanner are also necessary both to verify its physical characteristics and to ensure proper performance. The most basic of these phantoms is a cylindrical pool phantom. Cylindrical phantoms with various diameters are indispensable for examining sensitivity and S/N ratio and for performing routine cross-calibration procedures. A pie phantom is required to test linearity between pixel value and radioactivity concentration. Other phantoms can intuitively test the spatial resolution of the PET scanners. The Derenzo phantom is useful to evaluate transaxial resolution (see Figure

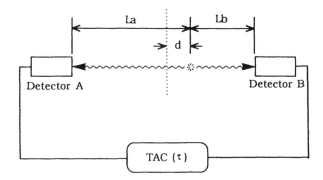

FIGURE 17. Principle of time-of-flight detection. The timing of two opposing fast detectors is fed into the time-amplitude converter (TAC). When the detectors determine a pair of annihilation gamma rays (at d from the center of the two detectors), the time difference (τ) of the arrival time is obtained and then converted to $\underline{d}$ ($d = $ (La-Lb)/2 $ = \tau C/2$). The state-of-the-art time resolution of the fast detector is 300 pico (10^{-12}) second, which corresponds to a distance of $3 \times 10^{12}/2/300 \times 10^{-12} \approx 5$ cm.

8). The Hoffman brain phantom is used to test the transaxial resolution that can actually be expected in the brain (see Figure 10). A finger phantom is used to test axial resolution.[31]

VII. PROSPECTS IN PET INSTRUMENTATION

A complete PET study comprises production of a radionuclide by the cyclotron, radiochemical synthesis, radioactivity distribution measurement by the PET instrument, and transformation of this measurement into physiologically and biochemically meaningful data based on a biological model. The role of PET instrumentation is to measure tissue concentration of tracers that are transformed into quantitative images representing physiological and biochemical variables.

Efforts to develop PET instrumentation focus on three areas. The first is further improvement of spatial resolution to improve quantitation as well as anatomical information. The second is widening of the axial FOV with fine axial sampling to permit reconstruction of three-dimensional images. The third is an attempt to increase the count-rate linearity, which is especially important in studies using ^{15}O-labeled tracers.

A. TIME-OF-FLIGHT PET

An alternative approach independent of the trend toward high-resolution PET is development of the time-of-flight (TOF) PET scanners. The principle of TOF is shown in Figure 17. TOF information currently available does not improve spatial resolution because even state-of-the-art TOF scanners provide a time resolution of 300 to 500 pico (10^{-12}) second, corresponding to only 5.0 to 7.5 cm in positioning an annhilation event.

TOF information does not improve spatial resolution of images. However, the TOF information results in improved S/N ratio. The reason for this is suppression of the noise dispersion in the reconstructed image. Since the TOF information can localize an annihilation event with a certain probability on a line, this information can be used to reduce noise. In addition, and important advantage of a TOF-PET system is its use of a narrow time window which reduces random coincidences especially useful in measurements with ^{15}O-labeled tracers (see Sections II.B and III.D, Coincidences and Signal-to-Noise Ratio).

B. ANIMAL PET

Several research groups now require PET cameras suitable for scanning animals.[56] Most scanners for animal use are designed to provide a spatial resolution of 2 to 3 mm in FWHM, which is close to the limit obtained with a BGO-PET system. The main framework of the animal PET system is no different from that used for human PET studies. The reduction in the number of detectors used in these systems does not substantially decrease the cost of the scanner because the detector assembly is only 10 to 20% of the cost of a current PET system.

C. NONINVASIVE PET

The goal of PET development is to realize a truly "noninvasive" study. Such a study would require neither arterial puncture nor immobilization of the subject, and would be much shorter than present scans. Work is already being carried out to eliminate the arterial puncture. Some cardiac studies are presently done completely in a noninvasive manner because they use the time radioactivity curves at the left atrium, the left ventricle, or the aorta. In the near future, software will be developed to obviate the necessity of rigid immobilization. The development of new tracers and of sophisticated biochemical and/or physiological models will shorten the study length. It is therefore quite possible that an entirely noninvasive PET study will be realized in the near future.

REFERENCES

1. **Phelps, M. E., Mazziotta, J. C., and Huang, S. C.,** Study of cerebral function with positron computed tomography, *J. Cereb. Blood Flow Metab.,* 2, 113, 1982.
2. **Phelps, M. E., Hoffman, E. J., Huang, S. C., and Ter-Pogossian, M. M.,** Effect of positron range on spatial resolution, *J. Nucl. Med.,* 16, 649, 1975.
3. **Cho, Z. H., Chan, J. K., Eriksson, L., Singh, M., Graham, S., MacDonald, N. S., and Yano, Y.,** Positron ranges obtained from biomedically important positron-emitting radionuclides, *J. Nucl. Med.,* 16, 1174, 1975.
4. **Iida, H., Kanno, I., Miura, S., Murakami, M., Takahashi, K., and Uemura, K.,** A simulation study of a method to reduce positron annihilation spread distributions using a strong magnetic field in positron emission tomography, *IEEE Trans. Nucl. Sci.,* NS-33, 597, 1986.
5. **Phelps, M. E., Hoffman, E. J., Huang, S. C., and Kuhl, D. E.,** ECAT: a new computerized tomographic imaging system for positron-emitting radiopharmaceuticals, *J. Nucl. Med.,* 19, 635, 1978.
6. **Bohm, C., Eriksson, L., Bergstrom, M., Litton, J., Sundman, R., and Singh, M.,** A computer assisted ring detector positron camera system for reconstruction tomography of the brain, *IEEE Trans. Nucl. Sci.,* NS-25, 624, 1978.
7. **Cho, Z. H. and Farukhi, M. R.,** Bismuth germanate as a potential scintillation detector in positron camera, *J. Nucl. Med.,* 18, 840, 1977.
8. **Thompson, C. J., Yamamoto, Y. L., and Meyer, E.,** Positome II: a high efficiency positron imaging device for dynamic brain studies, *IEEE Trans. Nucl. Sci.,* NS-26, 583, 1979.
9. **Nohara, N., Tanaka, E., Tomitani, T., Yamamoto, M., Murayama, H., Suda, Y., Endo, M., Iinuma, T., Teteno, Y., Shishido, F., Ishimatsu, K., Ueda, K., and Takami, K.,** Positologica: a positron ECT device with a continuously rotating detector ring, *IEEE Trans. Nucl. Sci.,* NS-28, 1128, 1980.
10. **Hoffman, E. J., Phelps, M. E., and Huang, S. C.,** Performance evaluation of a positron tomograph designed for brain imaging, *J. Nucl. Med.,* 24, 245, 1983.
11. **Litton, J., Bergstrom, M., Eriksson, L., Bohm, C., Blomqvist, G., and Kesselberg, M.,** Performance study of the PC-384 positron camera system for emission tomography of the brain, *J. Comput. Assist. Tomogr.,* 8, 74, 1984.
12. **Kanno, I., Miura, S., Yamamoto, S., Iida, H., Murakami, M., Takahashi, K., and Uemura, K.,** Design and evaluation of a positron emission tomograph: HEADTOME III, *J. Comput. Assist. Tomogr.,* 9, 931, 1985.
13. **Holte, S., Eriksson, L., Larsson, J. E., Ericson, T., Stjernberg, H., Hansen, P., Bohn, C., Kesselberg, M., Rota, E., Herzog, H., and Feinendegen, L.,** A preliminary evaluation of a positron camera system using weighted decoding of individual crystals, *IEEE Trans. Nucl. Sci.,* NS-35, 730, 1988.
14. **Spinks, T., Jones, T., Gilardi, M. C., and Heather, J. D.,** Physical performance of the latest generation of commercial positron scanner, *IEEE Trans. Nucl. Sci.,* NS-35, 721, 1988.

15. **Iida, H., Miura, S., Kanno, I., Murakami, M., Takahashi, K., Uemura, K., Hirose, Y., Amano, M., Yamamoto, S., and Tanaka, K.,** Design and evaluation of HEADTOME-IV, a whole-body positron emission tomogragh, *IEEE Trans. Nucl. Sci.,* NS-36, 1006, 1989.

16. **Laval, M., Moszynski, M., Allemand, R., Cormoreche, E., Guinet, P., Odru, R., and Vacher, J.,** Barium fluoride-inorganic scintillator for subnanosecond timing, *Nucl. Instrum. Methods,* 206, 169, 1983.

17. **Anderson, D. F.,** Properties of the high-density scintillator cerium fluoride, *IEEE Trans. Nucl. Sci.,* NS-36, 137, 1989.

18. **Murayama, H., Nohara, N., Tanaka, E., and Hayashi, T.,** A quad BGO detector and its timing and positioning discrimination for positron computed tomograph, *Nucl. Instrum. Methods,* 192, 501, 1982.

19. **Dahlbom, M. and Hoffman, E. J.,** An evaluation of a two-dimensional array detector for high resolution PET, *IEEE Trans. Med. Imag.,* MI-7, 264, 1988.

20. **Eriksson, L., Bohm, C., Kesselberg, M., Holte, S., Bergström, M., and Litton, J.,** Design studies of two possible detector blocks for high resolution positron emission tomography of the brain, *IEEE Trans. Nucl. Sci.,* NS-34, 344, 1987.

21. **Yamamoto, S., Miura, S., Iida, H., and Kanno, I.,** A BGO detector unit for a stationary high resolution positron emission tomograph, *J. Comput. Assist. Tomogr.,* 10, 851, 1986.

22. **Germano, G. and Hoffman, E. J.,** Investigation of count rate and deadtime characteristics of a high resolution PET system, *J. Comput. Assist. Tomogr.,* 12, 836, 1988.

23. **Ter-Pogossian, M. M., Ficke, D. C., Yamamoto, M., and Hood, J. T.,** Super PETT I: a positron emission tomograph utilizing photon time-of-flight information, *IEEE Trans. Med. Imag.,* MI-1, 179, 1982.

24. **Dereno, S. E., Budinger, T. F., Cahoon, J. L., Greenberg, W. L., Huesman, R. H., and Vuletich, T.,** The Donner 280-crystal high resolution positron tomograph, *IEEE Trans. Nucl. Sci.,* NS-26, 2790, 1979.

25. **Hoffman, E. J., Huang, S. C., Plummer, D., and Phelps, M. E.,** Quantitation in positron emission computed tomography. VI. Effect of nonuniform resolution, *J. Comput. Assist. Tomogr.,* 6, 987, 1982.

26. **Tanaka, E., Nohara, N., Tomitani, T., and Endo, M.,** Analytical study of the performance of a multilayer positron computed tomography scanner, *J. Comput. Assist. Tomogr.,* 6, 350, 1982.

27. **Huang, S. C., Hoffman, E. J., Phelps, M. E., and Kuhl, D. E.,** Quantitation in positron emission computed tomography. III. Effect of sampling, *J. Comput. Assist. Tomogr.,* 4, 819, 1980.

28. **Mullani, N., Ter-Pogossian, M. M., Higgins, C. S., Hood, J. T., and Ficke, D. C.,** Engineering aspects of PETT V, *IEEE Trans. Nucl. Sci.,* NS-26, 2703, 1979.

29. **Brooks, R. A., Sank, V. J., Talbert, A. J., and DiChiro, G.,** Sampling requirements and detector motion for positron emission tomograph, *IEEE Trans. Nucl. Sci.,* NS-26, 2760, 1979.

30. **Senda, M., Tamaki, N., Yonekura, Y., Tanada, S., Murata, K., Hayashi, N., Fujita, T., Saji, H., Konishi, J., Torizuka, K., Ishimatsu, K., Takami, K., and Tanaka, E.,** Performance characteristics of Positologica III: a whole-body positron emission tomograph, *J. Comput. Assist. Tomogr.,* 9, 940, 1985.

31. **Mullani, N. A.,** A phantom for quantitation of partial volume effects in ECT, *IEEE Trans. Nucl. Sci.,* NS-36, 983, 1989.

32. **Iida, H., Kanno, I., Takahashi, A., Miura, S., Murakami, M., Takahashi, K., Ono, Y., Shishido, F., Inugami, A., Tomura, N., Higano, S., Fujita, H., Sasaki, H., Nakamichi, H., Mizusawa, S., Kondo, Y., and Uemura, K.,** Measurement of absolute myocardial blood flow with $H_2^{15}O$ and dynamic positron emission tomography. Strategy for quantification in relation to the partial-volume effect, *Circulation,* 78, 104, 1988.

33. **Budinger, T. F., Derenzo, S. E., Greenberg, W. L., Gullberg, G. T., and Huesman, R. H.,** Quantitative potentials of dynamic emission computed tomography, *J. Nucl. Med.,* 19, 309, 1978.

34. **Phelps, M. E., Huang, S. C., Hoffman, E. J., Plummer, D., and Carson, R.,** An analysis of signal amplification using small detectors in positron emission tomography, *J. Comput. Assist. Tomogr.,* 6, 551, 1982.

35. **Dahlbom, M. and Hoffman, E. J.,** Problems in signal-to-noise ratio for attenuation correction in high resolution PET, *IEEE Trans. Nucl. Sci.,* NS-34, 288, 1987.

36. **Hoffman, E. J., Huang, S. C., and Phelps, M. E.,** Quantitation in positron emission tomography. I. Effect of object size, *J. Comput. Assist. Tomogr.,* 3, 299, 1979.

37. **Herscovitch, P. and Raichle, M. E.,** Effect of tissue heterogeneity on the measurement of cerebral blood flow with the equilibrium $C^{15}O_2$ inhalation technique, *J. Cereb. Blood. Flow Metab.,* 3, 407, 1983.

38. **Bergström, M., Litton, J., Eriksson, L., Bohm, C., and Blomqvist, G.,** Determination of object contour from projections for attenuation correction in cranial positron emission tomography, *J. Comput. Assist. Tomogr.,* 6, 365, 1982.

39. **Huang, S. C., Hoffman, E. J., Phelps, M. E., and Kuhl, D. E.,** Quantitation in positron emission computed tomography. II. Effect of inaccurate attenuation correction, *J. Comput. Assist. Tomogr.,* 3, 804, 1979.

40. **Daube-Witherspoon, M. E., Carson, R. E., and Green, M. V.,** Post-injection transmission attenuation measurements for PET, *IEEE Trans. Nucl. Sci.,* NS-35, 757, 1988.

41. **Ranger, N. T., Thompson, C. J., and Evans, A. C.,** The application of a masked orbiting transmission source for attenuation correction in PET, *J. Nucl. Med.,* 30, 1056, 1989.
42. **Bergström, M., Eriksson, L., Bohm, C., Blomqvist, G., and Litton, J.,** Correction for scattered radiation in a ring detector positron camera by integral transformation of the projections, *J. Comput. Assist. Tomogr.,* 7, 42, 1983.
43. **Chan, B., Bergström, M., Palmer, M. R., Sayre, C., and Pate, B. D.,** Scatter distribution in transmission measurements with positron emission tomography, *J. Comput. Assist. Tomogr.,* 10, 296, 1986.
44. **Thompson, C. J.,** The effects of collimation on scatter fraction in multi-slice PET, *IEEE Trans. Nucl. Sci.,* NS-35, 598, 1988.
45. **Digby, W. M. and Hoffman, E. J.,** An investigation of scatter in attenuation correction for PET, *IEEE Trans. Nucl. Sci.,* NS-36, 1038, 1989.
46. **Fox, P. T., Perlmutter, J. S., and Raichle, M. E.,** A stereotactic method of anatomical localization for positron emission tomography, *J. Comput. Assist. Tomogr.,* 9, 141, 1985.
47. **Friston, K. J., Passingham, R. E., Nutt, J. G., Heather, J. D., Sawle, G. V., and Frackowiak, R. S. J.,** Localization in PET image: direct fitting of the intercommissural (AC-PC) line, *J. Cereb. Blood. Flow. Metab.,* 9, 690, 1989.
48. **Frackowiak, R. S. J., Lenzi, G. L., Jones, T., and Heather, J. D.,** Quantitative measurement of regional cerebral blood flow and oxygen metabolism in man, using ^{15}O and positron emission tomography: theory, procedure and normal values, *J. Comput. Assist. Tomogr.,* 4, 727, 1980.
49. **Raichle, M. E., Martin, W. R. W., Herscovitch, P., Mintun, M. A., and Mahlkam, J.,** Brain blood flow measured with intravenous H$_2^{15}$O. II. Implementation and validation, *J. Nucl. Med.,* 24, 790, 1983.
50. **Kanno, I., Iida, H., Miura, S., Murakami, M., Takahashi, K., Sasaki, H., Inugami, A., Shishido, F., and Uemura, K.,** A system for cerebral blood flow measurement using H$_2^{15}$O autoradiographic method and positron emission tomography, *J. Cereb. Blood. Flow Metab.,* 7, 143, 1987.
51. **Huang, S. C., Phelps, M. E., Hoffman, E. J., Sideris, K., Selin, C. J., and Kuhl, D. E.,** Non-invasive determination of local cerebral metabolic rate of glucose in man, *Am. J. Physiol.,* 238, E69, 1980.
52. **Sokoloff, L., Reivich, M., Kennedy, C., Des Rosiers, M. H., Patlka, C. S., Pettigrew, K. D., Sakurada, O., and Shinohara, M.,** The (^{14}C)-deoxyglucose method for the measurement of local cerebral grucose utilization: theory, procedure, and normal values in the conscious and anesthetized albino rat, *J. Neurochem.,* 28, 987, 1977.
53. **Patlak, C. S., Blasberg, R. G., and Fenstermacher, J. D.,** Graphical evaluation of blood-to-brain transfer constants from multiple-time uptake data, *J. Cereb. Blood. Flow Metab.,* 3, 1, 1983.
54. **Kanno, I., Lammertsma, A. A., Heather, J. D., Gibbs, J. M., Rhodes, C. G., Clark, J. C., and Jones, T.,** Measurement of cerebral blood flow using bolus inhalation of C^{15}O$_2$ and positron emission tomography: description of the method and its comparison with the C^{15}O$_2$ continuous inhalation method, *J. Cereb. Blood. Flow Metab.,* 4, 224, 1984.
55. **Carson, R. E., Huang, S. C., and Phelps, M. E.,** BLD — a software system for physiological data handling and model analysis, Proc. 5th Annu. Symp. Comput. Appl. Medical Care, *IEEE Trans. Comput.,* p. 562, 1981.
56. **Tomitani, T., Nohara, N., Murayama, H., Yamamoto, M., and Tanaka, E.,** Development of a high resolution positron CT for animal studies, *IEEE Trans. Nucl. Sci.,* NS-32, 822, 1985.

Chapter 6

SPECT: STATE-OF-THE-ART SCANNERS AND RECONSTRUCTION STRATEGIES

Ronald J. Jaszczak

TABLE OF CONTENTS

I. NUCLEAR MEDICINE IMAGING

Nearly three decades ago, Kuhl et al.[1,2] developed the first tomographic device to image radiopharmaceuticals. However, this group employed reconstruction strategies that did not completely eliminate overlying and underlying source activities in the final image. In the early 1970s, Hounsfield[3] and Cormack[4] developed algorithms for X-ray computed tomography (CT) which effectively eliminated the superimposition artifacts. Progress in this area then led to renewed interest in emission tomography. The X-ray CT algorithms, when appropriately modified, could be adapted for single photon emission computed tomography (SPECT). Since 1970, many research groups throughout the world have contributed to the development of SPECT instrumentation and reconstruction techniques.[5-25]

Nuclear medicine imaging involves the detection and spatial mapping of the radiation (usually gamma rays) emitted by a radiopharmaceutical labeled with a specific radionuclide, such as ^{99m}Tc, ^{201}Tl, ^{123}I, ^{11}C, ^{13}N, ^{15}O, or ^{18}F. Objectives of a nuclear medicine scan of the brain may include, for example, the detection of lesions, the evaluation of regional cerebral blood flow, or the quantitative determination of a particular metabolic process, such as the rate of regional glucose utilization.

The first gamma ray imaging devices (nontomographic) were developed in the late 1940s and early 1950s. These used either Geiger counters[26] or scintillation counters[27] that mechanically scanned the detector over the patient while recording the observed count rate.[28,29] Most imaging systems for nuclear medicine use the principle of scintillation detection. This approach basically consists of a crystal, such as NaI(Tl), that emits light photons immediately after the absorption of a gamma photon, and a coupled photomultiplier tube (PMT) that converts the brief pulse of light to a pulse of electricity. The light pulse typically lasts less than a millionth of a second and, for certain crystals, may be less than a billionth of a second. A scintillation scanner, equipped with an appropriate lead collimator to determine the emission directon, maps the gamma ray intensity as a function of scanner position to produce a projectional, or planar, image.

The introduction of the scintillaton camera by Anger[30,31] in 1957 revolutionized the field of nuclear medicine imaging. The scintillation camera (Figure 1) is a positional-sensitive device that consists of a single, large-diameter (25 to 50 cm) NaI scintillation crystal viewed by an array of photomultiplier tubes (PMTs). When a gamma ray interacts with the crystal, the PMT in closest proximity to the point of interaction will generate the largest output signal, while the more distant tubes will have smaller outputs. These latter PMTs receive less light from the scintillator. By appropriately combining the output signals from all of the tubes, it is possible to compute the centroid of the light distribution and thus determine the point within the crystal where the gamma ray was absorbed. A projectional image of the gamma ray flux is obtained by placing a multi-hole collimator on the side of the scintillation crystal facing the patient. The collimator, typically about 2 or 3 cm in thickness, is made of lead and contains many small channels (1 to 2 mm in diameter) that allow gamma rays whose trajectories are appropriate to reach the crystal. The arrangement of the collimator holes determines the imaging geometry. For example, if the holes are all parallel to each other and perpendicular to the crystal, it is referred to as a parallel hole collimator. The resulting image will be a two-dimensional projection of the three-dimensional source distribution. The scintillation camera has the advantage of simultaneously viewing the entire organ of interest, and hence its sensitivity is much larger than a single scanning detector.

Nuclear medicine imaging also was impacted dramatically by the development of the ^{99}Mo/^{99m}Tc (gamma ray energy of 140 keV) generator by Harper et al. in 1962.[32,33] Having a physical half-life of 6 hr, ^{99m}Tc is well suited for routine clinical use, and the Anger scintillation camera tended to be optimized for the 140 keV photons emitted by ^{99m}Tc. Today, most nuclear medicine imaging is performed using ^{99m}Tc-labeled radiopharmaceuticals and scintillation cameras based on the Anger principle.

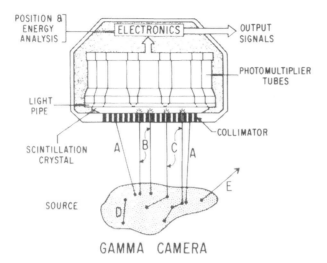

GAMMA CAMERA

FIGURE 1. Basic components of a gamma ray scintillation camera. Rays labeled A indicate gamma rays that strike the septa of the collimator. Rays labeled B pass through the collimator holes. Rays labeled C have been scattered within the source and then detected by the crystal. Rays labeled D and E are not detected. (From Jaszczak, R. J. and Coleman, R. E., *Invest. Radiol.*, 20, 897, 1985. With permission.)

II. EMISSION COMPUTED TOMOGRAPHY

The development of appropriate detectors, radiopharmaceuticals, and reconstruction strategies, coupled with rapid advancements in the areas of computers and electronics, made emission computed tomography (ECT) feasible.

ECT is an approach that attempts to determine the three-dimensional distribution of radionuclide concentrations within the organ by using a set of two-dimensional projection images acquired at many different angles around the patient. Since ECT eliminates overlying and underlying source activities, the image contrasts of lesions compared with surrounding normal tissue are markedly increased, which improves lesion detection, while the three-dimensional information improves localization. Perhaps even more importantly, ECT offers the potential for quantitative measurements of regional radiopharmaceutical concentrations. Hence, important physiologic and pathologic processes may be measured noninvasively and serially.

Instrumentation for ECT has developed in two complementary directions. These directions have been strongly influenced by the emission characteristics of the two types of radionuclides that are used. The approaches include (1) detection of the two coincident 511-keV photons resulting from positron-electron annihilation and (2) detection of gamma radiation from nonpositron photon emitters.

Positron emission tomography (PET) has the advantage of using physiologically important radionuclides that emit positrons, such as ^{11}C, ^{13}N, ^{15}O, and ^{18}F. Thus, it is possible to synthesize biochemically important radiopharmaceuticals for PET. However, the short half-lives of most PET isotopes require an expensive on-site accelerator for production and also increase the complexity in the synthesis of certain compounds.

SPECT (Figure 2) has the practical advantage of using readily available gamma-emitting radionuclides such as ^{99m}Tc, ^{201}Tl, and ^{123}I that are routinely employed in hospital nuclear medicine departments. The half-lives of these isotopes are sufficiently long that they (or the parent in the case of the ^{99}Mo/^{99m}Tc generator) can be produced in centralized nuclear reactors or accelerators and shipped directly to the hospital for use in diagnostic imaging procedures.

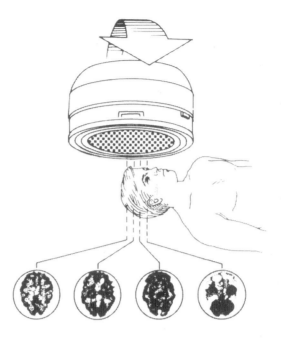

FIGURE 2.   Geometry used to acquire and reconstruct SPECT
images using a rotating gamma camera.

SPECT currently has several clinical applications that include the detection of space-occupying lesions within organs such as the liver and the brain, the evaluation of myocardial perfusion using ^{201}Tl, and the evaluation of regional blood perfusion using ^{123}I and ^{99m}Tc-labeled radiopharmaceuticals.

The evaluation of regional cerebral blood flow (rCBF) using single-photon radionuclide studies has been investigated using either freely diffusible compounds such as ^{133}Xe or lipophilic compounds,[34-47] such as *N*-Isopropyl-[^{123}I] *p*-iodoamphetamine (^{123}I IMP) and *N,N,N'*-trimethyl-*N*^{1}-(2-hydroxyl-3-methyl-5-[^{123}I] iodobenzyl)-1,3-propanediamine · HCl.

The ^{133}Xe method is based on measuring the regional cerebral clearance of inhaled ^{133}Xe. Typically a series of several 20- to 60-s scans must be acquired. To obtain an adequate number of detected photons within each scan requires very high sensitivity and high counting rate capabilities. However, it is not possible to attain simultaneously both the required high sensitivity and high spatial resolution (10 mm or less).

A principal advantage of the lipophilic brain agents, such as ^{123}I IMP and ^{123}I HIPDm, is the trapping that occurs after the agents enter the brain tissue. Thus, it is possible to use higher-resolution SPECT systems by increasing the scan time from several seconds to several minutes. Several investigators have demonstrated high-quality brain images of ^{123}I IMP and ^{123}I HIPDm using both discrete-detector and camera-based SPECT systems.[42-45] Further improvements in spatial localization could be obtained if a ^{99m}Tc compound were used. For example, a series of neutral and lipid soluble ^{99m}Tc bis-aminoethanethiol (BAT) derivatives have been developed as brain-imaging agents.[46] With ^{99m}Tc agents, it is possible to use SPECT systems having axial and transverse resolutions less than 10 mm full-width at half-maximum (FWHM), even though the sensitivity will be considerably less than that of lower-resolution configurations designed for ^{133}Xe clearance studies.

An example of a SPECT brain scan obtained using a ^{99m}Tc agent is shown in Figure 3. The SPECT scan was acquired with the Triad system using high-resolution collimation following intravenous injection of ^{99m}Tc-labeled exametazime and provides information on the regional blood perfusion within the brain.

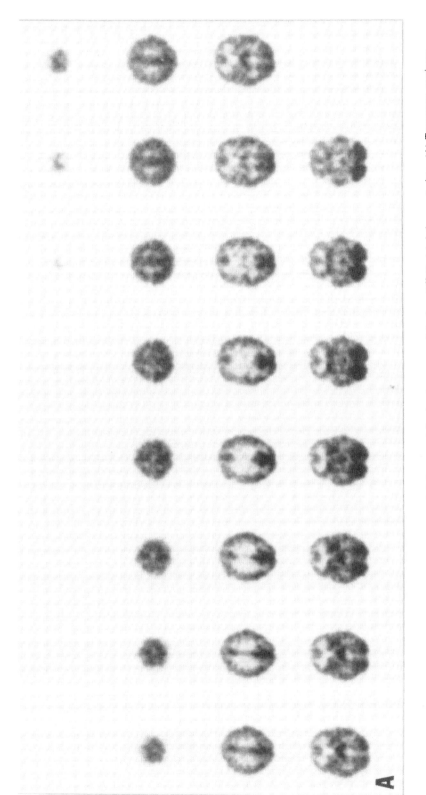

FIGURE 3. SPECT patient brain scan acquired using Duke Triad system following intravenous injection of ^{99m}Tc-labeled exametazime. (A) Transverse sections; (B) coronal sections; (C) sagittal sections.

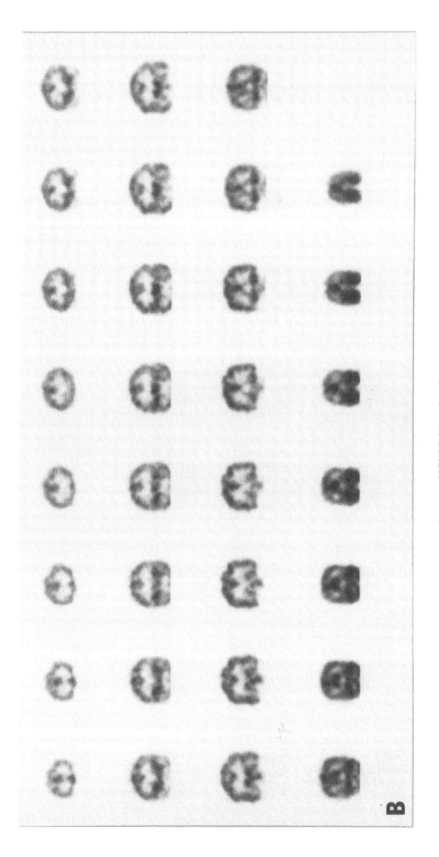

FIGURE 3 (continued).

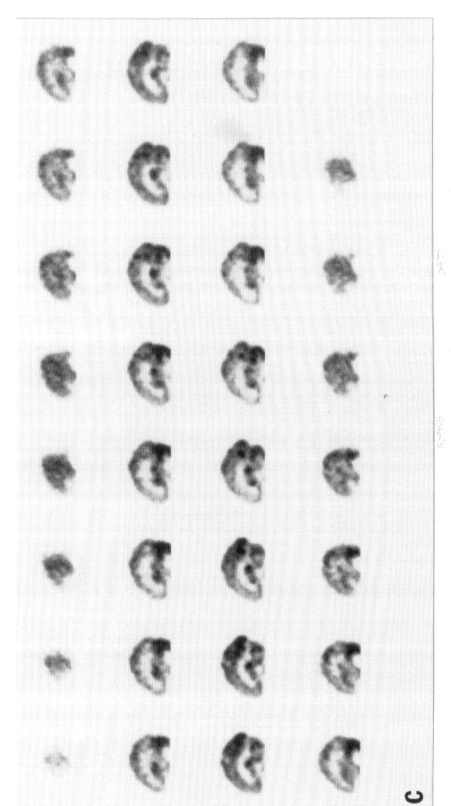

FIGURE 3 (continued).

III. SPECT INSTRUMENTATION

A. SENSITIVITY AND SPATIAL RESOLUTION CONSIDERATIONS

SPECT image quality is limited by the inability to acquire sufficient count densities within clinically useful imaging times. The statistical errors associated with the detection process are propagated through the reconstruction algorithm, resulting in both a qualitative and quantitative degradation of the SPECT image. Since the radionuclide within the patient emits photons isotropically, systems with a large active detector area are desirable.

The spatial resolution and sensitivity of a SPECT system are inversely related to each other and are largely determined by the total useful detector area and the geometric properties of the collimator. Spatial resolution may be quantitatively evaluated by scanning a line or point source. The source may be placed in air (or in a tissue equivalent medium such as water) at preselected locations within the field of view. Although complete profiles through the SPECT image of the line, or point, source may be used, usually only the FWHM and the full-width-at-tenth-maximum (FWTM) are determined. The spatial resolutions within a transaxial plane and parallel to the axis of rotation (i.e., in the axial direction) may or may not be the same. These resolutions will often be quite different. Furthermore, the point source response (resolution) can vary for different locations within the field of view. Thus, although resolution measurements are useful, system performance should also be evaluated using carefully selected phantoms that more effectively sample the complete field of view of the scanner and more closely simulate the specific clinical task.

Sensitivity measurements (i.e., the effectiveness of the scanner to detect gamma photons) also may be difficult to interpret. There are two general approaches to measuring sensitivity. The first approach is to determine the counting rate resulting from a point source of known activity that has been placed (in air) at a preselected location within the field of view. Alternatively, a water-filled cylindrical phantom, having a given diameter and length, may be filled with a given concentration of activity and scanned. This latter measurement is referred to as the volume sensitivity, and results may be given in terms of a single slice, the total phantom (all slices), or normalized on a "per axial centimeter" basis. The volume measurement usually includes detected scattered photons. Hence, a system with poor energy resolution may result in more detected scattered photons and appear more sensitive than another system; however, its imaging performance may be degraded. Similarly, one system may have a larger single slice volume sensitivity since its slice thickness may be large; however, its total volume sensitivity and its volume sensitivity per axial centimeter may be lower than another system.

SPECT system sensitivity can be increased by improving the collimator geometric efficiency (for example, by increasing the diameter of the individual holes) or by increasing the amount of active detector material that simultaneously views the organ of interest. Although both methods have been used, the former method obviously results in a corresponding degradation of spatial resolution. One must be careful in comparing SPECT devices on the basis of sensitivity, particularly in the absence of specific information on the corresponding spatial resolution and the detailed methodology used to determine the sensitivity and spatial resolution. In general it would be desirable to compare devices at the same spatial resolution and axial slice thickness; however, this is usually not possible because of differences in system design.

B. MULTI-DETECTOR SYSTEMS

Several acquisition approaches are being developed to improve SPECT brain imaging. Geometries having large active detector surfaces are being evaluated for both discrete detector- and camera-based SPECT systems (Figure 4). Improved collimator designs[15,48-51] are also being investigated to more effectively use the available detector surface (Figures 5 and 6).

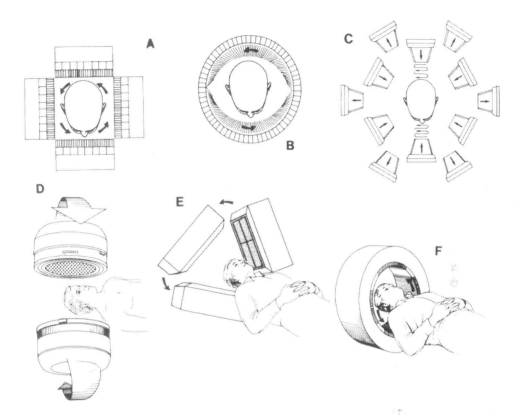

FIGURE 4.   Examples of several discrete detector (A to C) and camera-based (D to F) SPECT approaches.

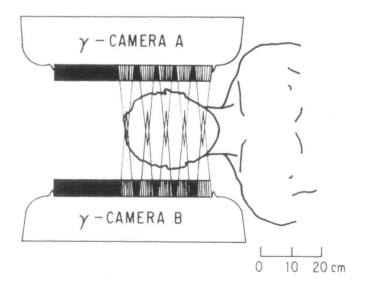

FIGURE 5.   Use of axially converging collimation to improve single slice sensitivity and uniformity.

Kuhl et al.[19] recognized several years ago that four banks of discrete detectors (see Figure 4A) could be used to improve SPECT performance. Discrete detector approaches are capable of high counting rates. These devices can be equipped with high sensitivity (poor spatial resolution) collimators to determine regional cerebral blood flow (rCBF), using [33]Xe

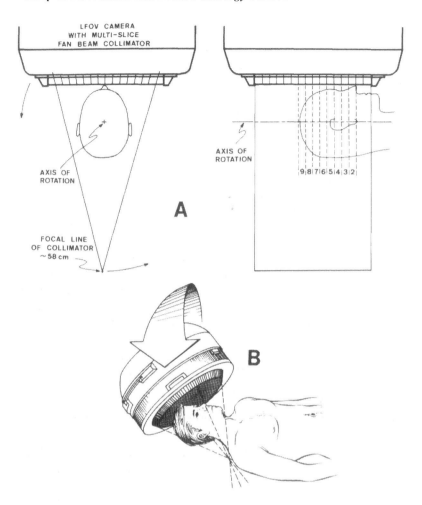

FIGURE 6.   Improved SPECT collimation. (A) Multislice fan beam collimation;[48] (B) cone, or converging, beam collimation.[49,50]

clearance measured from a series of 1-min SPECT scans. The useful spatial resolution of these configurations is typically greater than 2 cm (FWHM) in order to obtain adequate count densities from the 1-min scans. Hence, rCBF can be independently determined only for volume elements larger than about 4 cm on a side. Alternatively, discrete detector devices can be equipped with high resolution collimators (lower sensitivity) to obtain high quality images from static radiopharmaceutical distributions.

Multidetector systems typically produce only a few (one to three) sectional images from a single scan. The sections are obtained either by using multiple rings of discrete detectors or a set of one-dimensional bar cameras. Usually, axially converging collimation is used to increase the sensitivity per slice; however, the resulting sectional images may be noncontiguous. The slice thickness may be increased then (to as much as twice the in-plane resolution) to improve further the ''per slice'' sensitivity. Under these circumstances, the patient must be shifted and then rescanned in order to obtain a complete set of contiguous transaxial images. Thus, there exists a tradeoff between axial resolution, slice thickness, the number of slices, and contiguous vs. noncontiguous imaging. These characteristics make it difficult to compare SPECT devices simply on the basis of published sensitivity and resolution data.

Examples of multidetector systems include the Tomomatic[23] (Medimatic, Inc.), the

Headtome[21] SET-031 (Shimadzu Corp.), and the Harvard (Strichman Medical Equipment, Inc.) Model 810 scanner.[22,52] The Tomomatic is based on the configuration developed by Kuhl et al. (see Figure 4A) and uses four banks of one-dimensional bar cameras. Each bank contains 16 one-dimensional bar cameras. The one-dimensional bar cameras use NaI(Tl) crystals that are 16 cm long, 1.3 cm wide, and 2.5 cm thick. Three PMTs view each crystal in the 16-cm axial direction, and the signals from these PMTs are processed to localize the detected events. Typically three or five axial slices are imaged at once. Each crystal bar is equipped with three tapered-hole collimators that focus both axially and within the sectional plane. Spatial resolution ranges from 17 mm (FWHM) with a high sensitivity collimator to about 9 mm (FWHM) when a high or ultrahigh resolution collimator is used.

Although the original Tomomatic device was designed as a dedicated brain scanner, Lassen et al[53] have recently developed a novel whole body device that uses discrete detectors. By offsetting the four banks of detectors laterally, the detectors can maintain close proximity to the brain. The banks can also be moved radially outward (and shifted laterally) for whole-body imaging. Instead of bar detectors, this device uses 768 individual NaI(Tl) crystals (24 per bank). The system produces eight noncontiguous slices (1 cm thick), each separated by 2 cm. The Lassen device has several interesting design characteristics that are useful for brain (and body) imaging. First, as with most discrete detector approaches, it can be used for high counting rate studies such as ^{133}Xe wash-in or wash-out. The devices uses axially converging collimation of the type shown in Figure 5. Hence, 2 cm of crystal length (in the axial direction) is focused onto a 1-cm-thick axial slice of the brain. However, the resulting eight transaxial slices are not contiguous. To obtain a complete set of 16 contiguous slices, it is necessary to translate the patient and repeat the scan. A unique feature of the Lassen device is its capability to obtain SPECT scans of all parts of the body. This is possible since it uses four long (40 cm) detector banks that can be offset to maintain close proximity to either the brain or other parts of the body.

The Headtome[21] (see Figure 4B) consists of three circular rings of NaI(Tl) crystals (16 × 28 × 30 mm). Each ring has 64 individual detectors. The spacing between detector rings is 3.5 cm. An interesting characteristic of this device is that the detectors are stationary. The projection views are obtained using a vane-type collimator (see Figure 4B) that rotates in front of the stationary detectors. The reported in-plane resolutions and axial slice thicknesses can range from 10 to 20 and 16 to 25 mm, respectively, depending on the type of collimation. The volume sensitivity on a per slice basis ranges from 6 to 31 × 10^3 (counts/s)/(μCi/ml). However, both this system and the Tomomatic device use an energy window that may be more than twice as wide as camera-based systems. Hence, a larger number of scattered photons are detected and contribute to the volume sensitivity measurements.

The Strichman multidetector scanner (see Figure 4C) is a dedicated brain-only device originally developed in 1979 by Stoddart et al.[52] at Cleon (Union Carbide) and improved more recently by the group at Harvard.[22] This system uses a set of 12 large scintillators (20.3 × 12.7 × 2.5 cm), 12 PMTs, and 12 highly focused collimators. The detectors scan both radially and tangentially to provide a field of view that is 20.3 cm in diameter. The unique sampling characteristics of this multidetector scanner have required the development of special reconstruction techniques. The system has very high single-slice sensitivity since there are over 3000 cm^2 of active detector area focused on the slice. Since it can image only a single slice at a time a complete set of slices requires translating and rescanning the patient. Assuming 1- to 2-min acquisitions (per slice) are usually obtained, then the time required to obtain a complete set of scans is about the same as the time required by most other devices. An advantage of this system is the capability of obtaining high-count density, single-slice studies by scanning one section of the brain for several minutes. Also, the system can be used for high count rate studies. The volume sensitivity is 14 × 10^3 (counts/s)/ (μCi/ml), and the in-plane and axial resolutions are reported to be 10 and 15 mm (FWHM), respectively. However, this sensitivity value is affected by the large (5-in.), axially extended detectors and highly converging collimators that are used. The use of large detectors and converging collimation can result in sensitivity values that vary with the axial extent of the cylindrical source used for the measurement. For example, the volume sensitivity of a 15-mm-thick disk source would be much less than the volume sensitivity measured using a much longer (for example, 20-cm) cylindrical source of the same diameter. The detected photons from outside the 15-mm plane of interest may explain the large volume sensitivity. It is further reasonable to expect that source activity located in nearby planes would contribute noise in the reconstructed 15-mm-thick axial slice. The effect of this noise on image quality requires further investigation, and at the moment a comparison of this system's sensitivity with that of other SPECT devices is difficult.

Image reconstruction is obtained by deconvolving the line spread function of the system. With the current commercial system, a reconstruction software package is available for a standard Macintosh II workstation. More recently, iterative techniques have been implemented and have produced high-quality brain scans.

The discrete detection approaches described above have been used extensively for clinical imaging and are commercially available, and other discrete detector devices continue to be developed within several research laboratories.[20,24]

C. CAMERA-BASED SYSTEMS

Camera-based SPECT devices[7-11] (see Figures 4D to F) use one or several large-area two-dimensional scintillation cameras to acquire the projectional views. Presently, nearly all clinical SPECT scans are performed using a single large field of view (LFOV) camera that rotates about the patient (see Figure 2). The advantages of a single rotating gamma camera include:

1. It is suitable for both planar and SPECT imaging.
2. A single 360° scan provides a complete set of three-dimensional images of the entire organ volume.
3. In-plane and axial resolutions can be made to be nearly the same.
4. Both the head and the body can be scanned.
5. The cost is not significantly higher than the cost of a conventional planar camera equipped with a data processing computer.

It is for these reasons that most of the nuclear medicine systems purchased today consist of a single rotating gamma camera having SPECT capability.

However, single camera SPECT systems have a few significant disadvantages. The more important limitations include:

1. Single camera SPECT systems have suboptimal detection efficiency.
2. High counting rates may result in substantial count losses due to the deadtime of a single large-area detector.
3. Some camera-based SPECT systems are not able to clear the patient's shoulders because of their design and are therefore unsuitable for high-quality brain imaging.

For many years there have been camera-based SPECT systems that cleared the patient's shoulders and maintained close proximity to the brain.[8-10] Unfortunately, until relatively recently, some manufacturers underestimated the importance of this design consideration for SPECT imaging of the brain.

The second limitation, for example, prohibits the use of a camera-based SPECT system for ^{133}Xe clearance studies of the brain, although many other brain studies, such as those obtained using lipophilic agents, which result in low counting rates can be imaged with camera-based systems.

By far the most serious limitation of a single camera SPECT device is its low single-slice sensitivity, although its total organ (volume) sensitivity may be relatively high since the active detector area available is about 1300 cm². There are two methods available to improve the sensitivity of camera-based devices. First, it is possible to use collimator designs that focus the available detector surface onto the brain or other organ of interest. The second approach is to increase the active detector area, either by using multiple gamma cameras (see Figures 4D and E) or by using a single annular scintillation crystal (see Figure 4F). These configurations will be described in the following sections.

1. Improved Collimator Designs

One method to increase single-slice sensitivity and to improve the uniformity of the axial slice thickness is shown in Figure 5. The use of collimation that focuses in the axial direction about the midplane was proposed nearly a decade ago by our group.[15] This type of axially focusing collimation can increase single-slice sensitivity by a factor of two, and, in fact, many of the discrete detector SPECT systems, such as the Tomomatic[23] and the Harvard scanners,[52] use this concept to increase the single-slice sensitivity. This type of collimation does not increase total organ sensitivity since the slices are no longer contiguous. The sectional images are separated by gaps that would have to be scanned either by shifting the gamma camera axially or by shifting the patient and rescanning. Camera manufacturers have yet to implement this type of collimation, probably because the resulting slices are no longer contiguous.

An effective method that would significantly increase both single-slice and total organ sensitivity when imaging the brain is to use collimators that enable more of the available detector surface to view the brain. We have developed two collimator geometries to improve brain SPECT imaging using camera-based systems. In 1977 we built and evaluated[10,48] multi-slice fan beam collimators (see Figure 6, top) that have converging collimation perpendicular to the axis of rotation and parallel collimation along this axis. More recently, other investigators[54] have verified that this approach can improve SPECT image quality. Fan beam collimation can increase volume sensitivity by a factor of 1.3 to 1.6 as compared with a parallel hole collimator having similar spatial resolution.

However, even with multi-slice fan beam collimation, a large portion of the active detector surface is not used for brain imaging. Thus, we have recently[49,50] developed doubly converging, or cone beam collimation (see Figure 6, bottom) to further increase volume sensitivity. Similarly, Hawman et al.[51] have developed astigmatically focused collimators to optimize the camera field of view for brain imaging. These types of doubly converging collimators can increase volume sensitivity by as much as a factor of two, or perhaps even three, as compared with parallel hole collimaton having similar resolution characteristics. An example of the improvement in image quality that might be obtained using cone beam collimation is demonstrated in the patient scan shown in Figure 7.

To maintain the brain completely within the camera field of view when using cone beam collimators, it is necessary to tilt the camera with respect to the axis of rotation (see Figure 6, bottom). This increases the cone angle for locations near the top of the brain. Points located at large distances from the central ray are not completely sampled, and the filtered back projection algorithm currently used[55] cannot accurately reconstruct these locations. Hence, an axial blurring, or mispositioning of events, will occur near the superior part of the brain. As the cone angle increases (i.e., moving away from the central ray), the magnitude of the distortion increases. This distortion can be seen in the patient scan shown in Figure 7. We, and other investigators, are attempting to develop improved reconstruction approaches to minimize this distortion.

HIGH RESOLUTION PARALLEL HOLE

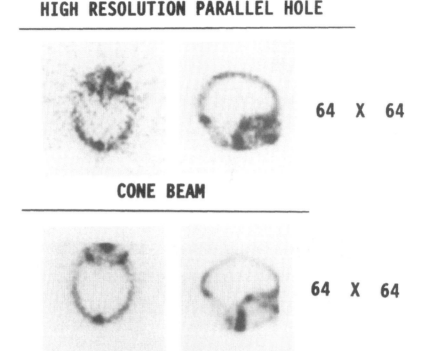

FIGURE 7.   Patient scan acquired using parallel hole (top) and cone beam collimation (bottom).

2. Multiple and Annular Crystal Camera Approaches

Several different camera-based SPECT devices are shown along the bottom row of Figure 4. While rotating single-head systems are suboptimal for brain imaging for the reasons discussed above, multiple head devices (see Figure 4D to E) using specially designed fan or cone beam collimation can result in imaging performance that is comparable to other dedicated SPECT devices.

The Triad system[16] (Trionix Research Laboratory, Inc.) (see Figure 4E and Figure 8) is one of several commercially available devices that uses three rectangular field of view scintillation cameras placed in a triangular pattern. It is capable of imaging either the body or the brain by moving the detectors radially outward or inward. Each scintillation camera has a 40 × 22-cm field of view, resulting in a total detector area of 2640 cm^2. Approximately 50 to 60% (1320 to 1584 cm^2) of this detector area can be focused onto the brain by using fan beam collimation. Cone beam collimators should result in a further improvement in sensitivity. Spatial resolutions range from 8 to 12 mm, depending on the type of collimation used. The volume sensitivity per axial centimeter (for 10-mm in-plane and axial resolution) is approximately 3 × 10^3 (counts/s)/(μCi/ml) using high-resolution fan beam collimation.

Another camera-based device that makes efficient use of the active detector area is the ASPECT system[18] (Digital Scintigraphics, Inc.) (see Figure 4F). The ASPECT is a dedicated brain scanner that uses a single fixed annular NaI(Tl) crystal (31-cm diameter, 13 cm wide) to completely encircle the patient's head. The total active detector area equals 1265 cm^2. Sixty-three PMTs are used to provide positional and energy information. Since the detector is stationary, a rotating annular collimator is used to generate the projection data necessary for SPECT. One of the available collimators preferentially focuses onto the central region of the brain. This configuration can provide, for the central region of the brain, a sensitivity that is six times that of single camera SPECT system using a parallel hole collimator. Typically, the sensitivity is about 3.5 to 4.0 times that of a comparable single rotating camera

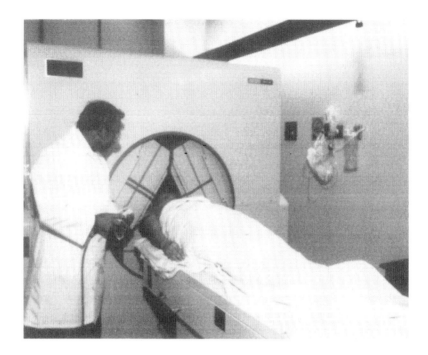

FIGURE 8. Triple camera (TRIAD) SPECT system (Duke University Medical Center, Durham, NC).

system when the brain is sampled completely. The reconstructed image resolution is approximately eight mm on the axis of rotation, depending on the collimator selected.

D. SYSTEM CALIBRATION AND IMAGING PERFORMANCE

Although high system sensitivity is an important factor affecting SPECT performance, other factors may influence reconstructed image quality. For example, regional nonuniformities may result in circular artifacts. Furthermore, relatively small errors (on the order of 1 or 2 mm) in the calibration of the center of rotation can significantly degrade the reconstructed spatial resolution.

Compensation for regional nonuniformities generally involves acquiring a high count density (greater than 30 to 60 million counts) image of a large, uniform sheet source that covers the entire field of view of the imaging system. This image is analyzed to determine the magnitudes and locations of the nonuniformities, and an algorithm is then used to compute a set of compensation factors. Subsequently, projection data are multiplied by these factors to remove the residual nonuniformities prior to reconstruction.

Center of rotation calibration is usually performed by scanning a line or point source placed within the field of view. The projection data of this source (i.e., the sinogram) are fitted as a function of angle to a sinusoidal function. The algorithm then determines the location of the physical center of rotation. This location is then used by the reconstruction algorithm to ensure that spatial resolution is not degraded and artifacts are not produced.

These calibration procedures and their associated software are essential components of any SPECT device. Unfortunately, the quality and capabilities of these subsystems are not always completely adequate and residual artifacts may result. Multiple camera systems, such as the TRIAD device, require very careful calibrations since each camera must be matched nearly perfectly.

Appropriately designed multifunctional phantoms can be used to evaluate the overall performance of a SPECT system. In Figure 9, for example, the high count density SPECT

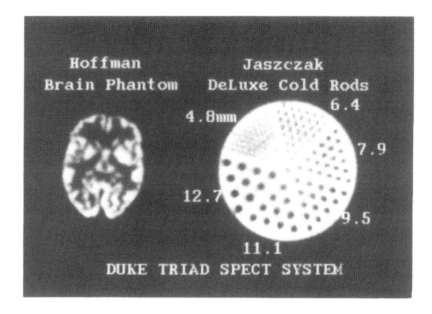

FIGURE 9.   High count density phantom images acquired with Triad SPECT system. Left, Hoffman brain phantom (Data Spectrum Model 8080); right, Jaszczak cold rod phantom (Data Spectrum Model 5000). Rod diameters range from 4.8 to 12.7 mm. SPECT scans of multi-function phantoms are used to ensure optimal system performance.

scans of the Hoffman brain phantom (Model 8080, Data Spectrum Corporation) and Jaszczak Deluxe cold rod phantom (Model 5000, Data Spectrum Corporation) acquired with the triple camera Triad system demonstrate acceptable image quality and proper calibration of all three cameras. Phantom evaluations should be performed routinely to ensure that SPECT clinical images are not being degraded by inaccurate calibration and compensation procedures.

IV. RECONSTRUCTION METHODS

SPECT is based on the acquisition of sets of line integrals or activity profiles measured from many angular views. Typically more than 100 angular samples are obtained with either continuous motion of the detector about the patient or in a "step-and-shoot" manner.

Several articles describe alternative reconstruction algorithms for computed tomography.[56-58] Basically, two general classes of solutions are used. One class involves iterative techniques whereby an initial trial solution is successively modified.[3,59-66] The second class is based on a direct analytic solution of the equations that relate the projection data with the transaxial image.[4,67-77] Presently, most X-ray CT and ECT systems use a filtered backprojection technique, which was first proposed by Bracewell and Riddle[68] and later by Ramachandran and Lakshminarayanan.[75]

A. BACKPROJECTION BY PIXEL-SUMMATION

The early single-photon transverse section images were reconstructed using a simple unfiltered backprojection algorithm.[1,2] This method is illustrated in the upper section of Figure 10 and consists of a summation of the unmodified projection data on a pixel-by-pixel basis. Shown in this example are the projection data for a point emission source measured at four angles. On the right, these and additional projection sets have been arranged as a function of increasing angular sampling values. For a point emission source located off the axis of rotation, the mean values for the distribution vary in a smooth, sinusoidal manner in this "projection space". The two-dimensional gray scale image formed by stacking the

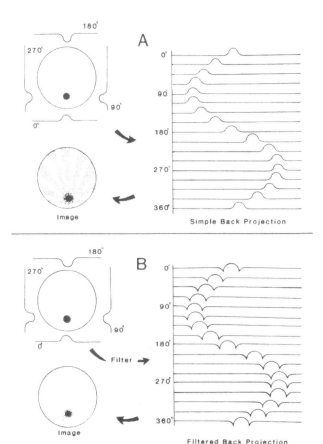

FIGURE 10.   SPECT image reconstruction using (A) simple and
(B) filtered back projection methods. Each source voxel describes a
unique sinusoid in projection space as indicated by the sinograms on
the right. By first filtering each projection (bottom) the ''star type''
artifact is eliminated. (From Jaszczak, R. J. and Coleman, R. E.,
Invest. Radiol., 20, 897, 1985. With permission.)

one-dimensional profiles obtained from many sequential angular views is referred to as a
''sinogram''. For each pixel within the field of view, there exists a unique (i.e., one and
only one) sinusoid of appropriate phase and amplitude in this projection space determined
by the data collection geometry. It is intuitively reasonable to attempt to reconstruct the
image on a pixel-by-pixel basis by summing the measured projection data along the appropriate sinusoidal curve. If necessary, a simple interpolation technique can be used to obtain
values at locations that do not exactly coincide with the spatially sampled projection data.
For example, the value for the pixel in the reconstructed image corresponding to the location
of the point source in the object shown in Figure 10 is obtained by summing the peak values
along the geometrically determined sinusoid for all projection sets. Unfortunately, since the
measured projection data can only have zero or positive values, other pixels in the reconstructed image also receive a net positive value when this simple algorithm is used even
though there is no activity located at these positions. For example, consider the pixel located
on the axis of rotation (i.e., at the center of the circular object). For this case, the path in
projection space is simply a vertical line passing through the center of the projection set.
The off-axis point source contributes positive values for projection sets near 0°, 180°, and
360°. This results in a point spread function that erroneously extends radially from its true

position. With very fine angular sampling, the point spread function for this unfiltered backprojection process will have smooth broad shoulders decreasing uniformly with the radial distance from the peak location. With fewer angular views a "star-type" artifact will result, as shown in Figure 10A.

This blurring or spreading of the point source activity over the entire image can be eliminated by appropriately modifying or "filtering" the original projection data prior to the summation process as shown in Figure 10B. Note that the filtered projections now have both positive and negative values. In this case, a correct value is obtained not only for the pixel corresponding to the actual off-axis point source location, but also for all other pixels in the reconstructed image. The center pixel, for example, will now have both positive and negative contributions during the summation process, and if the filter is chosen appropriately, these will cancel, yielding a net sum of zero. Of course, statistical variations are not eliminated completely and may result in the appearance of noise in regions away from the point source. A mathematical description of the reconstruction filter may be found in the literature.[58,60]

B. COMPENSATION FOR ATTENUATION

It is generally necessary to compensate for gamma photons that are not detected due to attenuation within the patient's body. Attenuation is caused mainly by Compton scattering within the source distribution and surrounding body tissue (see Figure 1). Photons, initially traveling toward the collimator apertures, are scattered in a direction such that the photons can no longer pass through the collimator holes and be detected. Attenuation can result in a large loss in detected photons. For example, less than about one in four of ^{99m}Tc gamma photons emitted near the center of a 20-cm-diameter, water-filled cylinder exit without scattering. In the presence of attenuation, sources located near the central region would "appear" to have much less activity when compared to similar sources located near the periphery (Figure 11A).

Most commercially available SPECT systems use a filtered backprojection algorithm, in conjunction with either a pre- or postprocessing procedure to compensate for photon attenuation within the patient. Various assumptions concerning the nature of the source activity and attenuating medium are required for each approach. Furthermore, the spatial extent of the medium (i.e., the patient's body contour) must be determined.

1. Body Contour Determination

Several methods have been proposed to estimate the body contour. For example, point marker sources can be placed on the patient's skin and imaged from several directions. These locations can then be fitted to an elliptical model of the patient's boundary. Over a decade ago[10] we developed and evaluated two other methods to measure this contour. One method used a ^{133}Xe transmission line source that was imaged at the same time that the ^{99m}Tc primary SPECT data was acquired. A threshold algorithm was used to determine the edge of the body. The second method used photons that were scattered within the patient, traversed the collimator holes, and were then detected within a second pulse-height window located below the primary photon energy. The method required no additional radiation dose to the patient, and the data could be acquired simultaneously with the primary SPECT data. This scattered photon method has been incorporated in several commercial SPECT systems. Some SPECT systems simply use the camera's orbit to determine the body contour; however, this approach can result in an inaccurate measurement and, hence, reconstruction errors.

2. Preprocessing Attenuation Compensation

Preprocessing attenuation compensation methods attempt to correct the measured projection data prior to image reconstruction.[78,79] Algorithms have been developed based on

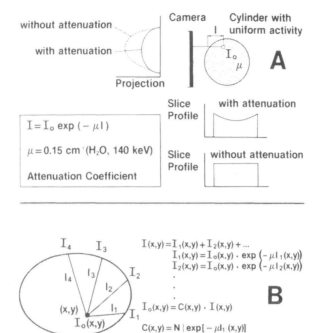

FIGURE 11. (A) Effect of attenuation on image of cylindrical source; (B) first-order multiplicative compensation method.[13]

the use of projection views that are 180° apart. Typically either the arithmetic or geometric mean is computed and a compensation equation is devised based on certain assumptions of uniform source activity and a constant attenuation coefficient. Even with these approaches, knowledge of the path length through the body is required. These methods may produce streak artifacts in the presence of noisy data and are considered generally to be less effective than other approaches.

3. Intrinsic Compensation

These methods are more sophisticated than the previous approaches and typically require more effort and computational time for implementation. Three basic approaches have been proposed. This first approach is based on including the effect of attenuation into an iterative reconstruction algorithm.[80-83] Typically, it is necessary to use an assumed or measured (perhaps with transmission CT) attenuation map as part of the reconstruction process. If an attenuation map measured with a ^{99m}Tc transmission source is used, this approach may provide quantitatively accurate results, even in the presence of nonuniform attenuation.

The second approach is to directly solve the attenuated Radon integral based on the assumption of a constant attenuation coefficient.[80,83] The final approach is to simultaneously solve for both the source activity and the attenuation map. Methods based on this sophisticated approach have been proposed by Censor et al.[83]

The intrinsic methods generally (except for certain iterative approaches) may have difficulty in the presence of noisy projection data. Although the better iterative approaches offer the potential for improved quantitative measurements, these approaches have not been widely used mainly as a result of the additional computational requirements.

4. Postprocessing Compensation

These approaches[10,12,13] to attenuation compensation first compute a reconstructed image using a conventional filtered backprojection algorithm and the assumption of no attenuation. A correction matrix is computed based on an independently measured body contour. Each element $C(x,y)$ for the correction matrix is equal to the reciprocal of average attenuation along all rays from the pixel (x,y) to the boundary of the attenuating medium (see Figure 11B). A constant linear attenuation coefficient is typically assumed. To obtain a first-order compensated image, the original reconstructed image is multiplied by the compensation matrix.

Although this first-order image provides an adequate correction for qualitative imaging situations, it is possible to obtain a second-order correction by applying an additional processing step.[10,13] A set of new projections is computed with a reprojection algorithm using the first-order compensated image, the measured body contour, and the assumed uniform linear attenuation coefficient. Error projectons are obtained by subtraction of the reprojected data from the original projections. An error image is then reconstructed using the error projections and multiplying by the compensation matrix. The second-order corrected image is finally obtained by adding the corrected error image to the first-order reconstructed image. The second-order correction may improve SPECT quantification; however, it has not been extensively utilized since the computational requirements are increased, and additional image noise may result. The first-order correction provides reasonable qualitative results when the attenuating media is uniform. Because this first-order method does not produce artifacts in the presence of noisy data and is simple to implement, it has become a standard, practical approach to compensate for attenuation.

C. COMPENSATION FOR DETECTED SCATTERED PHOTONS

Although attenuation results from primary photons that are scattered out of the collimator's acceptance angle, it is also possible for scattered photons to be directed into the collimator's acceptance angle and then be detected. Hence, the attenuation of photons and the inclusion of Compton-scattered photons within the primary energy window are closely related interactions. All gamma rays lose energy during the Compton scattering process. However, the amount of energy lost is often relatively small and the scattered photon may still have sufficient energy to be detected within the primary energy window. For example, even for a relatively large scattering angle of 45°, an incident 140-keV photon will result in a scattered photon having an energy that is equal to about 130 keV. Since NaI(Tl) scintillation cameras have an energy resolution of about 10%, this scattered photon is likely to be detected within the primary energy window.

Under typical clinical imaging situations, the ratio of detected scattered photons to nonscattered photons (i.e., the scatter fraction) may be as large as 40% or more for camera-based systems. Discrete detector devices often have poorer energy resolution and hence may have even larger scatter fractions.

The detection of scattered photons within the primary energy window reduces lesion contrast and results in quantitative inaccuracy. Several approaches are being investigated to compensate for scattered photons.[84-87]

One approach is based on convolutional methods.[85,86] In these techniques, scatter is modeled as a convolution of either the measured projection data or the nonscattered projection data with an empirically derived function. The resulting equations are then used to provide an estimate of the actual nonscatter projections. A limitation of these approaches is that the scattering function is not spatially invariant within the source distribution and changes shapes for different locations. Thus, convolutional techniques are not rigorously valid; however, these methods can be used to provide an approximate compensation for scatter.

Another approach to scatter compensation has been proposed by Jaszczak et al.[87] and is based on using the information that is contained in portions of the gamma ray energy spectrum outside the primary photopeak region. The method consists of subtracting a heuristically determined fraction of the image reconstructed using events recorded within a secondary (lower energy) pulse-height window from the image reconstructed from events recorded within the primary photopeak window. Alternatively, it is possible to directly compensate the projection data. This method is based on the assumption that the scatter image recorded from events collected within the secondary pulse height window is a close approximation to the true scatter component of the image recorded from the photopeak window.

We have evaluated the scatter subtraction technique using water-filled cylindrical phantoms and have found a significant improvement in SPECT quantification for the geometries investigated. Errors resulting from scatter were originally as great as 30%, while after compensation, these errors were reduced to less than 10%.

These, and similar, approaches to scatter compensation should be adequate for most SPECT studies of the brain, where the major errors will be caused by inadequate count densities and the effect of finite spatial resolution coupled with the very small structure sizes of the brain.

D. RECENT TRENDS IN RECONSTRUCTION TECHNIQUES

Iterative algebraic reconstruction techniques that make use of the statistical nature of the emission process and other *a priori* information are being investigated actively by many groups.[88-90] Although iterative approaches were first evaluated in the late 1960s and early 1970s as solutions to the reconstruction problems,[62,63,91] these methods were not used in commercial CT imaging systems. The wide acceptance of filtered backprojections methods is related to its computational efficiency. However, one would anticipate an improved reconstruction would result with the use of additional *a priori* information. Furthermore, the characteristics of a particular imaging device, and physical processes such as attenuation and scatter, may be directly incorporated into iterative algorithms. Custom dedicated processors offer the potential for practical reconstruction times. It is for these reasons that interest in iterative algebraic techniques has been revitalized.

Iterative algorithms start with an initial guess (typically a uniformly emitting distribution) for the source distribution. A set of projections is calculated using the initial source distribution and a known, or assumed, probability matrix. The calculated projections are compared with the measured projections, and using an appropriate recursion formula, a correction is applied to the estimated source distribution. This process is iteratively repeated until a solution is obtained that is optimal according to a preselected criteron. One criterion that has been proposed is the minimum mean squares error (MMSE) method[60] based on minimizing the discrepancies between the measured and calculated projections. Other criteria include the maximization of various conditional functions, for example, maximum entropy (ME) or maximum likelihood (ML),[92-95] or Bayesian methods such as maximum *a posteriori* approaches.[88-90,96] The ML method has generated considerable research interest after being proposed for ECT by Rockmore and Mackovski,[93] Shepp and Vardi,[94] and Lange and Carson.[95] Various approaches have been proposed to optimize its rate of convergence and stability.[97-99]

However, Bayesian formulism[88-90,96] is more general and perhaps may result in improved reconstructions since *a priori* information is explicitly included in the method. For emission imaging we may formally define the measurement process as follows (using the matrix notation suggested by Barrett):[96]

$$P = HF \tag{1}$$

where P is a J × 1 vector corresponding to the measured projection data, F is an I × 1 vector corresponding to the source distribution, and H is a J × I matrix (i.e., the probability or transfer matrix) characteristic of the acquisition geometry, attenuation, and scatter of photons by body tissue. The size of these matrices may be quite large. For example, consider a three-dimensional image volume of $128 \times 128 \times 128$ voxels and projection data acquired as $128 \times 128 \times 180$ (angles). The indices I and J are then approximately equal to 2.1×10^6 and 2.9×10^6, respectively. The matrix H could have over 10^{12} nonzero values, although for many imaging situations the number of nonzero elements may be much less.

Application of Bayes' law to the emission imaging problem gives

$$\text{Prob}[F|P] = \text{Prob}[P|F] \ \text{Prob}[F]/\text{Prob}[P]$$

where F is the source vector and P is the measured projection vector. One attempts to maximize the conditional *a posteriori* probability Prob[F|P] given the measured projections and constrained by any *a priori* information about feasible source distributions. The maximum *a posteriori* (MAP) method consists of determining the optimal source distribution F_{opt} which most likely resulted in the measured data, limited, or constrained, by any known *a priori* source information (contained in the probability distribution Prob[F]. Since Prob[P] is the probability distribution of projections for all possible sources, it is not dependent on F_{opt} and is equal to a constant. Hence, it will not affect the MAP solution. If there is no *a priori* information, then Prob[F] is a constant and the MAP solution becomes the maximum likelihood solution. However, *a priori* information is often available. For example, it is known that there cannot exist negative emission voxels. Also, correlations among neighboring voxels may exist. Other *a priori* information may also be available. Mathematically the optimal solution F_{opt} is obtained by maximizing the conditional Prob[F|P], or the logarithm of this function. An iterative expectation-maximization (EM) technique, as proposed by Dempster et al.[100] and others, may be used to determine the optimal F_{opt}, subject to appropriate constraints.

Although statistical reconstruction techniques are of great research interest, many practical issues remain to be resolved. For example, since the iterative approaches are computationally intense, methods need to be developed to accelerate the rate of convergence. Practical rules may be required to determine when the iteration process should be stopped,[88] if one is using a maximum likelihood algorithm, for example. The improvements that one gains using an iterative method must be compared with conventional reconstructions obtained using Metz, or other optimal filtering approaches,[101] and coupled with standard back projection methods. The latter approach is computationally fast compared with iterative methods and may result in similar improvements in image quality. These, and other issues, are currently under active investigation by several groups.

V. SUMMARY FUTURE TRENDS IN SPECT

The current trend to utilize detectors that completely encompass the patient's brain is likely to continue. For example, besides dual and triple camera SPECT systems, a dedicated brain device has recently been developed that uses four gamma cameras mounted in a square configuration. Most configurations that encircle the brain and have similar detector areas should have similar performance characteristics. The important considerations are (1) the total area of detector that is focused onto the organ of interest and (2) the proximity of the collimator to this organ. Thus, specially designed focusing collimators will continue to be utilized. It would be desirable to manufacture collimators more accurately than they are currently built. For example, intrinsic collimator inaccuracies have delayed the widespread utilization of fan and cone beam collimators. Overall system stability and performance should improve in the future as electronics and computers become more reliable. The availability of powerful computers should make the systems easier to use, permit more automated calibrations, and provide the capability to automatically monitor system malfunctions. The availability of these computers should also lead to the introduction within the clinical environment of sophisticated and more accurate, reconstruction techniques to compensate for attenuation, scatter, and collimator imperfections. These algorithms should improve both lesion detection and SPECT quantification. The improvements in SPECT instrumentation, coupled with newly developed radiopharmaceuticals, will ensure that SPECT remains an important modality for brain imaging.

ACKNOWLEDGMENTS

The author wishes to thank Wendy Painter for her excellent secretarial support. This work was supported in part by U.S. Public Health Services Grant R01-CA 33541 awarded by the National Cancer Institute, by Department of Energy Grant DE-FG05-89ER60894, and by National Science Foundation Grant CDR-8622201.

REFERENCES

1. **Kuhl, D. E. and Edwards, R. Q.**, Image separation radioisotope scanning, *Radiology*, 80, 653, 1963.
2. **Kuhl, D. E. and Edwards, R. Q.**, Cylindrical and section radioisotope scanning of the liver and brain, *Radiology*, 83, 926, 1964.
3. **Hounsfield, G. N.**, Computerized transverse axial scanning (tomography). I. Description of system, *Br. J. Radiol.*, 46, 1016, 1973.
4. **Cormack, A. M.**, Representation of a function by its line integrals, with some radiological applications. II, *J. Appl. Phys.*, 35, 2908, 1964.
5. **Kuhl, D. E., Edwards, R. Q., Ricci, A. R., and Reivich, M.**, Quantitative section scanning using orthogonal tangent correction, *J. Nucl. Med.*, 14, 196, 1973.
6. **Bowley, A. R., Taylor, C. G., Causer, D. A., Barber, D. C., Keyes, W. I., Undrill, P. E., Corfield, J. R., and Mallard, J. R.**, A radioisotope scanner for rectilinear, arc, transverse section and longitudinal section scanning. (ASS — the Aberdeen Section Scanner), *Br. J. Radiol.*, 46, 262, 1973.
7. **Keyes, W. I.**, A practical approach to transverse-section gamma-ray imaging, *Br. J. Radiol.*, 49, 62, 1976.
8. **Jaszczak, R., Huard, D., Murphy, P., and Burdine, J.**, Radionuclide emission computed tomography with a scintillation camera, *J. Nucl. Med.*, 17, 551, 1976.
9. **Jaszczak, R. J., Murphy, P. H., Huard, D., and Burdine, J. A.**, Radionuclide emission computed tomography of the head with ^{99m}Tc and a scintillation camera, *J. Nucl. Med.*, 18, 373, 1977.
10. **Jaszczak, R. J., Chang, L. T., Stein, N. A., and Moore, F. E.**, Whole-body single-photon emission computed tomography using dual, large-field-of-view scintillation cameras, *Phys. Med. Biol.*, 24, 1123, 1979.
11. **Keyes, Jr., J. W., Orlandea, N., Heetderks, W. J., Leonard, P. F., and Rogers, W. L.**, The humongotron — a scintillation-camera transaxial tomograph, *J. Nucl. Med.*, 18, 381, 1977.
12. **Walters, T. E., Simon, W., and Chesler, D. A. et al.**, Radionuclide axial tomography with correction for internal absorption, in *Information Processing in Scintigraphy*, Proc. 4th Int. Conf., Orsay, 1975, 333.
13. **Chang, L. T.**, A method for attenuation correction in radionuclide computed tomography, *IEEE Trans. Nucl. Sci.*, 25(2), 638, 1978.
14. **Moore, S. C., Brunelle, J. A., and Kirsch, C. M.**, An iterative attenuation correction for a single photon scanning multidetector tomography system, *J. Nucl. Med.*, 22, 65, 1981.
15. **Lim, C. B., Chang, L. T., and Jaszczak, R. J.**, Performance analysis of three camera configurations for single photon emission computed tomography, *IEEE Trans. Nucl. Sci.*, NS-27, 559, 1980.
16. **Lim, C. B., Gottschalk, S., Walker, R., and Schreiner, R. et al.**, Triangular SPECT system for 3-D total organ volume imaging. Design concept and preliminary imaging results, *IEEE Trans. Nucl. Sci.*, NS-32, 741, 1985.
17. **Logan, K. W. and Holmes, R. A.**, Missouri University multi-plane imager (MUMPI). A high sensitivity rapid dynamic ECT brain imager, *J. Nucl. Med.*, 25, 105, 1984.
18. **Genna, S. and Smith, A.**, The development of ASPECT, an annular single crystal brain camera for high efficiency SPECT, *IEEE Trans. Nucl. Sci.*, NS-35, 654, 1988.

19. **Kuhl, D. E., Edwards, R. Q., Ricci, A. R., Yacob, R. J., Mich, T. J., and Alavi, A.,** The Mark IV system for radionuclide computed tomography of the brain, *Radiology,* 121, 405, 1976.

20. **Cho, Z. H., Yi, W., Jung, K. J., Lee, B. U., and Min, H. B.,** Performance of single photon tomographic system-Gammatom-1, *IEEE Trans. Nucl. Sci.,* NS-29, 484, 1982.

21. **Hirose, Y., Ikeda, Y., and Higashi, Y. et al.,** A hybrid emission CT-HEADTOME II, *IEEE Trans. Nucl. Sci.,* NS-29, 520, 1982.

22. **Moore, S. C., Doherty, M. D., Zimmerman, R. E., and Holman, B. L.,** Improved performance from modifications to the multidetector SPECT brain scanner, *J. Nucl. Med.,* 25, 688, 1984.

23. **Stokely, E. M., Sveinsdottir, E., Lassen N. A., and Rommer, P.,** A single photon dynamic computer assisted tomograph (DCAT) for imaging brain function in multiple cross-sections, *J. Comput. Assist. Tomogr.,* 4, 230, 1980.

24. **Rogers, W. L., Clinthorne, N. H., and Stamos, J. et al.,** Performance evaluation of SPRINT, a single photon ring tomograph of brain imaging, *J. Nucl. Med.,* 25, 1013, 1984.

25. **Tretiak, O. J. and Metz, C.,** The exponential Radon transform, *SIAM J. Appl. Math.,* 39(2), 341, 1980.

26. **Geiger, H. and Muller, W.,** The electron counting tube, *Phys. Z.,* 29, 839, 1928.

27. **Hofstadter, R.,** Alkali halide scintillation counters, *Phys. Rev.,* 74, 100, 1948.

28. **Mayneord, W. V. and Newberry, S. P.,** An automatic method of studying the distribution of activity in a source of ionizing radiation, *Br. J. Radiol.,* 25, 589, 1952.

29. **Cassen, B., Curtis, L., Reed, C., and Libby, R.,** Instrumentation for I^{131} use in medical studies, *Nucleonics,* 9, 46, 1951.

30. **Anger, H. O.,** A new instrument for mapping gamma ray emitters, in Biology and Medicine Quarterly Report, UCRL-3653, 1957, 38.

31. **Anger, H. O.,** Scintillation camera, *Rev. Sci. Instrum.,* 29, 27, 1958.

32. **Harper, P. V., Andros, G., Lathrop, K., Siemens, W., and Weiss, L.,** Technetium-90 (sic), as a biological tracer, *J. Nucl. Med.,* 3, 209, 1962.

33. **Harper, P. V., Lathrop, K. A., Kiminez, F., Fink, R., and Gottschalk, A.,** Technetium-99m as a scanning agent, *Radiology,* 85, 101, 1965.

34. **Mallett, B. L. and Veall, N.,** Measurement of regional cerebral clearance rates in man using Xe-133 inhalation and extracranial recording, *Clin. Sci.,* 29, 179, 1965.

35. **Eichling, J. O. and Ter-Pogossian, M. M.,** Methodological shortcomings of the ^{133}Xe inhalation technique of measuring rCBF, *Acta Neurol. Scand.,* 56 (Suppl. 64), 464, 1977.

36. **Hazelrig, J. B., Katholi, C. R., Blauenstein, V. W., Halsey, J. H., Jr., Wilson, E. M., and Wills, E. L.,** Total curve analysis of regional cerebral blood flow with ^{133}Xe inhalation. Description of method and values obtained with normal volunteers, *IEEE Trans. Biomed. Eng.,* 28, 609, 1981.

37. **Obrist, W. D., Thompson, H. K., Wang, H. S., and Wilkinson, W. E.,** Regional cerebral blood flow estimated by ^{133}xenon inhalation, *Stroke,* 6, 245, 1975.

38. **Winchell, H. S., Baldwin, R. M., and Lin, T. H.,** Development of I-123 labeled amines for brain studies: localization of I-123 iodophenylalkyl amines in rat brain, *J. Nucl. Med.,* 21, 947, 1980.

39. **Winchell, H. S., Horst, W. D., Braun, L., Oldendorf, W. H., Hattner, R., and Parker, H.,** N-Isopropyl-[^{123}I] *p*-Iodoamphetamine. Single pass brain uptake and washout; binding to brain synaptosomes; and localization in dog and monkey brain, *J. Nucl. Med.,* 21, 947, 1980.

40. **Kung, H. F. and Blau, M.,** Regional intracellular pH shift. A proposed new mechanism for radiopharmaceutical uptake in brain and other tissues, *J. Nucl. Med.,* 21, 147, 1980.

41. **Kung, H. F., Tramposch, K. M., and Blau, M.,** A new brain perfusion imaging agent. [I-123]HIPDM:*N,N,N'*-Trimethyl-*N'*-[2-Hydroxy-3-methyl-5-iodobenzyl]-1,3 propanediamine, *J. Nucl. Med.,* 24, 66, 1983.

42. **Kuhl, D. E., Barrio, J. R., Huang, S. C., Selin, C., Ackermann, R. F., Lear, J. L., Wu, J. L., Lin, T. H., and Phelps, M. E.,** Quantifying local cerebral blood flow by *N*-isopropyl-*p*-[^{123}I] iodoamphetamine (IMP) tomography, *J. Nucl. Med.,* 23, 196, 1982.

43. **Hill, T. C., Holman, B. L., Lovett, R., O'Leary, F., Magistretti, P., Zimmerman, R. E., Moore, S., Clouse, M. E., Wu, J. L., Lin, T. H., and Baldwin, R. M.,** Initial experience with SPECT (single photon computerized tomography) of the brain using *N*-isopropyl I-123 *p*-iodoamphetamine. concise communication, *J. Nucl. Med.,* 23, 191, 1982.

44. **Drayer, B., Jaszczak, R., Friedman, A., Albright, R., Kung, H., Greer, K., Lischko, M., Petry, N., and Coleman, R. E.,** In vivo quantitation of regional cerebral blood flow in glioma and cerebral infarction. Validation of the HIPDm-SPECT method, *A. J. N. R.,* 4, 572, 1983.

45. **Drayer, B. P., Jaszczak, R. J., King, H. F., Friedman, A., Albright, R., Greer, K., Lischko, M., Petry, N., and Coleman, R. E.,** In vivo quantitation of regional cerebral blood flow. The SPECT-HIPD method, in *Functional Radionuclide Imaging of the Brain,* Magistretti, P. L., Ed., Raven Press, New York, 1983, 193.

46. **Hung, H. F., Yu, C. C., Billings, J., Molnar, M., Wicks, R., and Blau, M.,** New Tc-99m brain imaging agents, *J. Nucl. Med.,* 25, 16, 1984.

47. **Hung, H. F., Molnar, M., Billings, J., Wicks, R., and Blau, M.,** Synthesis and biodistribution of neutral lipid-soluble Tc-99m complexes that cross the blood-brain barrier, *J. Nucl. Med.,* 25, 326, 1984.

48. **Jaszczak, R. J., Chang, L. T., and Murphy, P. H.,** Single photon emission computed tomography using multi-slice fan beam collimators, *IEEE Trans, Nucl. Sci.,* 26, 610, 1979.

49. **Jaszczak, R. J., Floyd, C. E., Manglos, S. H., Greer, K. L., and Coleman, R. E.,** Cone beam collimation for single photon computed tomography. Analysis, simulation, and image reconstruction using filtered backprojection, *Med. Phys.,* 13, 484, 1986.

50. **Jaszczak, R. J., Greer, K. L., and Coleman, R. E.,** SPECT using a specially designed cone beam collimator, *J. Nucl. Med.,* 29, 1398, 1988.

51. **Hawman, E. G. and Hsieh, J.,** An astigmatic collimator for high sensitivity SPECT of the brain, *J. Nucl. Med.,* 27, 930, 1986.

52. **Stoddart, H. F. and Stoddart, H. A.,** A new development in single gamma transaxial tomography. Union Carbide focused collimator scanner, *IEEE Trans. Nucl. Sci.,* 26, 2710, 1979.

53. **Lassen, N. A., Holm, S., Petersen, R. K., Hansen, H. E., and Rommer, P.,** Whole body multidetector SPECT with variable aperture, *J. Nucl. Med.,* 30(5), 796 (abstract), 1989.

54. **Tsui, B. M. W., Gullberg, G. T., Edgerton, E. R., Gilland, D. R., Perry, J. R., and McCartney, W. H.,** Design and clinical utility of a fan beam collimator for SPECT imaging of the head, *J. Nucl. Med.,* 27, 810, 1986.

55. **Feldkamp, L. A., Davis, L. C., and Kress, J. W.,** Practical cone beam algorithm, *J. Opt. Soc. Am.,* 1, 612, 1984.

56. **Barrett, H. H.,** Perspective on SPECT, *SPIE J.,* 671, 178, 1986.

57. **Brooks, R. A. and DiChiro, G.,** Principles of computer assisted tomography (CAT) in radiographic and radioisotope imaging, *Phys. Med. Biol.,* 21, 689, 1976.

58. **Brooks, R. A. and DiChiro, G.,** Theory of image reconstructon in computed tomography, *Radiology,* 117, 561, 1975.

59. **Budinger, T. F., Derenzo, S. E., Gullberg, G. T., Greenberg, W. L., and Heusman, R. H.,** Emission computer assisted tomography with single-photon and positron annihilation photon emitters, *J. Comput. Assist. Tomogr.,* 1, 131, 1972.

60. **Budinger, T. F. and Gullberg, G. T.,** Transverse section reconstruction of gamma-ray emitting radio-nuclides in patients, in *Reconstruction Tomography in Diagnostic Radiology and Nuclear Medicine,* Ter-Pogossian, M. M., Phelps, M. E., Brownell, G. L., Cox, J. R., Davis, D. O., and Evens, R. G., Eds., University Park Press, Baltimore, 1977, 315.

61. **Budinger, T. F. and Gullberg, G. T.,** Three-dimensional reconstruction in nuclear medicine emission imaging, *IEEE Trans. Nucl. Sci.,* 21, 2, 1974.

62. **Gordon, R., Bender, R., and Herman, G. T.,** Algebraic reconstruction techniques (ART) for three-dimensional electron microscopy and X-ray photography, *J. Theor. Biol.,* 29, 471, 1970.

63. **Goitein, M.,** Three-dimensional density reconstruction from a series of two-dimensional projections, *Nucl. Instrum. Methods,* 101, 509, 1972.

64. **Bracewell, R. N.,** Strip integration in radio astronomy, *Aust. J. Phys.,* 9, 198, 1956.

65. **Kuhl, D. E., Edwards, R. Q., Ricci, A. R., and Reivich, M.,** Quantitative section scanning using orthogonal tangent correction, *J. Nucl. Med.,* 14, 196, 1973.

66. **Oppenheim, B. E.,** More accurate algorithm for iterative 3-dimensional reconstruction, *IEEE Trans. Nucl. Sci.,* 21, 72, 1974.

67. **Radon, J.,** Uber die Bestimmung von Funktionen durch ihre Integralwerte langs gewisser Mannigfaltig-keiten, *Ber. Verh. Saechs Adad. Wiss.,* 67, 262, 1917.

68. **Bracewell, R. N. and Riddle, A. C.,** Inversion of fan-beam scans in radio astronomy, *Astrophys. J.,* 150, 427, 1967.

69. **De Rosier, D. J. and Klug, A.,** Reconstruction of three dimensional structures from electron micrographs, *Astrophys. J.,* 150, 427, 1968.

70. **Berry, M. V. and Gibbs, D. F.,** The interpretation of optical projections, *Proc. R. Soc. London Ser. A.,* 314, 143, 1970.

71. **Shepp, L. A. and Logan, B. F.,** Reconstructing interior head tissue from X-ray transmissions, *IEEE Trans. Nucl. Sci.,* 21, 228, 1974.

72. **Herman, G. T., Lakshminarayanan, A. V., and Naparstek, A.,** Reconstruction using divergent-ray shadowgraph, in *Reconstruction Tomography in Diagnostic Radiology and Nuclear Medicine,* Ter-Pogossian, M. M. et al., Eds., University Park Press, Baltimore, 105, 1977.

73. **Bates, R. H. T. and Peters, T. M.,** Towards improvements in tomography, *N. Z. J. Sci.,* 14, 883, 1972.

74. **Smith, P. R., Peters, T. M., and Bates, R. H. T.,** Image reconstruction from finite numbers of projections, *J. Phys. A. Math. Nucl. Gen.,* 6, 361, 1973.

75. **Ramachandran, G. N. and Lakshminarayanan, A. V.,** Three dimensional reconstruction from radi-ographs and electron micrographs: application of convolutions instead of Fourier transforms, *Proc. Natl. Acad. Sci. U.S.A.,* 68, 2236, 1976.

76. **Tretiak, O. J. and Metz, C.,** The exponential Radon transform, *SIAM J. Appl. Math.,* 39(2), 341, 1980.

77. **Budinger, T. F., Gullberg, G. T., and Huesman, R. H.,** Emission computed tomography, in *Image Reconstruction from projections: Implementation and Applications,* Herman, G. T., Ed., Springer-Verlag, New York, 1979, 147.

78. **Kay, D. B. and Keyes, J. W., Jr.,** First order corrections for absorption and resolution compensation in radionuclide Fourier tomography, *J. Nucl. Med.,* 16, 540, 1975.

79. **Sorenson, J. A.,** Quantitative measurement of radiation in vivo by whole body counting, in *Instrumentation in Nuclear Medicine,* Vol. 2, Hine, G. H. and Sorenson, J. A., Eds., Academic Press, New York, 1984, 311.

80. **Cullberg, C. T. and Budinger, T. F.,** The use of filtering methods to compensate for constant attenuation in single-photon emission computed tomography, *IEEE Trans. Biomed. Eng.,* 28(2), 142, 1981.

81. **Bellini, S., Piacentini, M., Cafforio, C. et al.,** Compensation of tissue absorption in emission tomography, *IEEE Trans. Acoustics Speech Signal Process.,* 27, 213, 1979.

82. **Soussaline, F. P., Cao, A., and LeCoq, G. et al.,** An analytical approach to single photon emission computed tomography with attenuation effect, *Eur. J. Nucl. Med.,* 7, 487, 1982.

83. **Censor, Y., Gustafson, D. E., Lent, A., and Tuy, H.,** A new approach to the emission computed tomography problem: simultaneous calculation of attenuation and activity coefficients, Proc. IEEE Workshop on Physics and Engineering in Computerized Tomography, *IEEE Trans. Nucl. Sci.,* Newport Beach, CA, NS-26(2), 277S, 1979.

84. **Egbert, S. D. and May, R. S.,** An integral-transport method for Compton-scatter correction in emission computed tomography, *IEEE Trans. Nucl. Sci.,* 27, 543, 1980.

85. **Floyd, C. E., Jaszczak, R. J., and Harris, C. C. et al.,** Monte Carlo evaluating of Compton scatter by deconvolution in SPECT (abstract), *J. Nucl. Med.,* 25, 71, 1984.

86. **Axelsson, B., Msaki, P., and Israelsson, A.,** Subtraction of Compton-scattered photons in single-photon emission computerized tomography, *J. Nucl. Med.,* 25, 490, 1984.

87. **Jaszczak, R. J., Greer, K. L., and Floyd, C. E. et al.,** Improved SPECT quantification using compensation for scattered photons, *J. Nucl. Med.,* 25, 893, 1984.

88. **Levitan, E. and Herman, G. T.,** A maximum a posteriori probability expectation maximization algorithm for image reconstruction in emission tomography, *IEEE Trans. Med. Imag.,* 6, 185, 1987.

89. **Liang, Z. and Hart, H.,** Bayesian image processing of data from constrained source distribution. I. Nonvalued, uncorrelated and correlated constraints, *Bull. Math. Biol.,* 49, 51, 1987.

90. **Geman, S. and McClure, D. E.,** Bayesian image analysis: An application to single photon emission tomography, *Proc. of the Statistical Computing Section,* American Statistical Association, Washington, D.C., 1985, 12.

91. **Gilbert, P.,** Iterative methods for the three-dimensional reconstruction of an object from projections, *J. Theor. Biol.,* 36, 105, 1972.

92. **Minerbo, G.,** Maximum entropy reconstruction from cone beam projection data, *Comput. Biol. Med.,* 9, 29, 1979.

93. **Rockmore, A. J. and Macovski, A.,** A maximum likelihood approach to emission image reconstruction from projections, *IEEE Trans. Nucl. Sci.,* 23, 1428, 1977.

94. **Shepp, L. A. and Vardi, Y.,** Maximum likelihood reconstruction for emission tomography, *IEEE Trans. Med. Imag.,* 1, 113, 1982.

95. **Lange, K. and Carson, R.,** EM reconstruction algorithms for emission and transmission tomography, *J. Comput. Assist. Tomogr.,* 8, 306, 1984.

96. **Barrett, H. H.,** Perspective on SPECT, *SPIE J.,* 671, 178, 1986.

97. **Lewitt, R. M. and Muehllehner, C.,** Accelerated iterative reconstruction for positron emission tomography based on the EM algorithm for maximum likelihood estimation, *IEEE Trans. Med. Imag.,* 5, 16, 1986.

98. **Daube-Witherspoon, M. E. and Muehllehner, G.,** An iterative image space reconstruction algorithm suitable for volume ECT, *IEEE Trans. Med. Imag.,* 5, 61, 1986.

99. **Tanaka, E.,** A fast maximum likelihood algorithm for stationary positron emission tomography, *IEEE Trans. Med. Imag.,* 6, 98, 1987.

100. **Dempster, A., Laird, N., and Rubin, D.,** Maximum likelihood from incomplete data via the EM algorithm, *J. R. Stat. Soc. B,* 39, 1, 1977.

101. **King, M. A., Schwinger, R. B., and Penney, B. C.,** Variation of the count-density Metz filter with imaging system modulation transfer function, *Med. Phys.,* 13, 139, 1986.

Chapter 7

BIOCHEMISTRY OF THE BBB:
FUNCTIONAL ANALYSIS BY PET

L. E. Feinendegen and H. Herzog

TABLE OF CONTENTS

I. INTRODUCTION

The exchange of metabolic substrates between the blood and the brain is completely different from the exchange between blood and other organs. The first evidence for this difference was drawn from experiments with dyes which entered all organs of the body except the brain.[1-3] The obviously unique interface between the blood and the brain was first termed "Bluthirnschranke" or blood-brain barrier (BBB), by Lewandowsky in 1900.[3]

It took more than 60 years to establish the concept of the BBB and to prove that the cerebral capillary endothelium is the location of this barrier. Reese and Karnovsky were able to show that the low permeability of the BBB was caused by a completely closed endothelial layer with the cerebral capillary endothelial cells being connected by tight junctions (Figure 1).[4] The specific structure of this barrier permits the transfer of only a limited number of metabolic substrates into the brain. It was first pointed out by Krogh that lipophilic, in contrast to hydrophilic, molecules can readily pass the BBB since the endothelial membranes are constructed out of lipids.[6] Indeed, Fenstermacher and Rapoport found that the increasing permeability of soluble molecules was more correlated to the molecules' lipophilicity than to a decrease in the molecules' molecular weights.[7]

However, the major substrates for brain metabolism are hydrophilic, with the most important being glucose. For these substrates, several specific transport systems are present in the BBB.[8] Two systems ensure the supply of energy-delivering compounds: the hexose transport system[9,10] and the monocarboxylic transport system[11] which is necessary when the plasma glucose level is low. Three different systems are available for the transport of basic, acidic, and neutral amino acids.[10,12,13] Other systems, such as the choline and adenine transport systems, are involved in the passage of certain substrates required for neurotransmitter metabolism.[10] The important characteristics of all of these transport systems are saturability, competitive inhibition, and stereospecificity.

Specific features of the BBB, such as permeability for lipophilic compounds and transport systems for necessary nonlipophilic substrates, must be considered especially if drugs are to be supplied to the brain. These may be crucially altered in the presence of certain diseases. Thus, the BBB may become physically disrupted or change in concentration and/or function of transport systems. In any case, positron emission tomography (PET) allows the direct assessment of phenomena under consideration by using appropriate tracers labeled with positron emitters.

II. ION TRANSPORT

The exchange of information in the central nervous system (CNS) is based on the propagation of electrical action potentials along the neurons and chemical transmission at the synapses. The electrical phenomena are strongly dependent on the concentrations of ions, expecially K^+ and Na^+, inside and outside the neurons. Thus, it is very important for the homeostasis of these concentrations that the BBB is a strict barrier for ions with a permeability coefficient of about 10^{-7} cm/s.[14] This fact is also reflected by the high electrical resistance of the BBB, with values of 2000 Ω/cm^2 compared to normal epithelia with a resistance of about 100 Ω/cm^2.[15] Whereas the fenestra of normal epithelia allow the transfer of ions, the lipid membranes and the tight junctions between them effectively prevent the diffusion of ions through the BBB. Thus the brain is protected against fast changes in the concentrations of peripheral ions. There is, however, an exchange of ions between the brain and the plasma via the choroid plexus, which has a higher ion permeability but a much smaller surface than the BBB.[16,17] To a lesser extent, ion pumps may also be available in the BBB.[18] These limited transport mechanisms function to maintain cerebral homeostasis of ions.

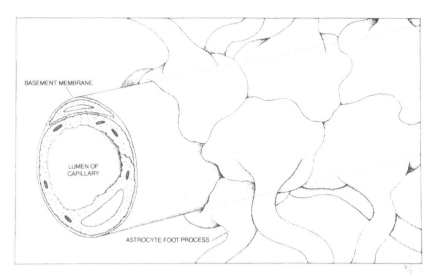

FIGURE 1. The cerebral capillary is constructed by a circle of endothelial cells which are connected by tight junctions and surrounded by a closed cover of astrocytes. (From Goldstein, G. W. and Betz, A. L., *Sci. Am.*, 255(3), 73, 1986. Copyright © 1986 by SCIENTIFIC AMERICAN, Inc. With permission.)

The difference between the low rate of transfer of ions across the BBB in normal brain tissue and the increased ion transfer in malignant brain tumors whose BBB is incomplete has been utilized by conventional scintigraphic imaging of radioactive ion tracers and also by PET, using ^{82}Rb ions to delineate tumors (see Section IV).[19-21] In addition to the imaging of brain tumors, PET studies used ^{82}Rb as a tool to assess alterations of the BBB produced during the treatment of brain tumors. Phillips et al. measured an increased rate constant of ^{82}Rb influx in normal brain issue after intravenous high-dose administration of the chemotherapeutic agent methotrexate.[22]

The improvement of the patient's state following steroid treatment with dexamethasone is assumed to be related to a changed permeability of the BBB. Jarden et al. investigated this question and found a decreased transfer rate constant k_1 of ^{82}Rb$^+$ influx.[23,24] They concluded that the antiedema effects of dexamethasone are mainly caused by this decrease of BBB permeability.

III. SPECIFIC TRANSPORT SYSTEMS

A. GLUCOSE TRANSPORT

Among the different transport systems for hydrophilic substrates present in the BBB, the glucose transport system has been investigated most extensively. The energy needed for neuronal activity is supplied by the oxidation of glucose, so that it is obvious that neuronal activity and cerebral glucose consumption are correlated.[25] As the cerebral glucose reserve is relatively low, a continuous transfer of glucose across the BBB is absolutely necessary for maintaining cerebral metabolism. The ratio between the velocity of unidirectional glucose influx (v_i) and the rate of glucose utilization (LCMRglu = local cerebral metabolic rate of glucose utilization) is, however, only 1.5:1 to 2:1.[26-29] These findings were usually obtained by experiments utilizing anesthetized rats.

Whereas nearly half of all PET papers ever published examined glucose utilization, only very few PET researchers reported glucose transport.[30-34] As in animal experiments, 3-O-methylglucose (CMG) was used in PET studies because this glucose analogue is not phosphorylated like glucose or deoxyglucose.[35] Labeled with ^{11}C, CMG was first synthesized by Kloster et al.[36] and used in man by Vyska et al.[30] In subjects without brain disease,

Vyska et al. obtained values for v_i of about 50 μmol/min/100 g in the cortex.[37] In another group of normal subjects, Feinendegen et al. found a mean cortical v_i of 67 μmol/min/100 g.[34] Both results are in agreement with the above-mentioned ratios between v_i and LCMRglu, if cortical LCMRglu ranges from 30 to 45 μmol/min/100 g as reported by several groups.[38] The values for v_i were obtained by assuming the plasma input curve as the sum of two exponential functions and then fitting the measured tissue curve to the sum of three exponential functions, where one exponent equals k_2^*, the rate constant of CMG efflux. Knowing k_2^*, v_i was calculated by:

$$v_i = c_P V_D^{CMG} k_2^* 1.11 \qquad (1)$$

where c_p is the plasma glucose concentration, $V_D^{CMG} = c_T^*/c_P^*$ is the distribution volume of CMG, and 1.11 equals the ratio between the transport rate constants of glucose and CMG. According to Pardridge and Oldendorf, the rates of CMG and glucose influx differ slightly from each other.[35] Based on these data, Vyska chose the above value of 1.11 to relate the transport rate constants of glucose k_1 (for influx) and k_2 (for efflux) to those of CMG k_1^*, k_2^*:[39]

$$1.11 = \frac{k_1}{k_1^*} = \frac{k_2}{k_2^*} \qquad (2)$$

Other values of v_i differed from those reported above when alternative approaches of model calculation were used.[32,33,40] Brooks et al. fitted the theoretically calculated tissue data of a three-compartment model to the tissue data of CMG measured by PET. In this way, the unidirectional rate of influx of CMG instead of glucose was determined and values of 22 μmol/min/100 g in normal controls were found.[33] Because of Pardridge's and Oldendorf's data,[35] the value of CMG influx obtained by Brooks et al. is close to the glucose influx. A similarly low value for v_i, i.e., μmol/min/100 g, was reported by the Jülich group in 1989; again direct fitting of the transport rate constants was applied.[40] These low values of v_i are smaller than the generally measured data of LCMRglu. Even lower values of v_i can be concluded from very small rate constants of CMG transport published by Herholz et al.,[41] who calculated parametric images by a fitting procedure on a pixel-by-pixel basis. Thus, different mathematical methods obviously yield different results of the rate of cerebral glucose transport which are concordant or discordant to the data of LCMRglu. Further investigations are obviously necessary to resolve these differences.

The unidirectional rate of glucose transport across the BBB, v, was found to be saturable and to depend on the glucose concentration, c, in the source compartment, e.g., the capillary plasma.[29,35] This relationship is usually described by Michaelis-Menten kinetics:

$$v = \frac{V_{max} \, c}{K_m + c} \qquad (3)$$

V_{max} denotes the maximum velocity of substrate transport and the Michaelis-Menten constant K_m equals the substrate concentration needed to yield a v of 0.5 V_{max}. Using Equation 3, the influx v_i is expressed as:

$$v_i = \frac{V_{max} \, c_P}{K_m + c_P} \qquad (4)$$

and the efflux v_e is expressed as:

$$v_e = \frac{V_{max}\, c_T}{K_m + c_T} \tag{5}$$

where c_p is the concentration of capillary plasma glucose and c_T is the concentration of cerebral tissue glucose. Here the BBB is assumed to act like a standard enzymatic reaction, although two cell membranes and the endothelial cytoplasma have to be crossed. The utilization of V_{max} and K_m in both Equations 4 and 5 is in agreement with the general assumption of a symmetric glucose carrier system.[35,42,43] This symmetric behavior may, however, be questioned by the finding of different carriers at the vascular and cerebral BBB membranes.[44] On the other hand, the glucose carrier system could be identified as a single membrane protein with a molecular weight of 53,000 daltons.[45,46]

The parameters V_{max} and K_m have been investigated by several animal studies, most of which were done in anesthetized rats. They yielded values of K_m between 6 and 11 μmol/g and values of V_{max} of about 2 μmol/min/g.[29,35,47] In awake rats, V_{max} was higher with values up to 7.9 μmol/min/g.[48] Similar data were also reported by PET studies in man. Brooks et al.[33] found K_m values of 3 to 11 μmol/g in a diabetic patient, Vyska et al.[37] reported a mean K_m of 6.4 μmol/g and a mean V_{max} of 2.5 μmol/min/g, and Feinendegen et al.[34] obtained a mean K_m of 3.8 μmol/g and a mean V_{max} of 2.0 μmol/min/g. Thus, it is concluded that K_m is in a range about equal for diabetic and normal subjects. Although the significance of this finding may be limited by the small number of subjects examined in these studies, it is in agreement with the fact that K_m is related to the chemical characteristics of the carrier. V_{max}, in contrast, mirrors the amount of available carrier molecules. There are speculations that the capacity of glucose transport might be changed by chronic changes in substrate concentration such as hypo- or hyperglycemia.[49] Until now this question of a V_{max} being changed under pathological conditions has not been studied with PET.

The investigations of the cerebral glucose influx by the Jülich group suggest that there are rather wide ranges of individual values of V_{max} for normal individuals and thus an individual capacity of glucose transport. The data of v_i published by Feinendegen et al. show a great variation between different subjects;[34] after the glucose level in plasma was raised (Figure 2), v_i increased individually as a function of the plasma glucose level, in agreement with the Michaelis-Menten equation for the kinetics of glucose transport. The changes of v_i due to the increased plasma glucose level were, however, within a much smaller range than were the values of v_i among the individual persons investigated. This observation leads to the assumption that each individual has his or her own specific capacity for glucose transport.

All PET studies which use a state of hyperglycemia to examine the characteristics of cerebral glucose transport apply normal Michaelis-Menten kinetics to the model description. Based on these studies, an influence of insulin on the glucose transport, which is obvious for other organs and which may be expected from the finding of insulin receptors in the BBB,[50] could not be verified. Several studies in animals gave contradictory results. Thus, whereas a non-PET investigation with the indicator dilution technique in normoglycemic humans by Hertz et al.[51] yielded a high increase of glucose transport at very high levels and a low increase at moderately high levels of insulin, an autoradiographic study with ^{14}C-methylglucose in normoglycemic conscious rats by Namba et al.[52] showed a decreased hexose transport. This discrepancy may be caused by the difference between species examined and methods used. This uncertainty suggests taking into account the possible effect of insulin in PET studies of glucose transport.

B. AMINO ACID TRANSPORT

Three different carrier systems are available for the transport of acidic, basic, and neutral

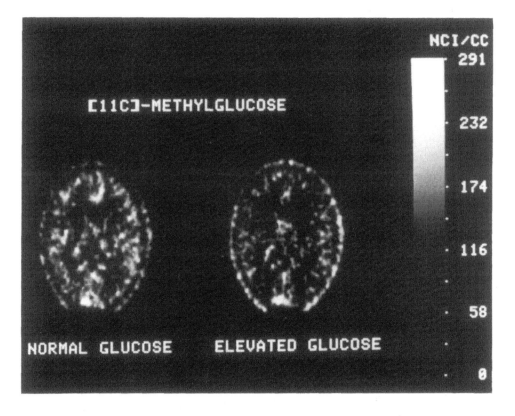

FIGURE 2. Images of cerebral uptake of ^{11}C-methylglucose (CMG). The right image demonstrates a decreased cerebral uptake of CMG in comparison to the left image due to competitive inhibition by plasma glucose the concentration of which was doubled using a glucose clamp technique.

TABLE 1
Classes of Amino Acids with Common Carrier Systems

Acidic amino acids:	Glutamate, aspartate
Basic amino acids:	Lysine, arginine, histidine, ornithine
Neutral amino acids:	Glycine, alanine, serine, threonine, valine, leucine, isoleucine, cysteine, methionine, phenylalanine, tyrosine, tryptophan

amino acids (Table 1). The latter system is responsible for the transport of the majority of the amino acids and can be divided into two subsystems,[12] one for large and the other for small neutral amino acids. The transfer of the large neutral amino acids (LNAA) into the brain tissue is especially important, because they are essential precursors for the intracerebral synthesis of neurotransmitters and proteins. In order to keep the concentration of a neurotransmitter at an appropriate level, the influx of the precursor substrate has to be controlled strictly by a dedicated carrier system which is governed according to Michaelis-Menten kinetics. Contrary to the influx into other tissues than the brain, there is competitive inhibition among the different LNAA at the BBB. This is easily explained because the K_m values of the LNAA are near to the plasma concentrations of the LNAA, so that the rate constants are sensitive to changes of these plasma concentrations.[13, 53]

Another feature of the amino acid transport is its asymmetry. The backflux of amino acids out of the brain is assumed to be much lower than the influx from the circulation.[54] This poses the difficulty in concluding from external measurements such as done with PET, whether a cerebral accumulation of labeled amino acids is due to incorporation into proteins or represents just a changed steady state with an increased tissue concentration of free amino acid caused by reduced backflux.

The uptake of amino acids into the human brain has been investigated in several PET studies mostly with ^{11}C-labeled LNAA, such as ^{11}C-leucine and ^{11}C-methylmethionine.[55-59] As tumors show an increased rate of protein synthesis, many of these PET studies reported tumor delineation in normal brain tissue in order to improve the differential diagnosis or find therapeutic effects. These studies are discussed in detail in Chapter 16. Here, they are of interest in regard to their results concerning the tranport across the BBB.

PET studies of the transport across the BBB and intracerebral metabolism of amino acids are nearly always confounded by the existence of labeled metabolites of the injected tracer both in blood and in cerebral tissue. In the case of ^{11}C-leucine, there is a rapid production of [11]CO_2, which is rapidly cleared into the blood, so that it can be neglected in comparison to the steadily increasing amount of labeled amino acids in the protein pool.[55] In the case of ^{11}C-methylmethionine, labeled metabolites in the blood have to be considered.[60] Moreover, in addition to their incorporation into proteins, amino acids enter other intracerebral pathways, i.e., for the production of other amino acids or neurotransmitters. Thus, a four-compartment model is necessary for describing the kinetics of labeled amino acids. Such models have been applied to the assessment of the rate of protein synthesis.[55,56,61] Indeed, not much attention has so far been directed toward the kinetics of amino acid transport across the BBB.

Although the calculation of transport rate constants is generally included in the solution of the kinetic model, this calculation may be unreliable. Whereas the model fit itself may be satisfactory, the accuracy of a single rate constant is doubtful, as the numerical values of the single rate constants are highly correlated so that errors in the determination of one rate constant affect results for all other rate constants.[62] Furthermore, the significance of the rate constants of amino acid transport is unclear even if they are exactly determined. Due to the competitive inhibition for transport across the BBB, the rate constants must be expected to be highly sensitive to changes in the plasma levels of all such amino acids, which are substrates of the same carrier system. Sometimes this dependency is not fully considered because only the cold "sister" of the examined labeled amino acid is taken into account.[63] The effect of cross-inhibition by all plasma amino acids, which are specific for the given carrier system, is especially important for studies of brain tumors. Here the BBB is at least partially destroyed so that the transport of amino acids into tumor is independent of their plasma levels and the Michaelis-Menten kinetics are inapplicable. Thus, the increased uptake of labeled amino acids in tumors may be due to simple diffusion which is unaffected by competitive amino acids, whereas the influx into normal tissue is governed by Michaelis-Menten kinetics. Furthermore, the comparison of the uptake ratio tumor/normal tissue, before and after therapy, may be influenced not only by therapeutic effects, but also by changed plasma levels with discordant consequences in tumor and normal tissue.

Despite the problem of differentiating between transport and metabolism of amino acids, PET succeeded in confirming findings of animal research. Using the D- and L-isomers of ^{11}C-methylmethionine, the stereospecificity of BBB transport of the LNAA preferring the L-isomer could be demonstrated.[57] There was a decrease of the accumulation rate of ^{11}C-methylmethionine due to an increased plasma level of unlabeled LNAA as shown by Bergström et al.[63] and O'Thuama et al.[59] Leenders et al. found a reduced uptake of ^{18}F-fluorodopa when the subject examined was loaded with cold amino acids, thus indicating the need to consider the interference between dietary amino acids and L-dopa given for therapy of Parkinson's disease.[64]

Recently, some amino acid analogues have been introduced which allow a better separation between rates of BBB transport and cerebral amino acid metabolism. As protein synthesis is a slow process, the choice of a tracer with a suitable half-life is important; thus,

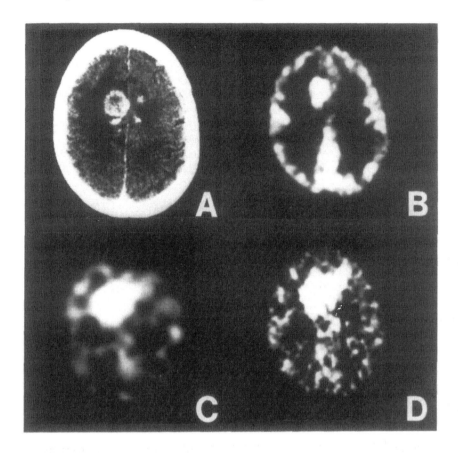

FIGURE 3.   Comparison of various approaches of imaging a malignant brain tumor. (A) CT shows a clear contrast enhancement in the tumor region due to the damaged BBB; (B) the increased metabolic activity of the tumor is shown by the FDG-scan. Using $^{123}I/^{124}I$-iodine-α-methyltyrosine and SPECT (C) or PET (D), increased amino acid transport was observed in a region greater than the one delineated by CT or FDG-PET. (From Langen, K.-J., Coenen, H. H., Roosen, N., Kling, P., Muzik, O., Herzog, H., Kuwert, T., Stöcklin, G., and Feinendegen, L. E., *J. Nucl. Med.*, 31, 281, 1990. With permission.)

incorporation of amino acids into proteins would be better traced with amino acids labeled with ^{18}F ($T_{1/2}$ = 109 min) than labeled with ^{11}C ($T_{1/2}$ = 20 min). When Coenen et al. synthesized L-[2-^{18}F]fluorotyrosine and tested its kinetics in mouse brain, they found that it is nearly exclusively incorporated into proteins without entering other pathways.[65] Thus, a three-compartment model equivalent to the one used for ^{18}F-flurodeoxyglucose (FDG) may be applied here. In contrast to L-[2-^{18}F]fluorotyrosine, the amino acid analogue L-3-iodo-α-methyltyrosine (IMT) was found not to be a tracer of protein synthesis but of amino acid transport.[66] Labeled with ^{123}I, it was utilized to image brain tumors by single photon emission computed tomography (SPECT)[67]; and when it was labeled with ^{124}I, Langen et al. employed PET for comparison (Figure 3).[68] The images were indeed similar, yet it remained unclear whether accumulation in tumor regions was caused by nonspecific trapping of tracer[54] or by interaction with tyrosine hydroxylase, for which IMT is known to be an inhibitor substrate.[69] Further investigations are presently underway.

IV. BBB IN TUMOR

Several imaging techniques are available for diagnosis of brain tumors, such as magnetic resonance imaging (MRI), (transmission) computed tomography (CT), conventional scintigraphy, SPECT, or PET. Whereas MRI and CT are mainly used to detect differences between tissue structures and qualities of normal brain and tumor, the tracer methods are able to assess the functional alteration of the BBB in tumors. Thus, the intact BBB is not permeable, for example, to contrast agents used in computed tomography (CT) because they are, in ionic or nonionic form, highly polar and hydrophilic. If they clearly delineate brain tumors, they must have penetrated from the blood vessels into the tumorous tissue where the BBB is absent.[70,71] Recently, however, several authors found that the BBB is more or less intact, at least in parts of the tumor.[72-74] The varying loss of BBB integrity is concordant with different levels of contrast enhancement. The partial absence of the BBB in tumors has been used for years by conventional scintigraphy and also by SPECT, e.g., with ^{99m}Tc-pertechnetate or ^{99m}Tc-DPTA as tracer.[75] Both conventional nuclear medicine methods and CT are more dedicated to qualitative imaging, whereas PET permits quantitation in examining the BBB damage in tumors more exactly; examples are the data with ^{11}C- and ^{18}F-labeled amino acids referred to above. Furthermore, ^{68}Ga-EDTA,[76-79] a hydrophilic complex, and ^{82}Rb ions[19-23] were used for probing BBB integrity with PET. Brooks et al. found an excellent correlation between extraction of ^{82}Rb and CT contrast enhancement, with extraction values of 29 $\pm$ 13% in tumors with contrast enhancement in CT and 1.3 $\pm$ 1.1% without contrast enhancement.[80]

Qualitative PET-imaging of brain tumors with ^{68}Ga-EDTA or ^{82}Rb does not offer advantages over CT or SPECT, unless it is linked with measuring the FDG (2-fluoro-2-deoxyglucose) uptake. In those cases where FDG uptake in tumors is not different from the surrounding normal tissue, the uptake of ^{68}Ga-EDTA or ^{82}Rb that is limited to tumors allows a better delineation of the region of interest.[21]

V. VITAMIN TRANSPORT INTO THE BRAIN

Vitamins are necessary for specific metabolic processes of the brain and insofar as they are lipophilic will cross the lipid membranes of the BBB easily. Little is known about the BBB transport of hydrophilic vitamins. Some investigators assume an indirect cerebrospinal transfer of hydrophilic vitamins as has already been described above for ions.[81] Although the permeability of the BBB is lower than that of the choroid plexus, its much larger surface eventually leads to an influx similar to the influx of LNAA.[82] Thus, it is possible that all vitamins are transferred directly through the BBB. As far as is known, PET studies with vitamins were performed only by the Jülich group,[83-85] who investigated the cerebral uptake of nicotinic acid and nicotinamide. A deficiency of these vitamins is responsible for pellagra, which is accompanied by various pathological symptoms of the CNS. First images of a study in man of the cerebral uptake of ^{11}C-nicotinic acid (NAC) suggested that this vitamin does not penetrate the BBB.[83] However, further studies with NAC and, in addition, with ^{11}C-nicotinamide (NAM) and quantitative analyses using Patlak plots revealed that both vitamins accumulated in small amounts in the cerebral tissue with similar values of k_3 of about 0.025 min^{-1}. Yet there were still differences in brain uptake; thus, at 30 min after injection, uptake was about 0.5% of the injected dose of NAC, but 2.6% of NAM. This discrepancy can be explained by the lower hydrophilicity and therefore better passage of NAM into the brain compared to NAC and by a selectively high uptake of NAC into the red blood cells.[84,85] Due to the accumulation of NAC in erythrocytes less substrate is available to be transferred from plasma to brain. Even if these data do not primarily distinguish between transport of vitamins via the BBB and the cerebrospinal space, in the case of NAM, the time sequence of tracer uptake and an uptake pattern similar to flow images allow the assumption that NAM enters the brain through the BBB.

VI. VARIOUS PET INVESTIGATIONS OF BBB TRANSPORT

Many compounds which are therapeutically effective in the body are prevented by the BBB from entering the brain. Thus, it would be attractive to open the BBB selectively and reversibly by partial disruption so that such compounds may enter the brain. Whereas nonphysiological conditions like hypercapnia, hypoxia, or ischemia yield a long-lasting or irreversible loss of physical integrity of the BBB,[86] the application of osmotic opening[87,88] or the administration of Metrazol[89] causes reversible short-term damage of the BBB. During such a transient period of high permeability, hydrophilic therapeutic compounds can be transferred into the brain without affecting the normal brain function. The method of osmotic opening is discussed in detail by Neuwelt and co-workers, who transferred their experiences in basic animal research to the clinical application to improve the uptake of chemotherapeutic agents in brain tumors.[87, 88] Recently PET was used in dogs to examine the effect of osmotic opening using an intracarotid infusion of hyperosmolar mannitol on the permeability of BBB for ^{11}C-quinidine and ^{11}C-morphine.[90] The results showed an increase of more than 100% of cerebral uptake of these drugs after the infusion of mannitol. The reversibility of the osmotic opening of BBB was, however, not investigated in this study.

The utility of positron emitting molecules to trace cerebral blood flow depends on their degree of diffusibility. The application of the Kety-Schmidt method, for example, requires a freely diffusible tracer. In numerous PET studies,[15]O-labeled water, either directly injected or produced in the lung after inhalation of $C^{15}O_2$, has been used for the measurement of cerebral blood flow (CBF).[91-93] Whereas this procedure is very attractive due to the convenience of tracer production, there is a methodological limitation because water crosses the BBB rapidly, but is not freely diffusible.[94] Herscovitch et al. used an intracarotid injection technique for comparing the permeability of $H_2^{15}O$ and of ^{11}C-butanol; they demonstrated that the lipophilic substrate ^{11}C-butanol diffuses across the BBB better, so that it is more appropriate for CBF measurements than $H_2^{15}O$.[95]

Damage of the BBB in multiple sclerosis is known from postmortem studies and from CT showing a contrast enhancement in multiple sclerosis plaques. Pozzilli and co-workers tried to quantify the permeability alteration with PET and ^{68}Ga-ETDA.[96] Their investigation yielded a moderate but significant increase of BBB permeability indicating the physical damage in the plaques. The authors suggested the use of PET to assess possible effects on BBB permeability caused by therapeutic interventions during and after acute exacerbation of multiple sclerosis.

VII. CONCLUDING REMARKS

In this chapter, some important features of the physical integrity and biochemistry of the BBB, such as the transfer of ions, and the facilitated transport of glucose and of amino acids, are briefly discussed and PET investigations concerning these subjects are reviewed. Comparing the PET publications related to the BBB with the whole of PET brain studies published, these papers are only a small fraction. This is in contrast to the importance of the BBB for the balance of cerebral metabolism and for the delivery of pharmaceutical agents. Various aspects of BBB biochemistry and its effects on cerebral metabolism were extensively reviewed by Pardridge in 1983,[13] who with co-workers[97] included a list of clinical problems in internal medicine which were assumed to affect the BBB biochemistry and consequently cerebral metabolism. Similar physiological and pathophysiological items of the BBB were also discussed by Suckling and co-workers in 1986.[98] Only recently, two extensive volumes on both basic and clinical aspects of the BBB were published by Neuwelt.[99] These reviews discuss a large variety of problems beyond those few already assessed by PET, which identify additional specific diagnostic applications of PET with its ability of measuring biochemistry *in vivo*.

REFERENCES

1. **Ehrlich, P.**, Über die Beziehungen von chemische Constitution, Verheilung, und Pharmakologischer Wirking, in *Collected Stud. in Immunity,* John Wiley & Sons, New York, 1906, 567 (reproduced and translated).
2. **Goldmann, E. E.**, Die äussere und innere Sekretion des gesunden und kranken Organismus im Licht der vitalen Färbung, *Beitr. Klin. Chir.,* 64, 192, 1909.
3. **Lewandowski, M.**, Zur Lehre der Cerebrospinal-flüssigkeit, *Z. Klin. Med.,* 40, 480, 1900.
4. **Reese, T. S. and Karnovsky, M. J.**, Fine structure localization of a blood-brain barrier to exogenous peroxidase, *J. Cell Biol.* 34, 208, 1967.
5. **Goldstein, G. W. and Betz, A. L.**, The blood-brain barrier, *Sci. Am.* 255(3), 74, 1986.
6. **Krogh, A.**, The active and passive exchanges of inorganic ions through the surfaces of living cells and through living membranes generally, *Proc. R. Soc. London Ser. B.* 133, 140, 1946.
7. **Fenstermacher, J. D. and Rapoport, S. I.**, Blood-brain barrier, in *Handbook of Physiology, Microcirculation,* Renkin, E. M. and Michel, C. C., Eds., American Physiological Society, Washington, D.C., 1984, 969.
8. **Pollay, M.**, Blood-brain barrier, *Contemp. Neurosurg.,* 9, 1, 1987.
9. **Crone, C.**, Facilitated transfer of glucose from blood to brain tissue, *J. Physiol. (London),* 181, 103, 1965.
10. **Oldendorf, W. H.**, Brain uptake of radiolabelled amino acids, amines, and hexoses after arterial injection, *Am. J. Physiol.,* 221, 1626, 1971.
11. **Oldendorf, W. H.**, Carrier-mediated blood-brain barrier transport of short chain monocarboxylic organic acids, *Am. J. Physiol.,* 224, 1450, 1973.
12. **Oldendorf, W. H. and Szabo, J.**, Amino acid assignment to one of three blood-brain barrier amino acid carriers, *Am. J. Physiol.,* 230, 94, 1976.
13. **Pardridge, W. M.**, Brain metabolism: a perspective from the blood-brain barrier, *Physiol. Rev.,* 63, 1481, 1983.
14. **Davson, H. and Welch, K.**, The permeation of several materials into the fluids of the rabbit's brain, *J. Physiol.* 218, 337, 1971.
15. **Crone, C. and Christensen, O.**, The electrical resistance of a capillary endothelium, *J. Gen. Physiol.,* 77, 349, 1981.
16. **Smith, Q. R., Woolbury, D. M., and Johanson, C. E.**, Kinetic analysis of Cl^{36}, Na^{22} and H^3-mannitol uptake into the in vivo choroid plexus-cerebrospinal fluid system; ontogeny of the blood-brain and blood-CSF barriers, *Dev. Brain Res.,* 3, 181, 1982.
17. **Sunderland, S.**, The nerve lesion in the carpal tunnel syndrome. *J. Neurol. Neurosurg. Psychiatry,* 39, 615, 1976.
18. **Bradbury, M. W. and Stulcova, B.**, Efflux mechanisms contributing to the stability of the potassium concentration in the cerebrospinal fluid, *J. Physiol.,* 208, 415, 1970.
19. **Brooks, D. J., Beaney, R. P., Lammertsma, A. A., Leenders, K. L., Horlock P. L., Kensett, M. J., Marshall, J., Thomas, D. G., and Jones, T.**, Quantitative measurement of blood-brain barrier permeability using rubidum-82 and positron emission tomography, *J. Cereb. Blood Flow Metab.,* 4, 535, 1984.
20. **Valk, P. E., Budinger, T. F., Levin, V. A., Silver, P., Gutin, P. H., and Doyle, W. K.**, PET of malignant cerebral tumors after interstitial brachytherapy. Demonstration of metabolic activity and correlation with clinical outcome, *J. Neurosurg.,* 69, 830, 1988.
21. **Doyle, W. K., Budinger, T. F., Valk, P. E., Levin, V. A., and Gutin, P. H.**, Differentiation of cerebral radiation necrosis from tumor recurrence by [^{18}F]FDG and ^{82}Rb positron emission tomography, *J. Comput. Assist. Tomogr.,* 11, 563, 1987.
22. **Phillips, P. C., Dhawan, V., Strother, S. C., Sidtis, J. J., Evans, A. C., Allen, J. C., and Rottenberg, D. A.**, Reduced cerebral glucose metabolism and increased brain capillary permeability following high-dose methotrexate chemotherapy: a positron emission tomographic study, *Ann. Neurol.,* 21, 59, 1987.
23. **Jarden, J. O., Dhawan, V., Poltorak, A., Posner, J. B., and Rottenberg, D. A.**, Positron emission tomographic measurement of blood-to-brain and blood-to-tumor transport of ^{82}Rb: the effect of dexamethasone and whole-brain radiation therapy, *Ann. Neurol.,* 18, 636, 1985.
24. **Jarden, J. O., Dhawan, V., Moeller, J. R., Strother, S. C., and Rottenberg, D. A.**, The time course of steroid action on blood-to-brain and blood-to-tumor transport of ^{82}Rb: a positron emission tomographic study, *Ann. Neurol.,* 25, 239, 1989.
25. **Sokoloff, L.**, Relationships among local functional activity, energy metabolism, and blood flow in the central nervous system, *Fed. Proc.,* 40, 2311, 1981.
26. **Cremer, J. E., Ray, D. E., Sarna, G. S., and Cunningham, V. J.**, A study of the kinetic behaviour of glucose based on simultaneous estimates of influx and phosphorylation in brain regions of rats in different physiological states, *Brain Res.,* 221, 331, 1981.
27. **Lund-Andersen, H. and Kjeldsen, C. J.**, Uptake of glucose analogues by rat brain cortex slices: membrane transport versus metabolism of 2-deoxy-D-glucose, *J. Neurochem.,* 29, 205, 1977.

28. **Pardridge, W. M., Crane, P. D., Mietus, L. J., and Oldendorf, W. H.,** Kinetics of regional blood-brain barrier transport and brain phosphorylation of glucose and 2-deoxyglucose in the barbiturate-anesthesized rat, *J. Neurochem.*, 38, 560, 1982.

29. **Gjedde, A. and Rasmussen, M.,** Pentobarbital anaesthesia reduces blood-brain glucose transfer in the rat, *J. Neurochem.*, 35, 1382, 1980.

30. **Vyska, K., Freundlieb, C., Höck, A., Becker, V., Feinendegen, L. E., Kloster, G., Stöcklin, G., Traube, H., and Heiss, W. D.,** The assessment of glucose transport across the blood brain barrier in man by use of 3-(^{11}C)-methyl-D-glucose, *J. Cereb. Blood Flow Metab.*, 1 (Suppl. 1), S42, 1981.

31. **Gjedde, A., Wienhard, K., Heiss, W. D., Kloster, G., Diemer, N. H., Herholz, K., and Pawlik, G.,** Comparative analysis of 2-fluorodeoxyglucose and methylglucose uptake in brain of four stroke patients. With special reference to the regional estimation of the lumped constant, *J. Cereb. Blood Flow Metab.*, 5, 163, 1985.

32. **Brooks, D. J., Beaney, R. P., Lammertsma, A. A., Herold, S., Turton, D. R., Luthra, S. K., Frackowiak, R. S. J., Thomas, D. G. T., Marshall, J., and Jones, T.,** Glucose transport across the blood-brain barrier in normal human subjects and patients with cerebral tumours studied using [^{11}C]3-O-methyl-D-glucose and positron emission tomography, *J. Cereb. Blood Flow Metab.*, 6, 230, 1986.

33. **Brooks, D. J., Gibbs, J. S. R., Sharp, P., Herold, S., Turton, D. R., Luthra, S. K., Kohner, E. M., Bloom, S. R., and Jones, T.,** Regional cerebral glucose transport in insulin-dependent diabetic patients studied using [^{11}C]3-O-methyl-D-glucose and positron emission tomography, *J. Cereb. Blood Flow Metab.*, 6, 240, 1986.

34. **Feinendegen, L. E., Herzog, H., Wieler, H., Patton, D. D., and Schmid, A.,** Glucose transfer from the circulation into metabolism in the human brain, analyzed by (C-11) methyl-glucose and positron emission tomography, *J. Nucl. Med.*, 27, 1867, 1986.

35. **Pardridge, W. M. and Oldendorf, W. L.,** Kinetics of blood-brain barrier transport of hexoses, *Biochim. Biophys. Acta*, 382, 377, 1975.

36. **Kloster, G., Müller-Platz, C., and Laufer, P.,** 3-^{11}-C-methyl-D-glucose: a potential agent for regional cerebral glucose utilization. Synthesis, chromatography, and tissue distribution in mice, *J. Lab. Comp. Radiopharm.*, 18, 855, 1981.

37. **Vyska, K., Magloire, J. R., Freundlieb, C., Höck, A., Becker, V., Schmid, A., Feinendegen, L. E., Kloster, G., Stöcklin, G., Schuier, F. J., and Thal, H. U.,** In vivo determination of the kinetic parameters of glucose transport in the human brain using ^{11}C-methyl-D-glucose (CMG) and dynamic positron emission tomography (dPET), *Eur. J. Nucl. Med.*, 11, 97, 1985.

38. **Mazziotta, J. C. and Phelps, M. E.,** Positron emission tomography studies of the brain, in *Positron Emission Tomography and Autoradiography: Principles and Applications for the Brain and Heart*, Phelps, M., Mazziotta, J. and Schelbert, H., Eds., Raven Press, New York, 1986, 498.

39. **Vyska, K., Profant, M., Schuier, F., Knust, E. J., Machulla, H. J., Mehdorn, H. M., Knapp, W. H., Spohr, G., von Seggern, R., Kimmling, I., Becker, V., and Feinendegen, L. E.,** In vivo determination of kinetic parameters for glucose influx and efflux by means of 3-O-^{11}C-methyl-D-glucose, ^{18}F-3-deoxy-3-fluoro-D-glucose and dynamic positron emission tomography: theory, method and normal values, in *Current Topics in Tumor Cell Physiology and Positron Emission Tomography*, Knapp, W. H. and Vyska, K., Eds., Springer-Verlag, Berlin, 1984, 37.

40. **Herzog, H., Lang, H. W., Langen, K. J., Rota Kops, E., Fulton, R., Kuwert, T., Muzik, O., and Feinendegen, L. E.,** Cortical perfusion, oxygen consumption, transport and consumption of glucose in symptomatic Huntington patients, in *Positron Emission Tomography in Clinical Research and Clinical Diagnosis*, Beckers, C., Goffinet, A., and Bol, A., Eds., Kluwer, Nijmwegen, 1989, 206.

41. **Herholz, K., Wienhard, K., Pietrzyk, U., Pawlik, G., and Heiss, W.-D.,** Measurement of blood-brain hexose transport with dynamic PET: comparison of [^{18}F]2-fluoro-2-deoxyglucose and [^{11}C]O-methylglucose, *J. Cereb. Blood Flow Metab.*, 9, 104, 1989.

42. **Betz, A. L., Gilboe, D. D., Yudilevich, D. L., and Drewes, L. R.,** Kinetics of unidirectional glucose transport into the isolated dog brain, *Am. J. Physiol.*, 225, 586, 1973.

43. **Hertz, M. M. and Paulson, O. B.,** Glucose transfer across the blood-brain barrier, in *CNS Regulation of Carbohydrate Metabolism*, Szabo, A. J., Ed., Academic Press, New York, 1983, 177.

44. **Betz, A. L., Firth, J. A., and Goldstein, G. W.,** Polarity of the blood-brain barrier: distribution of enzymes between the luminal and antiluminal membranes of the brain capillary endothelial cells, *Brain Res.*, 192, 17, 1980.

45. **Baldwin, S. A., Cairns, M. T., Gardiner, R. M., and Ruggier R.,** A D-glucose-sensitive cytocolasin B-binding component of cerebral microvessels, *J. Neurochem.*, 45, 650, 1985.

46. **Dick, A. P. K., Harik, S. I., Klip, A., and Walker, D. M.,** Identification and characterization of the glucose transporter of the blood-brain barrier by cytocholasin B binding and immunological reactivity, *Proc. Natl. Acad. Sci. U.S.A.*, 81, 7233, 1984.

47. **Bachelard, H. S., Daniel, P. M., Love, E. R., and Pratt, O. E.,** The transport of glucose into the brain of the rat in vivo. *Proc. R. Soc. London Ser. B.*, 183, 71, 1973.

48. **Cremer, J. E.**, Transport of glucose across the blood-brain barrier in relation to brain metabolism, in *The Blood-Brain Barrier in Health and Disease*, Suckling, A. J., Rumsby, M. G., and Bradbury, M. W. B., Eds., Ellis Horwood, Chichester, England, 1986, 77.

49. **Knudsen, G. M., Hertz, M. M., and Paulson, O. B.**, Metabolic disturbances of the blood-brain barrier with special emphasis on glucose and amino acid transport, in *Implications of the Blood-Brain Barrier and Its Manipulation, Vol. 2, Clinical Aspects*, Neuwelt, E. A., Ed., Plenum Medical, New York, 1989, 579.

50. **Van Houten, M. and Posner, B. I.**, Insulin binds to brain blood vessels in vivo, *Nature*, 282, 623, 1979.

51. **Hertz, M. M., Paulson, O. B., Barry, D. I., Christiansen, J. S., and Svendsen, P. A.**, Insulin increases glucose transfer across the blood-brain barrier in man, *J. Clin. Invest.*, 67, 597, 1981.

52. **Namba, H., Lucignani, G., Nehlig, A., Patlak, C., Pettigrew, K., Kennedy, C., and Sokoloff, L.**, Effects of insulin on hexose transport across blood-brain barrier in normoglycemia, *Am. J. Physiol. E. Baltimore*, 232, E299, 1987.

53. **Smith, Q. R., Momma, S. S., Aoyagi, M., and Rapoport S. I.**, Kinetics of neutral amino acid transport across the blood-brain barrier. *J. Neurochem.* 49, 1651, 1987.

54. **Paulson, O. B., Knudsen, G. M., Hertz, M. M. et al.**, Differences in characteristics of blood-brain barrier permeability by glucose and amino acids, *J. Cereb. Blood Flow Metab.*, 1, S73, 1985.

55. **Phelps, M. E., Barrio, J. R., Huang, S.-C., Keen, R. E., Chugani, H., and Mazziotta, J. C.**, Measurement of cerebral protein synthesis in man with positron computerized tomography: model, assumptions, and preliminary results, in *The Metabolism of the Human Brain Studied with Positron Emission Tomography*, Greitz T., Ingvar, D. H., and Widen, L., Eds., Raven Press, New York, 1985, 215.

56. **Bustany, P., Chatel, M., Derlon, J. M., Darcel, F. Sgouropoulos, P., Soussaline, F., and Syrota, A.**, Brain tumor protein synthesis and histological grades: a study by positron emission tomography (PET) with ¹¹C-L-methionine, *J. Neurooncol.*, 23, 397, 1986.

57. **Bergström, M., Lundqvist, H., Ericson, K., Lilja, A., Johnström, P., Langström, B., von Holst, H., Eriksson, L., and Blomqvist, G.**, Comparison of the accumulation kinetics of L-(methyl-¹¹C)-methionine and D-(methyl¹¹C)-methionine in brain tumours studied with positron emission tomography, *Acta Radiol.*, 28, 225, 1987.

58. **Schober, G., Duden, C., Meyer, G. J., Muller, J. A., and Hundeshagen, H.**, Nonselective transport of [¹¹C-methyl]-L- and D-methionine into a malignant glioma, *Eur. J. Nucl. Med.*, 13, 103, 1987.

59. **O'Thuama, L. A., Guilarte, T. R., Douglass, K. H., Wagner, H. N., Jr., Wong, D. F., Dannals, R. F., Ravert, H. T., Wilson, A. A., LaFrance, N. D., Bice, A. N., and Links, J. M.**, Assessment of [¹¹C]-L-methionine transport into the human brain, *J. Cereb. Blood Flow Metab.*, 8, 341, 1988.

60. **Lundqvist, H., Stalnacke, C.-G., Langström, B., and Jones, J.**, Labeled metabolites in plasma after intravenous administration of [¹¹CH₃]-L-methionine, in *The Metabolism of the Human Brain Studied with Positron Emission Tomography*, Greitz T., Ingvar, D. H., and Widen, L., Eds., Raven Press, New York, 1985, 233.

61. **Meyer, G. J., Schober, O., Gaab, M. R., Dietz, H., and Hundeshagen, H.**, Multi-parameter studies in brain tumors, in *Positron Emission Tomography in Clinical Research and Clinical Diagnosis*, Beckers, C., Goffinet, A., and Bol, A., Eds., Kluwer, Nijmwegen, 1989, 229.

62. **Carson, R. E.**, Parameter estimation in positron emission tomography, in *Positron Emission Tomography and Autoradiography: Principles and Applications for the Brain and Heart*, Phelps, M., Mazziotta, J., and Schelbert, H., Eds., Raven Press, New York, 1986, 354.

63. **Bergström, M., Ericson, K., Hagenfeldt, L., Mosskin, M., von Holst, H., Norén, G., Eriksson, K., Ehrin, E., and Johnström, P.**, PET study of methionine accumulation in glioma and normal brain tissue: competition with branched chain amino acids, *J. Comp. Assist. Tomogr.*, 11, 208, 1987.

64. **Leenders, K. L., Poewe, W. H., Palmer, A. J., Brenton, D. P., and Frackowiak, R. S.**, Inhibition of L-[¹⁸F]fluorodopa uptake into human brain by amino acids demonstrated by positron emission tomography, *Ann. Neurol.*, 20, 258, 1986.

65. **Coenen, H. H., Kling, P., and Stöcklin, G.**, Cerebral metabolism of L-[2-¹⁸F]fluorotyrosine, a new PET-tracer of protein synthesis, *J. Nucl. Med.*, 30, 1367, 1989.

66. **Kawai, K., Fujibayashi, Y., Saji, H., Konishi, J., and Yokoyama, A.**, New radioiodinated radiopharmaceutical for cerebral amino acid transport studies: 3-iodo-alpha-methyl-L-tyrosine, *J. Nucl. Med.*, 29, 778, 1988.

67. **Biersack, H. J., Coenen, H. H., Stöcklin, G., Reichmann, K., Bockisch, A., Oehr, P., Kashab, M., and Rollmann, O.**, Imaging of brain tumors with L-3[¹²³I]iodo-α-methyl-tyrosine and SPECT, *J. Nucl. Med.*, 30, 110, 198ᶜ

68. **Langen, K.-J., Coenen, H. H., Roosen, N., Kling, P., Muzik, O., Herzog, H., Kuwert, T., Stöcklin, G., and Feinendegen, L. E.**, SPECT studies of brain tumors with L-3-[¹²³I]iodo-α-methyl tyrosine (¹²³I-IMT): comparison with PET and ¹²⁴IMT and first clinical results, *J. Nucl. Med.*, 31, 281, 1990.

69. **Moore, K. E. and Dominic, J. A.**, Tyrosine hydroxylase inhibitors, *Fed. Proc.*, 30, 859, 1971.

70. **Long, D. M.**, Capillary ultrastructure and the blood-brain barrier in human malignant brain tumours, *J. Neurosurg.*, 32, 127, 1970.

71. **Vick, N. A., Khandeka, V. D., and Bigner, D. D.,** Chemotherapy of brain tumors: the "blood-brain barrier" is not a factor, *Arch. Neurol.,* 34, 523, 1977.

72. **Groothius, D. R., Molnar, P., and Blasberg, R. G.,** Regional blood flow and blood-to-tissue transport in five brain tumor models: implications for chemotherapy, *Prog. Exp. Tumor Res.,* 27, 132, 1984.

73. **Greig, N. H.,** Chemotherapy of brain metastases: current status, *Cancer Treat. Rev.* 11, 157, 1984.

74. **Neuwelt, E. A. and Frenkel, E. P.,** Osmotic blood-brain barrier modification in areas of barrier opening and progression in brain regions distant to barrier opening, *Neurosurgery,* 15, 362, 1984.

75. **Harper, P. V., Beck, R., Charleston, D., and Lathrop, R. A.,** Optimization of a scanning method using Tc-99m, *Nucleonics,* 22, 50, 1964.

76. **Mosskin, M., von Holst, H., Ericson, K., and Noren, G.,** The blood tumour barrier in intracranial tumours studied with X-ray computed tomography and positron emission tomography using ^{68}Ga-EDTA, *Neuroradiology,* 28, 259, 1986.

77. **Lilja, A., Lundqvist, H., Olson, Y., Spännare, B., Gullberg, P., and Langström, B.,** Positron emission tomography and computed tomography in differential diagnosis between recurrent or residual glioma and treatment-induced brain lesions, *Acta Radiol.* 30, 121, 1989.

78. **Iannotti, F., Fisschi, C., Alfano, B., Picozzi, P., Mansi, L., Pozzilli, C., Funzo, A., Del Vecchio, G., Lenzi, G. L., Salvatore, M. et al.,** Simplified noninvasive PET measurement of blood-brain barrier permeability, *J. Comput. Assist. Tomogr.,* 11, 390, 1987.

79. **Kessler, R. M., Goble, J. C., Bird, J. H., Girton, M. E., Doppman, J. L., Rapoport, S. I., and Barranger, J. A.,** Measurement of blood-brain barrier permeability with positron emission tomography and [^{68}Ga]EDTA, *J. Cereb. Blood Flow Metab.,* 4, 323, 1984.

80. **Brooks, D. J., Beaney, R. P., Lammertsma, A. A., Marshall, J., Thomas, D. G. T., Turton, D. R., Horlock, P., Kensett, M. J., Luthra, S. K., and Jonas, T.,** The use of positron emission tomography for studying blood-brain barrier function in human subjects, in *The Blood-Brain Barrier in Health and Disease,* Suckling, A. J., Rumsby, M. G., and Bradbury, M. W. B., Eds., Ellis Horwood Health Science Series, Chichester, England, 1986, 175.

81. **Spector, R.,** Development of vitamin transport systems in choroid plexus and brain, *J. Neurochem.,* 33, 1317, 1979.

82. **Pratt, O. E. and Greenwood, J.,** Movement of vitamins across the blood-brain barrier, in *The Blood-Brain Barrier in Health and Disease,* Suckling, A. J., Rumsby, M. G., Bradbury, M. W. B., Eds., Ellis Horwood Health Science Series, Chichester, England, 1986, 87.

83. **Hankes, L. V., Coenen, H. H., Rota, E., Langen, K.-J., Herzog, H., Stöcklin, G., and Feinendegen, L. E.,** Blood brain barrier blocks transport of carboxyl-^{11}C-nicotinic acid into human brain, *Eur. J. Nucl. Med.,* 14, 53, 1988.

84. **Hankes, L. V., Coenen, H. H., Rota, E., Langen, K. J., Herzog, H., Wutz, W., Stöcklin, G., and Feinendegen, L. E.,** Effect of Huntington's disease on the transport of nicotinic acid or nicotinamide across the human blood brain barrier, in *Proc. 6th Int. Meet. Int. Study Group for Tryptophan Research,* 1989, 67.

85. **Herzog, H., Hankes, L. V., Coenen, H. H., Rota, E., Langen, K.-J., Stöcklin, G., and Feinendegen, L. E.,** Cerebral uptake of nicotinic acid and nicotinamide in the human brain measured by PET in vivo, *Eur. J. Nucl. Med.,* 15, 577, 1989.

86. **Pardridge, W. M., Connor, J. D., and Crawford, I. L.,** Permeability changes in the blood-brain barrier: causes and consequences, *CRC Crit. Rev. Toxicol.* 3, 159, 1975.

87. **Neuwelt, E. A. and Barnett, P. A.,** The blood-brain barrier disruption in the treatment of brain tumors. Animal studies, in *Implications of the Blood-Brain Barrier and Its Manipulation,* Vol. 2, Clinical Aspects, Neuwelt, E. A., Ed., Plenum, New York, 1989, 107.

88. **Neuwelt, E. A. and Dahlberg, S. A.,** The blood-brain barrier disruption in the treatment of brain tumors. Clinical studies, in *Implications of the Blood-Brain Barrier and Its Manipulation,* Vol. 2, *Clinical Aspects,* Neuwelt, E. A., Ed., Plenum, New York, 1989, 195.

89. **Greig, N. and Cavanagh, J.,** Quantitative aspects of reversible opening of the blood-brain barrier by pentylenetetrazol, *J. Neuropathol. Appl. Neurobiol.,* 8, 245, 1982.

90. **Agon, P., Kaufman, J. M., Goethals, P., Van Haver, D., and Bogaert, M. G.,** Study with positron emission tomography of the osmotic opening of the dog blood-brain barrier for quinidine and morphine, *J. Pharm. Pharmacol.,* 40, 539, 1988.

91. **Huang, S. C., Phelps, M. E., Carson, R. E. et al.,** Tomographic measurement of local cerebral blood flow in man with 0-15 water, *J. Cereb. Blood Flow Metab.,* 1 (Suppl. 1), S31, 1981.

92. **Herscovitch, P., Markham, J., and Raichle, M. E.,** Brain blood flow measurement with intravenous H_2 ^{15}O. 1. Theory and error analysis, *J. Nucl. Med.,* 24, 782, 1983.

93. **Frackowiak, R. S. J., Lenzi, G. L., Jones, T., and Heather, J. D.,** Quantitative measurement of regional cerebral blood flow and oxygen metabolism in man using ^{15}O and positron emission tomography: theory, procedure and normal values, *J. Comput. Assist. Tomogr.,* 4, 727, 1980.

94. **Paulson O. B., Hertz, M. M., Bolwig, T. G., and Lassen, N. A.,** Filtration and diffusion of water across the blood-brain barrier in man, *Microvasc. Res.* 13, 113, 1977.

95. **Herscovitch, P., Raichle, M. E., Kilbourn, M. R., and Welch, M. J.,** Positron emission tomographic measurement of cerebral blood flow and permeability-surface area product of water using [^{15}O]water and [^{11}C]butanol. *J. Cereb. Blood Flow Metab.,* 7, 527, 1987.

96. **Pozzilli, C., Bernardi, S., Mansi, L., Picozzi, P., Iannotti, F., Alfano, B., Bozzao, L., Lenzi, G. I., Salvatore, M., Conforti, P., and Fieschi, C.,** Quantitative assessment of blood-brain barrier permeability in multiple sclerosis using 68-Ga-EDTA and positron emission tomography, *J. Neurol. Neurosurg. Neuropsychiat.,* 51, 1058, 1988.

97. **Pardridge, W. M., Oldendorf, W. H., Cancilla, P., and Frank, H. J. L.,** Blood-brain barrier: interface between internal medicine and the brain. *Ann. Intern. Med.,* 105, 82, 1986.

98. **Suckling, A. J., Rumsby, M. G., and Bradbury, M. W. B.,** *The Blood-Brain Barrier in Health and Disease,* Ellis Horwood Health Science Series, Chichester, Engl., 1986.

99. **Neuwelt, E. A., Ed.,** *Implications of the Blood-Brain Barrier and Its Manipulation,* Vols. 1 and 2, Plenum Press, New York, 1989.

Chapter 8

KINETIC ANALYSIS OF GLUCOSE TRACER UPTAKE AND METABOLISM BY BRAIN *IN VIVO*

Albert Gjedde and Hiroto Kuwabara

TABLE OF CONTENTS

1. INTRODUCTION

Brain energy metabolism is almost exclusively fueled by glucose. The work performed by the brain consists of the transfer of ions across cellular and mitochondrial membranes. Since the extrusion of sodium ions from the intracellular space in a single minute equals the total extracellular content of sodium ions, the work load is considerable. However, while both the oxygen reduction to water and oxidative phosphorylation of ADP are driven by cytochrome oxidase activity in the mitochondrion, glucose consumption is driven by phosphofructokinase activity outside the mitochondrion. Although well-correlated in the steady-state, these activities easily become uncoupled in the absence of steady-state. Hence, there are excellent reasons to measure both glucose and oxygen metabolism in the human brain.

This chapter (1) examines the interaction between tracer glucose uptake and phosphorylation in brain; (2) derives the equations for the use of labeled deoxyglucose (DG), fluorodeoxyglucose (FDG), or glucose itself to calculate brain glucose phosphorylation; and (3) estimates the value of the "lumped constant" as a function of the glucose phosphorylation rate, the plasma glucose concentration, the maximal transport capacity of the cerebral capillary endothelium, and the plasma flow of the brain.

This discussion is essentially a synthesis of kinetic models of this kind used or modified by Sokoloff et al.,[34] Gjedde,[16] and Lassen and Gjedde.[25] The discussion is not intended as a practical guide or critical review; both have been accomplished in recent works, e.g., Gjedde[17] for a critical review and Redies and Gjedde[29] for a practical guide. The discussion will present and derive the formulas necessary to compute the results of an experiment, according to one or more of the published versions of methods of measurement of glucose metabolism in brain *in* or *ex vivo*, but it will not attempt to evaluate the permissible interpretations critically.

II. COMPARTMENTAL ANALYSIS

A. *IN VITRO* and *EX VIVO* ANALYSIS

The fundamental differences between radiotracer studies *in vitro* and *in vivo* have spawned a science of the formulation and solution of kinetic models of tracer metabolism *in vivo*, of which only tomographic images can provide experimental verification.

"Residue detection" is the registration of the quantity of tracer left in the target organ at a given moment during circulation, in contrast to the methods of "inflow-outflow" detection in which arterial and venous concentrations are measured. In principle, the tracer residue can be detected both *in* and *ex vivo*. *Ex vivo*, the method is autoradiography. *In vivo*, positron emission tomography is one of several methods of residue detection.

B. DEFINITION OF COMPARTMENTS AS TRACER STATES

The principal target organs of positron emission tomography are brain, heart, liver, and lungs, but additional organs may soon be accessible to this method of investigation. In reality, organs are collections of cells that trap or process the tracer in different ways according to the physical and chemical properties of the tracer and the biochemical and physiological functions of the cells. The tracer and its metabolites assume different states, each of which must be well defined (within our imperfect understanding of nature), but all of which change and interact as functions of time. Eventually one or more of these states may become stationary and reach a "steady-state", often very far from equilibrium. The non-equilibrium steady-state can be maintained only in a thermodynamically open system. When no energy is provided and no energy is expended, true equilibrium irreversibly replaces the steady-state. Conversely, if a tracer steady-state of non-equilibrium continues indefinitely, the system absorbs energy.

A kinetic model consists of compartments that represent the different states of the tracer and its metabolites. Strictly speaking, the compartments have no formal relation to the structure of the target organ, except to the extent that the anatomy dictates the processes in which the tracer or its metabolites participate. Therefore, the compartments arise from a specific theory of the biochemistry of an organ and refer to quantities of tracer or its metabolites that may or may not be confined to separate physical subdivisions of the organ. In this sense, Sheppard[33] defined compartments as quantities of a tracer or its derivatives the concentrations of which remain the same "everywhere", each quantity having a single state that may vary in time, but not in space. A quantity is the number of molecules in units of mol ($6.0225 \cdot 10^{23}$).

The conventional definition of a tracer is a labeled analogue of a native molecule that is present in such low quantity that the characteristics of the processes in which the tracer participates do not change measureably. The reason for this requirement is the departure from steady-state that causes the concentrations of native molecules to change. The departure of the native system from steady-state may interfere with the first-order relaxation of tracer compartments, discussed below. There are acceptable solutions to this problem, but they do not, strictly speaking, belong to the realm of tracer kinetics.

Rescigno and Beck[32] interpreted Sheppard's definition of a compartment to mean that a given state of the tracer forms a separate compartment if the tracer in this state obeys the expression,

$$\frac{dM}{dt} = J(t) - kM \tag{1}$$

where M is the quantity ("mass") of tracer that belongs to the compartment (i.e., has the relevant state), k is the relaxation ("rate") constant, and J(t) is the flux of tracer molecules into the compartment as a function of time.

The equation requires that the escape of tracer from the particular state (the "relaxation" of the state) is a first-order process. It depends on the process responsible for the relaxation whether this requirement is met. The processes define the interfaces between compartments. In most cases, the interfaces are diffusion barriers like cell membranes or chemical reactions involving transporter, receptor, or enzyme proteins. The processes can be spatially well-defined, when the diffusion barrier or protein is associated with a cell membrane, or they can be distributed throughout the cells or tissue. Compartments of the latter kind represent virtual rather than real subdivisions of the tissue. Thus, compartments are states of the tracer, separated by chemical or physical barriers.

The driving force of the loss of tracer from a compartment is the aqueous concentration of the tracer in the compartment, if the tracer is in aqueous solution. The concentration is the ratio between M and the volume of tissue water in which the tracer is dissolved. Often, *in vivo*, it is not possible to determine the real volume of distribution directly. In other cases, the tracer is not dissolved in the literal sense of this word and hence does not have a concentration, e.g., when bound to receptors. Thus, although aqueous concentration is the key to the analysis of diffusion and chemical reactions, it is not always the key to compartmental analysis.

C. PURPOSE OF COMPARTMENTAL ANALYSIS *IN* OR *EX VIVO*

Compartmental analysis of tracer uptake *in vivo* serves to measure the size of the different compartments of the tracer, i.e., the relative abundance of its different states, and the magnitude of the transfer coefficient or relaxation constant of each compartment. The definition of each state of the tracer, relevant to the compartmental analysis, must be known before the tracer exchanges between compartments can be quantified, i.e., prior to the construction of a useful model of the compartments.

The transfer coefficients of a series of compartmental barriers can be distinguished only when their magnitudes are not too different because more resistant barriers tend to obscure less resistant barriers. Thus, by analyzing the brain uptake of a tracer as a function of time, only a limited number of compartments and transfer coefficients can be identified.

According to Equation 1, the basis of the compartmental analysis is the identification of transfer coefficients or relaxation constants, k. The transfer coefficients operate on masses or quantities of tracer in a given state, M, and frequently (but not always) dissolved in volumes of water, V. These basic variables define clearances $(K = kV)$, concentrations $(C = M/V)$, and fluxes $(J = kM)$. Although M symbolizes a mass, it can also conveniently be thought of as a measured quantity of tracer in a sample of brain. In most cases, it is not possible to determine tracer concentrations per unit water volume.

The net exchange of tracer between two compartments of a closed two-compartment system must be

$$-\frac{dM_1}{dt} = \frac{dM_2}{dt} = k_1 M_1 - k_2 M_2 \tag{2}$$

where M_2 is the net gain of tracer in compartment 2 as a function of time, and k_1 and k_2 are the relaxation constants or transfer coefficients of the two compartments. The interpretation of k_1 and k_2 depends on the particular process in question and must be based on the definition of the compartments analyzed in the study. The symbols M_1 and M_2 represent the tracer or its metabolites that retain the label in the respective compartments. The symbols, therefore, need not refer to the same chemical species. If the tracer or its metabolites initially exist only as M_1, it is convenient to think of M_1 as the independent variable and hence of M_2 as the dependent variable.

Experiments in which only a single *ex vivo* measurement of the dependent variable is performed are referred to as "autoradiographic" because they emulate the method of quantitative autoradiography in which the radioactivity in brain is determined only once, i.e., at the termination of the experiment. Experiments during which the dependent variable is determined several times are often referred to as "dynamic" and readily lend themselves to the transient analysis discussed below. Of course, only one coefficient can be estimated by autoradiography because the solution to Equation 2 has no degrees of freedom when only one measurement of the dependent variable is available.

The solution to Equation 2 is

$$M_2(T) = e^{-k_2 T}\left[M_2(0) + k_1 \int_0^T M_1 e^{k_2 t}\,dt\right] \tag{3}$$

Equation 3 is a transcendental equation which can be solved for k_2 only by iteration and may have zero, one, or two solutions. Special cases include $M_1 = 0$ ("wash-out") and $M_2(0) = 0$ ("fill-in").

In transient analysis, the independent and dependent variables are measured as functions of time and the desired coefficients are estimated from the equation by regression analysis, using computerized optimization. The unlimited application of regression analysis can be naive because the analysis almost always yields a result, but sometimes with no regard to the biological meaning of the estimate. Regression analysis is only meaningful when the validity of the model is independently established, and it is usually logically impossible to decide the validity of the model and obtain the best estimates of the coefficients in the same session. Several solutions presented below have been derived within the last few years to make multilinear regression possible.

III. TRANSIENT ANALYSIS OF BLOOD-BRAIN TRANSFER OF GLUCOSE TRACERS

A. BLOOD-BRAIN BARRIER

In many organs of the body, the capillary endothelial cells have leaky junctions and hence do not impede the escape of small polar solutes from the circulation. The capillary endothelial cells in the brain have particularly tight junctions that do form such a barrier, as noted above. Thus, in brain, the concentration difference of newly administered polar tracer solutes between the two sides of the endothelium initially is so great that the endothelium may be the only significant barrier to the distribution of such tracers in brain. For these tracers, the brain has two kinetic compartments, the vascular space and an extravascular space, separated by a blood-brain barrier. The two states of the tracer cannot be detected separately *in vivo* because the interface between the compartments is inside the brain tissue.

It is the purpose of the two-compartment transient analysis to estimate the tracer's rate of unidirectional clearance (K_1) from the vascular compartment to the extravascular compartment, as an indirect measure of the permeability of the blood-brain barrier to the tracer, and the tracer's volume of partition between the circulation and the extravascular tissue space (V_e). The clearance K_1 is the product of the transfer coefficient k_1 and the volume of the vascular space (V_a). The partition volume V_e is the ratio between K_1 and k_2, i.e., $V_a k_1/k_2$. In addition to the permeability of the blood-brain barrier, the magnitude of K_1 depends on the blood flow to the region in question and the binding of the tracer to proteins or other components of the vascular space, as discussed below.

B. DERIVATION OF BLOOD-BRAIN TRANSFER COEFFICENTS

The quantity of tracer transported from blood to brain in a given period of time is a function of the endothelial permeability (P, in units of cm s^{-1} or nm s^{-1}), the area of endothelial surface available for transport (S, in cm^2 g^{-1} or ml^{-1}), and the concentration gradient across the barrier.

Endothelial permeability is measurable only as an index of "clearance" (K_1) from the circulation. The clearance does not reflect the permeability-surface area product directly because the tracer concentration falls from the arterial to the venous end of the capillary. For this reason, the vascular space in brain may appear not to be a true compartment. However, it can be shown that the tracer in the vascular space (i.e., in the "intravascular" state) obeys Equation 3 and hence functions kinetically as a compartment.

The net rate of tracer transfer across the endothelium equals the difference between the unidirectional rates of transfer. Thus, the net transport of tracer across the endothelium is

$$\Delta J(t) = \frac{dM(t)}{dt} = P_1 S \frac{\overline{C}_c(t)}{\alpha} - P_2 S \frac{M_e(t)}{V_d} \tag{4}$$

in which M(t) is the net gain of radioactivity in the brain tissue, $M_e(t)$ is the amount of exchangeable, intact tracer accumulated until the time t, P_1 is the apparent permeability* of the tracer in the endothelium when passing from blood to brain, S is the cerebral capillary surface area, P_2 is the permeability of the tracer when passing from brain to blood, and V_d is the physical volume of distribution of the tracer in brain. The ratio V_d/α is identical to the partition volume V_e. The assumption is made that the gain of radioactivity in the brain equals the net transfer of tracer across the capillary endothelium. Hence, in the steady-state,

---

* Note that the tracer must pass both membranes of the capillary endothelium before entering the brain. It can be shown that the permeability of the entire cell is half of the permeability of each membrane if the two membranes are identical.

the net transport from the intravascular space also equals the product of the blood flow and the arteriovenous deficit,

$$\Delta J(t) = F[C_a(t) - C_v(t)] \tag{5}$$

The tracer variables are shown as functions of time after introduction of the tracer into the circulation. Note the use of $\overline{C}_c(t)$, the average capillary concentration of the tracer.

The concentration $\overline{C}_c(t)$ varies both with $C_a(t)$ and with the concentration of isotope in the brain interstitial fluid (precursor pool), $C_e(t)$, which in turn varies with $C_a(t)$. Generally speaking, therefore, $\overline{C}_c(t)$ relative to $C_a(t)$ is lowest initially when $C_e(t) \sim 0$ and later rises to its steady-state level just below $C_a(t)$.

Let $C_c(s)$ be the capillary tracer concentration in a small volume of plasma at the time s after its entry into the capillary. There are two populations of isotope in the capillary, represented by $C_p(s)$ and $C_b(s)$. $C_p(s)$ represents tracer originating in the arterial plasma and $C_b(s)$ represents tracer having diffused from the plasma into brain and back to the capillary. For simplicity, below, we set $P = P_1S/\alpha$, and $p = P_2S/V_d$,

$$V_c \frac{dC_p(s)}{ds} = -P\, C_p(s) \tag{6}$$

$$V_c \frac{dC_b(s)}{ds} = -P\, C_b(s) + pM_e \tag{7}$$

in which V_c is the volume of capillary plasma in brain. M_e and C_a are assumed not to change appreciably during the time of one capillary transit. This time, $\bar{s}$ equals V_c/F. Since $C_c(s) = C_b(s) + C_p(s)$ and $C_c(0) = C_p(0) = C_a$, integration from $s = 0$ to $s = u$ yields,

$$C_c(u) = C_a e^{-Pu/\{F\bar{s}\}} + M_e \frac{p}{P}(1 - e^{-Pu/\{F\bar{s}\}}) \tag{8}$$

Since $C_v(t) = C_c(\bar{s})$,

$$C_v(t) = C_a(t)e^{-P/F} + M_e(t)\frac{p}{P}(1 - e^{-P/F}) \tag{9}$$

After insertion of this expression, Equation 5 changes to

$$\Delta J(t) = P\left(\frac{1 - e^{-P/F}}{P/F}\right) C_a(t) - p\left(\frac{1 - e^{-P/F}}{P/F}\right) M_e(t) \tag{10}$$

The term $1 - e^{-P/F}$ is the extraction fraction, $E(o)$, for unidirectional transfer of tracer into brain, and $p(1 - e^{-P/F})/P$ is the ratio m/V_d of Kety.[21,22] Thus, $m \neq E(o)$ when $P_2 \neq P_1$.

It is now possible to introduce the constants K_1^* and k_2. Equation 10 shows that these are simple exponential forms of the apparent permeabilities of the capillary endothelium to the movement of the tracer in question from blood to brain, and from brain to blood,

$$K_1 = P\left[\frac{1 - e^{-P/F}}{P/F}\right] = F(1 - e^{-P/F}) = E(o)F \tag{11}$$

---

* Note that throughout the chapter, the symbol K is used for "clearances" (except in the case of the affinity constants K_t and K_e), while the symbol k is reserved for proper rate constants. Therefore, K's are expressed in units of ml g^{-1} min^{-1} and k's are expressed in units of min^{-1}. "Clearances" include the net steady-state clearance $K [= E(\infty) \times F]$ and the unidirectional clearance $K_1 [= E(o) \times F]$.

and

$$k_2 = \frac{mF}{V_d} = p\left[\frac{1 - e^{-P/F}}{P/F}\right] = \frac{F}{V_e}(1 - e^{-P/F}) = \frac{E(o)F}{V_e} \tag{12}$$

From these definitions, it follows that $K_1/k_2 = P/p = V_e$. K_1 is the unidirectional clearance of the tracer (clearance at zero time), measured in units of ml g^{-1} min^{-1}. With these substitutions, Equation 10 reads,

$$\Delta J(t) = \frac{dM(t)}{dt} = K_1 C_a(t) - k_2 M_e(t) \tag{13}$$

This fundamental equation, developed by Kety,[21,22] and Johnson and Wilson[20] for inert substances, shows that the vascular compartment fulfills the definition of a compartment. From this equation, it is possible to obtain the equation derived by Bohr[4] for the relationship between clearance, perfusion (F), and the permeability-surface area produce (P_1S) of the endothelium. When $P_2 = P_1$ (simple diffusion), the partition volume V_e is defined as $P/p = V_d/\alpha$. In this case,

$$\frac{P}{F} = \frac{P_1S}{\alpha F} = -\ln\frac{\Delta C_v}{\Delta C_a} \tag{14}$$

where α is the solubility of the tracer in the sample of blood relative to that in tissue and plasma water. ΔC_v is the concentration difference between tissue and blood at the venous end of the capillary ($= C_v - M_e/V_e$), and ΔC_a is the concentration difference at the arterial end ($= C_a - M_e/V_e$). When the clearance is determined so soon after the tracer administration that tissue concentrations can be assumed to be negligible, and hence $C_v = \Delta C_v$ and $C_a = \Delta C_a$, Crone[7] completed the circle by showing that the equation changed to:

$$P = P_1S/\alpha = -F\ln(1 - E(o)) \tag{15}$$

where $E(o)$, the so-called "first-pass" extraction fraction, is the K_1/F ratio. Hence, the equation is valid only for extraction fractions measured in the presence of negligible tissue concentration of the tracer. Only tracers with relatively low permeability of the blood-brain barrier qualify because the extravascular accumulation of tracer must remain negligible during the passage of the tracer bolus. Hence, "first pass extraction" means negligible tissue concentration and no backflux, relative to the circulation. $E(t)$ is the fraction of the tracer extracted at the time t. Since it equals $1 - C_v(t)/C_a(t)$,

$$E(t) = (1 - e^{-P/F})\left(1 - \frac{M_e(t)}{V_e C_a(t)}\right) = E(o)\left(1 - \frac{M_e(t)}{V_e C_a(t)}\right) \tag{16}$$

Equation 16 confirms that $E(t) = E(o)$ for $M_e = 0$. With this equation, it is possible to calculate the relative decline of the extraction fraction with time,

$$\frac{E(t)}{E(o)} = 1 - \frac{M_e(T)}{V_e C_a(T)} \tag{17}$$

Integration of Equation 9 makes it possible to determine the actual degree of backflux of tracer from brain to blood. For this purpose, it is necessary to introduce the term $E(T)$ for

TABLE 1
Transport and Phosphorylation Ratios

Species	Tracer	τ	φ	Ref.
Rat	Deoxyglucose	1.35	0.37	8
	Deoxyglucose	1.75		16
	Deoxyglucose	1.39	0.38	5a
	Deoxyglucose	1.52		12
	Fluorodeoxyglucose	1.67	0.55	5a
	Fluorodeoxyglucose	1.65		12
Man	Fluorodeoxyglucose[a]	1.10	0.30	23
	Fluorodeoxyglucose[b]	1.10	0.33	24

[a] k_3 model.
[b] k_4 model.

the fraction of tracer extracted net *in* (rather than *at*) the time T. By definition, it equals $1 - [\int_0^T C_v(t)dt / \int_0^T C_a(t)dt]$ and hence also the ratio $M(T)/[F\int_0^T C_a(t)dt]$. By integration of Equation 9 and substitution,

$$\frac{E(T)}{E(o)} = 1 - E(T) \frac{F\int_0^T M_e(t)\,dt}{V_e M(T)} \tag{18}$$

which expresses the relative reduction of the "net" extraction with time. For low values of T, Equation 18 changes to:

$$\frac{E(T)}{E(o)} \cong 1 - E(T) \frac{FT}{2V_e} \tag{19}$$

which describes the decrease of E(T) with time after the tracer injection, and hence describes the degree of backflux in the period after the injection. If the "net" extraction is less than 50% in 1 min, if blood flow is less than 0.5 ml min^{-1}, and if the partition volume is greater than 0.5 ml, the extent of backflux is less that 25% in the 1 min.

The ratio, henceforth called τ, between the unidirectional blood-brain clearances of two glucose tracers can be computed from Equation 11,

$$\tau = \frac{K_1^*}{K_1} = \frac{1 - e^{P*/F}}{1 - e^{P/F}} \tag{20}$$

which is approximately a constant for relevant values of P* (for DG or FDG), P (for glucose), and F. Estimated values of τ from the literature are listed in Table 1.

C. BLOOD-BRAIN TRANSFER COEFFICIENTS FOR FACILITATED DIFFUSION

The measured permeabilities of the blood-brain barrier to hydrophilic substances are low, of the order of 1 to 10 nm s^{-1}. The permeabilities of the blood-brain barrier to lipophilic substances are 100 to 1000 nm s^{-1}. Lipophilic substances have low solubilities in plasma water relative to whole blood because they tend to be bound to plasma proteins. The consequent large values of α limit the clearance of lipophilic tracers, including oxygen. Most substrates of brain metabolism are hydrophilic, but facilitated diffusion allows the hydrophilic nutrients to cross the blood-brain barrier more readily than inert polar non-electrolytes.

The Michaelis-Menten formula for the association of a solute with a transporter will be derived in the chapter on radioligand binding (see below). Gjedde[14] and Reith et al.[30] derived, for all practical purposes, the following expression of the apparent permeability of the blood-brain barrier to true tracers subject to facilitated diffusion,

$$P = \frac{P_1 S}{\alpha} = \frac{T_{max}}{K_t \left(1 + \Sigma \dfrac{C_{a_j}}{K_{t_i}} \right)} \tag{21}$$

and, assuming symmetrical transport,

$$p = \frac{P_2 S}{V_d} = \frac{T_{max} \alpha}{V_d K_t \left(1 + \Sigma \dfrac{C_{e_i}}{V_d K_{t_i}} \right)} \tag{22}$$

in which K_t and K_{t_i} are the half-saturation concentrations in whole-blood or plasma for the endothelial transport of the tracer and its inhibitor(s), and C_{a_i} and C_{e_i}, are the concentration of the inhibitor(s) in whole-blood and tissue water. The potential inhibitors are native substrates of the transporter responsible for the facilitated diffusion. Division of Equation 21 by Equation 22 yields the P/p ratio V_e,

$$V_e = \frac{P}{p} = \frac{V_d}{\alpha} \left[\frac{1 + \Sigma \dfrac{C_{e_j}}{K_{t_i}}}{1 + \Sigma \dfrac{C_{a_j}}{K_{t_i}}} \right] \tag{23}$$

According to this equation, the partition volume of a tracer subject to facilitated diffusion depends not only on the solubility in blood and brain tissue but also on the concentrations of inhibitors. In fact, this ratio is independent of the specific tracer used (when V_d and α do not vary between the individual tracers). For example, Equation 23 predicts that the partition volume must be the same for all glucose analog tracers when the T_{max} values are identical for the individual analogs and the transport is symmetrical. A similar prediction can be made for large neutral amino acid analog tracers. For tracer hexoses for which $\alpha = 1$, and for which there is only a single native competitor, glucose, the equation reduces to

$$V_e = \frac{V_d K_t + M_e}{K_t + C_a} \tag{24}$$

It is important to note that the magnitude of V_e equals the distribution volume of the extravascular space (V_d) only in the absence of protein or other nonspecific binding of the tracer in whole-blood or plasma, i.e., only when $\alpha = 1$, and in the absence of facilitated diffusion of the tracer between blood and brain, i.e., only when $P_2 = P_1$.

D. DETERMINATION OF BLOOD-BRAIN TRANSFER COEFFICIENTS

Standard solution — The transfer coefficients K_1 and k_2 can be determined by fitting the solution of Equation 13 to simultaneously determined values of t, M(t), the quantity of tracer in brain, and $C_a(t)$, the concentration of tracer in whole blood or plasma. The quantity of tracer present in the vascular space, M_a, equals the product of the relevant concentration (C_a) and the vascular volume. The quantity of tracer present in the compartment corresponding to the extravascular space, M_e, can be calculated from Equation (3). If $M(T) = M_e(T) + V_a C_a(T)$, the complete operational equation giving the sum of M_a and M_e for $M_e(0) = 0$ is

$$M(T) = K_1 e^{-k_2 T} \int_0^T C_a(t) e^{k_2 t} dt + V_a C_a(T) \tag{25}$$

where V_a is the vascular volume of brain. The values of K_1 and k_2 can be converted to values of P and p by means of Equations 11 and 12 when the plasma flow of brain is known. Alternatively, the parameter k_2 can be replaced by K_1/V_e since $K_1/k_2 = P/p = V_e$. The three parameters, V_a, K_1, and k_2 or V_e, can be determined only by nonlinear regression analysis, on the basis of at least three separate observations. If the magnitude of one or more of the parameters is independently known, the number of required observations can be reduced proportionately. In experiments in which a single observation of M is available, Equation 13 can be solved directly for K_1,

$$K_1 = \frac{M(T) - V_a C_a(T)}{\int_0^T C_a(t) e^{k_2(T-t)} dt} \tag{26}$$

Equation 26 indicates that the calculation of K_1 from a single measurement of M requires both negligible concentration of the tracer in the brain (integral of $M(T)$ relative to V_e) and a known or negligible product of V_a and C_a. The product of V_a and C_a must be determined with a second tracer which remains in the vascular compartment. Negligible extravascular concentration (i.e., mininal "backflux") cannot be ruled out in a single measurement and often must be inferred from indirect evidence. The general condition is short periods of circulation,

$$\lim_{T \to 0} K_1 = \frac{M(T) - V_a C_a(T)}{\int_0^T C_a(t) dt} \tag{27}$$

Multivariate linear solution — Statistically, linear regression analysis may be preferable to nonlinear regression. By integration of Equation 13 for $M_e(0) = 0$,

$$M = V_a C_a(T) + K_1 \left(1 + \frac{V_a}{V_e}\right) \int_0^T C_a(t) dt - \left(\frac{K_1}{V_e}\right) \int_0^T M(t) dt \tag{28}$$

where $C_a(t)$ is the vascular concentration, measured in arterial whole-blood or a subcompartment of blood, and V_a is the vascular volume of distribution of the tracer in brain, approximately (but not exactly) equal to the plasma water volume in brain when the tracer is distributed and measured only in plasma water. This equation is of the form,

$$Y = p_1 X_1 + p_2 X_2 + p_3 X_3 \tag{29}$$

where p_i denotes the coefficient of the independent variable X_i. The variables include three independent variables (i.e., C_a and the two integrals) and one dependent variable (i.e., M). The coefficients of Equation 29 can be obtained by multiple linear regression in a single step.

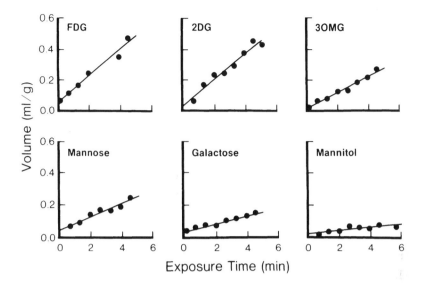

FIGURE 1. Initial slope method of determining blood-brain transfer coefficients in neonatal rats: apparent distribution volume in brain of five glucose analogs and mannitol (a plasma marker) as function of time. The five hexoses include fluorodeoxyglucose (FDG), deoxyglucose (2DG), and 3-O-methylglucose (3OMG). Abscissae: blood integrals, normalized against concentration, or weighted exposure time (minutes); ordinates: brain/blood radioactivity ratios. Each point is one observation. (Redrawn from Fuglsang, A., Lomholt, M., and Gjedde, A., *J. Neurochem.*, 46, 1417, 1986.)

Normalized solution — In many cases, both the vascular tracer content and the extent of backflux can be inferred by consideration of the normalized solution of the operational equation. Normalization is the process of dividing the radioactivity in brain by the arterial radioactivity concentrations of tracer obtained in individual experiments,

$$V(T) = \frac{M(T)}{C_a(T)} = V_a + K_1 \left[\left(1 + \frac{V_a}{V_e}\right) \frac{\int_0^T C_a(t)\,dt}{C_a(T)} - \left(\frac{1}{V_e}\right) \frac{\int_0^T M(t)\,dt}{C_a(T)} \right] \quad (30)$$

where the $M(T)/C_a(T)$ ratio is the actual volume of distribution. When V_e is sufficiently large, the equation shrinks to

$$V(T) = V_a + K_1\Theta(T) \quad (31)$$

where $\Theta(T)$ is the normalized integral $\int_0^T C_a(t)dt/C_a(T)$ that has unit of time and represents a modified time variable. $V(T)$ is then a linear function of $\Theta(T)$ when the $\int_0^T M(t)dt/V_e$ ratio remains negligible compared to $\int_0^T C_a(t)dt$. Equation 31 underlies the "slope-intercept" or "Patlak" plot (Gjedde,[15,16] Patlak et al.[27]). Continued linearity of this plot signifies absent backflux and a slope of K_1. Examples of this plot are shown in Figure 1. Values of K_1 for deoxyglucose in rat are listed in Table 2.

IV. TRANSIENT ANALYSIS OF GLUCOSE TRACER METABOLISM

A. NATIVE GLUCOSE METABOLISM

The net rate of native glucose transfer into brain in the steady state can be described in

TABLE 2
Deoxyglucose Transfer Coefficients For Rat Somatosensory Cortex

Year	K_1^* ml g^{-1} min^{-1}	k_2^* min^{-1}	k_3^* min^{-1}	V_e ml g^{-1}	K^* ml g^{-1} min^{-1}	Ref.
1977	0.19	0.21	0.049	0.93	0.037	34
1981	0.27	0.45	0.152	0.58	0.066	6
1982	0.35	1.09	0.090	0.32	0.027	16
1986	0.28	0.55	0.077	0.51	0.035	12
1990	0.23	0.14	0.030	1.64	0.040	26

two different ways. First, it can be calculated from the arteriovenous deficit and the plasma flow, as prescribed by the principle of Fick,[11]

$$J_{net} = FC_a - FC_v = KC_a \qquad (32)$$

in which J_{net} is the net rate of glucose uptake, F is the plasma flow, and K is the net clearance of glucose into brain in the steady state. In the steady state, the net rate of glucose phosphorylation must also equal the difference between the unidirectional rates of transfer across the cerebral capillary endothelium. In the case of appreciable net transfer, a difference persists between the mean capillary concentration $(\overline{C}_c)$ and the interstitial concentration of the mother substance (C_e),

$$J_{net} = k_3 C_e V_d = P\overline{C}_c - pM_e = K_1 C_a - k_2 M_e \qquad (33)$$

where k_3 is the net phosphorylation coefficient of glucose, C_e is the glucose concentration of the precursor pool in brain, P is the apparent permeability of glucose in the endothelium when passing from blood to brain, and p is the permeability of glucose when passing from brain to blood. Equations 32 and 33 combine to yield,

$$k_3 M_e = K_1 C_a - k_2 M_e = KC_a \qquad (34)$$

where $M_e = C_e V_d$. From this equation, it follows that K, the net clearance of native glucose in the steady state, is the following combination of the individual transfer coefficients,

$$K = k_3 \left[\frac{K_1}{k_2 + k_3} \right] = E_{net}F \qquad (35)$$

where E_{net} is the fraction of net extraction of glucose.

B. GLUCOSE TRACER METABOLISM

When a tracer of glucose is introduced in the blood stream, the movement of glucose tracer molecules can be described by two differential equations,

$$\frac{dM_m(t)}{dt} = k_3^* M_e(t) - k_4^* M_m'(t) - k_5^* M_m''(t) \qquad (36)$$

and

$$\frac{dM_e(t)}{dt} = \Delta J(t) - k_3^* M_e(t) + k_4^* M_m'(t) \qquad (37)$$

in which $M_m(t)$ symbolizes the amount of tracer metabolized in the time t (amount of product accumulated until time t), k_3^* is the rate of phosphorylation of the tracer, k_4^* is the rate of (hypothetical) dephosphorylation, $M_m'(t)$ is the quantity of tracer metabolites available for dephosphorylation (if any), and $M_m''(t)$ is the quantity of glucose tracer metabolites (also if any) that escape directly from brain at the rate of k_5^*. Note that the tracer transfer coefficients are identified by an asterisk, and that tracer variables are shown as functions of time. With the substitution of Equation 13, Equation 37 yields,

$$\frac{dM_e(t)}{dt} = K_1^* C_a(t) - (k_2^* + k_3^*) M_e(t) + k_4^* M_m'(t) \qquad (38)$$

For steady state (t→∞), Equation 13 yields the net steady-state rate of tracer transfer,

$$\Delta J(t) = K_1^* C_a(t) - k_2^* M_e(t) \qquad (39)$$

The solutions of these equations depends on the significance assigned *a priori* to the process described by k_4. The user must decide whether to deal with a "k_4 brain" or a "non-k_4-brain" when the k_3-model is used. This dependence has caused considerable confusion in the past, and the two solutions frequently have been mixed completely. Below, these two "brains" will be dealt with separately.

C. DEFINITION OF TRACER METABOLISM COEFFICIENTS AND LUMPED CONSTANT

1. k_3-Model Applied to "Non-k_4-Brain"

In analogy with Equation 34, in the steady state and in the absence of loss of labeled metabolites of the tracer,

$$\Delta J(\infty) = k_3^* M_e(\infty) = K^* C_a(\infty) \qquad (40)$$

the combination of Equations 39 and 40 yields,

$$K^* = k_3^* \left[\frac{K_1^*}{k_2^* + k_3^*} \right] = E(\infty)F \qquad (41)$$

Equation 43 states that the net steady state of uptake is equal to the rate of phosphorylation multiplied by the apparent steady-state volume of distribution of unphosphorylated glucose analog. Several solutions of the fundamental Equations 36 and 38 have been published and used in the past; they will be discussed below. By introducing a term for a "metabolism" index,

$$\beta = \frac{k_2^*}{k_2^* + k_3^*} = 1 - \frac{K^*}{K_1^*} \qquad (42)$$

it is possible to further simplify Equation 41:

$$K^* = E(\infty)F = k_3^* \beta V_e \qquad (43)$$

Metabolism Coefficients—The metabolism coefficient k_3^* equals the ratio between the metabolic flux through the enzymatic step in question and the quantity of tracer in the precursor state. Glucose is subject to phosphorylation by hexokinase as the first step in its metabolism. The transfer coefficient defined for this unidirectional process is

$$k_3^* = \frac{v^*}{M_e(t)} = \frac{\alpha V_{max}^*}{V_d K_m^* \left(1 + \Sigma \dfrac{C_e}{K_m}\right)} = \frac{\alpha V_{max}^* K_m}{K_m^*(V_d + M_e)} \tag{44}$$

where v^* is the metabolic flux and K_m^* and K_m are the Michaelis half-saturation concentrations for tracer and native inhibitor(s). Since glucose is the only native inhibitor under normal circumstances,

$$\frac{V_d \Sigma C_e}{K_m} = \frac{M_e}{K_m} \tag{45}$$

For native glucose itself, the coefficient of unidirectional phosphorylation can be similarly defined,

$$k_3 = \frac{v}{M_e} = \frac{\alpha V_{max}}{V_d K_m \left(1 + \Sigma \dfrac{C_e}{K_m}\right)} = \frac{\alpha V_{max}}{V_d + M_e} \tag{46}$$

where V_{max} is the maximal rate of glucose phosphorylation. When native glucose is the only inhibitor of tracer metabolism, it follows that the phosphorylation ratio between tracer and native glucose is

$$\varphi \equiv \frac{k_3^*}{k_3} = \frac{V_{max}^* K_m}{V_{max} K_m^*} \tag{47}$$

which is a constant, to be used below to determine the lumped constant. Values of φ are listed in Table 1.

Lumped constant — The "lumped constant", Λ, is the ratio between K^* and K. It converts the steady-state rate of tracer phosphorylation to rate of glucose phosphorylation. Although it is by definition an operational constant, it can be given several exact descriptions, of which the one due to Sokoloff et al.[34] is only one.

In the k_3^*-model, the lumped constant (Λ) is the ratio between the net clearances of DG or FDG and glucose. In the k_4^*-model, which has no net clearance in the steady state, we define Λ as the ratio between the unidirectional rate of phosphorylation of DG or FDG ($K^* = K_1^* k_3^*/(k_2^* + k_3^*)$) and the net clearance of glucose ($K = K_1 k_3/(k_2 + k_3)$). The unidirectional rate of phosphorylation reflects the fraction of DG or FDG which, once transported into tissue, is further phosphorylated. This rate is given by the same equation that defines the net clearance of the k_3^*-model.

Since $J_{net} = K^* C_a/\Lambda$, Equations 35 and 41 yield the following expression for what we choose to call the "true" lumped constant,

$$\Lambda = \frac{K^*}{K} = \frac{K_1^* k_3^* (k_2 + k_3)}{K_1 k_3 (k_2^* + k_3^*)} \tag{48}$$

Equation 48 is the basis for the determination of the lumped constant for DG or FDG as the ratio between the steady-state extraction fractions. Equation 48 shows that the lumped constant is the product of the ratio of phosphorylation rates and the ratio of the apparent distribution spaces in brain. The regional variation of the lumped constant can be assessed by the following rearrangement of Equation 48,

$$\Lambda = \tau - \beta (\tau - \varphi) \tag{49}$$

in which φ is the k_3^*/k_3 ratio between the hexokinase affinities for the tracer and native glucose, assumed to be approximately constant throughout brain, τ the K_1^*/K_1 and k_2^*/k_2 ratios of the blood-brain barrier transporter affinities for deoxyglucose and glucose, also assumed to be constant throughout brain, β is the nontrapping index, equal to the $k_2^*/(k_2^* + k_3^*)$ index that indicates the fraction of tracer that is not phosphorylated (trapped).

The constrained method of Kuwabara et al.[23,24] (see below) indicated how this relationship may be used to estimate the lumped constant regionally,

$$\Lambda = \frac{K^*}{K} = \varphi + (\tau - \varphi) \frac{K^*}{K_1^*} \tag{50}$$

where τ and φ are the "universal" transport and phosphorylation ratios between DG or FDG and glucose. The glucose metabolic rate is calculated as,

$$J_{net} = \frac{K^* C_a}{\Lambda} \tag{51}$$

The lumped constant determined by Sokoloff et al.[34] in the manner of Equation 48 averaged 0.483 in rat.

2. k_3-Model Applied to "k_4-Brain"

In the presence of loss of labeled metabolites by dephosphorylation,

$$\Delta J(t) - \frac{dM_e(t)}{dt} = k_3^* M_e(t) - k_4^* M_m'(t) = k_3'^*(t) M_e(t) = K'^*(t) C_a(t) \tag{52}$$

in which some coefficients are actual variables. If $\frac{dM_e(t)}{dt} \ll \Delta J(t)$ (quasi-steady-state), the combination of Equations 39 and 52 yields,

$$K'^*(t) = k_3'^*(t) \left[\frac{K_1^*}{k_2^* + k_3'^*(t)} \right] = E(t) F \tag{53}$$

By introducing the term for the "metabolism" index,

$$\beta'(t) = \frac{k_2^*}{k_2^* + k_3'^*(t)} = 1 - \frac{K'^*(t)}{K_1^*} \tag{54}$$

it is possible also to simplify Equation 53,

$$K'^*(t) = E(t) F = k_3'^*(t) \beta'(t) V_e \tag{55}$$

which is a steadily declining function in the presence of dephosphorylation.

Metabolism Coefficients — If the brain has the necessary phosphatase activity to dephosphorylate glucose and glucose tracers, this enzymatic step is assigned the following transfer coefficients for the tracer,

$$k_4^* = \frac{\alpha V_{max}'^*}{V_d K_m'^* \left(1 + \Sigma \dfrac{C_m'}{K_m'}\right)} = \frac{\alpha V_{max}'^* K_m'}{K_m'^* (V_d + M_m')} \tag{56}$$

and for native glucose,

$$k_4 = \frac{\alpha V_{max}'}{V_d K_m' \left(1 + \Sigma \dfrac{C_m'}{K_m'}\right)} = \frac{\alpha V_{max}'}{V_d + M_m'} \tag{57}$$

where M_m' is the brain content of glucose-6-phosphate, and V_{max}' and $V_{max}'^*$ are the maximal rates of dephosphorylation of native glucose and tracer 6-phosphate. Hence, in the presence of signigicant phosphatase activity, the apparent phosphorylation coefficient is

$$k_3'^*(t) = k_3^* \left(1 - \frac{k_4^*}{k_3^*} \frac{M_m'(t)}{M_e(t)}\right) \tag{58}$$

where $M_m'(t)$ is the glucose tracer-6-phosphate content. The k_4^*/k_3^* ratio is the inverse of a "metabolic trapping potential", akin to the "binding potential" of radioligands. This potential, p_M, is

$$p_M = \frac{k_3^*}{k_4^*} = \frac{V_{max}^* K_m'^* K_m (V_d + M_m')}{V_{max}'^* K_m^* K_m' (V_d + M_e)} \tag{59}$$

Thus, when $k_4^* = 0$, it follows that $k_3'^* = k_3^*$. The metabolic trapping potential is close to 10 (Table 3). When the brain, as a consequence of this finding, is assigned a significant value of k_4^*, the $k_3'^*/k_3^*$ ratio is a variable, the magnitude of which decreases as a function of time,

$$\frac{k_3'^*(t)}{k_3^*} = 1 - \frac{1}{p_M} \left(\frac{M_m'(t)}{M_e(t)}\right) \tag{60}$$

where $M_m'(t)$ is the brain content of tracer-6-phosphate. For native glucose, the net phosphorylation coefficient in the steady state must also take dephosphorylation into account. Sokoloff et al.[34] introduced the fraction Φ to indicate this effect, although they argued that it was likely to be negligible,

$$\Phi = \frac{k_3 M_e}{k_3 M_e + k_4 M_m'} \tag{61}$$

which is close to unity because $M_m' \ll M_e$ for glucose and because $k_4 \ll k_3$. In this equation, k_3 remains the net rate of native glucose phosphorylation in the steady state whether in the presence or absence of glucose-6-phosphate hydrolysis. When $k_4 = 0$, $\Phi = 1$, of course.

TABLE 3
Fluorodeoxyglucose Transfer Coefficients For Human Cerebral Cortex

Author(s)	K_1^*	K^*	k_2^*	k_3^*	k_4^*	p_M	$V(\infty)$	Ref.
	(ml g^{-1} min^{-1})		(min^{-1})			(ratio)	(ml g^{-1})	
k_3^*-Model								
Reivich et al. (1985)	0.105	0.035	0.148	0.074	—	—	—	31a
Lammertsma et al. (1987)	0.064	0.027	0.066	0.047	—	—	—	24a
Kuwabara and Gjedde (1990)								24
Constrained method	0.087	0.032	0.203	0.127	—	—	—	
Conventional method	0.068	0.026	0.092	0.054	—	—	—	
k_4^* Model: 45-min studies								
Friedland et al. (1983)	0.131	0.042	0.225	0.106	0.010	11	6.9	11a
Lammertsma et al. (1987)	0.069	0.035	0.109	0.113	0.020	6	4.2	24a
Kuwabara and Gjedde (1990)								24
Constrained method	0.085	0.035	0.197	0.149	0.018	8	4.0	
Conventional method	0.077	0.031	0.146	0.097	0.014	7	4.2	
k_4^*-model: longer studies ($\sim$3 h)								
Phelps et al. (1979)	0.102	0.033	0.130	0.062	0.0068	9	7.4	28
Hawkins et al. (1983)†								19a
Young adults	0.084	0.025	0.166	0.071	0.0051	14	7.6	
Old adults	0.100	0.024	0.253	0.079	0.0060	13	5.6	
Reivich et al. (1985)	0.095	0.034	0.125	0.069	0.0055	13	10.3	31a
Hawkins et al. (1986)	0.041	0.014	0.119	0.064	0.0075	8	3.3	19a

† Single dagger indicates no corrections for the radioactivity within vessels in tissue.

Equation 61 can be solved for k_3,

$$k_3 = k_4 \left[\frac{\Phi}{1 - \Phi} \right] \frac{M_m'}{M_e} \tag{62}$$

where M_m' is the tissue content of glucose-6-phosphate. Hence, both Φ and k_3 are true constants. The apparent phosphorylation ratio is

$$\varphi'(t) \equiv \frac{k_3'^*(t)}{k_3} = \frac{k_3^*}{k_3} \left(1 - \frac{1}{p_M} \left[\frac{M_m'(t)}{M_e(t)} \right] \right) = \frac{V_{max}^* K_m}{\Phi V_{max} K_m^*} \left(1 - \frac{1}{p_M} \left[\frac{M_m'(t)}{M_e(t)} \right] \right) \tag{63}$$

which shows that the phosphorylation ratio for the k_3-model is lower in a time-dependent fashion when the model is applied to a "k_4-brain" than when it is applied to a "k_3-brain."

Lumped constant—The "apparent" lumped constant, introduced by Sokoloff et al.,[34] was measured as the ratio,

$$\Lambda'(t) = \frac{E(t)}{E_{net}} = \frac{K'^*(t)}{K} = \frac{K_1^* k_3'^*(t) (k_2 + k_3)}{K_1 k_3 (k_2^* + k_3'^*(t))} = \tau - \beta'(t)[\tau - \varphi'(t)] \tag{64}$$

which shows that the "apparent" or measured lumped constant is a function of time if $k_4^* \neq 0$. When $k_4^* = 0$, $\Lambda'(t) = \Lambda$. Since the "true" and "apparent" lumped constants are related,

$$\frac{K'^*(t)}{\Lambda'(t)} = \frac{K}{\Lambda} \tag{65}$$

it does not, in theory, matter which of the two "brains", "k_3" or "k_4", underlies the determination of glucose metabolic rate, as long as they are used consistently. The glucose phosphorylation rate is therefore correctly calculated in both ways,

$$J_{net} = \frac{K'^*(t)C_a}{\Lambda'(t)} = \frac{K^*C_a}{\Lambda} \tag{66}$$

Reivich et al.[31] measured the apparent lumped constants for both DG and FDG in the human brain. They averaged 0.52 and 0.56.

e. k_4-Model Applied to "k_4-Brain"

When the k_4-model is applied to a hypothetical k_4-brain, the time-dependencies disappear. The definitions of k_3, k_3^*, k_4, and k_4^* are as indicated in Equations 44, 46, 56, and 57. The same is true of the lumped constant, defined as in Equation 50. However, the phosphorylation ratio is not numerically identical to the ratio defined in Equation 47.

$$\varphi \equiv \frac{k_3^*}{k_3} = \frac{V_{max}^* K_m}{\Phi V_{max} K_m^*} = \frac{\varphi'(t)}{\left(1 - \frac{1}{p_M} \left[\frac{M_m'(t)}{M_e(t)} \right] \right)} \tag{67}$$

The glucose phosphorylation rate is again calculated as

$$J_{net} = \frac{K^*C_a}{\Lambda} \tag{68}$$

Thus, the value of J_{net} is independent of the model when applied consistently. The identical results were confirmed experimentally by Kuwabara and Gjedde.[23] The two values, φ and $\varphi'(t)$, estimated by Kuwabara and Gjedde,[24] are listed in Table 1. The resulting lumped constants averaged 0.66 and 0.60.

D. DETERMINATION OF METABOLISM COEFFICIENTS

Attempts to determine the metabolism coefficients of glucose have been made with labeled deoxyglucose, labeled fluorodeoxyglucose, and labeled glucose itself. The two most wanted transfer coefficients are k_3 and the net clearance K.

1. Tracer Deoxyglucose

Although the first use of a labeled tracer of glucose was the use of labeled glucose itself by Gaitonde,[13] the first examination of the kinetics of glucose tracer metabolism was carried out with labeled deoxyglucose by Sokoloff et al.[34]

Sokoloff's solution ("k_3-Brain") — Sokoloff et al.[34] solved Equations 36 and 38 for tracer deoxyglucose by first assuming that $k_4^* = k_5^* = 0$, i.e., that brain allows no significant dephosphorylation of deoxyglucose, because of the absence or sequestration of the phosphatase required for this process, and no significant direct loss of tracer deoxyglucose metabolites, because of little continued metabolism to compounds that cross cellular membranes. Under these circumstances, the solution to Equation 38 is

$$M_e(T) = K_1^* e^{-(k_2^* + k_3^*)T} \int_0^T C_a(t) e^{(k_2^* + k_3^*)t} dt \tag{69}$$

With these assumptions, the total radioactivity in brain, $M(T)$, is the sum $M_m(T) + M_e(T) + M_a(T)$, where $M_a(T)$ is the quantity in the intravascular space. Since $M_m(T) = k_3^* \int_0^T M_e(t)\,dt$, this sum equals,

$$M(T) = V_a C_a(T) + K_1^* e^{-(k_2^*+k_3^*)T} \int_0^T C_a(t) e^{(k_2^*+k_3^*)t}\,dt$$

$$+ k_3^* K_1^* \int_0^T \left[e^{-(k_2^*+k_3^*)T'} \int_0^{T'} C_a(t) e^{(k_2^*+k_3^*)t}\,dt \right] dT' \tag{70}$$

which, with the exception of the vascular component, is identical to Equation 11 of Sokoloff et al.[34] Sokoloff et al.[34] used this equation to estimate the individual transfer coefficients by nonlinear regression analysis in rats, listed in Table 2. However, since the equation was intended for use with autoradiography in which only a single observation of $M(T)$ is available, the authors solved the equation for the net clearance of the tracer:

$$K^* = \frac{M(T) - M_e(T) - V_a C_a(T)}{\int_0^T C_a(t)\,dt - \dfrac{M_e(T)}{K_1^*}} \tag{71}$$

where K^* is the net clearance of the tracer. Except for the vascular and "lumped constant" terms, the equation is identical to Equation 30 of Sokoloff et al.[34] Since only a single observation of $M(T)$ is available, the additional terms of $M_e(T)$, $M_e(T)/K_1^*$, and $V_a C_a(T)$ must be determined separately or ignored. Sokoloff et al.[34] advocated (1) calculating $M_e(T)$ and $M_e(T)/K_1^*$ from the separate estimates of the transfer coefficients by means of Equation 69, (2) circulating the tracer for 45 min to minimize the magnitude of $M_e(T)$ relative to $M(T)$, and (3) ignoring the contribution of the vascular space. At first, Sokoloff et al.[34] did not attempt to assign any physiological meaning to the transfer coefficients. This position created a problem because it was later shown (see Cunningham's solution below) that the solution was overparameterized; in reality there may be only two independent parameters (transfer coefficients). The presence of three parameters in the original solution meant that the three estimates (K_1^*, k_2^*, and k_3^*) could be uncertain.

An additional problem will be dealt with below. Comparison of Equations 35 and 41 reveals that K^* and K are not identical when the transfer coefficients differ. The ratio between these is the "lumped constant" which must be known to determine K from K^*.

Cunningham's solution ("k_3-Brain") — Cunningham and Cremer[8] very significantly improved the understanding of the deoxyglucose method by showing that the solution has only two real parameters (K_1^* and k_3^*, or K_1^* and K^*) because the third transfer coefficient can be deduced from the data itself, provided the relationship between plasma and brain glucose is known. By introducing the ratios between K_1^* and K_1 (τ) and between k_3^* and k_3 (φ), the authors solved Equations 36 and 38 for deoxyglucose for the transfer coefficients of glucose:

$$\frac{dM_e(t)}{dt} = \tau K_1 C_a(t) - (\tau K_1 - \tau K + \varphi K)\frac{M_e(t)}{V_f} \tag{72}$$

where V_f is the distribution volume of native glucose, determined separately, where the transfer coefficients both refer to glucose; and

$$\frac{dM_m(t)}{dt} = \varphi K \frac{M_e(t)}{V_f} \tag{73}$$

$C_a(t)$ and $M_e(t)$ refer to deoxyglucose while the parameters refer to native glucose. Interestingly, these equations eliminate the problem of the lumped constant because the estimated transfer coefficients refer to glucose rather than the tracer deoxyglucose. The disadvantage is the need to know the ratios τ and φ and the native brain glucose content M_e. Note that the approach originally was believed to require separation of unphosphorylated and phosphorylated tracer. However, measurements of total activity actually suffice since $M_m(T)$ and $M_e(T)$ can be estimated separately from the transfer coefficients. A linear version is given in Equation 74:

$$\frac{M_m(T)}{M_e(T)} = K^* \frac{\int_0^T C_a(t)\,dt}{M_e(T)} - \frac{K^*}{K_1^*} \tag{74}$$

which describes a straight line of slope K^* and y-intercept $-K^*/K_1^*$. Cunningham's equation was used by Cremer et al.[6] to obtain estimates of transfer coefficients on the basis of known concentrations of glucose in plasma and brain. In Table 2, they have been compared with other previous and subsequent estimates. It is fairly apparent by this table that the estimates of K_1^* and K^* are considerably more robust (variation approximately twofold) than the estimates of k_2^* (variation approximately tenfold), k_3^* (variation approximately fivefold), and V_e (variation approximately eightfold). Since the estimates of Cremer et al.[6] and Fuglsang et al.[12] were both constrained by known values of the relationship between plasma and brain glucose, they are *a priori* more likely to be robust.

Normalized solution (''k_3-Brain'')—Gjedde[16] showed that Equation 71 rearranges to a simpler expression,

$$K^* = \frac{M(T) - \beta M_e(T) - V_a C_a(T)}{\int_0^T C_a(t)\,dt} \tag{75}$$

where $\beta = k_2^*/(k_2^* + k_3^*)$. This equation may be advantageous in cases in which the integral is known with certainty while the individual concentrations $C_a(t)$ are less well known. The equation was further simplified by dividing it by the final glucose tracer concentration in arterial plasma, $C_a(T)$. This division introduces the concept of ''apparent'' (virtual) volumes of distribution, which is experimentally useful because it normalizes values arising from the use of different absolute amounts of tracers in different animals. Following such a division, Equation 75 changes to

$$K^* = \frac{V(T) - \beta \dfrac{M_e(T)}{C_a(T)} - V_a}{\Theta(T)} \tag{76}$$

where $V(T) = M(T)/C_a(T)$, and $\Theta(T) = \int_0^T C_a(t)\,dt/C_a(T)$. The term $\Theta(T)$ is expressed in unit of time and represents the duration of the experiment when a constant unit $C_a(T)$ is used. Thus, after bolus injection, this term ''stretches'' the duration of the experiment. An example of the use of this equation is shown in Figure 2. It may well be asked what error is incurred by ignoring the amount of unphosphorylated deoxyglucose in the precursor pool at the time of termination of an experiment with the deoxyglucose method. The answer to

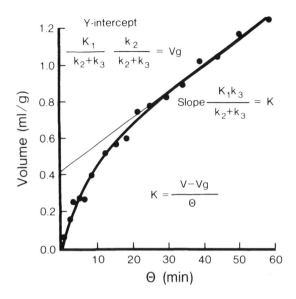

FIGURE 2. Normalized solution of deoxyglucose equation: apparent volume of distribution as function of the normalized blood concentration-time integra. Abscissa. $\Theta(T)$, equal to the time integral of blood concentration, normalized against concentration (minutes); ordinate: $V(T)$, equal brain radioactivity, normalized against blood concentration of [^{18}F]fluorodeoxyglucose. Fine curve shows steady-state asymptote of accumulation of [^{18}F]fluorodeoxyglucose; heavy line shows actual curve of accumulation. Formulas indicate steady-state slope and ordinate intercept in terms of the three-compartment model and the transfer constants K_1^*, k_2^*, and k_3^*. Normalized solution from Gjedde et al. 1985.[15a]

this question depends on the duration of the experiment, be it 45 min, 30 min or less. Equations 75 and 76 describe the magnitude of this error at steady state, i.e., $T \rightarrow \infty$,

$$K^* = \frac{V(\infty) - (\beta^2 V_e + V_a)}{\Theta(\infty)} = \frac{V(\infty) - (V_g + V_a)}{\Theta(\infty)} \qquad (77)$$

in which the equilibration delay (i.e., the separate error incurred by assumption of instant tracer equilibration between plasma and precursor pool) equals $1/(k_2^* + k_3^*)$, and the apparent precursor pool volume, $\beta^2 V_e$, equals $K_1^* k_2^*/(k_2^* + k_3^*)^2$. The error incurred by ignoring these terms can be significant, particularly when the glucose phosphorylation rate is low. In one study, Crane et al.[5a] equated K^* with $M_m(T)/[C_a(T)\Theta(T)]$ in experiments lasting no longer than 4 min. Not only could steady-state not have been reached at this early time, but an error of 25 to 50% (see later) may have been incurred by not correcting the plasma integral for the equilibration delay time. The latter criticism also applies to the original work of Gaitonde.[13]

2. Tracer Fluorodeoxyglucose

Phelps' solution ("k_4-Brain")—Both Phelps et al.[28] and Reivich et al.[31] published extensions of the DG method to the study of brain glucose metabolism in humans by positron tomography and labeled FDG. However, while Reivich et al. used the method of Sokoloff et al.[34] in its original form, including the transfer coefficients determined in rat, Phelps et al. judged the value of k_4^* to be sufficiently significant in human brain to merit its inclusion in the operational equation. Thus, the solution of Equations 36 and 38, for $k_5^* = 0$ and $M_m'(t) = M_m(t)$, was extended to:

$$M_e(T) = \frac{K_1^*(k_4^* - \alpha_1)}{\alpha_2 - \alpha_1} e^{-\alpha_1 T} \int_0^T C_a(t)e^{\alpha_1 t}dt + \frac{K_1^*(\alpha_2 - k_4^*)}{\alpha_2 - \alpha_1} e^{-\alpha_2 T} \int_0^T C_a(t)e^{\alpha_2 t}dt$$

and

$$M_m(T) = \frac{K_1^* k_3^*}{\alpha_2 - \alpha_1} e^{-\alpha_1 T} \int_0^T C_a(t)e^{\alpha_1 t}dt - \frac{K_1^* k_3^*}{\alpha_2 - \alpha_1} e^{-\alpha_2 T} \int_0^T C_a(t)e^{\alpha_2 t}dt \tag{78}$$

where α_1 and α_2 are the following expressions:

$$\alpha_1 = \frac{k_2^* + k_3^* + k_4^* - \sqrt{(k_2^* + k_3^* + k_4^*)^2 - 4k_2^* k_4^*}}{2}$$

and

$$\alpha_1 = \frac{k_2^* + k_3^* + k_4^* + \sqrt{(k_2^* + k_3^* + k_4^*)^2 - 4k_2^* k_4^*}}{2}$$

Equation 78 was used to determine all four transfer coefficients, by nonlinear regression analysis of multiple time points, for use in the operational equation. These estimates are listed in Table 3. The operational equation was different in principle from Equation 69 in that it was not simply the solution of equation 78 for K*. Rather, the desired clearance of the tracer was determined as the ratio between measured values and values calculated from the predetermined transfer coefficients,

$$K^* = K^{calc} \left[\frac{M(T) - M_e^{calc}(T)}{M_m^{calc}(T)} \right] \tag{79}$$

Again, the operational equation was designed to allow the determination of the deoxyglucose clearance from a single observation of M(T). Thus, the values of K^{calc}, $M_e^{calc}(T)$, and $M_m^{calc}(T)$ were calculated from the predetermined transfer coefficients. This approach ignored the additional information that can be gained from positron tomography studies by executing the studies dynamically, i.e., by registration of the time course of radioactivity accumulation in brain in individual studies. This approach was facilitated by the multiple linear solution discussed below.

Both assumptions, $k_4^* \neq 0$ and $M_m'(t) = M_m(t)$, i.e., that the entire pool of FDG metabolites is available for dephosphorylation to FDG, have been challenged fiercely by the advocates of the DG method in its original form (see Sokoloff's solution above), mostly on biochemical grounds, while the opposite assumptions, i.e., $k_4^* = 0$ and $M_m'(t) \neq M_m(t)$, likewise have been disputed in very heated arguments, mostly on mathematical, statistical grounds (see Hawkins' solution below). The discussion is somewhat sterile, however, since neither pair of assumptions affect the resulting estimates of glucose metabolic rate, if they are treated correctly (see constrained solution below) and since nature undoubtedly cannot be completely represented by three or four coefficients (Figure 3).

Multivariate linear solution ("k_4-Brain")—In order to obtain estimates of the transfer coefficients, we extended the linear, multivariate, least-squares regression method of

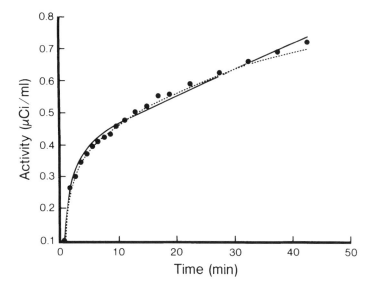

FIGURE 3. Comparison of k_3- and k_4-model solutions.[24] The abscissa is time of circulation, ordinate radioactivity registered in representative region of interest in brain. Full curve represents k_3-model solution; broken curve represents k_4-model solution. Estimates are listed in Table 3.

Blomqvist[1] to include the coefficient of dephosphorylation of FDG-6-phosphate (k_4^*), and the correction for the cerebral vascular volume in tissue (V_a) (Evans[10]):

$$M(T) = \alpha_1 C_a(T) + \alpha_2 \int_0^T C_a(t)\,dt + \alpha_3 \int_0^T \int_0^u C_a(t)\,dt\,du$$
$$+ \alpha_4 \int_0^T M(t)\,dt + \alpha_5 \int_0^T \int_0^u M(t)\,dt\,du \tag{80}$$

where $M(T)$ and $C_a(t)$ are radioactivities observed per gram of brain tissue and per milliliter of arterial plasma, respectively, and

$$\begin{cases} \alpha_1 = V_a \\ \alpha_2 = K_1^* + (k_2^* + k_3^* + k_4^*)V_a \\ \alpha_3 = K_1^*(k_3^* + k_4^*) + k_2^* k_4^* V_a \\ \alpha_4 = -(k_2^* + k_3^* + k_4^*) \\ \alpha_5 = -k_2^* k_4^* \end{cases} \tag{81}$$

and where K_1^* is the unidirectional clearance from blood to brain, k_2^* the fractional clearance from brain to blood, and k_3^* the phosphorylation coefficient. In principle, this solution is similar to Equation 28.

Constrained solution ("k_3-Brain") — From the derivation of Cunningham's solution above, it was clear that the three-parameter solution only has two real parameters when the relationship between plasma and brain glucose is known. Fundamentally, these two parameters are the T_{max} of hexose transfer across the blood-brain barrier, and the V_{max} of hexokinase in brain, when the half-saturation concentrations K_t and K_m of the Michaelis-Menten processes are known, and provided the facilitated diffusion of the blood-brain is symmetrical. Since brain glucose cannot be measured in the human brain *in vivo*, the application of the Cunningham solution to the human brain required a different approach by which the relationship was established by introducing the Michaelis half-saturation constants. From the preceding discussion, it is clear that (1)

$$\frac{K_1}{k_2} = \frac{K_1^*}{k_2^*} = V_e \tag{82}$$

(2) ($\alpha = 1$ in plasma)

$$V_e = \frac{V_d K_t + M_e}{K_t + C_a} \tag{83}$$

and (3)

$$J_{net} = \frac{K^* C_a}{\Lambda} = k_3 M_e \tag{84}$$

On this basis, Kuwabara et al.[23] redefined k_2^* and k_3^* in terms of K_1^* and K^*,

$$k_2^* = \frac{K_1^* + \mu K^*}{V_d} \quad \text{and} \quad k_3^* = \frac{K^*}{K_1^* - K^*} \left[\frac{K_1^* + \mu K^*}{V_d} \right] \tag{85}$$

where μ is ($\tau C_a / [\Lambda K_t]$), K_t is the Michaelis constant for glucose transport, V_d is the brain water volume, and C_a is the arterial plasma glucose concentration. For the k_3-model of FDG metabolism for which $\Lambda'(t) \cong \Lambda$ (Equation 64), the introduction of the constraint had the following consequence for the operational equation:

$$M(T) = (K_1^* - K^*) e^{-\gamma T} \int_0^T C_a(t) e^{\gamma} dt + K^* \int_0^T C_a(t) dt + V_a C_a(T) \tag{86}$$

where $M(T)$ is the [^{18}F] radioactivity recorded in a region-of-interest at time T, V_a is the cerebral vascular volume, and

$$\gamma = k_2^* + k_3^* = \frac{K_1^*}{K_1^* - K^*} \left[K_1^* + \left(\frac{\tau K^*}{\varphi + (\tau - \varphi)(K^*/K_1^*)} \right) \frac{C_a}{K_t} \right] \frac{1}{V_d} \tag{87}$$

From this equation, K^*, K_1^*, and V_a can be directly estimated by multiple linear least-squares regression. The values of K_1^* - k_3^*, calculated from these estimations, are listed in Table 3.

Constrained solution ("k_4-Brain")—For the k_4^* model, Equation 70 was exchanged by Equation 80 and yielded estimates of K_1^*, K^*, and k_4^* (Kuwabara and Gjedde[24]). Transfer coefficients obtained with the k_3^* and the k_4^*-models are compared in Table 3.

3. Tracer Glucose

Hawkins' solution ("k_5-Brain") — Hawkins et al.[18] proposed the use of tracer glucose itself rather than tracer DG to avoid having to know the value of the lumped constant and to avoid making the assumption that $k_4^* = 0$ since this assumption, in the workers' opinion, is invalid. The significant magnitude of k_4^* would not affect tracer glucose because the continued metabolism of tracer glucose-6-phosphate would tend to maintain $M_m'(T)$ at a low level and hence minimize the dephosphorylation reaction, if it exists. Of course, the use of glucose means that it is no longer valid to assume $k_5 = 0$ and hence the solution must include an additional term, $M_\tau(T) = k_5 \int_0^T M_m''(t)dt$,

$$K = \frac{M(T) - M_e(T) + M_\tau(T) - V_aC_a(T)}{\int_0^T C_a(t)\,dt - \dfrac{M_e(T)}{K_1}} \tag{88}$$

in which $M_\tau(T)$ is the amount of label lost from brain in the time T due to loss of metabolites. To reduce the loss of labeled metabolites from brain, the authors advocated (1) shortening the circulation of the tracer to 10 min and (2) using glucose labeled in the C1 or, preferably, in the C6 position where little labeled carbon dioxide is generated during the first passage through Krebs' cycle (Hawkins et al.[19]). However, the shortening of the circulation increased $M_e(T)$ relative to $M(T)$ and necessitated a correction for intact tracer glucose and delayed equilibration ($M_e(T)/K_1$). These corrections were accomplished by either measuring the ratio between native glucose in brain and plasma and using this ratio to determine $M_e(T)$ (Duckrow and Bryan[9]) or using predetermined values of K_1, k_2, and k_3 to calculate $M_e(T)$ according to Equation 34 (Hawkins et al.[19]).

Blomqvist's solution — Tracer glucose itself has been used in humans as well as in rats. In humans, using dynamic positron emission tomography, the quantity of unmetabolized tracer glucose can be determined kinetically but the problem of the egress of labeled metabolites remains. Blomqvist et al.[2] devised the following modification to Equation 88,

$$K = \frac{M(T)\left(1 + \dfrac{M_\tau(T)}{M(T)}\right) - M_e(T) - V_aC_a(T)}{\int_0^T C_a(t)\,dt - \dfrac{M_e(T)}{K_1}} \tag{89}$$

and represented the ratio $M_\tau(T)/M(T)$ by the formula,

$$g = \frac{M_\tau(T)}{M(T)} = aT(1 - e^{-bT}) \tag{90}$$

For a given value of T, g is a constant applied uniformly to all values of $M(T)$. It is based on the assumption that the egress of carbon dioxide is proportional to the accumulated radioactivity and hence to the glucose metabolic rate. The values used for normal brain were $a = 0.021$ and $b = 0.63$ for uniformly labeled glucose (Blomqvist et al.[3]).

Brøndsted's solution—In rats, Brøndsted and Gjedde[5] chose to solve Equation 88 slightly differently, using the normalized form

$$K = \left[\frac{V(T) - V_g}{\Theta(T)}\right] e^{\frac{K}{V_k}\Theta(t)} \tag{91}$$

where V_g is the product of β^2V_e for glucose, and V_k is an apparent pool of glucose metabolites. The constants were determined to be 0.21 ml g^{-1} (V_g), 0.100 ml g^{-1} min^{-1} (K), and 12.9 ml g^{-1} (V_k) for glucose labeled in the C6 position. The value of V_g was close to the value of $M_e(T)/C_a(T)$ determined for Hawkins' solution above. This usage implies that the rate constant K/V_k must change in proportion to any change of K. In actual practice, K/V_k is used as a constant to facilitate the solution of Equation 91. Figure 4 shows the application of Brøndsted's solution to the rat brain.

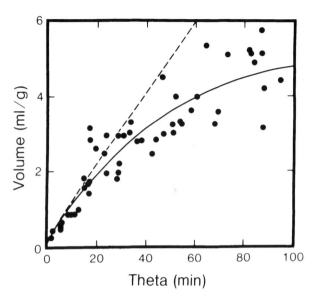

FIGURE 4. Use of radioactive glucose instead of radioactive deoxyglucose, as tracer of glucose metabolism: volume of distribution of glucose-derived radioactivity in rat parietal cortex vs. normalized time integral of labeled glucose concentration in arterial plasma. Abscissa: normalized integral (Θ) of concentration of labeled glucose in plasma (min); ordinate: apparent volume of distribution of radioactivity concentration in parietal cortex (ml/g), calculated as ratio between radioactivity in brain and plasma. (Redrawn from Brøndsted, H. E. and Gjedde, A., *Am. J. Physiol.*, 254, E443, 1988.)

V. GLOSSARY

α Solubility of glucose tracer in sample of blood or plasma (source fluid) relative to water (ratio)

β "Backflux" fraction of tracer subject to sequestration in tissue; the closer to unity β is, the higher the fraction of tracer that escapes sequestration; equal to $k_2/(k_2 + k_3)$ (ratio)

C_a Native glucose concentration in arterial sample of source fluid (plasma or blood) (mM)

$C_a(t)$ Glucose tracer concentration in arterial sample of source fluid (plasma or blood) (μCi ml^{-1})

C_v Native glucose concentration in venous sample (mM)

E_{net} Steady-state fraction of native glucose extracted from the circulation by brain (ratio)

$E(0)$ Fraction of glucose tracer extracted during "first pass", i.e., in the absence of backflux (ratio)

F Source fluid (plasma or blood) flow to brain sample (ml g^{-1} min^{-1})

φ Ratio between phosphorylation coefficients of glucose tracer and native glucose (ratio)

$\varphi'(t)$ Ratio between apparent phosphorylation coefficients of glucose tracer and native glucose in presence of significant dephosphorylation of phosphorylated glucose tracer (ratio)

Φ Ratio between net and unidirectional rates of phosphorylation of native glucose in presence of significant dephosphorylation of native glucose-6-phosphate (ratio)

J Native glucose flux (μmol g^{-1} min^{-1} or μmol ml^{-1} min^{-1})

J_{net} Net native glucose flux (μmol g^{-1} min^{-1} or μmol ml^{-1} min^{-1})

$\Delta J(t)$ Net glucose tracer flux across blood-brain barrier (μmol g^{-1} min^{-1} or μmol ml^{-1} min^{-1})

K   Coefficient of steady state, net volume transfer (clearance) of native glucose into brain tissue (ml g^{-1} min^{-1} or ml ml^{-1} min^{-1})

K*   Coefficient of steady state, net volume transfer (clearance) of glucose tracer into brain tissue (ml g^{-1} min^{-1} or ml ml^{-1} min^{-1})

K$'$*(t) Coefficient of apparent net volume transfer (clearance) of glucose tracer into brain tissue in presence of significant dephosphorylation of tracer-6-phosphate to exchangeable tracer (ml^{-1} min^{-1} or ml ml^{-1} min^{-1})

K_1 Coefficient of unidirectional volume transfer (clearance) of native glucose into brain tissue (ml g^{-1} min^{-1} or ml ml^{-1} min^{-1})

K_1^* Coefficient of unidirectional volume transfer (clearance) of glucose tracer into brain tissue (ml g^{-1} min^{-1} or ml ml^{-1} min^{-1})

k_2 Relative clearance of native glucose from brain (min^{-1})

k_2^* Relative clearance of glucose tracer from brain (min^{-1})

k_3 Relative rate of phosphorylation of native glucose in brain (min^{-1})

k_3^* Relative rate of phosphorylation of glucose tracer in brain (min^{-1})

$k_3'^*(t)$ Apparent, relative rate of net phosphorylation of glucose tracer in brain in presence of significant dephosphorylation (min^{-1})

k_4 Relative rate of dephosphorylation of glucose-6-phosphate (min^{-1})

k_4^* Relative rate of dephosphorylation of glucose tracer-6-phosphate in brain (min^{-1})

k_5^* Relative rate of direct loss of glucose tracer metabolites from brain (min^{-1})

K_m Michaelis half-saturation concentration for phosphorylation of native glucose by brain hexokinase (mM)

K_m^* Michaelis half-saturation concentration for phosphorylation of glucose tracer by brain hexokinase (mM)

K_m' Michaelis half-saturation concentration for dephosphorylation of native glucose-6-phosphate by brain phosphatase (mM)

$K_m'^*$ Michaelis half-saturation concentration for dephosphorylation of glucose tracer-6-phosphate by brain phosphatase (mM)

K_t Michaelis half-saturation concentration for blood-brain transfer of native glucose (mM)

K_t^* Michaelis half-saturation concentration for blood-brain transfer of glucose tracer (mM)

Λ Ratio between steady-state extraction fractions of glucose tracer and native glucose (ratio)

$\Lambda'(t)$ Time-dependent ratio between apparent steady-state extraction fractions of glucose tracer and native glucose in presence of significant dephosphorylation of phosphorylated glucose tracer (ratio)

M(t)   Total quantity of tracer and all labeled materials derived from tracer, in a sample of brain tissue (μCi g^{-1} or ml^{-1})

M_a Total quantity of native glucose in the vascular compartment of a sample of brain tissue (μmol g^{-1} or ml^{-1})

$M_a(t)$ Total quantity of tracer and all labeled materials derived from tracer, in the vascular compartment of a sample of brain tissue (μCi g^{-1} or ml^{-1})

M_e Quantity of native glucose in brain (μmol g^{-1} or ml^{-1})

$M_e(t)$ Quantity of exchangeable intact tracer in brain (μCi g^{-1} or ml^{-1})

M_m' Quantity of native glucose-6-phosphate in brain (μmol g^{-1} or ml^{-1})

$M_m(t)$ Quantity of glucose tracer metabolites in brain (μCi g^{-1} or ml^{-1})

$M'_m(t)$ Quantity of glucose tracer metabolites subject to dephosphorylation to exchangeable glucose tracer in brain (μCi g^{-1} or ml^{-1})

$M''_m(t)$ Quantity of glucose tracer metabolites subject to direct loss from brain (μCi g^{-1} or ml^{-1})

$M_\tau(t)$ Quantity of glucose tracer metabolites lost from brain (μCi g^{-1} or ml^{-1})

$p(_i)$ Coefficient of independent variable $X_{(i)}$

p_m "Metabolic trapping potential" of tracer in brain, equivalent to k_3^*/k_4^* (ratio)

t Time during experiment (min)

T Time of termination of experiment (min)

T_{max} Maximal transport capacity of blood-brain barrier glucose transporter (μmol g^{-1} or ml^{-1} min^{-1})

$\Theta(T)$ Normalized time-integral of $C_a(t)$ (min)

V Measured volume of distribution; ratio between M and C_a (ml g^{-1} or ml^{-1})

V_a Actual vascular volume of distribution of tracer in sample of brain; ratio between M_a and C_a (ml g^{-1} or ml^{-1})

V_d Volume of distribution of tracer in sample of brain relative to water; ratio between M_e and C (ml g^{-1} or ml^{-1})

V_e Partition volume of exchangeable tracer in sample of brain relative to source fluid (plasma or blood); ratio between $M_e(\infty)$ and $C_a(\infty)$ (ml g^{-1} or ml^{-1})

V_f Steady-state nonequilibrium volume of distribution of substrate of metabolism in sample of brain; equal to βV_e (ml g^{-1} or ml^{-1})

V_{max} Maximal phosphorylation capacity of brain hexokinase towards native glucose (μmol g^{-1} or ml^{-1} min^{-1})

V_{max}^* Maximal phosphorylation capacity of brain hexokinase towards glucose tracer (μmol g^{-1} or ml^{-1} min^{-1})

V'_{max} Maximal dephosphorylation capacity of brain phosphatase towards native glucose-6-phosphatase (μmol g^{-1} or ml^{-1} min^{-1})

V'^*_{max} Maximal dephosphorylation capacity of brain phosphatase towards glucose tracer-6-phosphatase (μmol g^{-1} or ml^{-1} min^{-1})

$X_{(i)}$ Independent variable i

Y Dependent variable

REFERENCES

1. **Blomqvist, G.,** On the construction of functional maps in positron emission tomography, *J. Cereb. Blood Flow Metab.* 4, 629, 1984.

2. **Blomqvist, G., Bergstrom, K., Bergstrom, M., Elvin, E., Eriksson, L., Garmelius, B., Lindberg, B., Lilja, A., Litton, J. E., Lundmark, L., Lundqvist, H., Malmberg, P., Mostrom, U., Nilsson, L., Stone-Elander, S., and Widén, L.,** Models for ^{11}C-glucose, in *The Metabolism of the Brain Studied With Positron Emission Tomography*, Greitz, T., Ingvar, D. H., and Widen, L., Eds., Raven Press, New York, 1985, 185.

3. **Blomqvist, G., Gjedde, A., Gutniak, M., Grill, V., Widén, L., Stone-Elander, S., and Hellstrand, E.,** Facilitated transport of glucose from blood to brain in man and the effect of moderate hypoglycemia on regional cerebral glucose utilization, *Eur. J. Neurosci.*, in press.

4. **Bohr, C.,** Über die spezifische Tätigkeit der Lungen bei der respiratorischen Gasaufnahme und ihr Verhalten zu der durch die Alveolarwand stattfindenden Gasdiffusion, *Skand. Arch. Physiol.*, 22, 221, 1909.

5. **Brøndsted, H. E., and Gjedde, A.,** Measuring brain glucose phosphorylation with labeled glucose, *Am. J. Physiol.*, 254, E443, 1988.

5a. **Crane, P. D., Pardridge, W. M., Braun, L. D., and Oldendorf, W. H.,** Kinetics of transport and phosphorylation of 2-fluoro-2-deoxy-D-glucose in rat brain, *J. Neurochem.*, 40, 160, 1983.

6. **Cremer, J. E., Ray, D. E., Sarna, G. S., and Cunningham, V. J.**, A study of the kinetic behavior of glucose based on simultaneous estimates of influx and phosphorylation in brain regions of rats in different physiological states, *Brain Res.*, 221, 331, 1981.

7. **Crone, C.**, The permeability of capillaries in various organs as determined by use of the "indicator diffusion" method, *Acta Physiol. Scand.*, 58, 292, 1963.

8. **Cunningham, V. J. and Cremer, J. E.**, A method for the simultaneous estimation of regional rates of glucose influx and phosphorylation in rat brain using radio-labelled 2-deoxy-glucose, *Brain Res.*, 221, 319, 1981.

9. **Duckrow, R. B. and Bryan, R. M., Jr.**, Regional cerebral glucose utilization during hyper-glycemia, *J. Neurochem.* 48, 989, 1987.

10. **Evans, A. C.**, A double integral form of the three-compartmental, four-rate-constant model for faster generation of parameter maps, *J. Cereb. Blood Flow Metab.*, 7 (Suppl.1), S453, 1987.

11. **Fick, A.**, *Sitzungsber. Phys. Med. Ges. Würzburg*, 36, 1870.

11a. **Friedland, R. P., Budinger, T. F., Vano, Y., Huesman, R. H., Knittel, B., Derenzo, S. E., Koss, B., and Ober, B. A.**, Regional cerebral metabolic alterations in Alzheimer-type dementia: kinetic studies with 18-fluorodeoxyglucose, *J. Cereb. Blood Flow Metab.*, 3(Suppl. 1), S510, 1983.

12. **Fuglsang, A., Lomholt, M., and Gjedde, A.**, Blood-brain transfer of glucose and glucose analogs in newborn rats, *J. Neurochem.*, 46, 1417, 1986.

13. **Gaitonde, M. K.**, Rate of utilization of glucose and "compartmentation" of α-oxoglutarate and glutamate in rat brain, *Biochem. J.*, 95, 803, 1965.

14. **Gjedde, A.**, Rapid steady-state analysis of blood-brain glucose transfer in rat, *Acta Physiol. Scand.*, 108, 331, 1980.

15. **Gjedde, A.**, High- and low-affinity transport of D-glucose from blood to brain, *J. Neurochem.*, 36, 1463, 1981.

15a. **Gjedde, A., Wienhard, K., Heiss, W.-D., Kloster, G., Diemer, N. H., Herholz, K., and Pawlik, G.**, Comparative regional analysis of 2-fluorodeoxyglucose and methylglucose uptake in brain of four stroke patients. With special reference to the regional estimation of the lumped constant, *J. Cereb. Blood Flow Metab.*, 5, 163, 1985.

16. **Gjedde, A.**, Calculation of glucose phosphorylation from brain uptake of glucose analogs *in vivo*: a re-examination, *Brain Res. Rev.*, 4, 237, 1982.

17. **Gjedde, A.**, Does deoxyglucose uptake in the brain reflect energy metabolism?, *Biochem. Pharmacol.*, 36, 1853, 1987.

18. **Hawkins, R. A., Hass, K., and Ransohoff, J.**, Measurement of regional brain glucose utiliztion *in vivo* using [2-14C]glucose, *Stroke*, 10, 690, 1979.

19. **Hawkins, R. A., Mans, À. M., Davis, D. W., Vina, J. R., and Hubbard, L. S.**, Cerebral glucose use measured [14C]glucose labeled in the 1, 2, or 6 position, *Am. J. Physiol.*, 248, C170, 1985.

19a. **Hawkins, R. A., Phelps, M. E., and Huang, S. C.**, Effects of temporal sampling, glucose metabolic rates, and disruptions of the blood-brain barrier on the FDG model with and without a vascular compartment: studies in human brain tumor with PET, *J. Cereb. Blood Flow Metab.*, 6, 170, 1986.

20. **Johnson, J. A. and Wilson, T. A.**, A model for capillary exchange, *Am. J. Physiol.* 210, 1299, 1966.

21. **Kety, S. S.**, The theory and application of the exchange of inert gas at the lungs and tissues, *Pharmacol. Rev.*, 3, 1, 1951.

22. **Kety, S. S.**, Measurements of local blood flow by the exchange of an inert diffusible substance, in *Methods in Medical Research*, Vol. 8, Year Book Medical Publishers, Chicago, 1960, 223.

23. **Kuwabara, H., Evans, A. C., and Gjedde, A.**, Michaelis-Menten constraints improved cerebral glucose metabolism and regional lumped constant measurements with [18F]fluorodeoxy-glucose. *J. Cerebr. Blood Flow Metab.*, 10, 180, 1990.

24. **Kuwabara, H. and Gjedde, A.**, Measurements of glucose phosphorylation with FDG, *J. Nucl. Med.*, in press.

24a. **Lammertsma, A. A., Brooks, D. J., Frackowiak, S. J., Beaney, R. P., Herold, S., Heather, J. D., Palmer, A. J., and Jones, T.**, Measurement of glucose utilization with [18F]2-fluoro-2-deoxy-D-gluclose: a comparison of different analytical methods, *J. Cereb. Blood Flow Metab.*, 7, 161, 1987.

25. **Lassen, N. A. and Gjedde, A.**, Kinetic analysis of the uptake of glucose and some of its analogs in the brain, using the single capillary model: comments on some points of controversy, in *Lecture Notes in Biomathematics 48: Tracer Kinetics and Physiologic Modeling*, Lambrecht, R. M., Rescigno, A., Eds., Springer-Verlag, New York, 1983, 387.

26. **Mori, K., Schmidt, K., Jay, T., Palombo, E., Nelson, T., Lucigniani, G., Pettigrew, K., Kennedy, C., and Sokoloff, L.**, Optimal duration of experimental period in measurement of local cerebral glucose utilization with the deoxyglucose method, *J. Neurochem.*, 54, 307, 1990.

27. **Patlak, C., Blasberg, R. G., and Fenstermacher, J. D.**, Graphical evaluation of blood-to-brain transfer constants from multiple-time uptake data, *J. Cereb. Blood Flow Metab.*, 3, 1, 1983.

28. **Phelps, M. E., Huang, S. C., Hoffman, E. J., Selin, C., Sokoloff, L., and Kuhl, D. E.,** Tomographic measurement of local cerebral glucose metabolic rate in humans with [^{18}F]2-fluoro-2-deoxy-D-glucose: validation of method, *Ann. Neurol.* 6, 371, 1979.

29. **Redies, C. and Gjedde, A.,** Double-label and conventional deoxyglucose methods: a practical guide for the user, *Cereb. Vasc. Brain Metab. Rev.,* 1, 319, 1989.

30. **Reith, J., Ermisch, A., Diemer, N. H., and Gjedde, A.,** Saturable retention of vasopressin by hippocampus vessels *in vivo,* associated with inhibition of blood-brain transfer of large neutral amino acids, *J. Neurochem.,* 49, 1471, 1987.

31. **Reivich, M., Kuhl, D., Wolf, A., Greenberg, J., Phelps, M., Ido, T., Casella, V., Fowler, J., Hoffman, E., Alavi, A., Som, P., and Sokoloff, L.** The [^{18}F] fluoro-deoxyglucose method for the measurement of local cerebral glucose utilization in man, *Circ. Res.,* 44, 127, 1979.

31a. **Reivich, M., Alavi, A., Wolf, A., Fowler, J., Russell, J., Arnett, C., MacGregor, R. R., Shiue, C. Y., Atkins, H., Anand, A., Dann, R., and Greenberg, J. H.,** Glucose metabolic rate kinetic model parameter determination in humans: the lumped constants and rate constants for [^{18}F]fluorodeoxyglucose and [^{11}C]deoxyglucose, *J. Cereb. Blood Flow Metab.,* 5, 179, 1985.

32. **Rescigno, A. and Beck, J. S.,** Compartments, in *Foundations of Mathematical Biology,* Vol. 2, Rosen, R., Ed., Academic Press, New York, 1972, 255.

33. **Sheppard, C. W.,** The theory of the study of transfers within a multi-compartmental system using isotopic tracers, *J. Appl. Phys.* 19, 70, 1948.

34. **Sokoloff, L., Reivich, M., Kennedy, C. des Rosiers, M. H., Patlak, C. S., Pettigrew, K. D., Sakurada, O., and Shinohara, M.,** The [^{14}C]deoxyglucose method for the measurement of local cerebral glucose utilization: theory, procedure, and normal values in the conscious and anesthetized albino rat, *J. Neurochem.,* 28, 897, 1977.

Chapter 9

¹⁵O STUDIES WITH PET

Ernst Meyer

TABLE OF CONTENTS

I. INTRODUCTION

Among the positron emitting radioisotopes frequently used in positron emission tomography (PET), ^{15}O is the one with the shortest physical half-life ($T_{1/2} = 2.035$ min) and the most energetic positron ($E_{max_{B^+}} = 1.72$ MeV). Both of these facts have their consequences for the use of ^{15}O in PET imaging. The short physical half-life makes the on-site production of ^{15}O by a medical cyclotron mandatory. On the other hand, it allows multiple ^{15}O studies to be performed on the same subject within a short time with limited radiation exposure and negligible interference from residual radioactivity. This possibility has given rise to some exciting applications of ^{15}O (see Section VII). The large positron energy of ^{15}O results in a resolution broadening effect of $\geqslant 1$ mm (FWHM) which is much larger than that for ^{18}F (~ 0.22 mm).[1] The major applications of ^{15}O in PET imaging, at present, are centered on the study of brain function.[2] The compounds used include $C^{15}O$ for the measurement of cerebral blood volume (CBV), $CO^{15}O$ and $H_2^{15}O$ for the measurement of cerebral blood flow (CBF), and $O^{15}O$ which, together with the former studies, allows estimation of the cerebral oxygen extraction fraction (OEF) and oxygen utilization (CMRO$_2$). This chapter describes current physiological models that form the basis for the estimation of these parameters together with selected examples. Some practical aspects of data analysis are discussed as well.

II. CEREBRAL BLOOD VOLUME

The measurement of regional cerebral blood volume (rCBV) was one of the first tasks successfully accomplished with PET. In this application, the circulating blood pool is visualized via the labeling of the intravascular agent carboxyhemoglobin. Although ^{11}C ($T_{1/2} = 20.3$ min) was initially used for that purpose,[3,4] the method has recently been validated for ^{15}O-carboxyhemoglobin.[5] The shorter physical half-life ($T_{1/2} = 2.035$ min) of this tracer reduces the radiation dose to the subject (see Table 3) and allows further PET studies, such as the ones required for the measurement of CMRO$_2$ by the sequential bolus approach of Mintun et al.,[6] to be performed rapidly without interference from residual radioactivity.

A. THEORY AND PROCEDURE

Labeling of carboxyhemoglobin is usually achieved *in vivo* by inhalation of a mixture of air and trace amounts of $C^{15}O$ in the form of one (bolus inhalation) or several deep breaths. *In vitro* labeling by withdrawing and reinjecting a sample of the subject's own blood may be performed, too, but this procedure has been mainly used with the longer-lived ^{11}CO label.[7] Following an equilibration period of approximately 2 min which is required to obtain a uniform labeling of the entire blood pool, characterized by equal ^{15}O activities in arterial and venous blood samples,[5] a PET scan, lasting typically 5 min, is performed. During this scan, one or several arterial or venous blood samples are withdrawn from a peripheral vessel and their ^{15}O activity, C_b(cps/g), assayed in a well counter that is calibrated with respect to the sensitivity of the tomograph. This "cross-calibration" is expressed by a factor, X, whose units are (Bq/cps) or equivalent. Based on simple tracer dilution, rCBV(ml/100 g) is then calculated as:

$$rCBV = \frac{\overline{C} \cdot 100}{\overline{C}_b \cdot X \cdot R \cdot d_t \cdot d_b} \tag{1}$$

where $\overline{C}$ (Bq/cm^3) is the average regional tissue concentration obtained from the equilibrium tomographic image, $\overline{C}_b$(cps/g) is the blood ^{15}O concentration averaged over all blood samples, R is the cerebral small-to-large-vessel hematocrit ratio, d_t is the density of cerebral tissue,

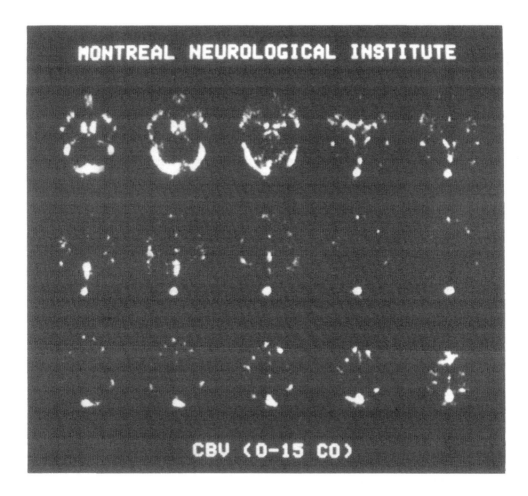

FIGURE 1.   CBV study on a normal volunteer. Data were collected for 5 min, following the inhalation of $C^{15}O$ and an equilibration period of 2 min. The 15 slices are 6.8 mm thick (FWHM) with an in-plane reconstructed spatial resolution of 6 mm (FWHM). The large blood vessels and vascular spaces of the brain such as the superior and inferior sagittal sinuses, the vein of Galen, the straight sinus, the internal carotid arteries, the anterior, middle and posterior cerebral arteries, the transverse and sigmoid sinuses as well as the jugular bulb can be distinguished. (Scanditronix PC-2048B).

and d_b is the density of blood. A value of R = 0.85 has been frequently used,[3,8,9] although more recent *in vivo* estimates of cerebral hematocrit suggest a smaller value of R ≈ 0.7.[10,11] For both d_b and d_t, a value of 1.05 g/cm^3 is normally used.[3,9] Exclusive labeling of red cells is assured by assaying plasma samples whose ^{15}O activities should be negligible.

B. APPLICATIONS

The *in vivo* measurement of rCBV by PET is useful mainly for two reasons. First, its assessment together with measurements of regional CBF and oxygen utilization has an important part in the full evaluation of cerebral hemodynamics in cerebrovascular disease.[12-17] Second, knowledge of rCBV is frequently required to correct other PET data for intravascular ^{15}O activity such as in the measurement of regional CMRO$_2$.[6,18-20] Figure 1 shows a $C^{15}O$ PET CBV study of a normal human brain which clearly illustrates the major vascular spaces.

III. CEREBRAL BLOOD FLOW

The measurement of regional CBF is, together with ^{18}F-deoxyglucose studies, one of the most frequently performed PET procedures. Frequently, PET CBF investigations are carried out as independent studies in their own right such as, for instance, during physiological blood flow activation studies (see Section VII). CBF determinations are, however, also part of the sequence of measurements required for the assessment of cerebral oxygen utilization by current PET techniques.

Most present PET CBF methods use $H_2^{15}O$ as a tracer and are based on the Kety-Schmidt single compartment model for diffusible inert tracers.[21,22]

A. THEORY

Consider a volume element of tissue, V_t, perfused with an inert freely diffusible tracer which enters the tissue compartment through the arterial blood stream at a flow F(ml/min) with concentration C_a(Bq or μCi/ml) (Figure 2A). Loss of tracer from the compartment occurs on the one hand through venous outflow with concentration C_v and, on the other hand, via radioactive decay (physical decay constant λ for ^{15}O is 0.34 min^{-1}). It is assumed that equilibrium between venous blood and tissue is achieved upon a single capillary transit, i.e., the tracer is perfectly diffusible and its delivery to tissue only limited by the blood flow. In this case, the change per unit time in the amount of tracer, Q, in the tissue compartment is given by:

$$\frac{dQ}{dt} = F(C_a - C_v) - \lambda Q \tag{2}$$

Equation 2 is based on the Fick principle expressing conservation of mass[21] which has been adapted to account for physical decay of the tracer. We define the tissue tracer concentration $C = Q/W_t$ (Bq/g) and the venous, or compartmental, tracer concentration $C_v = Q/V_d$(Bq/ml) where W_t is the weight of the tissue element (grams) and V_d is the tracer distribution volume (milliliters). With this notation, the equilibrium tissue-blood partition coefficient of the tracer, defined as $p = C/C_v$, is equal to the distribution volume per tissue weight ($p = V_d/W_t$). With these definitions, Equation 2 becomes:

$$\frac{dC}{dt} = fC_a - \left(\frac{f}{p} + \lambda\right) C \tag{3}$$

where $f = F/W_t$ is the tissue blood flow [ml/(min·g)]. With C and C_a being variable with time, integration of Equation 3 with the initial condition $C(0) = 0$ yields:

$$C(t) = fe^{-(f/p + \lambda)t} \int_0^t C_a(\tau)e^{(f/p + \lambda)\tau}d\tau \tag{4}$$

Here, C(t) and $C_a(t)$ represent instantaneous tissue and arterial blood tracer activities that have not been corrected for radioactive decay. By defining $\tilde{C}(t) = C(t)\,e^{\lambda t}$ and $\tilde{C}_a(\tau) = C_a(\tau)\,e^{\lambda\tau}$, we can rewrite Equation 4 in terms of the decay-corrected quantities $\tilde{C}(t)$ and $\tilde{C}_a(\tau)$:

$$\tilde{C}(t) = fe^{-k_2 t} \int_0^t \tilde{C}_a(\tau)e^{k_2\tau} \, d\tau \tag{5}$$

A

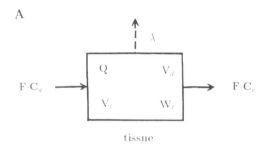

tissue

B

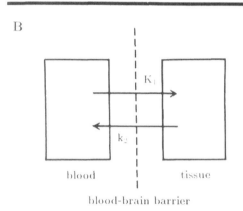

blood-brain barrier

FIGURE 2. Current compartmental models for the measurement of regional cerebral blood flow (rCBF) with ^{15}O-labeled compounds and PET. (A) One-compartment model for a freely diffusible inert tracer. Shown is a tissue element with volume V_t, weight W_t, tracer content Q, and tracer distribution volume V_d. The tissue element is homogeneously perfused with blood flow F at arterial and venous tracer concentrations C_a and C_v. Lambda (λ) is the physical decay constant of [15]O. (B) Two-compartment model representing intravascular (blood) and extravascular (tissue) spaces separated by the blood-brain barrier for a diffusion-limited tracer with a first pass capillary extraction fraction of E<1. The unidirectional clearance constant, $K_1 =$ E·f, represents the extraction rate of tracer from blood into tissue. The fractional clearance, or rate constant, $k_2 =$ E·f/p, describes the washout of tracer from tissue. Notice that f = F/W_t, and p = $K_1/k_2 = V_d/W_t$ is the tissue-blood partition coefficient.

with $k_2 = f/p$. Equation 5 may also be written in the form of a convolution integral:

$$\tilde{C}(t) = f \cdot \int_0^t \tilde{C}_a(\tau)e^{-k_2(t-\tau)}d\tau = f \cdot \tilde{C}_a(t) * e^{-k_2t} \qquad (6)$$

where the asterisk (*) denotes the convolution operation.

The above derivation is valid only for a perfectly diffusible inert tracer with a capillary first pass extraction fraction of E = 1. It has long been known, however, that water is not freely diffusible.[23] In his original derivation, Kety[21] has accounted for this possibility by introducing an equilibrium constant, m, with a value between 0 and 1, that represents the extent to which the tracer reaches diffusion equilibrium between blood and tissue during a single capillary transit. With this modification, Equation 2 is written in a more general form:

$$\frac{dQ}{dt} = m \cdot F(C_a - C_v) - \lambda Q \tag{7}$$

After division by W_t and assuming all parameters to be corrected for radioactive decay ($\lambda = 0$), we get:

$$\frac{d\tilde{C}}{dt} = K_1 \tilde{C}_a - k_2 \tilde{C} \tag{8}$$

where we have introduced the unidirectional clearance constant $K_1 = E \cdot f$ [ml/(min·g)] characterizing the extraction of tracer from the vascular space into tissue and the fractional clearance (or rate constant) $k_2 = E \cdot f/p$ [min^{-1}] describing the washout of tracer from the extravascular space (Figure 2B). It follows that $E = K_1/f$ and $K_1/k_2 = p$. It can be shown that m is equivalent to the first pass extraction fraction, E, for a diffusible tracer[21] which, according to the Renkin-Crone model,[24,25] is related to tissue blood flow f [ml/(min·g)], the capillary surface area S (cm^2/g) and its permeability P (cm/min) as follows:

$$E = 1 - e^{-\frac{PS}{f}} \tag{9}$$

Although this expression was derived for a single capillary under constant tracer infusion, Raichle and Larson[26] have shown that it also applies to an entire tissue element during a finite tracer injection with an arbitrary time course. In a more general formulation of Equation 6, f is therefore replaced by $E \cdot f$ and the new definition of $k_2 = E \cdot f/p$ is used (see also Equation 18).

Unfortunately, Equation 6 cannot be solved for f (or $E \cdot f$) explicitly. Also, current positron emission tomographs do not provide instantaneous tissue count rates, $\tilde{C}(t)$. Therefore, special scanning procedures and/or approximations have been developed in order to account for, or circumvent, these difficulties. This has led to the practical CBF methods in current use which are described in the following paragraphs.

B. STEADY-STATE METHOD

One effort to avoid the problems posed by Equation 6 has resulted in the so-called steady-state techniques which are still being used to measure cerebral blood flow and oxygen utilization.[27-30] When CO^{15}O is inhaled, the ^{15}O label is transferred from the CO$_2$ molecule to an H$_2$O molecule in the pulmonary alveolar capillaries under the action of carbonic anhydrase[31] which results in the *in vivo* labeling of circulating water. If a subject is inhaling CO^{15}O continuously at a steady rate, a dynamic equilibrium is attained within a few tracer half-lives (typically 10 min for CO^{15}O) at which point the continuous arrival of H$_2$^{15}O to the brain is balanced by its rate of washout in the blood stream and by radioactive decay. At this dynamic equilibrium or "steady state", the change in regional brain ^{15}O concentration per unit time is zero and regional CBF may be calculated from Equation 3 by setting dC/dt = 0 and solving for f:

$$rCBF_0 = \frac{\lambda \cdot 100}{\dfrac{\overline{C}_a \cdot (X \cdot d_b \cdot d_t)}{\overline{C}} - \dfrac{1}{p}} \tag{10}$$

The calculated steady state flow, $rCBF_0$, has units of [ml/(min·100 g)], and $\overline{C}_a$ and $\overline{C}$ are the constant arterial blood [cps/g] and tissue [Bq/cm^3] tracer concentrations measured during the PET scan at steady state, p the tissue-blood partition coefficient of water and d_b and d_t the densities of blood and tissue as previously defined. Again, $X(Bq/cps)$ is the cross-calibration factor that accounts for the different sensitivities of the well counter and the tomograph. (When $\overline{C}_a$ and $\overline{C}$ are measured in the units indicated in brackets, a new cross-calibration factor $X' = X \cdot d_b \cdot d_t$ may be defined).

In practice, the subject inhales a mixture of air and trace amounts of $CO^{15}O$ supplied from a medical cyclotron at a well-controlled steady flow of ^{15}O activity via a tight fitting face mask. Care is taken to vent any excess ^{15}O activity in order to avoid contamination of personnel and equipment. Upon reaching a stable brain ^{15}O level after an inhalation period of ~ 10 min, scanning is initiated and data accumulated for typically 5 min. During this time, several arterial blood samples are withdrawn, assayed in a well counter, decay-corrected to the time of withdrawal and averaged to yield an estimate of $\overline{C}_a$. Since the arterial ^{15}O activity fluctuates in synchrony with the subject's respiratory pattern, arterial samples should be withdrawn over at least one entire respiratory cycle (~ 15 s). If the CBF study is part of an oxygen utilization determination, plasma ^{15}O activity is assayed as well and used in the calculation of the oxygen extraction fraction (see below). CBF_0 is calculated pixel-by-pixel according to Equation 10 using $\lambda_{15_O} = 0.34$ min^{-1} and a tissue-blood equilibrium partition coefficient for water of $p = 0.9$ ml/g.[32]

The steady-state CBF and $CMRO_2$ method was adapted for routine clinical use by Frackowiak et al.[30] It has since been used in numerous clinical studies (see this volume).

1. Error Considerations

The limitations and sources of error inherent to the steady-state method in general and of its CBF component in particular have been discussed by several authors,[33-37] and only two important points are briefly discussed here. An error analysis of Equation 10 shows that the calculated value of $rCBF_0$ is extremely sensitive to errors in the estimate of the steady state arterial tracer concentration such that a 10% error in $\overline{C}_a$ translates into a 20 to 30% error in $rCBF_0$.[33] Similar errors result from an unstable tracer supply.[36] Multiple arterial sampling helps to reduce these errors and is highly recommended with this method.[36] Senda et al.[38] have expanded the steady-state method to include an explicit correction for variations in arterial tracer concentration while Jones et al.[39] have adapted it for the continuous intra-venous infusion of $H_2^{15}O$. This makes CBF measurements possible on patients who are unable to cooperate during tracer inhalation. Another error arises from the fact that the steady state tissue concentration, $\overline{C}$, is related to $rCBF_0$ in a strongly nonlinear fashion (Figure 3A). As a consequence, the blood flow for a heterogeneous volume element consisting of varying mixtures of gray and white matter tissue may be underestimated by as much as 20% compared to the true arithmetic mean flow (Figure 4).[40]

C. AUTORADIOGRAPHIC METHOD

A frequently used technique for the measurement of regional CBF with PET is an adaptation of the tissue autoradiographic blood flow method for animals. This method is based on the principles of inert gas exchange between blood and tissue developed by Kety and colleagues,[21,22,41] and it was successfully adapted and applied to laboratory animals.[42-44] Its theoretical basis is summarized in Equations 2 to 9. Briefly, a biologically inert, freely diffusible radioactive tracer is administered intravenously for a short period of time, t, followed by decapitation of the animal. The instantaneous regional tissue tracer concentration, $\tilde{C}(t)$, is determined by quantitative autoradiography and $\tilde{C}_a(\tau)$ by arterial blood sampling. With the equilibrium tissue-blood partition coefficient of the tracer, p, determined from separate studies, regional blood flow of the animal's brain may be calculated by numerically solving Equation 6 for f. Raichle[9] has proposed to adapt the tissue autoradiographic method for *in vivo* rCBF measurements with PET. This proposal was developed and implemented

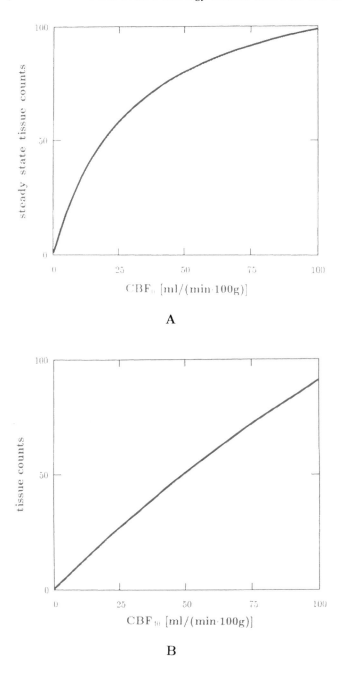

FIGURE 3.   Relationship between PET tissue counts (arbitrary units) and calculated cerebral blood flow. (A) Pronounced nonlinear behavior in the case of the steady-state method; (B) nearly linear relationship for the PET adaptation of the tissue autoradiographic method with a scan duration of 40 s.

by Herscovitch et al.[45] and Raichle et al.[46] who initially used intravenously administered ^{15}O-labeled water as a tracer.

Since current positron emission tomographs do not provide instantaneous tissue tracer concentrations or count rates, $\tilde{C}(t)$, but rather the sum of radioactive decay events occurring during a given scan interval, one has to integrate Equation 6 over the data accumulation

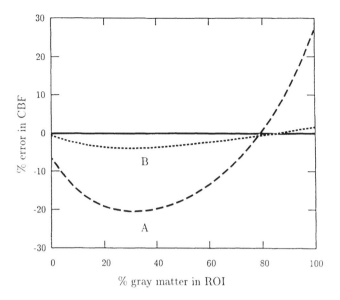

FIGURE 4. Percentage errors in calculated rCBF with respect to the true weighted flow for a region of interest (ROI) with varying gray-white matter mixtures. Tissue curves were generated using gray and white matter flows of 80 and 20 [ml/(min·100g)] and gray and white matter tissue partition coefficients of 0.98 and 0.82 (ml/g), respectively. CBF was calculated using an average brain-blood partition coefficient of p = 0.9 (ml/g). (A) Predominant underestimation reaching ~20% with the steady-state method. (B) Error within approximately ± 5% for the PET adaptation of the tissue autoradiographic method.

interval T_1 to T_2 in order to obtain the PET equivalent operational equation of the tissue autoradiographic method:

$$\tilde{C} = \int_{T_1}^{T_2} \tilde{C}(t)\,dt = \int_{T_1}^{T_2} f e^{-k_2 t} \int_0^t \tilde{C}_a(\tau) e^{k_2 \tau}\,d\tau\,dt \tag{11}$$

$\tilde{C}$ is the total number of radioactive events per unit weight of brain tissue detected by the tomograph during the scan and the arterial tracer concentration, $\tilde{C}_a(\tau)$, is available from arterial blood samples. Both the tissue and the blood data are corrected for radioactive decay of the tracer, and a value for the equilibrium tissue-blood partition coefficient of water has to be specified.[32] Equation 11 can then be numerically solved for f by one of several proposed methods[46] and a tomographic image of the regional distribution of CBF constructed (Figure 5). It is worth noticing that, contrary to the animal version of the tissue autoradiographic method (Equation 6) where multiple solutions for the calculation of f from a given tissue activity may exist,[47] this situation does not occur in the PET adaptation due to the use of the double integral form (Equation 11) and short scan intervals, T_1 to T_2, of typically 40 to 60 s.[45,48]

1. Error Considerations

The PET adaptation of the tissue autoradiographic method for the *in vivo* measurement of rCBF was validated for scan intervals ≤ 60 s by comparison with results from intracarotid injections of $H_2^{15}O$ in baboons using standard residue detection techniques.[46] Due to the known diffusion limitation of water, CBF is progressively underestimated for flow values

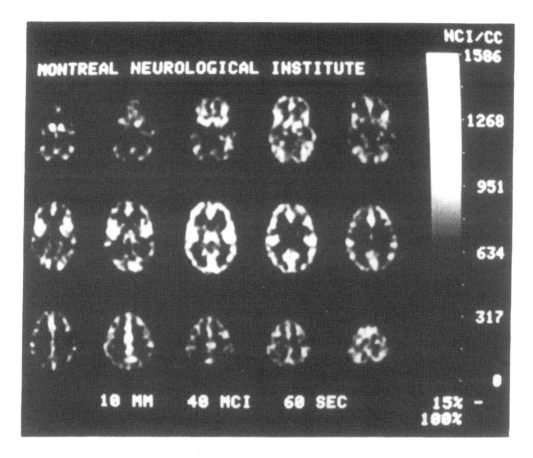

FIGURE 5. $H_2{}^{15}O$ CBF study on a normal volunteer. Images represent tissue counts which reflect regional CBF distribution. Data were collected over 60 s, following intravenous injection of 40 mCi of $H_2{}^{15}O$. The 15 slices are 6.8 mm thick (FWHM) with an in-plane reconstructed spatial resolution of 10 mm (FWHM). (Scanditronix PC-2048B).

above ~65 [ml/(min·100 g)] by this technique. This problem may be overcome by using ^{11}C-butanol[49] or, more recently, ^{15}O-butanol[50] which is a freely diffusible tracer (E = 1).

Another possible error arises from the fact that the calculation of absolute rCBF requires knowledge of the brain tissue-blood partition coefficient of water which may not be the same for normal and pathological tissue.[51] Contrary to the steady-state technique, there exists a nearly linear relationship between total tissue counts, $\tilde{C}$, and rCBF with the autoradiographic method, provided the scan intervals are kept short (≤60 s) (see Figure 3B).[45] Errors in calculated rCBF due to tissue heterogeneity, therefore, are smaller (~5%) with this method than with the steady-state technique (see Figure 4).

Since absolute rCBF calculation requires measurement of the arterial [15]O input function, which is commonly achieved by rapid manual or automatic sampling from the radial artery, appropriate corrections are required for (1) the systematic time difference between the tracer arrival time in the brain relative to the peripheral sampling site (delay correction) and (2) the difference in the degree of distortion in the shape of the input function due to dispersion of the tracer bolus in the blood vessels between the heart and the brain on the one hand and between the heart and the sampling site on the other hand (dispersion correction). The critical dependence of calculated CBF on these two corrections has been demonstrated.[45,52-55] In particular, a delay correction error as small as 2 s may translate into a 10% error in calculated rCBF.[45] Rather than matching the upslopes of the tissue and blood curves (Figure 6B),[55] the time delay correction is more accurately performed by fitting the decay-corrected total

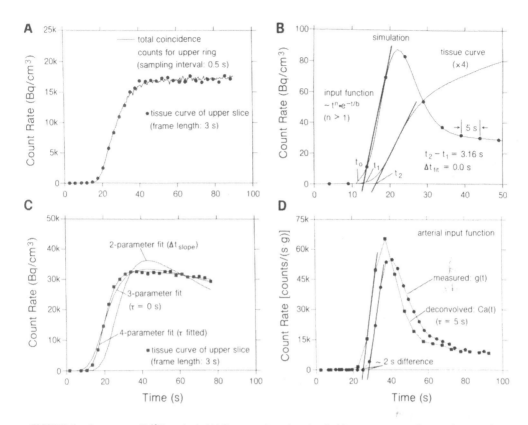

FIGURE 6. Intravenous H$_2$^{15}O method. (A) Decay-corrected total coincidence count rate of upper detector ring sampled at 0.5-s intervals (continuous line) superposed onto corresponding whole slice tissue curve (solid black circles) reconstructed with a frame length of 3 s; (B) slope method vs. global fit method for determination of tracer arrival time differences, Δt, between peripheral arterial blood and brain. This simulation demonstrates that the time intercepts of the visually drawn upslopes give a wrong Δt of $t_2 - t_1 = 3.16$ s as compared to the theoretically correct value of $\Delta t_{th} = 0$ s which was also found with the global fitting method[56] (Δt_{fit}). (C) Whole slice tissue time-activity curve (solid black squares) together with three fitted tissue curves. For the two-parameter fit, a fixed time delay determined by the slope method, Δt_{slope}, was used. For the three-parameter fit, the time delay, Δt, was fitted as a third parameter. In the four-parameter fit, the dispersion correction time constant, τ, was included as a fourth fitting parameter.[59] The significantly better three- and four-parameter fits are apparent. (D) Effect of dispersion correction on shape of input function. Shown are a measured, g(t), and the corresponding dispersion corrected or deconvolved input function, C$_a$(t), using a dispersion correction time constant of $\tau = 5$ s. (The measured data shown on this figure were acquired with Positome IIIp[141]). (From Meyer, E., *J. Nucl. Med.*, 30, 1069, 1989. With permission.)

slice coincidence count rate, $\tilde{X}$(t), to the right-hand side of Equatin 6 (Figure 6A) where, in addition to the arbitrary parameters A and B, the global time delay, Δt, applying to the entire slice is introduced as a fitting parameter:[56]

$$\tilde{X}(t) = \tilde{C}_a(t + \Delta t) * Ae^{-Bt} \tag{12}$$

The double integral form (Equation 11) is not necessary here since $\tilde{X}$(t) is usually sampled at least once a second. Iida et al.[56] have shown that time delays as long as ± 2 s exist between individual brain structures in addition to the global delay, Δt.

Dispersion correction has been treated by several authors in a number of models.[53,57,58] Iida et al.[53] have used a monoexponential dispersion function of the form:

$$d(t) = \frac{1}{\tau} \exp\left(\frac{-t}{\tau}\right) \tag{13}$$

where τ is the dispersion time constant. The peripherally measured dispersed input function, $\tilde{g}(t)$, and the true, dispersion corrected, input function, $\tilde{C}_a(t)$, are related by:

$$\tilde{g}(t) = \tilde{C}_a(t) * d(t) \qquad \text{or} \qquad \tilde{C}_a(t) = \tilde{g}(t) + \tau\frac{d\tilde{g}}{dt} \tag{14}$$

with the asterisk standing for the convolution operation (Figure 6D).

With these definitions, Equation 11 may be written in terms of the observed input function, $\tilde{g}(t)$, as:[53,55,56]

$$\tilde{C} = \int_{T_1}^{T_2} \tilde{C}_i(t)\,dt = \int_{T_1}^{T_2} [\tau pk_2\tilde{g}(t) + (1 - \tau k_2)\tilde{g}(t) * pk_2 e^{-k_2 t}]\,dt \tag{15}$$

As demonstrated by Meyer,[59] time delay and dispersion may be corrected for simultaneously by extending the global fitting approach of Iida et al.[56] to include the dispersion time constant, τ, as a fourth fitting parameter (Figure 6C). By reformulating Equation (15) in its single integral version and adapting it to the form of Equation (12) we get:

$$\tilde{X}(t) = \tau A\tilde{g}(t + \Delta t) + (1 - \tau B)\tilde{g}(t + \Delta t) * Ae^{-Bt} \tag{16}$$

This expression can be rearranged into a form suitable for nonlinear least-squares fitting by substituting u for $t + \Delta t$ and explicitly expanding the convolution operation:

$$\tilde{X}(t) = \tau A\tilde{g}(t + \Delta t) + (1 - \tau B)Ae^{-B(t + \Delta t)} \int_{\Delta t}^{t + \Delta t} \tilde{g}(u)e^{Bu}\,du \tag{17}$$

The value of $\tilde{g}(t + \Delta t)$ may be obtained by cubic spline interpolation of the discretely measured peripheral input function. Dispersion time constants of $\tau \sim 5$ s are typically found with manual radial artery sampling.[53,59] Neglecting to account for dispersion leads to a significant overestimation of calculated rCBF, particularly for short scans (typically $+ 15\%$ error for a 40-s scan).[53]

It has been repeatedly noted that gradually extending the scan interval results in a progressive underestimation of calculated rCBF with the autoradiographic method in its present form.[46,60] The contribution to this observation of factors such as tissue heterogeneity,[61] arterial tracer dispersion,[53] and the use of a single compartment model has been investigated.[62]

Errors in calculated rCBF may also occur if the intravascular ^{15}O activity is neglected as shown by Kanno et al.[63] and Koeppe et al.[54] The effect of arterial ^{15}O activity is particularly important in the early phase after $H_2{}^{15}$O injection.[54]

Most of the correction factors discussed above may be taken into account by expanding Equation 11 to include the partition coefficient (p), the extraction fraction (E), the tracer delay (Δt) and the fractions of the total volume of the brain element occupied by arterial (V_a) and venous (V_v) blood and by tissue (V_t):

$$\tilde{C} = \int_{T_1}^{T_2} [V_t + V_v/p)Ef\tilde{C}_a(t + \Delta t) * e^{-(Ef/p)t}$$

$$+ V_v(1 - E)\tilde{C}_a(t + \Delta t) + V_a\tilde{C}_a(t + \Delta t)] dt \qquad (18)$$

$\tilde{C}$ is the total number of counts recorded from the brain element and $V_a + V_v + V_t = 1$.

D. DYNAMIC METHODS

Methods described so far derive functional information such as rCBF from a single tomographic tissue count image and from the appropriate blood data. Another approach is to acquire a sequence of images (dynamic scan) and to calculate the desired parameters by means of a regression, or curve fitting, procedure that matches the measured tissue time-activity curve with that predicted by a particular model. As an example, Equation 11 may be used to estimate f and p by nonlinear least squares fitting from a series of scans with different start and end times T_1 and T_2. However, statistical restrictions together with the limited timing resolution of positron emission tomographs and the computationally intensive task of serial image reconstruction and pixel-by-pixel curve fitting make the generation of functional images by this approach a tedious task. It has, nevertheless, been used to estimate CBF, partition coefficient and tracer delay for extended brain regions of interest.[64-66]

A computationally more efficient scheme of parameter estimation, the so-called integrated projection technique, was proposed by Tsui and Budinger[67] and implemented by Carson[68] and Huang[69,70] for the pixel-by-pixel determination of rCBF and partition coefficient. This method takes advantage of the linearity of the tracer compartment model with regard to f and f/p (Equation 3) and the image reconstruction algorithm. The two tissue time integrals appearing in their operational equations for CBF and p, therefore, are calculated on the projection data rather than on individually reconstructed dynamic images. These modified projection sums are then used to reconstruct the two "integral" images required for the calculation of CBF and p.

The weighted integration method described by Alpert,[71] and refined by Carson,[68] is a more general version of the integrated projection technique. With this method, weighted images, instead of the ordinary integral images, are generated by means of weighting functions that may be selected such that the variance of the desired parameter estimate, e.g., rCBF, is minimized. This method has been successfully used for the simultaneous determination of pixel-by-pixel images of regional CBF and distribution volume of water.[64,68-70]

With the growing interest in dynamic PET methods, a number of papers have been published dealing with the analysis of dynamic PET data.[72-75] Also, distributed parameter models, as opposed to compartmental models, have recently been proposed in an attempt to provide a more adequate description of experimental dynamic data.[76,77]

IV. CEREBRAL OXYGEN UTILIZATION

The *in vivo* determination of regional cerebral oxygen utilization, or cerebral metabolic rate of O_2 ($CMRO_2$), was pioneered by Ter-Pogossian and associates[78,79] who used the intracarotid bolus injection of ^{15}O-labeled water and ^{15}O-oxyhemoglobin together with conventional external monitoring techniques to measure CBF and the first pass extraction fraction, E, of oxygen, respectively. $CMRO_2$ was then calculated as the product of CBF, the extraction fraction, E, and the arterial oxygen content, $[O_2]_a$ (see Equation 33). These results were validated by comparison with hemispheric $CMRO_2$ values obtained from direct measurements of arteriovenous oxygen content differences and assuming E and the net extraction fraction at steady state, $E_0 = (C_a^{O_2}-C_v^{O_2})/C_a^{O_2}$, to be equal. $C_a^{O_2}$ and $C_v^{O_2}$ are the arterial and venous oxygen contents at steady state.

Current PET techniques for the measurement of $CMRO_2$, which include the continuous inhalation or steady-state method validated by Frackowiak and associates[30] and the sequential bolus approach of Mintun and colleagues,[6] are based on the same principles. With these techniques, $CMRO_2$ is calculated from a series of three PET studies which includes measurements of CBF, CBV and oxygen extraction fraction, OEF. Such a multitracer study may last from 30 min (bolus method) to over an hour (steady-state method) during which the subject's head has to be comfortably but effectively restrained. Faster methods to measure $CMRO_2$ have been proposed[80-82] and are under continuous investigation.[83]

A. OXYGEN EXTRACTION FRACTION

Unlike in the original experiment by Ter-Pogossian et al.[78] where ^{15}O-labeled oxyhemoglobin was intravenously administered to measure the extraction fraction of oxygen (OEF), in the PET adaptation, *in vivo* labeling is achieved by administering molecular oxygen, $O^{15}O$, by inhalation.

1. Steady-State Method

The model underlying the steady-state approach is depicted in Figure 7A.[29,30] Upon inhalation, $O^{15}O$ initially appears in the blood stream in the form of ^{15}O-oxyhemoglobin. The amount of physically dissolved labeled O_2 is small and assumed to be negligible. Upon capillary transit, a certain fraction, E, of $O^{15}O$ is extracted from arterial blood into brain tissue and immediately metabolized to water, forming ^{15}O labeled water of metabolism. This model assumes that the tissue concentration of molecular oxygen is negligible. As $H_2^{15}O$ is washed out from the tissue, it reappears in the arterial blood in the form of recirculating water and is eventually redeposited in tissue. In the tissue compartment, the ^{15}O label therefore only appears in the form of $H_2^{15}O$ (water of metabolism plus deposited recirculating water) whereas in blood both $H_2^{15}O$ ($C_a^{H_2O}(t)$) and hemoglobin-bound $O^{15}O$ ($C_a^{O_2}(t)$) exist. In analogy to the notation used for the CBF model (see Equation 3), the change in tissue tracer concentration, C(t), per unit time is given by:

$$\frac{dC}{dt} = f \cdot C_a^{H_2O}(t) + f \cdot E \cdot C_a^{O_2}(t) - \left(\frac{f}{p} + \lambda\right) \cdot C(t) \qquad (19)$$

where f, p and λ are defined as before.

The quantity of interest is the oxygen extraction fraction, E, which may be derived on the basis of Equation 19.

In practice, $O^{15}O$ is inhaled in a controlled continuous fashion, as in the case of $CO^{15}O$, and a dynamic equilibrium, or steady state, eventually develops where the tissue ^{15}O concentration is constant over time, i.e., dC/dt = 0. Under this condition, Equation 19 is easily solved for E:

$$E = \frac{\hat{C} \cdot \left(\frac{f}{p} + \lambda\right) - f \cdot \hat{C}_a^{H_2O}}{f \cdot \hat{C}_a^{O_2}} \qquad (20)$$

Here, $\hat{C}$, $\hat{C}_a^{H_2O}$, and $\hat{C}_a^{O_2}$ are the constant steady state tissue and whole blood tracer concentrations. By replacing f with the "steady state" expression previously derived for $rCBF_0$ (Equation 10) and rearranging, we get:

A

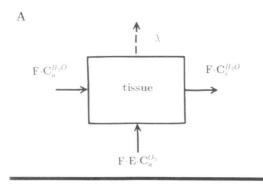

B

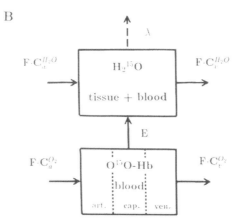

FIGURE 7. Current compartmental models for the measurement of regional cerebral oxygen extraction fraction (rOEF = E) and oxygen metabolic rate ($CMRO_2$) with ^{15}O-labeled compounds and PET. (A) one-compartment model representing the tissue water space only with inputs from ^{15}O-labeled molecular oxygen that is extracted from arterial blood and rapidly metabolized to $H_2^{15}O$ and from recirculating ^{15}O-labeled water delivered to tissue by the blood flow F;[29,30] (B) two-compartment model representing the water space in the tissue and blood, and the vascular space, subdivided into an arterial, capillary and venous segment, which contains hemoglobin-bound molecular oxygen.[6] Lambda (λ) is the physical decay constant of ^{15}O.

$$E = \frac{\dfrac{\hat{C}}{\overline{\overline{C}}} \cdot \dfrac{\overline{C}_a^{H2O}}{\hat{C}_a^{H2O}} - 1}{\dfrac{\hat{C}_a^O}{\hat{C}_a^{H2O}} - 1} \qquad (21)$$

Notice that the quantities $\overline{C}$ and $\overline{C}_a^{H_2O}$ were determined from the $CO^{15}O$ inhalation study and $\hat{C}_a^O$ is the total ^{15}O activity in whole blood which is the sum of the activities in the form of hemoglobin-bound molecular oxygen and labeled water:

$$\hat{C}_a^O = \hat{C}_a^{O_2} + \hat{C}_a^{H_2O} \qquad (22)$$

Unfortunately, therefore, $\hat{C}_a^{H_2O}$ cannot be measured directly due to the presence in arterial blood of ^{15}O-labeled hemoglobin-bound molecular oxygen. However, the ratio of water in arterial whole blood to that in plasma is the same regardless of whether $O^{15}O$ or $CO^{15}O$ is being inhaled and the whole blood to plasma water ratio measured during the continuous $CO^{15}O$ inhalation study may be used:

$$\hat{C}_a^{H_2O} = A \cdot \hat{C}_p^{H_2O} \tag{23}$$

where $A = \overline{C}_a^{H_2O}/\overline{C}_p^{H_2O}$ with $\overline{C}_p^{H_2O}$ and $\hat{C}_p^{H_2O}$ representing the plasma $H_2^{15}O$ concentrations measured during the $CO^{15}O$ and the $O^{15}O$ inhalation study, respectively. With these substitutions, we obtain the operational equation for the calculation of the cerebral net oxygen extraction fraction, $E_0 = E = rOEF$, where all variables can be directly measured:

$$rOEF = \frac{\dfrac{\hat{C}}{\overline{C}} \cdot \dfrac{\overline{C}_a^{H_2O}}{\hat{C}_p^{H_2O}} - A}{\dfrac{\hat{C}_a^O}{\hat{C}_p^{H_2O}} - A} \tag{24}$$

As suggested by Frackowiak et al.,[30] A may also be calculated from the packed cell volume (PCV) of the arterial blood as $A = 1 - 0.245 \cdot PCV$.

Since in the above expression the tissue and blood or plasma activities appear as ratios only, the calculation of OEF by the steady-state method is independent of the cross-calibration factor, X.

Lammertsma et al.[18] have noted that normal OEF values calculated according to Equation 24 were higher than those obtained by arteriovenous sampling.[84,85] They showed that this overestimation, which is around 10 to 15% in normal volunteers but may reach several times this value in pathological cases,[19] was the result of vascular[15]O activity which was neglected in the derivation of the above expression for rOEF. These authors, therefore, introduced a correction factor that accounts for intravascular ^{15}O activity.[18,20] Its calculation requires, in addition to CBF, the measurement of regional CBV (blood volume correction) which has since become an integral part of any OEF determination by the steady-state method. The corrected oxygen extraction fraction, rOEF, is calculated as follows:

$$rOEF = \frac{E' - Y}{1 - Y} \tag{25}$$

with

$$Y = \frac{d_t \cdot [rCBF/100] + \lambda}{[rCBF/(R \cdot rCBV)] + \lambda} \tag{26}$$

E' is the uncorrected regional oxygen extraction fraction, d_t is the brain tissue density ($\sim 1.05 g/cm^{-3}$), λ is the physical decay constant of ^{15}O and R is the small-to-large-vessel hematocrit ratio. Regional CBF and CBV are measured in milliliters per minute per 100 g and milliliters per 100 g, respectively. It is worth noting the Y is independent of the regional hematocrit since rCBV is inversely proportional to R (see Equation 1). Further error discussions of the steady-state method can be found elsewhere.[33,34,37]

2. Bolus Inhalation Method

A method similar to the autoradiographic approach for the measurement of rCBF has been developed for the calculation of the regional oxygen extraction fraction of the brain.[6] With this method, a mixture of air and $O^{15}O$ is administered as a bolus in one or several deep inhalations ($\sim$70 mCi) followed by a brief period of breath holding ($\sim$15 s) to facilitate the exchange of the label between the inhaled gas and the blood. A PET scan of typically 40 s duration and arterial blood sampling are initiated at the same time. The underlying model (see Figure 7B) consists of two compartments corresponding to the two states of the tracer, namely, hemoglobin-bound $O^{15}O$ and $H_2{}^{15}O$. The $O^{15}O$ compartment corresponds to the cerebral vascular space which, in this model, is explicitly accounted for in contrast to the steady-state method where a correction factor was introduced retrospectively.[18,20] A fraction, E, of the arterial $O^{15}O$ diffuses into the brain tissue where it is immediately metabolized to $H_2{}^{15}O$. Upon venous drainage, this water of metabolism eventually reappears in the arterial blood as recirculating labeled water. The $O^{15}O$ content in tissue is assumed to be zero. The second compartment corresponds to the distribution space of free water in brain, including its vascular spaces. It receives its input from locally generated water of metabolism, which is equal to the amount of extracted $O^{15}O$, and through arterial recirculating $H_2{}^{15}O$ redistributed from all tissues. Similar to the one-compartment model for the measurement of blood flow (see Figure 2A), except for the additional input from extracted $O^{15}O$, the differential equation describing the flow of tracer through this second compartment is

$$\frac{d\tilde{C}_2}{dt} = f \cdot \tilde{C}_a^{H_2O}(t) + f \cdot E \cdot \tilde{C}_a^{O_2}(t) - \frac{f}{p} \cdot \tilde{C}_2(t) \tag{27}$$

which, upon integration, becomes:

$$\tilde{C}_2(t) = f \cdot \tilde{C}_a^{H_2O}(t) * e^{-k_2 t} + f \cdot E \cdot \tilde{C}_a^{O_2}(t) * e^{-k_2 t} \tag{28}$$

The vascular compartment is subdivided into an arterial, a capillary, and a venous segment. Assuming the venous tracer concentration to be the arterial $O^{15}O$ concentration times the unextracted fraction 1-E, and the capillary concentration the average of arterial and venous concentration, we obtain for the concentration of the entire vascular compartment:

$$\tilde{C}_1(t) = CBV \cdot R \cdot \tilde{C}_a^{O_2}(t) \cdot (1 - 0.835 \cdot E) \tag{29}$$

Literature values of 0.83 and 0.01 were used for the venous and the capillary fractions of total blood volume, CBV.[86] Furthermore, CBV was corrected for effects of hematocrit dilution occurring at the tissue level by multiplication with the small-to-large-vessel hematocrit ratio, R. This is necessary since $O^{15}O$ is tightly bound to hemoglobin.

In analogy to Equation 11, the total number of radioactive events for both compartments detected by the tomograph in the scan interval T_1 to T_2 becomes:

$$\tilde{C}_{PET} = \int_{T_1}^{T_2} [\tilde{C}_1(t) + \tilde{C}_2(t)] \, dt \tag{30}$$

or explicitly:

$$\begin{aligned}
\tilde{C}_{PET} = {} & CBV \cdot R \cdot (1 - 0.835 \cdot OEF) \cdot \int_{T_1}^{T_2} \tilde{C}_a^{O_2}(t) dt \\
& + CBF \cdot \int_{T_1}^{T_2} \tilde{C}_a^{H_2O}(t) * e^{-(CBF/p)t} \, dt \\
& + CBF \cdot OEF \cdot \int_{T_1}^{T_2} \tilde{C}_a^{O_2}(t) * e^{-(CBF/p)t} \, dt
\end{aligned} \tag{31}$$

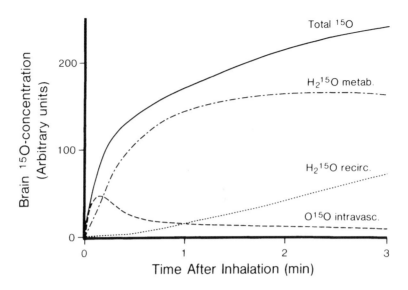

FIGURE 8.   O^{15}O bolus inhalation. Simulated tissue time-activity curve with its principal components (see Equation 31) according to the model of Mintun et al.[6] The total brain ^{15}O concentration is the sum of activities due to intravascular hemoglobin-bound molecular oxygen, recirculating H$_2$^{15}O and water of metabolism. (From Meyer, E., Tyler, J. L., Thompson, C. J., Redies, C., Diksic, M., and Hakim, A. M., *J. Cereb. Blood Flow Metab.*, 7, 403, 1987. With permission.)

The three components of this sum are depicted in Figure 8. By solving Equation 31 for OEF, we obtain the operational equation for the calculation of the local oxygen extraction fraction:

$$OEF = \frac{\tilde{C}_{PET} - CBF \cdot \int_{T_1}^{T_2} \tilde{C}_a^{H_2O}(t) * e^{-(CBF/p)t}\, dt - CBV \cdot R \cdot \int_{T_1}^{T_2} \tilde{C}_a^{O_2}(t)dt}{CBF \cdot \int_{T_1}^{T_2} \tilde{C}_a^{O_2}(t) * e^{-(CBF/p)t}\, dt - CBV \cdot R \cdot (0.835) \cdot \int_{T_1}^{T_2} \tilde{C}_a^{O_2}(t)\, dt} \tag{32}$$

Local CBF and CBV are determined from separate PET studies as previously described. The time course of whole blood and plasma ^{15}O activity is obtained by rapid arterial sampling during the scan. Under the assumption that O^{15}O only exists in the form of oxyhemoglobin, all plasma ^{15}O activity, $\tilde{C}_p(t)$, may be considered to be H$_2$^{15}O and the arterial concentration of recirculating water is therefore given by $\tilde{C}_a^{H_2O}(t) = \tilde{C}_p(t) \cdot FWC_b/FWC_p$ where $FWC_b = 0.80$ g/g$_{blood}$ and $FWC_p = 0.92$ g/g$_{plasma}$ are the fractional water contents of blood and plasma.[87] $\tilde{C}_a^{O_2}(t)$ is then obtained by subtracting $\tilde{C}_a^{H_2O}(t)$ from the whole blood activity curve. An example of measured tissue and blood activity curves is shown in Figure 9. Tracer delay and dispersion effects may be accounted for as described under the H$_2$^{15}O bolus CBF method.[57] A detailed error discussion of the method is found in Mintun's original paper.[6]

B. CEREBRAL METABOLIC RATE OF OXYGEN

1. Steady-State and Bolus Inhalation Methods

With both the steady-state[30] and the bolus inhalation method,[6] the calculation of cerebral metabolic rate of oxygen, CMRO$_2$, is based on a sequence of three PET studies that includes determinations of CBF, CBV and OEF with the CBV study being used to account for blood-borne radioactivity in the calculation of OEF. Once these quantities are known, CMRO$_2$ is calculated from the relationship:

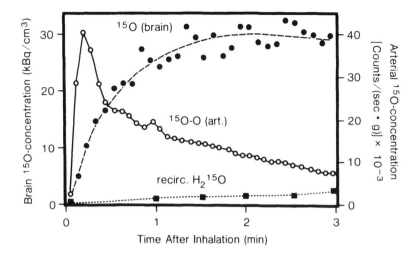

FIGURE 9. O^{15}O bolus inhalation. Measured tissue (filled black circles) and blood ^{15}O activity curves. The dashed line represents the tissue curve obtained by a dynamic three-parameter (rCBV, rCBF, and rOEF) least squares fit for a 2.3-cm-diameter ROI according to Equation 31. (Notice the different vertical scales for the brain and the blood data). (From Meyer, E., Tyler, J. L., Thompson, C. J., Redies, C., Diksic, M., and Hakim, A. M., *J. Cereb. Blood Flow Metab.*, 7, 403, 1987. With permission.)

$$CMRO_2 = CBF \cdot OEF \cdot [O_2]_a \tag{33}$$

The total oxygen content of arterial blood, $[O_2]_a$, is obtained from arterial samples and is typically around 20 ml of O_2 per 100 ml of blood. With OEF being a dimensionless parameter (typically ~ 0.4) and CBF measured in [ml/(min·100 g)], the units of $CMRO_2$ are [ml$_{O_2}$/(min·100 g)] or, upon multiplication by 44.64, [μmol$_{O_2}$/(min·100 g)].

2. Dynamic Methods

Methods described above for the calculation of $CMRO_2$ require three separate studies that take between 30 and 60 min for completion. The results are, therefore, vulnerable to errors arising from subject movement and/or from a change in the physiological state of the subject during the imaging period. Efforts have therefore been, and are being, made to estimate $CMRO_2$ from a single short PET study. Huang et al.[80,88] and Lammertsma et al.[81] have done some ground work in this field. Meyer et al.[82] successfully estimated regional $CMRO_2$ from a single 3-min dynamic O^{15}O bolus inhalation study by means of nonlinear regression analysis based on the model proposed by Mintun et al.[6] Holden et al.[89] have described a similar approach. The observation by Fox and Raichle[90] and Fox et al.[91] that oxygen consumption increased much less than blood flow and glucose consumption during somatosensory stimulation of the brain has triggered new efforts to study cerebral oxygen metabolism by simple though accurate methods. From H$_2$^{15}O and O^{15}O studies performed pairwise in a series of volunteers at normo- and hypocapnia, Gjedde et al.[92] conclude that the washout of radioactive water generated from labeled oxygen (water of metabolism) bears no simple relation to blood flow and that the use of separate measurements for CBF in the calculation of the fraction of extracted oxygen may lead to errors. The same authors are presently evaluating a method that should allow the generation of $CMRO_2$ maps on a pixel-by-pixel basis from a single short dynamic O^{15}O bolus inhalation study.[83]

V. DATA ANALYSIS

This section discusses some aspects of data treatment and analysis that are particularly relevant to PET studies with ^{15}O-labeled compounds. More general aspects related to the quantification of PET data such as correction for random and scattered coincidence events, attenuation and dead time are treated in other chapters of this volume. Also, some error considerations specific to a particular model or application have already been included in the respective sections.

A. DECAY CORRECTION

Due to the short physical half-life for ^{15}O of $T_{1/2}$ = 2.035 min, correction for physical decay of this tracer requires particular attention. While the PET tissue data in the steady-state models do not require such a correction, it is important for the dynamic methods as well as for the tissue autoradiographic approaches for the calculation of CBF and OEF. Raichle et al.[46] introduced an average decay correction based on a constant tissue function, except for its decrease over time as a result of radioactive decay alone:

$$A = \frac{1}{T_s} \cdot \int_0^{T_s} e^{-\lambda T_s} \, dt = \frac{1 - e^{-\lambda T_s}}{\lambda T_s} \tag{34}$$

Inversion of the average decay, A, yields an average decay correction factor, 1/A, which, for scan durations of $T_s \leq 60$ s, was shown by simulations to be accurate within approximately 4%.[46] Videen et al.[93] have proposed the inclusion of the decay constant in the model by defining $k_2 = (f + \lambda)/p$ which gives slightly more accurate results ($\sim$3%).

Manual determination of the arterial input function, $C_a(t)$, requires accurate timing of sample withdrawal and counting times. Assaying the blood samples in reverse order of their withdrawal assures more uniform counting statistics in tracer bolus studies. Time constraints due to the rapid tracer dacay have led to the development of automated continuous blood sampling systems.[55,94,95] Careful correction for tracer delay and dispersion effects is important with such equipment. When, in addition to whole blood, serial plasma samples are required as during OEF measurements by the O^{15}O bolus inhalation method, on-line plasma separators have been used with partial success to avoid the time consuming operation of centrifuging individual blood samples.[96,97]

B. FUNCTIONAL IMAGE GENERATION AND ANALYSIS

The results of a PET study are usually displayed in the form of functional tomographic images with the picture elements representing color coded numerical values of rCBV, rCBF, rOEF, or rCMRO$_2$, for example. The generation of these functional images is based on the pixel-by-pixel evaluation of operational equations such as Equations 1, 10, 11, 24, 32, or 33. While some expressions can be evaluated by means of simple mathematical operations (e.g., Equations 1, 10, 24, and 33), others (e.g., 11 and 32) require more time-consuming procedures which, when repeated for the many thousand pixels of each image and each tomographic slice, may result in significant processing times. The operational equation for the evaluation of rCBF by the autoradiographic approach is such an example (Equation 11). Since this equation cannot be solved explicitly for blood flow, numerical approximations have to be used. One method consists in the evaluation and tabulation of the right-hand side of the equation for a number of finely spaced values of the unknown parameter f. Measured tissue counts, $\bar{C}$, are then compared pixel-by-pixel to these tabulated theoretical tissue counts and the corresponding value of f assigned to the appropriate pixel in the functional image (table look-up method).[46] This approach was further rationalized by Raichle et al.,[46] who have shown that, for short scan intervals ($T_2 - T_1 \leq 60$ s) the relationship between blood flow (f) and tissue counts (C) (Equation 11) may be approximated with good accuracy (less than 1% error) (see Figure 3B) by a second-order polynomial of the form:

$$f = a \cdot \tilde{C} + b \cdot \tilde{C}^2 \tag{35}$$

The same principle can be used to calculate rOEF according to Equation 32.[98] More generally applicable algorithms for the rapid generation of functional maps suitable for a variety of models have been proposed.[99,100]

With regard to the autoradiographic CBF method, a further simplification has been introduced.[101] Due to the nearly linear relationship between tissue counts and blood flow with this method (see Figure 3B), the tissue count image alone is sufficient for the calculation of relative regional blood flow changes. Such changes in the blood flow pattern, and not absolute rCBF values, are of prime interest in so-called physiological blood flow activation studies for example (see Section VII). Images are usually normalized with regard to total brain tissue counts ($\sim$ global CBF) to correct for global blood flow changes that may occur due to experimental or physiological factors such as variations in arterial carbon dioxide tension. In situations where CBF studies are repeated on the same subject under different physiological conditions, focal CBF changes (activation foci) may be better visualized on a subtraction image representing the difference between two normalized tissue count or CBF images (see Plate 2*). Furthermore, following the signal averaging principle, the definition of activation foci is significantly improved by adding up a number of subtraction images obtained either from a single subject (intrasubject averaging) or from several different subjects (intersubject averaging).[102] In the latter case, a method to account for the anatomical variability between different brains has to be available.[103,104] Statistical analysis of averaged subtraction images is performed by means of the gamma-2 statistic that tests the distribution of local maxima for outliers which may then be identified with statistically significant activation foci.[102,105] Other methods have been proposed for the analysis of functional patterns in PET data.[106]

The activation and averaging principle described above also provides a method for the functional or physiological definition of regions of interest (ROIs) as opposed to their structural or anatomical definition discussed in the next paragraph.

C. ANATOMICAL-FUNCTIONAL CORRELATION AND ATROPHY CORRECTION

Anatomical-functional correlation and atrophy correction are not specific to PET studies with $O^{15}O$-labeled compounds. However, in view of recent developments using intersubject averaging for the enhancement of subtle regional CBF changes in response to physiological activation,[102,105] the use of CT and/or MRI for anatomical standardization and subsequent superposition of functional information has become a topic of prime interest.[103,104] Furthermore, since PET studies provide functional images that contain limited or sometimes no structural information (e.g., rOEF), analysis of subtle metabolic changes, particularly in small structures, requires the precise localization and identification of the anatomy underlying a functional region of interest or vice versa. Several attempts at correlating PET with CT or MRI information have been made in the past.[103,104,107-111] The stereotactic approach of Fox et al.[103] has been used successfully. The more recent approach by Evans et al.[104] is noteworthy as it allows nonlinear, individualized matching based on a computerized deformable brain atlas (see Plate 2*). With CT and/or MRI becoming increasingly part of PET investigations, methods to correct functional images for atrophy effects have been recently

---

* Plate 2 follows page 198.

proposed and should become part of routine image analysis software in the near future.[112-116]

VI. NORMAL VALUES

Normal values for CBV, CBF, OEF, and $CMRO_2$ obtained on healthy human volunteers by means of ^{15}O labeled compounds and current PET methods are given in Table 1. For comparison, whole brain values calculated by Kety's original method[117] using nitrous oxide and arteriovenous differences are included. Also, CBV values determined with ^{11}C-labeled carbon monoxide are shown.[3,4] When necessary, the units used in the original reference were converted so that all parameters listed in Table 1 are expressed on a per 100 g of brain tissue basis (tissue density $d_t = 1.05$ g/cm³). Also, μmol_{O_2} were used for $CMRO_2$ (1 $ml_{O_2} = 44.64$ μmol_{O_2} at 0°C and 760 mmHg).

Interpretation and comparison of normal values is not easy and several points, including both hardware and software as well as physiological factors, have to be kept in mind. The tomograph itself together with the procedure used to correct for random and scattered events as well as for attenuation and tracer decay have to be considered. Physiological factors such as the degree of sensory deprivation of the subject during data acquisition[118] and the level of arterial carbon dioxide tension, a potent determinant of CBF, are important. Parameters may be measured by different methods which yield different results. As an example, CBF values calculated by the steady-state method generally are lower than those obtained with the autoradiographic method due to the stronger tissue heterogeneity effect with the former.[40,45] Size, shape, and position of the region of interest used to extract functional information from the tomographic image are important. For bolus methods, effects of tracer delay and dispersion[45,52-55] as well as the scan length[46,60] have to be considered. Corrections for intravascular tracer activity, so-called blood volume corrections, may significantly change the value of a parameter such as OEF values obtained by the steady-state method.[19] With regard to the effect of age and gender, an extensive review by Hoyer[119] concludes that CBF and $CMRO_2$ of the normally aging brain are maintained unchanged from the third to the seventh decade of life. Thereafter, these parameters may decrease. Statistically significantly higher global CBF values (~11%), however, have been observed in females as compared to males.[120] Data regarding the short- and long-term reproducibility of CBV, CBF, OEF, and $CMRO_2$ PET measurements are given in Table 2. For the chronic studies, part of the observed changes between repeat measurements may be explained by head repositioning errors. Notice that $CMRO_2$ shows a smaller variability than both CBF and OEF. It might be argued that $CMRO_2$ which is closely related to cerebral function is kept fairly constant by appropriate inverse adjustments of CBF and OEF.

VII. SELECTED APPLICATIONS

Many applications of the PET methods discussed in this chapter are described elsewhere in this volume. We therefore present just two illustrations selected from an area of research that has recently gained considerable interest, namely, the mapping of functional activity of the brain by means of so-called blood flow activation studies and PET.

The application of this technique to cognitive neuroscience is particularly promising and has already provided support for new hypotheses such as the one advanced by Posner et al.[121] proposing that cognitive human tasks are performed by a number of strictly localized elementary operations rather than by any particular single area of the brain alone. PET imaging during the execution of a cognitive task, broken down according to a carefully designed paradigm, together with sophisticated image analysis methods such as described in Section V allows to describe and anatomically localize these component elementary operations.

TABLE 1
Normal Values for CBV, CBF, OEF, and CMRO$_2$[a]

Method[b]	Age (years)	N	Structure	CBV	CBF	OEF	CMRO$_2$	Tomograph	Ref.
N$_2$O ci	—	14	Global	—	54 ± 12	—	147 ± 18		117
^{11}CO bi	22—27	3	Global	4.2 ± 0.4	—	—	—	ECAT II	4
^{11}CO bi	—	10	Global[c]	4.3 ± 0.4	—	—	—	PETT III	3
C^{15}O bi	26 ± 6	7	Global[d]	3.3 ± 0.5				Positome IIIp	Our data
			Gray	4.0 ± 0.7	—	—	—		
			White	1.9 ± 0.4	—	—	—		
ss[e]	26—74	14	Gray	—	62 ± 7	49 ± 8	251 ± 25	ECAT II	30
			White	—	20 ± 2	48 ± 4	77 ± 9		
ss[e]	50 ± 12	27	Gray	—	45 ± 10	49 ± 10	166 ± 34	ECAT II	142
			White	—	24 ± 5	46 ± 8	81 ± 17		
ss[f]	26—63	13	Gray	3.9 ± 0.6	43 ± 10	40 ± 6	136 ± 13	ECAT II	143
			White	2.3 ± 0.8	26 ± 5	38 ± 6	77 ± 9		
ss[f]	19—25	11	Gray	4.3 ± 0.3	50 ± 12	42 ± 7	183 ± 42	Therascan 3128	Our data
			White	2.0 ± 0.5	20 ± 5	47 ± 16	81 ± 22		
	62—72	6	Gray	3.5 ± 0.3	42 ± 4	44 ± 6	160 ± 13		
			White	2.2 ± 0.4	28 ± 8	40 ± 7	105 ± 16		
ss	—	4	Gray		41			Headtome III	55
			White		16				
H$_2$^{15}O i.v.	—	4	Gray		51				
			White		23				
H$_2$^{15}O i.v.	26 ± 6	13	Gray		58 ± 8			Positome IIIp	Our data
			White		29 ± 7				
CO$_2$ ci and H$_2$^{15}O i.v.	27 ± 7	7	Gray		58 ± 11			NeoroECAT	70
			White		20 ± 4				
			Global		42 ± 8				
H$_2$^{15}O cont. i.v. infusion	24 ± 4	8	Global	4.7 ± 1.1	62 ± 13			PETT V	39
Seq. bolus	18—84	24	MCA territory		49 ± 9	37 ± 10	129 ± 27	PETT VI	144
Seq. bolus	28 ± 7	4	Global	4.0 ± 0.6	41 ± 6	44 ± 6	160 ± 39	Positome IIIp	Our data

TABLE 1 (continued)
Normal Values for CBV, CBF, OEF, and CMRO$_2$[a]

a   CBV in ml/100g; CBF in ml/(min·100g); OEF in %, CMRO$_2$ in μmol$_{O2}$/(min·100g). Errors are ±SD.

b   Abbreviations: ci, continuous inhalation; bi, bolus inhalation; ss, steady-state method; H$_2^{15}$O i.v., intravenous H$_2^{15}$O bolus; seq. bolus, sequential bolus method consisting of C^{15}O bi, H$_2^{15}$O i.v., and O^{15}O bi.

c   Level 4 cm above orbito-meatal line.

d   Level 8 cm above orbito-meatal line.

e   ss, steady-state method without CBV correction for OEF.

f   ss, steady-state method with CBV correction for OEF.

TABLE 2
Reproducibility of CBV, CBF, OEF, and CMRO$_2$ PET Measurements in Normal Volunteers

Method	Age (years)	N	Interval	Structure	Mean absolute % change between measurements				Ref.
					CBV	CBF	OEF	CMRO$_2$	
ss	52 ± 18	4	150 ± 25 d	Gray	—	27 ± 30	20 ± 18	7 ± 6	30
ss	—	4	<1 h	Global	—	11 ± 11	12 ± 10	7 ± 5	145
	—	9	180 ± 60 d	Gray	—	16 ± 16	17 ± 14	12 ± 11	
ss	23 ± 2	5	41 ± 29 d	Gray	6 ± 4	28 ± 37	14 ± 15	16 ± 7	Our data
H$_2$^{15}O i.v.	26 ± 7	8	<30 min	Gray	—	11 ± 10	—	—	Our data

Note: Abbreviations: ss, steady-state method; H$_2$^{15}O i.v., intravenous H$_2$^{15}O bolus method. Errors are ± SD.

The basis of the CBF activation method is the assumption that, for the normal brain, neuronal activity, glucose utilization, and cerebral blood flow (CBF) are closely coupled;[122] in other words, regional CBF may be considered as an indicator of local neuronal activity. The CBF activation method has allowed to map the response of the human brain to a variety of stimuli, including sensory activation,[101,123-126] motor tasks,[127] as well as higher mental functions.[128-131] The development of the intravenous H$_2$^{15}O bolus technique for the repeated, rapid measurement of CBF in humans[45,46] was instrumental for the realization of such studies. In order to detect the often subtle CBF changes which may be of the order of only a few percent when studying more complex mental processes,[128] the signal averaging[102,132] and image analysis[105] techniques described in Section V have proved to be powerful tools in this exciting field of research.

Our first example (Plate 1*) shows the CBF response to vibrotactile stimulation in the human primary sensory cortex.[133] This strong regional blood flow response was elicited by sweeping a hand-held vibrator (frequency 110 Hz, amplitude 2 mm) across the fingertips of the right hand of a young healthy volunteer at a rate of one sweep per second. An increase in regional CBF around 30% reflecting the increased neuronal activity in the primary cortex due to this potent stimulus is consistently observed without any particular image analysis strategies.[124] A region of interest template derived from a matched MRI study[104] is superposed onto the functional PET image, confirming the activated area to belong to the postcentral gyrus. Several applications of this stimulation paradigm have been described or proposed.[124] They are based on two aspects of the CBF response, namely, its magnitude, reflecting functional status, or its location, reflecting functional topography.

Recently, we have successfully explored the topographical potential of the method for preoperative functional mapping in connection with surgical planning for the treatment of patients suffering from arteriovenous malformations located in particularly eloquent cortical areas.[134]

The second example (Plate 2**) illustrates a subtraction image (color), representing the difference between two normalized H$_2$^{15}O bolus tissue activity images, that has been merged with its matched MRI image (black and white) upon which a customized ROI template has been superposed.[104] The baseline image was acquired while the subject was fixing a pulsating (0.5 Hz) cross-hair on a computer screen. The activation image was obtained during the passive viewing of images of animals presented at a rate of one image every 2 s. The PET images were not translated into functional CBF units, making use of the proportionality

* Plate 1 follows page 198.
** Plate 2 follows page 198.

TABLE 3
Radiation Absorbed Dose Estimates for ^{15}O PET Studies in Adults[a]

		Radiation absorbed dose estimate (mrad/mCi)[b]					
	i.v. Bolus	**Bolus inhalation**				**Continuous inhalation**	
Organ	**($H_2{}^{15}O$)**	**CO^{15}O**	**O^{15}O**	**^{11}CO**	**C^{15}O**	**CO^{15}O**	**O^{15}O**
Adrenals	1.9	1.9	6.7	29.0	4.4	1.9	2.3
Bladder	1.3	1.2	1.5	13.0	1.6	1.2	0.5
Blood	10.0	5.5	10.0	75.0	14.0	5.5	3.4
Brain	0.6	0.6	1.2	4.5	0.6	0.6	1.6
Heart	2.1	2.2	1.9	33.0	4.5	2.2	1.8
Kidneys	2.1	2.1	1.6	28.0	4.3	2.1	2.2
Liver	2.0	2.0	1.3	21.0	3.0	2.0	1.1
Lungs	2.0	4.9	12.0	55.0	10.0	6.2	5.2
Muscle	2.0	2.0	0.5	6.7	1.0	2.0	0.4
Ovaries	2.1	2.1	1.7	27.0	4.2	2.1	1.1
Red marrow	1.2	1.2	1.7	10.0	1.5	1.2	1.1
Spleen	2.1	2.1	2.7	91.0	15.0	2.1	1.4
Testes	2.2	2.1	1.7	8.5	1.0	2.1	1.1
Thyroid	1.9	1.9	6.3	19.0	3.1	1.9	2.1
Whole body	1.6	1.6	1.7	19.0	1.6	1.6	0.6

[a]   Data derived mainly from Kearfott[135] and Bigler and Sgouros.[136]
[b]   Absorbed dose per total administered activity.

between tissue counts and relative rCBF.[132] The residual tissue count rate observed in the occipital visual areas is thought to reflect the increased neuronal activity due exclusively to the process of passive picture processing. This example illustrates how, by way of appropriate subtractions, unwanted background activity may be eliminated leaving as a result the information related to a very specific process. Petersen et al.[128] have used this method to identify cerebral areas involved in higher mental processes such as in the formation of semantic associations.

VIII. RADIATION DOSIMETRY

The dosimetry for some current PET methods involving ^{15}O-labeled compounds plus ^{11}C-labeled carbon monoxide is summarized in Table 3. It gives radiation absorbed doses in millirads per millicurie of administered radioactivity. Conversion to SI units may be performed using the relation 100 rad = 1 gray (Gy). The tabulated values are based on a number of references[135,136] and calculations,[70,137] and they may vary depending on the assumptions made. The reader is therefore advised to consult the pertinent references in addition to the table. Most calculations are based on the absorbed fraction method and the Medical Internal Radiation Dosimetry (MIRD) publications.[138] With this approach, the average absorbed dose, D_t [rad], to a target organ is calculated as:

$$D_t = \sum_s A_s \cdot S_{t \leftarrow s} \qquad (36)$$

where A_s is the cumulated activity in the source organ number s [μCi·h] and $S_{t \leftarrow s}$ is the corresponding S-factor [rad/(μCi·h)] representing the dose absorbed by this organ per cumulated activity. S-factors for selected radionuclides and organs are tabulated in the MIRD publications. Typical amounts of tracer radioactivity administered for the various PET procedures are as follows: 10 to 15 mCi for ^{11}CO bolus inhalation,[3,4] 50 to 100 mCi for C^{15}O bolus inhalation,[5] 40 to 80 mCi for O^{15}O bolus inhalation,[6,82] 30 to 100 mCi for CO^{15}O bolus inhalation,[63,70] and 20 to 100 mCi for intravenous $H_2{}^{15}O$ bolus administration.[46,55,82] For the continuous inhalation of $CO^{15}O$ and $O^{15}O$, activity is typically supplied for 20 min (10-min build-up to equilibrium followed by a 5- to 10-min scan) at a rate of 5 to 10 mCi/min for a total of 100 to 200 mCi and an average arterial ^{15}O concentration of 1 to 2 μCi/g.[30,137] These are guideline figures only that might have to be modified to satisfy a particular experimental situation. Powers et al.[139] give detailed absorbed dose estimates for $H_2{}^{15}O$, $C^{15}O$, and $O^{15}O$ studies in newborn infants. The absorbed dose to the upper airways during the continuous inhalation of $CO^{15}O$ and $O^{15}O$ has also been carefully evaluated.[140]

IX. OUTLOOK

A category of ^{15}O PET studies that have been particularly fruitful and innovative in the recent past, and that will continue to produce first-hand basic information on higher human brain function, are physiological blood flow activation studies performed with the intravenous $H_2{}^{15}O$ bolus method. Further exciting progress is expected in this field, helped by the rapidly developing methodology concerned with the three-dimensional matching of functional and anatomical information. The inclusion of nonlinear deformation algorithms (warping) should, together with the improved spatial resolution of present tomographs and the support of careful simulation studies, allow to refine the process of intersubject image averaging for improved identification of activation foci in these studies. In addition to their continued application to the field of cognitive neuroscience with normal control populations, such studies will likely be extended to the investigation of cognitive function in pathologies such as Alzheimer's disease and anxiety disorders, to name just a few.

The advances in the field of anatomical-functional correlation should lead to the development of software for the routing correction of functional PET images for effects of atrophy observed on MRI.

The search for methods allowing the measurement of cerebral oxygen utilization by means of a single PET study, comparable in its duration and simplicity to an $H_2{}^{15}O$ bolus blood flow study, will continue in an attempt to contribute to the investigation of the degree of coupling between cerebral blood flow, oxygen and glucose consumption during physiological activation. Progress in the accurate automated acquistion of blood and plasma data together with advanced dynamic data processing and modeling should lead to some interesting new approaches in quest of the simultaneous determination of blood flow and oxygen utilization.

ACKNOWLEDGMENTS

This work was partly supported by Grand PG-41 from the Medical Research Council of Canada and the Isaac Walton Killam Fellowship Fund of the Montreal Neurological Institute. The contributions of Dr. S. Ohta, the discussions with Drs. A. Gjedde and A. C. Evans, as well as the help of Mr. S. Marrett and Mr. P. Neelin with image analysis are gratefully acknowledged. Special thanks are due to the personnel of the Positron Imaging Laboratories, the Medical Cyclotron Radiochemistry Unit, the Radiology and Neurophotography Departments, and to Mrs. F. Alberico for preparation of the references.

REFERENCES

1. **Hoffman, E. J. and Phelps, M. E.,** Positron emission tomography: principles and quantitation, in *Positron Emission Tomography and Autoradiography: Principles and Applications for the Brain and Heart,* Phelps, M., Mazziotta, J., and Schelbert, H., Eds., Raven Press, New York, 1989, 237.
2. **Mazziotta, J. C. and Phelps, M. E.,** Positron emission tomography studies in the brain, in *Positron Emission Tomography and Autoradiography: Principles and Applications for the Brain and Heart,* Phelps, M., Mazziotta, J., and Schelbert, H., Eds., Raven Press, New York, 1989, 493.
3. **Grubb, R. J., Jr., Raichle, M. E., Higgins, C. S. et al.,** Measurement of regional cerebral blood volume by emission tomography, *Ann. Neurol.,* 4, 322, 1978.
4. **Phelps, M. E., Huang, S. C., Hoffman, E. J., and Kuhl, D. E.,** Validation of tomographic measurement of cerebral blood volume with C-11-labeled carboxyhemoglobin, *J. Nucl. Med.,* 20, 328, 1979.
5. **Martin, W. R. W., Powers, W. J., and Raichle, M. E.,** Cerebral blood volume measured with inhaled $C^{15}O$ and positron emission tomography, *J. Cereb. Blood Flow Metab.,* 7, 421, 1987.
6. **Mintun, M. A., Raichle, M. E., Martin, W. R. W., and Herscovitch, P.,** Brain oxygen utilization measured with O-15 radiotracers and positron emission tomography, *J. Nucl. Med.,* 25, 177, 1984.
7. **Lars, J., Nilsson, G., Stone-Elander, S., Ehrin, E., Garmelius, B., and Johnström, P.,** [^{11}C]-Labeled compounds for the study of cerebral blood volume, blood flow, and energy metabolism, in *Metabolism of the Human Brain Studied with Positron Emission Tomography,* Greitz, T., Ingvar, D. H., and Widén, L., Eds., Raven Press, New York, 1985.
8. **Grubb, R. L., Phelps, M. E., and Ter-Pogossian, M. M.,** Regional cerebral blood volume in humans. X-ray fluorescence studies, *Arch. Neurol.,* 28, 38, 1973.
9. **Raichle, M. E.,** Quantitative *in vivo* autoradiography with positron emission tomography, *Brain Res. Rev.,* 1, 47, 1979.
10. **Lammertsma, A. A., Brooks, D. J., Beaney, R. P. et al.,** *In vivo* measurement of regional cerebral haematocrit using positron emission tomography, *J. Cereb. Blood Flow Metab.,* 4, 317, 1984.
11. **Sakai, F., Nakazawa, K., Tazaki, Y. et al.,** Regional cerebral blood volume and hematocrit measured in normal human volunteers by single-photon emission computed tomography, *J. Cereb. Blood Flow and Metab.,* 5, 207, 1985.
12. **Powers, W. J. and Martin, W. R. W., Herscovitch, P., Raichle, M. E., and Grubb, R. L.,** Extracranial-intracranial bypass surgery: hemodynamic and metabolic effects, *Neurology (Cleveland),* 34, 1168, 1984.
13. **Powers, W. J. and Raichle, M. E.,** Positron emission tomography and its application to the study of cerebrovascular disease in man, *Stroke,* 16, 361, 1985.
14. **Leblanc, R., Yamamoto, Y. L., Tyler, J. L., Diksic, M., and Hakim, A. M.,** Borderzone ischemia, *Ann. Neurol.,* 22, 707, 1987.
15. **Hakim, A. M., Pokrupa, R. P., Villaneuva, J., Diksic, M., Evans, A. C., Thompson, C. J., Meyer, E., Yamamoto, Y. L., and Feindel, W. H.,** The effect of spontaneous reperfusion on metabolic function in early human cerebral infarcts, *Ann. Neurol.,* 21, 279, 1987.
16. **Leblanc, R., Tyler, J. L., Mohr, G., Meyer, E., Diksic, M., Yamamoto, Y. L., Taylor, L., Gauthier, S., and Hakim, A.,** Hemodynamic and metabolic effects of cerebral revascularization, *J. Neurosurg.,* 66, 529, 1987.
17. **Tyler, J. L., Leblanc, R., Meyer, E., Dagher, A., Yamamoto, Y. L., Diksic, M., and Hakim, A.,** Hemodynamic and metabolic consequences of cerebral arteriovenous malformations studied by positron emission tomography, *Stroke,* 20, 890, 1989.
18. **Lammertsma, A. A. and Jones, T.,** Correction for the presence of intravascular oxygen-15 in the steady-state technique for measuring regional oxygen extraction ratio in the brain. I. description of the method, *J. Cereb. Blood Flow Metab.,* 3, 416, 1983.
19. **Lammertsma, A. A., Wise, R. J. S., Heather, J. D., Gibbs, J. M., Leenders, K. L., Frackowiak, R. S. J., Rhodes, C. G., and Jones, T.,** Correction for the presence of intravascular oxygen-15 in the steady-state technique for measuring regional oxygen extraction ratio in the brain. II. Results in normal subjects and brain tumour and stroke patients, *J. Cereb. Blood Flow Metab.,* 3, 425, 1983.
20. **Lammertsma, A. A., Baron, J.-C., and Jones, T.,** Correction for intravascular activity in the oxygen-15 steady-state technique is independent of the regional hematocrit, *J. Cereb. Blood Flow Metab.,* 7, 372, 1987.
21. **Kety, S. S.,** The theory and application of the exchange of inert gas at the lungs and tissues, *Pharmacol. Rev.,* 3, 1, 1951.
22. **Kety, S. S.,** Measurement of local blood flow by the exchange of an inert, diffusible substance, *Methods Med. Res.,* 8, 228, 1960.
23. **Eichling, J. O., Raichle, M. E., Grubb, R. L., Jr. et al.,** Evidence of the limitations of water as a freely diffusible tracer in brain of the Rhesus monkey, *Circ. Res.,* 35, 358, 1974.
24. **Renkin, E. M.,** Transport of potassium-42 from blood to tissue in isolated mammalian skeletal muscles, *Am. J. Physiol.,* 197, 1205, 1959.

25. **Crone, C.**, Permeability of capillaries in various organs as determined by use of the indicator diffusion method, *Acta Physiol. Scand.*, 58, 292, 1964.

26. **Raichle, M. E. and Larson, K. B.**, The significance of the NH_3-NH_4^+ equilibrium on the passage of ^{13}N-ammonia from blood to brain, *Circ. Res.*, 48, 913, 1981.

27. **Russ, G. A., Bigler, R. E., and Tilbury, R. S.**, Whole body scanning and organ imaging with oxygen-15 at the steady-state, in *Proc. 1st World Congr. Nuclear Medicine*, Tokyo, Japan, 1974, 904.

28. **Jones, T., Chesler, D. A., and Ter-Pogossian, M. M.**, The continuous inhalation of oxygen-15 for assessing regional oxygen extraction in the brain of man, *Br. J. Radiol.*, 40, 339, 1976.

29. **Subramanyan, R., Alpert, N. M., Hoop, B., Jr., Brownell, G. L., and Tavera, J. M.**, A model for regional cerebral oxygen distribution during continuous inhalation of $^{15}O_2$, $C^{15}O$ and $C^{15}O_2$, *J. Nucl. Med.*, 19, 48, 1978.

30. **Frackowiak, R. S. J., Lenzi, G. L., Jones, T. et al.**, Quantitative measurement of regional cerebral blood flow and oxygen metabolism in man using O-15 and positron emission tomography. Theory, procedure, and normal values. *J. Comput. Assist. Tomogr.*, 4, 727, 1980.

31. **West, J. B. and Dollery, C. T.**, Uptake of oxygen-15 labeled CO_2 compared with carbon-11-labeled CO_2 in the lung, *J. Appl. Physiol.*, 17, 9, 1962.

32. **Herscovitch, P. and Raichle, M. E.**, What is the correct value for the brain-blood partition coefficient for water?, *J. Cereb. Blood Flow Metab.*, 5, 65, 1985.

33. **Lammertsma, A. A., Frackowiak, R. S. J., Lenzi, G. L., Heather, J. D., Pozzilli, C., and Jones, T.**, Accuracy of the oxygen-15 steady state technique for measuring rCBF and $rCMRO_2$, *J. Cereb. Blood Flow Metab.*, 1, S3, 1981.

34. **Lammertsma, A. A., Jones, T., Frackowiak, F. S., and Lenzi, G. L.**, A theoretical study of the steady-state model for measuring regional cerebral blood flow and oxygen utilization using oxygen-15, *J. Comput. Assist. Tomogr.*, 5, 544, 1981.

35. **Jones, S. C., Greenberg, J. H., and Reivich, M.**, Error analysis for the determination of cerebral blood flow with the continuous inhalation of ^{15}O-labeled carbon dioxide and positron emission tomography, *J. Comput. Assist. Tomogr.*, 6, 116, 1982.

36. **Meyer, E. and Yamamoto, Y. L.**, The requirement for constant arterial radioactivity in the $C^{15}O_2$ steady-state blood-flow model, *J. Nucl. Med.*, 25, 455, 1984.

37. **Correia, J. A., Alpert, N. M., Buxton, R. B., and Ackerman, R. H.**, Analysis of some errors in the measurement of oxygen extraction and oxygen consumption by the equilibrium inhalation method, *J. Cereb. Blood Flow Metab.*, 5, 591, 1985.

38. **Senda, M., Buxton, R. B., Alpert, N. M., Correia, J. A., Mackay, B. C., Weise, S. B. and Ackerman, R. H.**, The ^{15}O steady-state method: correction for variation in arterial concentration, *J. Cereb. Blood Flow Metabol.*, 8, 681, 1988.

39. **Jones, S. C., Greenberg, J. H., Dann, R., Robinson, G. D., Jr., Kushner, M., Alavi, A., and Reivich, M.**, Cerebral Blood flow with the continuous infusion of oxygen-15-labeled water, *J. Cereb. Blood Flow Metab.*, 5, 566, 1985.

40. **Herscovitch, P. and Raichle, M. E.**, Effect of tissue heterogeneity on the measurement of cerebral blood flow with the equilibrium $C^{15}O_2$ inhalation technique, *J. Cereb. Blood Flow Metab.*, 3, 407, 1983.

41. **Landau, W. M., Freygang, W. H., Jr., Rowland, L. P., et al.**, The local circulation of the living brain; values in the unanesthetized and anesthetized cat, *Trans. Am. Neurol. Assoc.*, 80, 125, 1955.

42. **Freygang, W. H., Jr. and Sokoloff, L.**, Quantitative measurement of regional circulation in the central nervous system by the use of radioactive inert gas, *Adv. Biol. Med. Phys.*, 6, 263, 1959.

43. **Reivich, M., Jehle, J., Sokoloff, L., and Kety, S. S.**, Measurement of regional cerebral blood flow with antipyrine-^{14}C in awake cats, *J. Appl. Physiol.*, 27, 296, 1969.

44. **Sakurada, O., Kennedy, C., Jehle, J., Brown, J. D., Carbin, G. L., and Sokoloff, L.**, Measurement of local cerebral blood flow with iodo(^{14}C)antipyrine, *Am. J. Physiol.*, 234, H59, 1978.

45. **Herscovitch, P., Markham, J., and Raichle, M. E.**, Brain blood flow measured with intravenous H_2O ^{15}O. I. Theory and error analysis, *J. Nucl. Med.*, 24, 782, 1983.

46. **Raichle, M. E., Martin, W. R. W., Herscovitch, P. et al.**, Brain blood flow measured with intravenous H_2O ^{15}O. II. Implementation and validation, *J. Nucl. Med.*, 24, 790, 1983.

47. **Ginsberg, M. D., Lockwood, A. H., Busto, R., Finn, R. D., Butler, C. M., Cendan, I. E., and Goddard, J.**, A simplified *in vivo* autoradiographic strategy for the determination of regional cerebral blood flow by positron emission tomography: theoretical considerations and validation studies in the rat, *J. Cereb. Blood Flow Metab.*, 2, 89, 1982.

48. **Howard, B. E., Ginsberg, M. D., Hassel, W. R., Lockwood, A. H., and Freed, P.**, On the uniqueness of cerebral blood flow measured by the *in vivo* autoradiographic strategy and positron emission tomography, *J. Cereb. Blood Flow Metab.*, 3, 432, 1983.

49. **Herscovitch, P., Raichle, M. E., Kilbourne, M. R., and Welch, M. J.**, Positron emission tomographic measurement of cerebral blood flow and permeability — surface area product of water using [^{15}O] water and [^{11}C]butanol, *J. Cereb. Blood Flow Metab.*, 7, 527, 1987.

50. **Berridge, M. S. and Cassidy, E.,** An improved, routine, synthesis of oxygen-15 labeled butanol for positron tomography, *J. Nucl. Med.,* 30, 927, 1989.

51. **Takagi, H., Shapiro, K., Marmarou, A., and Wisoff, H.,** Microgravimetric analysis of human brain tissue. Correlation with computerized tomography scanning, *J. Neurosurg.,* 54, 797, 1981.

52. **Dhawan, V., Conti, J., Mernyk, M., et al.,** Accuracy of PET rCBF measurements: effect of time shift between blood and brain radioactivity curves, *Phys. Med. Biol.,* 31, 507, 1986.

53. **Iida, H., Kanno, I., Miura, S., et al.,** Error analysis of a quantitative cerebral blood flow measurement using O-15 H₂O autoradiography and positron emission tomography with respect to the dispersion of the input function, *J. Cereb. Blood Flow Metab.,* 6, 536, 1986.

54. **Koeppe, R. A., Hutchins, G. D., Rothley, J. M., and Hichwa, R. D.,** Examination of assumptions for local cerebral blood flow studies in PET, *J. Nucl. Med.,* 28, 1695, 1987.

55. **Kanno, I., Iida, H., Miura, S., et al.,** A system for cerebral blood flow measurement using an H₂ ¹⁵O autoradiographic method and positron emission tomography, *J. Cereb. Blood Flow Metab.,* 7, 143, 153, 1987.

56. **Iida, H., Higano, S., Tomura, N., Shishido, F., Kanno, I., Miura, S., Murakami, M., Takahashi, K., Sasaki, H., and Uemura, K.,** Evaluation of regional differences of tracer appearance time in cerebral tissue using [¹⁵O] water and dynamic positron emission tomography, *J. Cereb. Blood Flow Metab.,* 8, 285, 1988.

57. **Meyer, E., Tyler, J. L., Thompson, C. J., et al.,** The time dependence of the O-15 bolus model for CMRO₂ measurement by PET with respect to dispersion of the input function, *J. Nucl. Med.,* 28, 699, 1987.

58. **Senda, M., Nishizawa, S., Shibata, T., et al.,** Effect of arterial blood dispersion on the measurement of cerebral blood flow using PET and O-15 water, *J. Nucl. Med.,* 28, 656, 1987.

59. **Meyer, E.,** Simultaneous correction for tracer arrival delay and dispersion in CBF measurements by the H₂ ¹⁵O autoradiographic method and dynamic PET, *J. Nucl. Med.,* 30, 1069, 1989.

60. **Eklöf, B., Lassen, N. A., Nilsson, L., et al.,** Regional cerebral blood flow in the rat measured by the tissue sampling technique; a critical evaluation using four indicators C¹⁴-antipyrine, C¹⁴-ethanol, H³-water and xenon¹³³, *Acta Physiol. Scand.,* 91, 1, 1974.

61. **Schmidt, K., Sokoloff, L., and Kety, S. S.,** Effects of tissue heterogeneity on estimates of regional cerebral blood flow, *J. Cereb. Blood Flow Metab.,* 9(Suppl.1), S242, 1989.

62. **Gambhir, S. S., Huang, S. C., Hawkins, R. A., et al.,** A study of the single compartment tracer kinetic model for the measurement of local cerebral blood flow using O-15 water and positron emission tomography, *J. Cereb. Blood Flow Metab.,* 7, 13, 1974.

63. **Kanno, I., Lammerstma, A. A., Heather, J. D., Gibbs, J. M., Rhodes, C. G., Clark, J. C., and Jones, T.,** Measurement of cerebral blood flow using bolus inhalation of ¹⁵O CO₂ and positron emission tomography: description of the method and its comparison with the O-15 CO₂ continuous inhalation method, *J. Cereb. Blood Flow Metab.,* 4, 224, 1984.

64. **Koeppe, R. A., Holden, J. E., Polcyn, R. E., et al.,** Quantitation of local cerebral blood flow and partition coefficient without arterial sampling: theory and validation, *J. Cereb. Blood Flow Metab.,* 5, 214, 224, 1985.

65. **Iida, H., Kanno, I., Miura, S., Murakami, M., Takahashi, K., Inugami, A., Shishido, F., and Uemura, K.,** An accurate determination of regional brain/blood partition coefficient of water using dynamic positron-emission tomography: validation of Kety-Schmidt single compartment model for H₂ ¹⁵O based on measurement, *J. Cereb. Blood Flow Metab.,* 7, S576, 1987.

66. **Lammertsma, A. A., Frackowiak, R. S. J., Hoffman, J. M., Huang, S. C., Weinberg, I. N., Dahlbom, M., MacDonald, N. S., Hoffman, E. J., Mazziotta, J. C., Heather, J. D., Forse, G. R., Phelps, M. E., and Jones, T.,** The C¹⁵O₂ build-up technique to measure regional cerebral blood flow and volume distribution of water, *J. Cereb. Blood Flow Metab.,* 9, 461, 1989.

67. **Tsui, E. and Budinger, T. F.,** Transverse section imaging of mean clearance time, *Phys. Med. Biol.,* 23, 644, 1978.

68. **Carson, R. E., Huang, S. S., and Green, M. V.,** Weighted integration method for local cerebral blood flow measurements with positron emission tomography, *J. Cereb. Blood Flow Metab.,* 6, 245, 1986.

69. **Huang, S. S., Carson, R. E., and Phelps, M. E.,** Measurement of local blood flow and distribution volume with short-lived isotopes: a general input technique, *J. Cereb. Blood Flow Metab.,* 2, 99, 1982.

70. **Huang, S. S., Carson, R. E., Hoffman, E. J., Carson, J., MacDonald, N., Barrio, J. R., and Phelps, M. E.,** Quantitative measurement of local blood flow in humans by positron computed tomography and ¹⁵O-water, *J. Cereb. Blood Flow Metab.,* 3, 141, 1983.

71. **Alpert, N. M., Eriksson, L., Chang, J. Y., Bergström, M., Litton, J. E., Correia, J. A., Bohm, C., Ackerman, R. H., and Taveras, J. M.,** Strategy for the measurement of regional cerebral blood flow using short-lived tracers and emission tomography, *J. Cereb. Blood Flow Metab.,* 4, 28, 1984.

72. **Mazoyer, B. M., Huesman, R. H., Budinger, T. F., and Knittel, B. L.** Dynamic PET data analysis, *J. Comp. Assist. Tomogr.,* 10, 645, 1986.

73. **Huesman, R. H. and Mazoyer, B. M.,** Kinetic data analysis with a noisy input function, *Phys. Med. Biol.,* 32, 1569, 1987.

74. **Herholz, K. and Patlak, C. S.,** The influence of tissue heterogeneity on results of fitting nonlinear model equations to regional tracer uptake curves: with an application to compartmental models used in positron emission tomography, *J. Cereb. Blood Flow Metab.,* 7, 214, 1987.

75. **Jovkar, S., Evans, A. C., Diksic, M., Nakai, H., and Yamamoto, Y. L.,** Minimization of parameter estimation errors in dynamic PET: choice of scanning schedules, *Phys. Med. Biol.,* 34, 895, 1989.

76. **Larson, K. B., Markham, J., and Raichle, M. E.,** Tracer-kinetic models for measuring cerebral blood flow using externally detected radiotracers, *J. Cereb. Blood Flow Metab.,* Vol. 7, 443, 1987.

77. **Sawada, Y., Sugiyama, Y., Iga, T., and Hanano, M.,** Tracer disposition kinetics in the determination of local cerebral blood flow by a venous equilibrium model, tube model, and distributed model, *J. Cereb. Blood Flow Metab.,* 7, 433, 1987.

78. **Ter-Pogossian, M. M., Eichling, J. O., Davis, D. O., and Welch, M. J.,** The measure *in vivo* of regional cerebral oxygen utilization by means of oxyhemoglobin labeled with radioactive oxygen-15, *J. Clin. Invest.,* 49, 381, 1970.

79. **Raichle, M. E., Grubb, R. L., Jr., Eichling, J. O., and Ter-Pogossian, M. M.,** Measurement of brain oxyten utilization with radioactive oxygen-15: experimental verification, *J. Appl. Physiol.,* 40, 638, 1976.

80. **Huang, S. C., Feng, D. G., and Phelps, M. E.,** Model dependency and estimation reliability in measurement of cerebral oxygen utilization rate with oxygen-15 and dynamic positron emission tomography, *J. Cereb. Blood Flow Metab.,* 6, 105, 1986.

81. **Lammertsma, A. A., Frackowiak, R. S. J., and Huang, S. C. et al.,** The simultaneous measurement of regional cerebral blood flow and oxygen utilization, *J. Nucl. Med.,* 27, 913, 1986.

82. **Meyer, E., Tyler, J. L., Thompson, C. J., Redies, C., Diksic, M., and Hakim, A. M.,** Estimation of cerebral oxygen utilization rate by single-bolus O-15 O_2 inhalation and dynamic positron emission tomography, *J. Cereb. Blood Flow Metab.,* 7, 403, 1987.

83. **Ohta, S., Meyer, E., and Gjedde, A.,** $CMRO_2$ measurement in PET by weighted integration with CBV correction, *J. Nucl. Med.,* 31, 861, 1990.

84. **Frackowiak, R. S. J., Jones, T., Lenzi, G. L., and Heather, J. D.,** Regional cerebral oxygen utilization and blood flow in normal man using oxygen-15 and positron emission tomography, *Acta Neurol. Scand.,* 62, 336, 1982.

85. **Baron, J. C., Steinling, M., Tanaka, T., et al.,** Quantitative measurement of CBF, oxygen extraction fraction (OEF) and $CMRO_2$ with ^{15}O continuous inhalation techniques and positron emission tomography (PET): experimental evidence and normal values in man, *J. Cereb. Blood Flow Metab.,* Suppl. 1, S5, 1981.

86. **Rushmer, R. F.,** *Cardiovascular Dynamics,* 4th ed., W. B. Saunders, Philadelphia, 1976, 8.

87. **Davis, F. E., Kenyon, K., and Kirk, J.,** A rapid titrimetic method for determining the water content of human blood, *Science,* 118, 276, 1953.

88. **Huang, S. C., Carson, R. E., and Phelps, M. E.,** A tomographic technique for simultaneous determination of LCBF and local metabolic rate of oxygen with O-15 oxygen, *J. Cereb. Blood Flow Metab.,* 3(Suppl. 1), S17, 1983.

89. **Holden, J. E., Eriksson, L., Roland, P. E., Stone-Elander, S., Widén, L., and Kesselber, M.,** Direct comparison of single-scan autoradiographic with multiple-scan least-squares fitting approaches to PET $CMRO_2$ estimation, *J. Cereb. Blood Flow Metabol.,* 8, 671, 1988.

90. **Fox, P. T. and Raichle, M. E.,** Focal physiological uncoupling of cerebral blood flow and oxidative metabolism during somatosensory stimulation in human subjects, *Proc. Natl. Acad. Sci. U.S.A.,* 83, 1140, 1986.

91. **Fox, P. T., Raichle, M. E., Mintun, M. A., and Dence, C.,** Nonoxidative glucose consumption during focal physiologic neural activity, *Science,* 241, 426, 1986.

92. **Gjedde, A., Meyer, E., and Ohta, S.,** Direct determination of cerebral oxygen consumption by PET, *J. Nucl. Med.,* 31, 882, 1990.

93. **Videen, T. O., Perlmutter, J. S., Herscovitch, P., and Raichle, M. E.,** Brain blood volume, flow, and oxygen utilization measured with ^{15}O radiotracers and positron emission tomography: revised metabolic computations, *J. Cereb. Blood Flow Metab.,* 7, 513, 1987.

94. **Hutchins, G. D., Hichwa, R. D., and Koeppe, R. A.,** A continuous flow input function detector for H_2O blood flow studies in positron emission tomography, *IEEE Trans. Nucl. Sci.,* 33, 546, 1986.

95. **Eriksson, L., Holte, S., Bohm, C., Kesselberg, M., and Hovander, B.,** Automated blood sampling systems for positron emission tomography, *IEEE Trans. Nucl. Sci.,* NS-35, 703, 1988.

96. **Lammertsma, A. A., Frackowiak, R. S. J., Hoffman, J. M., Huang, S. C., Weinberg, I., Phelps, M. E., and Jones, T.,** Simultaneous measurement of regional cerebral blood flow and oxygen metabolism: a feasibility study, *J. Cereb. Blood Flow Metab.,* 7(Suppl. 1), S587, 1987.

97. **Daghighian, F., Huang, S. C., Weinberg, I., Hoffman, E. J., Hoffman, J. M., Grafton, S., Sumida, R., Digby, W., Mazziotta, J. C., and Phelps, M. E.,** On-line plasma separation and measurement of the arterial input function for dynamic PET oxygen studies, *J. Cereb. Blood Flow Metab.,* 9(Suppl. 1), S421, 1989.

98. **Herscovitch, P., Mintun, M. A., and Raichle, M. E.,** Brain oxygen utilization measured with oxygen-15 radiotracers and positron emission tomography: generation of metabolic images, *J. Nucl. Med.,* 26, 416, 1985.

99. **Blomqvist, G.,** On the construction of functional maps in positron emission tomography, *J. Cereb. Blood Flow Metabol.,* 4, 629, 1984.

100. **Evans, A. C.,** A double integral form of the three-compartmental, four-rate-constant model for faster generation of parameter maps, *J. Cereb. Blood Flow Metabol.,* 7(Suppl. 1), S453, 1987.

101. **Fox. P. T., Mintun, M. A., Raichle, M. E., et al.,** A non-invasive approach to quantiative functional brain mapping with $H_2^{15}O$ and positron emission tomography, *J. Cereb. Blood Flow Metab.,* 4, 329, 1984.

102. **Fox, P. T., Mintun, M. A., Reiman, E. M., and Raichle, M. E.,** Enhanced detection of focal brain responses using intersubject averaging and change-distribution analysis of subtracted PET images, *J. Cereb. Blood Flow Metabol.,* 8, 642, 1988.

103. **Fox, P. T., Perlmutter, J. S., and Raichle, M. E.,** A stereotactic method of anatomical localization for positron emission tomography, *J. Comput. Assist. Tomogr.,* 9, 141, 1985.

104. **Evans, A. C., Beil, C., Marrett, S., Thompson, C. J., and Hakim, A.,** Anatomical-functional correlation using an adjustable MRI-based region of interest atlas with positron emission tomography, *J. Cereb. Blood Flow Metab.,* 8, 513, 1988.

105. **Mintun, M. A., Fox, P. T., and Raichle, M. E.,** A highly accurate method of localizing regions of neuronal activation in the human brain with the positron emission tomography, *J. Cereb. Blood Flow Metabol.,* 9, 96, 1989.

106. **Moeller, J. R., Strother, S. C., Sidtis, J. J., and Rottenberg, D. A.,** Scaled subprofile model: a statistical approach to the analysis of functional patterns in positron emission tomographic data, *J. Cereb. Blood Flow Metab.,* 7, 649, 1987.

107. **Bajcsy, R., Lieberson, R., and Reivich, M.,** A computerized system for the elastic matching of deformed radiographic images to idealized atlas images, *J. Comput. Assist. Tomogr.,* 7, 618, 1983.

108. **Bohm, C., Greitz, T., Kingsley, D., Berggren, B., and Olsson, L.,** Adjustable computerized stereotactic brain atlas for transmission and emission tomography, *AJNR,* 4, 731, 1983.

109. **Bohm, C., Greitz, T., Kingsley, D., Berggren, B. M., and Olsson, L.,** A computerized individually variable stereotactic brain atlas, in *The Metabolism of the Brain Studied with Positron Emission Tomography,* Greitz, T., Ed., Raven Press, New York, 1985, 85.

110. **Herholz, K., Pawlik, G., Weinhard, K., and Heiss, W. D.,** Computer assisted mapping in quantitative analysis of cerebral positron emission tomograms, *J. Comput. Assist. Tomogr.,* 9, 154, 1985.

111. **Friston, K. J., Passingham, R. E., Nutt, J. G., Heather, J. D., Sawle, G. V., and Frackowiak, R. S. J.,** Localisation in PET images: direct fitting of the intercommissural (AC-PC) line, *J. Cereb. Blood Flow Metab.,* 9, 690, 1989.

112. **Herscovitch, P., Auchus, A. P., Gado, M., Chi, D., and Raichle, M. E.,** Correction of positron emission tomography data for cerebral atrophy, *J. Cereb. Blood Flow Metab.,* 6, 120, 1986.

113. **Condon, B., Patterson, J., Wyper, D., Hadley, D., Teasdale, G., Grant, R., Jenkins, A., Macpherson, P., and Rowan, J.,** A quantitative index of ventricular and extraventricular CSF fluid volumes using MR imaging, *J. Comput. Assist. Tomogr.,* 10, 784, 1986.

114. **Condon, B., Patterson, J., Wyper, D. J., Hadley, D. M., Grant, R., Teasdale, G., and Rowan, J.,** Intracranial CSF volumes determined using magnetic resonance imaging, *Lancet,* 8494, 1355, 1986.

115. **Clark, C., Hayden, M., Hollenberg, S., Li, D., and Stoessl, A. J.,** Controlling for cerebral atrophy in positron emission tomography data, *J. Cereb. Blood Flow Metab.,* 7, 510, 1987.

116. **Videen, T. O., Perlmutter, J. S., Mintun, M. A., and Raichle, M. E.,** Regional correction for the effects of brain atrophy in positron emission tomography, *J. Nucl. Med.,* 29, 773, 1988.

117. **Kety, S. S. and Schmidt, C. F.,** The nitrous oxide method for the quantitative determination of cerebral blood flow in man: theory, procedure and normal values, *J. Clin. Invest.,* 27, 476, 1948.

118. **Mazziotta, J. C. and Phelps, M. E.,** Human sensory stimulation and deprivation. PET results and strategies, *Ann. Neurol.,* 15(Suppl. 1), S50, 1984.

119. **Hoyer, S.,** Senile dementia and Alzheimer's disease. Brain blood flow and metabolism, *Prog. Neuro Psychopharmacol. Biol. Psychiatry,* 10, 447, 1986.

120. **Rodriguez, G., Warkentin, S., Risberg, J., and Rosadini, G.,** Sex differences in regional cerebral blood flow, *J. Cereb. Blood Flow Metab.,* 8, 783, 1988.

121. **Posner, M. I., Petersen, S. E., Fox, P. T., and Raichle, M. E.,** Localization of cognitive operations in the human brain, *Science,* 240, 1627, 1988.

122. **Sokoloff, L.,** Relationships among local functional activity, energy metabolism, and blood flow in the central nervous system, *Fed. Proc.,* 40, 2311, 1981.

123. **Fox, P. T., Miezin, F. M., Allman, J. M., Van Essen, D. C., and Raichle, M. E.,** Retinotopic organization of human visual cortex mapped with positron-emission tomography, *J. Neurosci.,* Vol. 7, 913, 1987.

124. **Fox, P. T., Burton, H., and Raichle, M. E.,** Mapping human somatosensory cortex with positron emission tomography, *J. Neurosurg.,* 67, 34, 1987.

125. **Lauter, J. L., Herscovitch, P., Formby, C., and Raichle, M. E.,** Tonotopic organization in human auditory cortex revealed by positron emission tomography, *Hearing Res.,* 20, 199, 1985.

126. **Roland, P. E., Skinhøj, E., and Lassen, N. A.,** Focal activations of human cerebral cortex during auditory discrimination, *J. Neurophys.,* 45, 1139, 1981.

127. **Roland, P. E., Meyer, E., Shibasaki, T., Yamamoto, Y. L., and Thompson, C. J.,** Regional cerebral blood flow changes in cortex and basal ganglia during voluntary movements in normal human volunteers, *J. Neurophysiol.,* 48, 467, 1982.

128. **Petersen, S. E., Fox, P. T., Posner, M. I., Mintun, M., and Raichle, M. E.,** Positron emission tomographic studies of the cortical anatomy of single-word processing, *Nature,* 331/6157, 585, 1988.

129. **Bryan, R. M., Jr. and Lehman, A. W.,** Cerebral glucose utilization after aversive conditioning and during conditioned fear in the rat, *Brain Res.,* 444, 17, 1988.

130. **Roland, P. E., Eriksson, L., Widén, L., and Stone-Elander, S.,** Changes in regional cerebral oxidative metabolism induced by tactile learning and recognition in man, *J. Cereb. Blood Flow Metab.,* 8, 642, 1988.

131. **Lassen, N. A. and Friberg, L.,** Physiological activation of the human cerebral cortex during auditory perception and speech revealed by regional increases in cerebral blood flow, *Scand. Audiol. Suppl.,* 30, 173, 1988.

132. **Fox, P. T. and Mintun, M. A.,** Noninvasive functional brain mapping by change-distribution analysis of averaged PET images of $H_2^{15}O$ tissue activity, *J. Nucl. Med.,* 30, 141, 1989.

133. **Meyer, E., Ferguson, S., Zatorre, R. J., Alivisatos, B., Marrett, S., Evans, A. C., and Hakim, A. M.,** Attention modulates somatosensory CBF response to vibrotactile stimulation as measured by PET, *Ann. Neurol.,* in press.

134. **Leblanc, R. and Meyer, E.,** Functional PET scanning in the treatment of AVMs, *J. Neurosurg.,* in press.

135. **Kearfott, K. J.,** Absorbed dose estimates for positron emission tomography (PET):$C^{15}O$, ^{11}CO, and $CO^{15}O$, *J. Nucl. Med.,* 23, 1031, 1982.

136. **Bigler, R. E. and Sgouros, G.,** Biological analysis and dosimetry for ^{15}O-labeled O_2, CO_2 and CO gases administered continuously by inhalation, *J. Nucl. Med.,* 24, 431, 1983.

137. **Rhodes, C. G., Wise, R. J. S., Gibbs, J. M., Frackowiak, R. S. J., Hatazawa, J., Palmer, A. J., Thomas, D. G. T., and Jones, T.,** *In vivo* disturbance of the oxidative metabolism of glucose in human cerebral gliomas, *Ann. Neurol.,* 14, 614, 1983.

138. **Snyder, W., Ford, M., Warner, G., and Watson, S.,** "S", Absorbed Dose per Unit Cumulated Activity for Selected Radionuclides and Organs, MIRD Pamphlet No. 11, Society of Nuclear Medicine, New York, 1975.

139. **Powers, W. J., Stabin, M., Howse, D., Eichling, J. O., and Herscovitch, P.,** Radiation absorbed dose estimates for oxygen-15 radiopharmaceuticals (H_2O, $C^{15}O$, $O^{15}O$) in newborn infants, *J. Nucl. Med.,* 29, 1961, 1988.

140. **Meyer, E., Yamamoto, Y. L., Evans, A. C., Tyler, J. L., Diksic, M., and Feindel, W.,** Radiation dose to upper airways from inhaled oxygen-15 carbon dioxide, *J. Nucl. Med.,* 28, 234, 1987.

141. **Thompson, C. J., Dagher, A., Meyer, E., et al.,** Imaging performance of a dynamic positron emission tomograph: Positome IIIp, *IEEE Trans. Med. Imag.,* 5, 183, 1986.

142. **Pantano, P., Baron, J.-C., Lebrun-Grandié, P., Duquesnoy, N., Bousser, M.-G., and Comar, D.,** Regional cerebral blood flow and oxygen consumption in human aging, *Stroke,* 15, 635, 1984.

143. **Brooks, D. J.,** Studies on regional cerebral function in multiple sclerosis using positron emission tomography, in *Current Problems in Neurology,* Vol. 3, Symp. Multiple Sclerosis, Rose, F. C. and Jones, R., Eds., John Libbey Eurotext, London, 1987, 129.

144. **Powers, W. J., Grubb, R. L., Jr., Darriet, D., and Raichle, M. E.,** Cerebral blood flow and cerebral metabolic rate of oxygen requirements for cerebral function and viability in humans, *J. Cereb. Blood Flow Metab.,* 5, 600, 1985.

145. **Lenzi, G. L., Gibbs, J. M., Frackowiak, R. S. J., and Jones, T.,** Measurement of cerebral blood flow and oxygen metabolism by positron emission tomography and the ^{15}O steady-state technique: Aspects of methodology reproducibility, and clinical application, in *Functional Radionuclide Imaging of the Brain,* Magistretti, L. Ed., Raven Press, New York, 1983, 291.

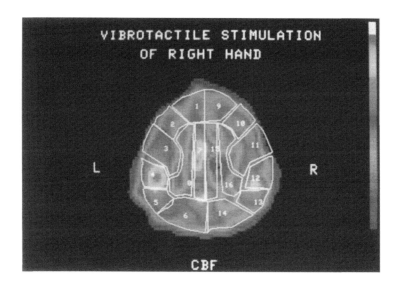

CHAPTER 9, PLATE 1.   Cerebral blood flow image (intravenous $H_2{}^{15}O$ bolus method) acquired on a normal volunteer during vibrotactile stimulation of the fingers of his right hand showing increased cerebral blood flow (CBF) in left somatosensory cortex.[133] A region of interest (ROI) template derived from a matched magnetic resonance image (MRI) of the same subject is superposed, confirming activated region (4) to be postcentral (sensory) gyrus.[104] The additional medial response corresponds to the supplementary motor area which was consistently activated by vibrotactile stimulation of the fingers.[124,133] (PET data acquired with Positome IIIp.[141])

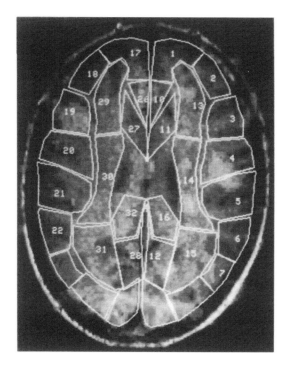

CHAPTER 9, PLATE 2.   Subtraction image (color), representing the difference between two normalized $H_2{}^{15}O$ bolus tissue activity images, that has been merged with its matched MRI image (black and white) upon which a customized ROI template has been superposed using the method of Evans et al.[104] The base line image was acquired while the subject was fixing a pulsating (0.5 Hz) cross hair on a computer screen. The activation image was obtained during passive viewing of images of animals presented on the screen at a rate of one image every 2 s. The residual tissue count activity observed in the occipital visual areas reflects increased neuronal activity due to passive picture processing (Scanditronix PC-2048B).

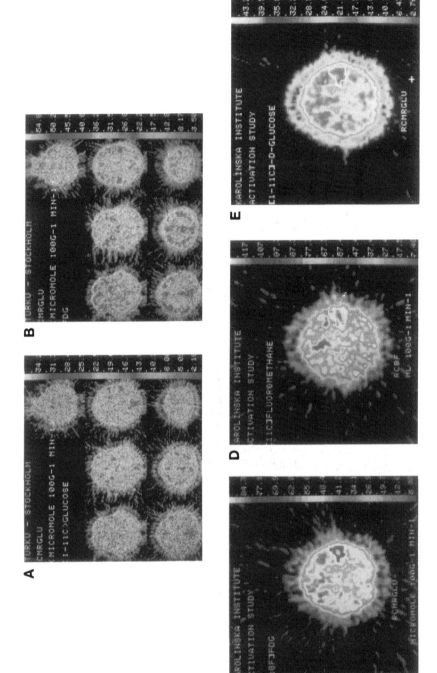

CHAPTER 10, PLATE 1.   rCMRgl measured in seven slices of the brain of a resting subject. (A) With 1-[^{11}C]-D-glucose (1-[^{11}C]glc); correction was made for loss of ^{11}C-CO$_2$, but not other labeled metabolites; (B) with ^{18}F-fluorodeoxyglucose (FDG). Higher rCMR values and more marked contrast between different areas with FDG. (C to E) Measurements during movements of the fingers of the right hand. (C) and (E) show rCMRglc measured with FDG and 1-[^{11}C]glc, respectively. In (D) rCBF was measured with CH$_3$F. Note that the activation of sensorimotor areas is much more pronounced in (C) and (D). The primary sensorimotor area is delineated with the aid of the data base of the computerized stereotactic atlas. (Courtesy of Dr. G. Blomqvist.)

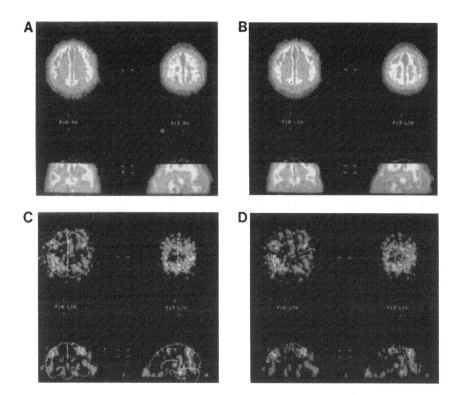

CHAPTER 10, PLATE 2. (A) Average image of rCBF during rest, measured withCH₃F. PET data have been reformated into a standardized anatomy. Mean of nine healthy subjects at rest. In (A to D) the slices of the upper panels are transverse; lower left slices are coronal and lower right slices are sagittal. (B) Average image from the same subjects during motor learning; (C) mean of difference images from the experiments of (A) and (B). In the color scale, red indicates the largest increase in rCBF. (D) "Significance images". Weighted differences are divided by their standard deviations pixel by pixel. Relevant anatomical areas are called in from the data base. They are identified as the precentral gyri (mainly activated on the left), left postcentral gyrus, bilateral SMA, right angular and supramarginal gyri (areas 39 and 40), and posterior part of right cuneus (area 7). (From Greitz, T., Bohm, Ch., Eriksson, L., Mogard, J., Roland, P. E., Seitz, R. J., and Wiesel, F. A., in *Visualization of Brain Functions*, Ottoson, D. and Rosène, W., Eds., Macmillan, London, 1989, 137. With permission.)

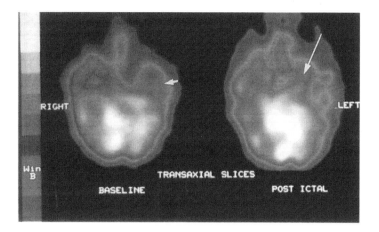

CHAPTER 13, PLATE 1. Transaxial interictal (base line) and postictal SPECT scans using ⁹⁹Tc-HMPAO in a patient with left temporal lobe epilepsy. Details of SPECT acquisition are described elsewhere.[64] The base line study shows *hypoperfusion* of the left temporal region (small arrow). The postictal study shows *hyperperfusion* in the left mesial temporal region (long arrow) with extensive lateral *hypoperfusion*.

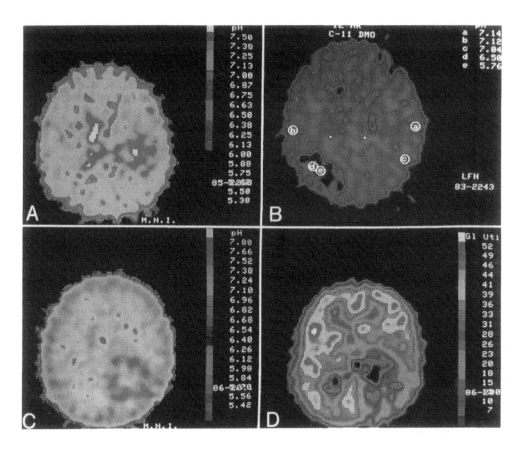

CHAPTER 14, PLATE 1.   The data follow convention that the patient's left is to the right of the page and that anterior is toward the top of the page. (A) pH measurement in subject with a malignant glioma, illustrating increased ^{11}C-DMO in the tumor consistent with tissue alkalosis. (B) pH measurement in subject, performed 12 h postonset of right MCA stroke. Samples of tissue pH marked a and b illustrate the normal pH in homologous regions, while samples marked d and e are in the affected region and show reduced (acidotic) tissue pH compared to c on the contralateral side. (C and D) Measurement of pH with ^{11}C-DMO and glucose metabolism with [18]FDG in a newly diagnosed glioma, with alkalosis and low glucose utilization in the solid part of the tumor. (Courtesy of M. Diksic, McGill University, Montreal.)

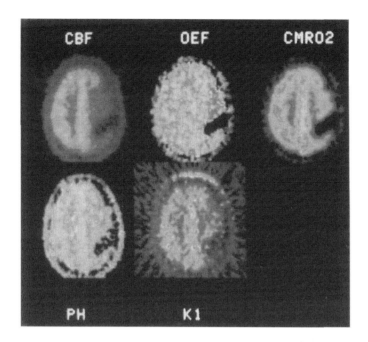

CHAPTER 14, PLATE 2. The data follow convention that the patient's left is to the right of the page and that anterior is toward the top of the page. Shown are five images, all of the same slice, indicating the distribution of cerebral blood flow (CBF), oxygen extraction fraction (OEF), oxygen consumption (CMRO$_2$), tissue pH (PH), and the blood-to-tissue rate constant for CO$_2$ in a case of acute stroke measured 11 h after the ictus. The CBF image shows an extensive, virtually hemispheric hypoperfusion, with a smaller region of nearly absent perfusion in the posterior quadrant of the left MCA territory. The OEF image shows increased OEF (in red) in the hypoperfused region and decreased (nearly zero) OEF in the region with "no flow". Oxygen consumption is effectively absent in the "no flow region", but preserved in the region of hypoperfusion. The corresponding pH data show normal pH as yellow, with tissue acidosis (yellow-green-blue) in much of the region with increased OEF. It should be noted that pH cannot be inferred from regions with very low or no flow because tracer is not delivered to such tissue. The blood-to-tissue rate constant correlates well with the distribution of CBF. (Courtesy of Massachusetts General Hospital, Boston.)

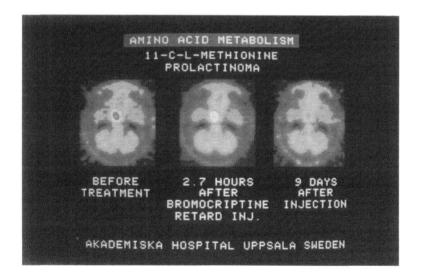

CHAPTER 16, PLATE 1. Prolactinoma analyzed with methionine before and at several time intervals after bromocryptine treatment. (Courtesy of Prof. M. Bergström, Upsalla, Sweden.)

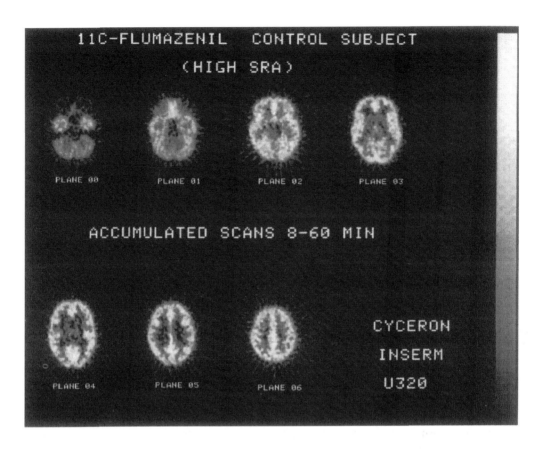

CHAPTER 18, PLATE 1. Brain images obtained with the LETI-TTV 03 PET camera in a young untreated healthy volunteer following intravenous bolus injection of ^{11}C-flumazenil at high specific radioactivity. The images shown are accumulated over the 8- to 60-min post-injection interval. The scans shown have an axial resolution (slice thickness) and a lateral resolution of 9 and 5.5 mm, respectively; seven planes are studied with interplane void of 3 min and start at the cerebellum level (plane 00) up to the upper hemispheric convexity (plane 06). The pseudocolor scale ranges from black (no radioactivity) to white (maximal pixel value). The distribution of ^{11}C-flumazenil at these late time intervals represents essentially tracer specifically bound to the central benzodiazepine receptors (cBZR). Maximum binding occurs in the cerebral cortex (especially visual cortex), intermediate levels occur in the cerebellum, and low binding occurs in the basal ganglia and thalamus; in hemispheric white matter, consistent with the known distribution of cBZR in the human brain, the concentration of ^{11}C-FLU should represent unbound tracer only.

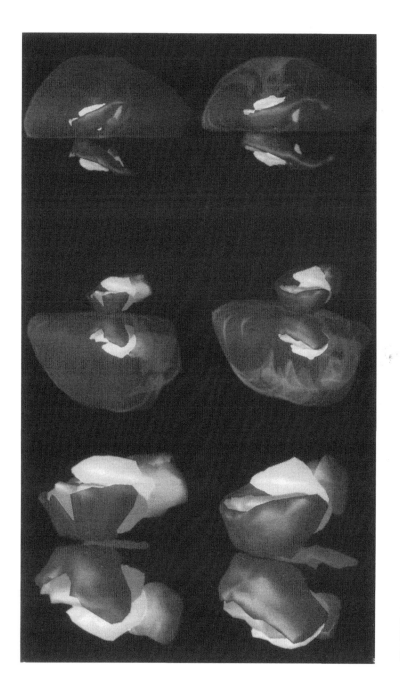

CHAPTER 20, PLATE 1.   Three-dimensional region of interest (ROI) reconstructions. The figure depicts the corresponding three-dimensional (3D) reconstructions of selected ROIs of a normal individual (top panel) and a schizophrenic patient (bottom panel) whose MRI sections are shown in Figure 1. The left-hand set of patient-control images depicts the lateral and third ventricles (blue), the caudate nucleus (orange), the lenticular nucleus (yellow), and the amygdaloid complex as seen from in front. This computer-generated 3D image can be used to visually check the accuracy with which specific ROIs were defined on two-dimensional (2D) images. The bottom row of images shows considerable lateral and third ventricular dilatation in the schizophrenic patient. The same computer program also gives volumetric estimates of the structures outlined in Figure 1. The center and right-hand sets of patient-control scans depict an anterior and posterior view of the hemispheres and the central cerebral gray matter structures.

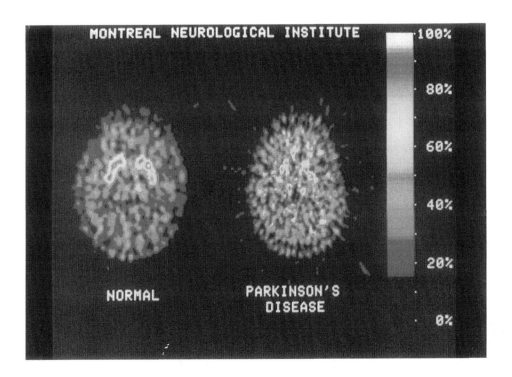

CHAPTER 21, PLATE 1. 6-FD PET studies comparing a normal subject and a patient with Parkinson's disease. Note the reduced radioactivity in the regions corresponding to the striatum.

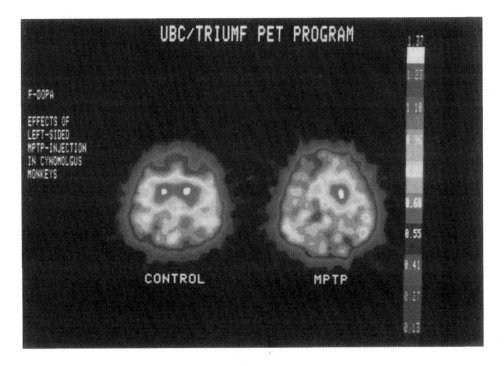

CHAPTER 21, PLATE 2. 6-FD PET studies of a normal cynomolgus monkey (left) and a hemiparkinsonian monkey after the unilateral administration of MPTP into the left internal carotid artery. Note the reduction of radioactivity in the striatal region on the lesioned side.

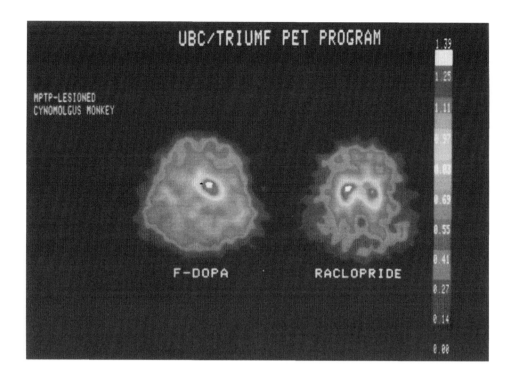

CHAPTER 21, PLATE 3. PET studies of a hemiparkinsonian monkey with 6-FD (left) and raclopride (right) showing reduced presynaptic uptake and increased D-2 receptor binding on the lesioned side.

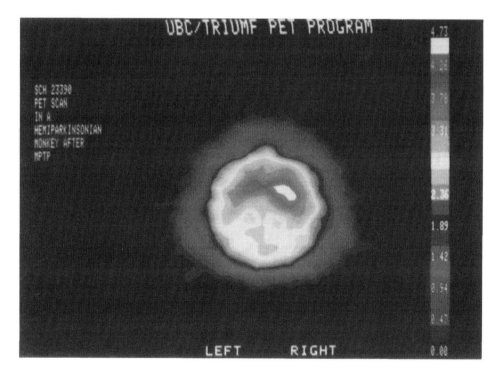

CHAPTER 21, PLATE 4. PET study in a hemiparkinsonian monkey with [^{11}C]SCH-23390 to study D-1 receptors. There is reduced binding on the lesioned side after MPTP administration.

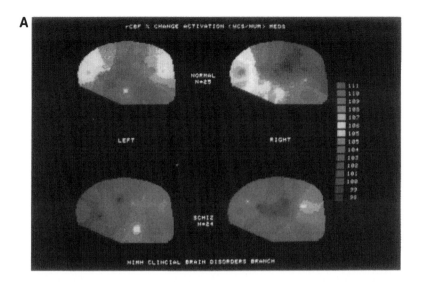

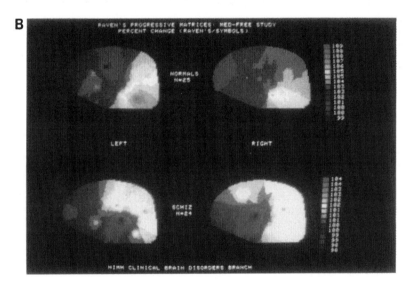

CHAPTER 22, PLATE 1. Lateral maps of group mean gray matter regional cortical blood flow for normal subjects and patients with schizophrenia during two different complex mental tasks. (A) Wisconsin Card Sorting Test (WCS); (B) Raven's Progressive Matrix Test (RPM). The anterior pole is at the left in all images. rCBF values for the reasoning tasks are expressed as a percentage of flow during a matched sensorimotor control task. Note that although both tasks involve abstract reasoning and problem solving, the patterns of activation for the normal subjects are quite different with maximal activation frontally during the WCS, but posteriorly during RPM. Moreover, differences between the patients and the controls were demonstrated only during the frontal lobe task (the WCS). Data are from References 33 and 41.

Chapter 10

PET STUDIES OF THE HUMAN BRAIN DURING MENTAL ACTIVITY: SOME THEORETICAL AND PRACTICAL CONSIDERATIONS

Lennart Widén

TABLE OF CONTENTS

I. INTRODUCTION

Even though certain aspects of cognition like elementary learning and certain types of memory can be studied in animals, the rather stereotyped experimental conditions unavoidably require both sensory stimulation and motor response. Only in man can pure cognition be studied. This can be done with essentially two methods: psychological and neurophysiological, or a combination of both. Electrophysiological methods are not suitable for this purpose since it is impossible to record simultaneously the electrical activitity of the whole brain. In noninvasive electrophysiolgicial investigations of the human brain (EEG), only about one fifth of the surface area of the cerebral cortex is accessible to study. Future development of magnetoencephalography (MEG) may provide us with a method of investigating the electrophysiology of some subcortical structures.

However, neuronal activation can also be studied by measurements of changes in the rate of regional cerebral energy metabolism (rCMR) or blood flow (rCBF). Pioneering studies using the Lassen-Ingvar technique of measuring the rCBF[1] were performed by Ingvar and Risberg and by Roland et al. They showed that ''mental effort''[2,3] or pure mental activity without stimulation or motor response[4-6] is paralleled by changes in the rCBF patterns. These were important observations that stimulated continued research in this field.

However, there are some limitations inherent in the nontomographic measurements of rCBF using external detection of gamma-emitting tracers administered by intracarotid injection (for a more detailed discussion see, e.g., References 7 and 8). Data are mainly recorded from the surface of the cerebral cortex. Furthermore, the occipital lobe and parts of the temporal lobe of the hemisphere ipsilateral to the site of injection and the entire contralateral hemisphere are not seen because they are not reached by the injected radiotracer. With the more recently developed intravenous or inhalation techniques, all parts of both hemispheres are accessible to measurements. The accurate quantification of data is then, however, further complicated by extracerebral radioactivity in the field of view. Therefore, subtle changes during physiological stimulation may be difficult to interpret.

The advent of positron emission tomography (PET), therefore signified a remarkable breakthrough for human brain research in general and for the study of brain-mind relationships in particular. The PET technique[9] has the following advantages compared to the earlier techniques referred to above:

1. It is a tomographic method and consequently all regions of the brain are accessible to measurements.
2. Modern PETs have a spatial resolution of 4 to 5 mm, i.e., much superior to the nontomographic methods.
3. With adequate corrections of the positron camera data for dead time, random coincidences and attenuation (correction for events lost due to scatter and absorption of the radiation) accurate quantification can be achieved.[9]
4. No only rCBF but also rCMR can be measured (and in addition a number of other important physiological parameters).

The detection of photon pairs emitted by the positron annihilation makes the PET technique superior to single photon emission computed tomography (SPECT).[10] Since bioisotopes are used in PET, the possible applications are many more than for SPECT. The short half-lives of these isotopes are useful assets, permitting administration of large doses of activity with tolerable radiation exposure and enabling repeated studies to be performed.

The purpose of this chapter is to review briefly some practical and technical considerations when performing PET studies of mental activity and to present a few examples illustrating the potential of PET in such studies. The aim is not to catalogue the literature regarding the results obtained in various studies. For more comprehensive reviews, see for example, References 8 and 11.

A comparatively large section of this chapter is devoted to methodological problems for a number of reasons. Mental events are certainly an elusive materia and it is apparent that studies of brain-mind relationships with PET make special demands on the design of the experiments. Therefore, some space is devoted to this problem. Further, the choice of the proper parameter for measurements of regional "brain work" is important, and this matter is discussed in the light of recent findings concerning the relations between rCBF—the classic indicator of brain work—and regional oxygen and glucose consumption. Some problems in the interpretation of data are pointed out and new methods of selection of regions of interest are briefly described. They aim both at accurate identification of anatomical structures and minimization of observer bias.

Only a few examples of PET studies of mental activity are given. The selection is highly subjective. The purpose is to demonstrate the potential of PET in investigations of various aspects of mental activity and in addition to present some new information on its neurobiological substrate. Whenever possible, attempts are made to relate the PET findings to results obtained from animal experiments, clinical and neuropsychological studies and theories. PET studies on aging and abnormal mental activity are not included in this brief review.

II. METHODOLOGICAL CONSIDERATIONS

A. BEHAVIORAL CONTROL

In PET studies of the effects of various types of mental activities on the rCBF or rCMR patterns, it is imperative to have the functional states of the experimental subjects as well defined as possible. This is not an easy task. In the numerous tomographic and nontomographic studies of the effects of sensory stimuli or mental activation, it has been customary to compare the activated state with the "resting state". This reference state implies that the subjects are immobile and seemingly relaxed though not asleep. All external stimuli that are not inherent in the basic measurement situation are avoided.

Even in this resting state, however, a PET measurement involves unavoidable external stimuli. There are certainly interindividual differences concerning the response to these stimuli, including affective reactions to the experimental situation as a whole. Even though experience shows that most subjects seem to overcome quickly their initial apprehension of having their brains examined, a varying degree of anxiety may remain in a few subjects. This may influence the results since it has been shown that apprehensiveness or anxiety increases the average CMR and CBF[12] and that these changes show regional differences indicating special involvement of limbic structures.[13-15]

Furthermore, even if the external environment and the behavior of the experimental subjects can be reasonably well controlled with close supervision, recording of EEG, EMG, eye movements, etc., nevertheless, no one can control what is going on in their brains. They are instructed to try to be relaxed and not to think of anything and yet avoid falling asleep. It is not unlikely, however, that in spite of this, some kind of thinking is pursued. It has been shown that the rCBF and $rCMRO_2$ patterns are influenced by the type of thinking going on.[16] Consequently, interindividual differences in the type of thinking as well as intraindividual differences between different occasions of examination would result in differences in cerebral metabolic patterns, which may contribute to the wide variation in the rCMR and rCBF values of the resting state found in most studies.

It may therefore be advantageous to have instead of a passive resting state an operationally well-defined control state of minimal stimulation, which makes the experimental subjects focus their attention on the stimulus and thereby to a certain extent constraining the content of their mental activity. The PET images of this control state are then subtracted from the activated state pixel-by-pixel. When the object of study is complex cognitive operations, the experimental design should be a multiple-level subtractive hierarchy, in which each task state adds a small number of adequately defined operations to those of the preceding, subordinate control state (cf. Reference 17).

B. STIMULATION PROCEDURES

In studies of the effects of pure mental activation, such as focusing of attention, mental arithmetic, silent speech, mental image processing, semantic processing, etc., the important thing is to avoid all kinds of external environmental change and motor activation as the experimental subjects switch from the "resting state" to mental activation. Accordingly, they have to be carefully trained, supervised, and accustomed to the experimental situation. Other types of investigations, e.g., of sensory or motor learning, word-processing, etc., unavoidably involve both sensory stimuli and motor activity. In these cases, it is imperative to have adequate control of factors such as rate and intensity of the stimuli which are known to be quantitatively related to the metabolic response.[18-20]

In a recent study of rCMRO$_2$ during tactile learning and recognition of complicated geometrical "nonsense" objects for instance, it was found that the frequency of manipulatory exploratory movements during tactile recognition was twice as high as that during tactile learning which significantly influenced the PET measurements due to the higher influx of information to the somatosensory receiving areas.[21]

C. MEASUREMENT OF REGIONAL "BRAIN WORK"
1. Relations Between rCBF, rCMRglc, and rCMRO$_2$

The interpretation of the results of the numerous nontomographic rCBF studies of cerebral responses to different types of physiological activation is based upon the assumption that there is a close coupling between the rate of cerebral blood flow and the rate of energy metabolism, i.e., oxygen consumption, so that the rCBF is regulated by the metabolic demands of functional activity. (For a recent review, see Reference 22.) Though this dogma can be dated to the well-known paper by Roy and Sherrington in 1890,[23] it has not been substantiated by animal experimental work for the obvious reason that there is no technique available for regional determination of the rCMRO$_2$. However, measurements of rCBF and deoxyglucose (DG) uptake in anesthetized animals[24] have shown excellent regional correlation between CBF and the rate of glucose consumption (rCMRglc) as measured with the deoxyglucose method.[25] Similar regional correlations between CBF, CMRglc, and CMRO$_2$ were found in PET studies of resting humans.[26] Later it was shown in experiments on rats with double label autoradiography[27] that there was an overall preservation of the coupling between local CBF (lCBF) and lCMRglc during somatosensory activation. Furthermore, the percentage increments of lCBF and lCMRglc closely matched.

However, in a PET study of the effect of somatosensory stimulation on rCMRO$_2$ and rCBF in human subjects, Fox and Raichle,[28] while confirming the "excellent" regional correlation between the two variables in the resting state, made the surprising observation of a regional uncoupling during neuronal activation. The focal augmentation of CBF (mean, 29%) induced by somatosensory stimulation far exceeded the concomitant local increase in CMRO$_2$ (mean, 5%). In a later PET study using visual stimulation,[29] a similar uncoupling during neural activation was found. The additional observation was made that there was a proportionate augmentation of rCBF and rCMRglc (50 and 51%, respectively), whereas rCMRO$_2$ increased only by 5%. It was concluded that transient neural activation causes a glucose utilization in excess of that consumed by oxidative metabolism and regulates rCBF for purposes other than oxidative metabolism.

These findings of the St. Louis group were confirmed in essence by the Stockholm PET

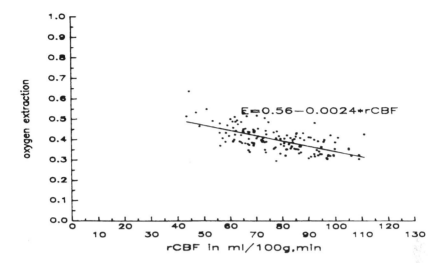

FIGURE 1. Representative example of one of the rOEF vs. rCBF plots. The solid line is the result of a least squares fit.

group.[30,31] They measured the rCBF and the regional oxygen extraction fraction (rOEF) in the resting state and during somatosensory stimulations of different strengths. A linear relationship was found between rCBF and rOEF in the examined flow range of 20 to 120 ml/100 g/min (Figure 1). The higher the rCBF, the lower was the corresponding rOEF. This relationship was independent of rest or activation. If upon activation CBF increases in a region, the rOEF and hence the rCMRO$_2$ directly follows from the flow-extraction dependence found. Quantitatively the results indicated a 25 to 30% CMRO$_2$ increase for a 50% flow increase, i.e., substantially higher than the one reported by Fox et al.[29]

According to the above-mentioned findings of Fox et al.[29] the fraction of nonoxidatively metabolized glucose—generally estimated to be up to 15% of the total cerebral glucose uptake[32]—is larger during transient increases in neural activity. This fraction can be directly determined by a comparison between the value for rCMRglc measured according to the DG method[25] or the Fick principle, which measure the total glucose uptake, and the value obtained with [11C]-glucose selectively labeled in the carbon 1 position. With the last mentioned tracer, only the oxidative glucose consumption is measured, provided the loss of [11C]-CO$_2$ is adequately corrected for.[33]* The studies of Blomqvist et al. have shown that the difference between regional DG and 1-[11C]-D-glucose uptake increases with increasing total CMRglc. Consequently, it shows regional variations and changes between rest and activation (see Plate 1**).[34]

Blomqvist et al. further showed that there is an egress of nonoxidatively produced acid metabolites from the brain, the accumulated average value of which amounts to 5% of the tissue radioactivity during the first 15 min after bolus injection of 1-[11C]-glucose. The observation that increased cerebral glucose uptake implies an increased fraction of nonoxidative glycolysis is supported by animal experiments[35,36] showing increases in tissue lactate during activation of somatosensory and visual systems.

The recent findings concerning the relationship between cerebral blood flow and metabolism undeniably have somewhat upset the conventional conceptions and raise questions

---

* By application of the proper kinetic model, the loss of labeled CO$_2$ can be estimated without actual measurement of the cerebral arteriovenous difference.

** Plate 1 follows page 198.

concerning the proper way of measuring changes in regional brain work during mental activation. Supposedly, the aim of most earlier studies has been to measure regional changes in the rate of cerebral "energy metabolism" in response to physiological stimulation, either directly via $rCMRO_2$ or rCMRglc or indirectly via rCBF. The concept of "energy metabolism" is less well defined but probably often considered synonymous with the rate of oxidative metabolism. It is evident, then, that only measurements of $rCMRO_2$ or rCMRglc, using l-[^{11}C]-glucose as a tracer, give an estimate of the regional rate of energy metabolism in this strict sense.*

The role of the nonoxidative fraction of glucose uptake in regional brain work is not quite clear. It is a priori unlikely that it serves the energy production, since as is well known the anaerobic glycolysis produces only one-sixteenth of the energy yield of aerobic glucose breakdown. It might at least partially be incorporated into glycogen.[37] This part of it, however, must be very small because if we compensate for the loss of radioactivity from the brain due to egress of labeled CO_2 and acid metabolites, then the value of CMRglc obtained using the tracer l-[^{11}C]-glucose is very close to the one we get with the deoxyglucose model.[34]

2. Practical Considerations

In addition to the basic problems discussed above, there are practical considerations to be given to the choice of technique of measuring neuronal activation during mentation. The DG method in its application for PET[38-40] requires 40 to 45 min of steady state which obviously is difficult to achieve in many psychophysiological studies. With l-[^{11}C]-glucose as a tracer, the measurement period does not need to be longer than 12 to 15 min[33] but even this may be too long for many types of experiments. For the calculation of $rCMRO_2$ with the autoradiographic single scan method of Mintun et al.[41] measurements of oxygen uptake, rCBF, and the regional blood volume, rCBV, are needed. The regional oxygen extraction fraction (rOEF) is first calculated from the oxygen uptake, rCBF, and rCBV data. The $rCMRO_2$ is then obtained from the relationship:

$$CMRO_2 = rOEF \times rCBF \times CaO_2$$

where CaO_2 is the total oxygen content of arterial blood. For the computation of rOEF or rCBF, a total scanning time of only 40[41] or 80[42] s is needed and for rCBV a couple of minutes. This is a great advantage in experiments using various cognitive tasks. A drawback, however, is that three measurements are required for each one of the resting and test situations. It is of course difficult to make sure that the functional state of the brain is exactly the same on the three different occasions. A further disadvantage is the amount of radiation exposure which precludes comparison of more than two different states in the same individual.[21]

It seems therefore at present that the most suitable method for PET studies of regional brain responses to mental activation is in fact measurement of the rCBF changes. It is true that these not only reflect the oxidative metabolism, but are also regulated by other, partially unknown mechanisms.[43,44] Nevertheless, rCBF changes serve as excellent indicators of changes in the degree of neuronal activation. Their measurements are fast and can be repeated several times without exceeding acceptable levels of radiation exposure.

Common tracers for PET measurements of rCBF are ^{15}O-water[45] and [^{11}C]-CH_3F.[46,47] Water has the advantages of easy administration (bolus intravenous injection) and very short half-life of the isotope, making repeated measurements possible at short intervals. It is, however, diffusion limited[48] which leads to a certain underestimation of rCBF. The magnitude

---

* With U[^{11}C]-glucose as a tracer, the loss of labeled CO_2 is so large and rapid that an adequate correction is difficult to achieve.[26]

of this error increases with the rate of blood flow. Another disadvantage is the relatively large positron range of ^{15}O which contributes to unsharpness of the ^{15}O-water images. Fluoromethane, on the other hand, is freely diffusible and gives accurate measurements over the whole range of rCBF, but has to be inhaled, at least in its present mode of application, which may cause some inconveniences. The drawback of the longer half-life of ^{11}C can be overcome by subtraction of the cerebral background activity remaining after a preceding PET measurement. The freely diffusible tracer butanol, labeled either with ^{11}C,[49] or ^{15}O,[50] is an interesting alternative to fluroromethane.

D. PROBLEMS OF INTERPRETATION

It is generally agreed that the processes requiring most energy in the central nervous system (CNS) are the ionic pumps. Furthermore, it has been shown that the neuropil accounts for most of the metabolic activity. During stimulation an increase in CMRglc occurs in the region of the nerve terminals due to their synaptic activity, whereas there is little or no change in metabolic rate of the cell bodies.[51,52] These observations have two important but rarely discussed implications for the interpretation of the results of mental activation.

Firstly, an alteration—augmentation or decrease—of the local metabolism somewhere in the brain may be sign of a change of excitability of cell bodies situated at a distance from the locus of the observed change. This may lead to faulty conclusions concerning the localization of functions tested by psychophysiological experiments.

Secondly, processes related to synaptic activity in nerve terminals releasing inhibitory transmitters are assumed to be as energy consuming as those of excitatory neurons.[53] If this is true, there does not seem to be at present a way of distinguishing between excitatory and inhibitory processes in PET studies. The importance of this uncertainty is hard to evaluate. "Neuronal activation" always implies a mixture of excitation and inhibition at the very local level. There is perhaps not a single nerve cell in the CNS that does not send collaterals to inhibitory interneurons in addition to the main fiber, participating in excitatory pathways. The interaction between adjacent cortical modules is largely inhibitory. "Nowhere is there uncontrolled excitation."[54]

It is evident, therefore, that even in a single voxel in PET there is a mixture of excitatory and inhibitory events. The same is of course true for the large fields that according to Roland[55] constitute units of cortical activation, at least as revealed by PET and nontomographic rCBF studies. It seems, however, reasonable to assume that within a given field either excitatory or inhibitory processes predominate but it is not possible to distinguish between these two opposite functional states with the aid of PET. Several studies of cerebral information processing, for instance, have shown bilateral activation of homologous cortical fields. By the same reasoning as above, caution should be exercised in the interpretation of the results. It can not be excluded—even though it may seem less likely—that in some of these cases there is a unilateral excitation and concomitant contralateral inhibition.

The primary aim of PET studies of the effect of mental activity on the brain is of course to locate and quantitate changes in rCMR or rCBF in activated areas. Spontaneous fluctuations of these variables in nonactivated regions influencing the value of the global CBF (gCBF) would be a confounding factor. Even though it has been shown that the CBF in nonactivated brain areas is quite stable over time,[20] a method of linear normalization has been introduced, intended to negate the effects of gCBF fluctuations.[56] It consists of multiplying every pixel of each CBF scan by a correction factor equal to the mean standard gCBF [50 ml/(100 g × min)] divided by the individual scan gCBF, before making rCBF comparisons. It has been claimed, however, that this normalization procedure introduces errors (over- and under-estimation) in the quantification of rCBF.[57] For a discussion of these problems, the reader is referred to the original papers.

E. ANATOMICAL LOCALIZATION AND SELECTION OF ROIs

It is a truism to point out that in studies of cerebral physiology including effects of mental activation, it is of fundamental importance to have as accurate localization and identification of the structures involved as possible. The topic of the procedures for anatomical localization is further dealt with in Chapter 9. Suffice it to say here that the combined use of a stereotactic head fixation system,[58] adaptable to both PET, CT and MRI, and a computerized stereotactic atlas has considerably improved the accuracy of anatomical localization and also substantially reduced observer bias in the selection of regions of interest.[59]

In addition, it provides excellent facilities for comparisons between groups, e.g., of resting and stimulated subjects, respectively. MR images are taken of each subject equipped with the stereotactic helmet. Then the individual MR images are fitted to the computerized brain atlas, i.e., they are transformed into a standard brain. By applying the inverse atlas transformations to the PET data volume, it is possible to relate the PET information to the reference atlas instead of the anatomy of the experimental subject or patient. In this way, all brains will be anatomically alike and all differences observed will be functional. Changes in rCBF or rCMR are calculated by pixel-by-pixel subtractions of intrasubject pairs (task state minus control state) of PET images. Population mean, variance, and significance are calculated and demonstrated pictorially. Finally, anatomical structures are projected over these images. This procedure is not only accurate, but also highly objective (see Plate 2*).

One of the limitations of the PET method—often pointed out by its critics—is the poor spatial resolution, hampering detailed identification of structures involved in, for example, internal brain work. It is well known that two point sources of radioactivity occurring simultaneously in the field of view of a PET machine cannot be resolved unless they are separated by at least one FWHM. In functional studies, however, in which some kind of activation paradigm is employed, one can make use of subtraction methods, as discussed above, to improve the effective spatial resolution. This was first pointed out by the St. Louis group and applied by them in a series of studies.[60,61] Loci of neuronal activation separated by less than one FWHM can be distinguished from one another by sequential activation and determination of the location of each response within subtraction-format images. It is claimed that in rCBF studies using this procedure a localization accuracy of 0.5 to 1.5 mm can be attained with an image resolution of 18 mm (FWHM) and rCBF changes of 20 to 30%. Since modern PET instruments have a spatial resolution of 4 to 5 mm, the precision achieved should be still better.

III. PET STUDIES OF BRAIN/MIND RELATIONSHIPS

According to general textbooks in psychology, mental activity includes various kinds of psychic phenomena such as sensation, perception, recognition, focusing attention, learning, recall, various types of thinking from passive day dreaming to active problem solving, etc. All these types of mental activity may have their special emotional and volitional qualities. In the following, a few PET studies will be reviewed in which attempts have been made to investigate the neurobiological substrate of some of the above components of human consciousness.

A. VISUAL IMAGERY

The first question to be answered is whether pure mental activity without any external stimulation influences the rCMR pattern. This is to be expected from earlier two-dimensional rCBF studies.[4-6,62] Roland et al.[47] recently investigated this problem in a study in which the rCMRO$_2$ was measured (according to the technique of Mintun et al.[41]) in ten healthy vol-

---

* Plate 2 follows page 198.

unteers at rest and during visual imagery, implying that they imagined walking around in the familiar surroundings of their home following the streets according to instructions given prior to the experiment. They should concentrate on the appearance of the environs and try to recall it in color. The rationale behind this test was that, once started, the visual imagery went on in the mind of the subjects until they were stopped after 180 s. The results showed that this particular kind of mental activity raised the $rCMRO_2$ in 25 cortical fields of homotypical cortex, ranging in size from 2 to 10 cm.[3] Subcortically, the $rCMRO_2$ increased in the neostriatum and posterior thalamus. These focal metabolic increases were so strong that the $CMRO_2$ of the whole brain increased by 10%.

Sokoloff et al.[63] measured $CMRO_2$ with the Kety-Schmidt technique when subjects were at rest and when they tried to solve mathematical problems. No change of the $CMRO_2$ was found. As a tentative explanation of this negative finding, Sokoloff et al. suggested that $CMRO_2$ increased in some active regions while it decreased in others such that the net change in metabolism was zero. Rolant et al.,[47] however, did not find any decreases in $rCMRO_2$ during visual imagery.

During the imagined walk, the subjects did not perform any visible movements, and there were no $rCMRO_2$ increases in the cortical motor areas. Neither was there a change of metabolism in the calcarine cortex or the immediately surrounding cortex. The largest increases were found in the posterior, superior parietal cortex as well as in the superior and lateral prefrontal cortex.

What new information was acquired from these experiments? First of all they demonstrated that pure thinking is energy consuming which is probably in agreement with the subjective impressions of most people. The particular type of thinking studied implied retrieval of stored visual information. Since the only change that occurred in connection with the increases in rCMR was the switch from "resting state" to visual imagery, it is reasonable to assume that the alterations of the CMR pattern were in some way related to the mobilization of visual memories. Some of the activated areas might then be the loci of memory storage, others might be involved in the process of retrieval.

There is evidently not one single cortical area, the activation of which is required for mental imagery. Instead, there are a number of activated cortical fields, the pattern of which is characteristic of this particular type of mental activity. This finding lends support to current theories of the neurophysiological basis of memory which state that "no single memory center exists and many parts of the nervous system participate in the representation of a single event" (see Reference 64, page 123).

It is further assumed that "the processing systems that analyze information also participate in and influence the representation of that information" (Reference 64, page 129) and that "the neural plasticity that underlies long-term memory storage probably occurs preferentially in the higher reaches of modality-specific pathways and in polymodal areas as opposed to early-stage, cortical processing areas" (Reference 64, page 126). It is striking how well these assumptions, based on rather indirect evidence, acquired from animal experiments and psychological studies agree with our PET findings. There was no activation of the primary visual area or the immediate surrounding cortex whereas a maximum of activation was found in the posterior superior parietal cortex. Experiments on monkeys and a few clinical studies of humans with cerebral lesions have led to the conclusions that there are at least two major pathways from the striate cortex, one of which follows a dorsal route into the parietal lobe. This is the so-called motion pathway conveying information of movement and spatial awareness.[65,66] The activated area of the parietal lobe may therefore very well be the storehouse for certain aspects of the total visual information that has been received by the experimental subjects during their walks in the surroundings of their homes.

The other major visual pathway, the ventral one, is concerned with object recognition and leads into the temporal lobe. There were significant increases in $CMRO_2$ in the posterior inferior temporal cortex though not as great as in the parietal lobe. (This part of the temporal cortex, however, was subject to partial volume effects.)

It was shown in early two-dimensional rCBF studies[67] that memorization activates prefrontal regions in addition to activating posterior, modality specific areas.[68] The prefrontal areas are assumed to participate in the intrinsic organization of brain work.[69] The lateral prefrontal cortex (area 46 in particular) seems to participate in cortical networks subserving spatial mnemonic processes in the macaque. In this animal, there are heavy anatomical connections between prefrontal and parietal association areas, characterized by topographic relationships, reciprocity and parallellism.[70] They are also unified by their input from the pulvinar. In the experiments by Roland et al.[47] there was a highly significant increase in $CMRO_2$ in the posterior thalamus, which must at least in part have originated from the pulvinar. According to Goldman-Rakic,[70] the medial pulvinar, which is particularly prominent in primates, is in position to recruit a neural system defined by cortico-cortical connectivity and possibly dedicated to the function of being oriented in time and space. The coactivation of the prefrontal and parietal areas as well as the posterior thalamus during the experimental subjects' imagined walk in a three-dimensional visual space was thus in agreement with previous experimental findings on the macaque.

Traditionally it was believed that whereas the left hemisphere is involved with language processing, the right is engaged in the production of mental imagery. However, we did not find any significant differences between the right and left hemispheres during visual imagery. Accordingly, both hemispheres seem to be involved in the production of mental images. This is in agreement with results of neuropsychological tests, recently presented by Kosslyn et al. They conclude[71] that imagery is carried out by multiple processes in the two hemispheres even though not all of these processes are implemented equally effectively in both hemispheres.

B. LEARNING AND RECOGNITION

It is generally agreed that there are multiple memory systems and also different functional components of memory. One of these is the process of storing information, i.e., learning; another one is recognition, which is perhaps a special form of retrieval of information, possibly subserved by different brain mechanisms from recall.[72] In a recent paper by Roland et al.,[21] attempts were made to identify cerebral structures, differentially activated by tactile learning and recognition of complicated geometrical objects which were manipulated and examined with the right hand. The $rCMRO_2$ was measured in these two functional states as well as during rest. Video-recordings were made during all PET measurements and it was found that the frequency of exploratory finger movements during tactile recognition was twice that of tactile learning.

The experimental situation involved both motor activity (finger movements of the right hand) and somatosensory stimulation which by themselves caused widespread activation of cortical and subcortical structures and made the analysis of the results more complicated. There were no differences between the structures that were activated during learning and recognition. $rCMRO_2$ increased in identical areas in the prefrontal cortex, motor and somatosensory cortices, insula, lingual gyrus, hippocampus, basal ganglia, thalamus, parasagittal anterior cerebellum, and neocerebellum. There were, however, quantitative differences: in tactile recognition, during which the frequency of movements of the right hand was higher than during learning, the $rCMRO_2$ increases were correspondingly larger in the left premotor, supplementary, and somatosensory hand areas. In spite of the higher sensorimotor activity during tactile recognition, the $rCMRO_2$ increases in the lateral neocerebellum were larger during learning, which was an unexpected finding (see below).

Since the cortical structures activated in this series of experiments have been shown in earlier studies (see in Reference 21) to participate in motor and sensory control of exploratory movements and tactile discrimination we could not point out specific cortical sites of storage and retrieval of the tactile information.

However, the high rCMRO$_2$ in neocerebellum during tactile learning deserves some comments. It can hardly be due to the motor activity per se, since this was lower than during recognition. Furthermore, it has been shown that movements of the fingers of one hand elicit rCBF increases bilaterally, parasagittally in the anterior part of the cerebellum and not in the lateral part of the neocerebellum.[73] There were no learning effects on the exploratory finger movements, i.e., no signs of motor learning. Moreover, motor learning has recently been shown to activate the parasagittal, not the lateral cerebellum.[74] It seems, therefore, that the extra oxygen consumption in the lateral neocerebellum is most easily explained as due to energy-demanding processes, related to storage of the tactile information. It is noteworthy that these parts of the neocerebellum are not activated by visual learning.[75]

The involvement of the neocerebellum in somatosensory learning has been demonstrated experimentally (for reviews, see References 76 and 77). The basic mechanism proposed is that the climbing fibers depress the excitability of the simultaneously activated parallel fiber synapses on the Purkinje cells. In this way, the synaptic effectiveness of the parallel fibers is selected in accordance with the active climbing fiber territory. It was shown by Gilbert and Thach[78] that climbing fiber responses were much more frequent during the learning phase of a sensorimotor task than after the information was stored.

A single impulse in a climbing fiber elicits a prolonged and complex EPSP in the Purkinje cell, inducing a large ionic transfer in the hundreds of synapses on the cell and concomitant energy requirement of the ionic pumps. The increased rCMRO$_2$ in the lateral neocerebellum during tactile learning might thus be explained by the increased climbing fiber activity during the learning phase.

This seems to be the first time that the involvement of cerebellum in learning has been demonstrated in humans.

C. SINGLE WORD PROCESSING

Speech and language—essential characteristics of the human species—have been studied with two-dimensional rCBF as well as with PET methods. It is striking that both speech production and listening to spoken words or reading elicit bilateral responses. Asymmetries are often reported, but the degree of asymmetry as well as the location of areas of maximal activation differ between investigators. The interpretation of data is often difficult because the experimental situations have been complex, involving vocalization, visual, or auditory stimulation. In a recent two-dimensional rCBF study of "silent speech" (counting), however, activated areas were mainly confined to the dominant hemisphere.[79] For recent reviews, see References 8, 11, and 80.

A model PET study of the cortical anatomy of single-word processing was recently published by Petersen et al.[17] In this work, there is a clearly defined problem and a flawless, step-by-step analysis of three levels of single-word processing. The aim of the study was to determine which one of two different theories of lexical processing is in accordance with experimental results. One of the theories, the cognitive model, states that visually and auditorily perceived words involve separate, modality-specific codes with parallel, independent pathways to shared meaning and output codes. In contrast, the "neurological" model implies a serial processing, with recoding of visual signals into an auditory-based code—assumed to take place in the angular gyrus—before they are relayed and recoded in the semantic and output networks.

rCBF was measured in 17 normal volunteers using $^{15}O_2$-labeled water as tracer. Four task states formed a three-level subtractive hierarchy, each state adding a small number of operations to those of its subordinate or control state. By subtracting the control state from the stimulus state, areas of activation were identified and related to those mental operations that were present in the stimulated state, but not in the control state. In this way, areas involved in semantic processing could be distinguished from those involved in speech, involuntary word-form processing and passive sensory input.

Task-state minus control-state gave images of rCBF changes related to the operations of each cognitive level. Intersubject averaging was used with the aid of a stereotactic method of anatomical localization.

For a detailed description of activated areas, the reader is referred to the original paper.[17] It is sufficient to mention here that at the lowest levels of the hierarchically organized tasks, those of passive sensory processing and of word-form processing, modality-specific primary and nonprimary sensory regions were activated, i.e., two different computional levels. These were separate for visual and auditory activation.

The next step was to identify areas for articulatory coding and motor output. These were activated when words were repeated aloud and were similar for visual and auditory presentation. The "highest level" in this context was activated by an association task (saying a use for each word presented). The area in question was located in the left anterior inferior frontal lobe and considered to be specific to semantic language tasks and related to a semantic network involved in this type of word associations.

The results of this careful PET study were not consistent with the "neurological", serial model according to which visual information has to be phonologically recoded in the angular gyrus and then semantically processed in Wernicke's area before output coding. Instead, they strongly support the cognitive model of separate modality-specific coding networks with common semantic and output codes.

A few other observations in this study should be mentioned since they illustrate the potential of PET in the analysis of the neurobiological substrate of cognitive processes. The specific response to visually presented words was a bilateral activation of a small set of areas in the prestriate region as far anterior as the occipital-temporal boundary. These areas have so far only been activated by these types of visual stimuli.[81] It is concluded that they may represent a network coding for visual word form. This hypothesis is supported by the fact that cases of pure alexia have been described, caused by lesions of the occipital-temporal boundary.

Similarly, an area was found that responded specifically to word auditory stimuli, namely, the temporoparietal region near the angular and supramarginal gyri. It was left-lateralized. It is well known that lesions of this area cause disturbances of the comprehension of spoken language. Based on this fact and the PET findings it can be considered a phonological coding region.

Evidently, complex visual and phonological analyses are processed in the areas mentioned above. These PET findings are in agreement with recently published evidence for modality-specific meaning systems in the brain, acquired from studies of patients with discrete lesions of the cerebral cortex.[82,83]

D. ATTENTION SYSTEMS

Some experiments in the study by the St. Louis group[17] were designed to distinguish between structures subserving semantic processes and those related to "attention for action". The attention system is assumed not to be related to any particular sensory or cognitive content but is involved in selecting operations controlling output systems. In this study it was activated in experiments requiring detection of a great number of targets. Whereas semantic processing involved the anterior inferior frontal lobe (on the left side), vigilance tasks with multiple targets activated the anterior cingulate gyrus which, therefore, is believed to be part of an anterior system of attention for action. Again, PET findings are congruent with clinical observations. It has been shown that lesions of the anterior cingulate produce akinetic mutism.

E. EMOTIONS

Most "normal" emotions are difficult to study experimentally for obvious reasons. One exception is anticipatory and performance anxiety which has been shown to influence the cerebral metabolic pattern.[13-15] Recently, Reiman et al. extended their interesting PET studies on panic disorders to include also normal anticipatory anxiety.[84] Patients with a certain type of panic disorder have an abnormal hemispheric asymmetry of parahippocampal CBF and $CMRO_2$ in the nonpanic state.[85] During the production of lactate-induced anxiety attacks in these patients, there are rCBF increases in a number of areas, among them the temporal poles. Also in a state of "normal" anxiety in healthy volunteers expecting an unpleasant electric shock, bilateral rCBF increases were demonstrated in the temporal poles. The findings are interpreted as indicating a final common pathway involving the temporal poles both for normal anticipatory and lactate-induced anxiety. It is interesting that also two-dimensional rCBF studies have shown involvement of fronto-temporal areas in anxiety states in healthy male volunteers.[15]

IV. CONCLUDING REMARKS

PET is still a relatively new technique for investigations of human brain function. Much time and effort have been dedicated to technical development, improvement of the scanners, production of new tracers, tracer kinetic modeling, and diverse clinical studies. Only a minor part of PET research has been devoted to the neurobiological basis of mental operations. Naturally, early studies in this field were hampered by diverse technical shortcomings. However, important technical advances have been made, particularly concerning scanners, identification of brain structures and objectivity of observations through the use of stereotactic methods of anatomical localization and use of stereotactic atlases. The development of new tracers enables us to perform several measurements with short scanning times on the same subject and apply subtraction image methods. In this way the rCMR or rCBF responses to different task states in a hierarchy can be isolated and the interpretation of the results is made much less ambiguous.

PET is at present better suited for studies of gross structure-function relationships than any other available technology. Important new information on the functional anatomy of some types of mental operations have already been acquired. Theories and hypotheses based on results from animal experiments, neuropsychological investigations and clinical studies can be tested. The field is in a stage of rapid and very promising development. Knowledge of the normal pattern of rCBF/rCMR during various forms of mental activity enables us to disclose deviations caused by functional disturbances or gross organic lesions of the brain.[11]

"Regional brain activation" in the sense this term is used in PET studies is synonymous with increases in rCMR or rCBF. In the future, other physiological parameters may be investigated, like synaptic transmission in various neurotransmitter systems or the functional states of ion channels.

REFERENCES

1. **Lassen, N. A. and Ingvar, D. H.,** The blood flow of the cerebral cortex determined by radioactive krypton-85, *Experientia,* 17, 42, 1961.
2. **Ingvar, D. H. and Risberg, J.,** Influence of mental activity upon regional cerebral blood flow in man, *Acta Neurol. Scand.,* 41(Suppl. 14), 93, 1965.
3. **Risberg, J. and Ingvar, D. H.,** Patterns of activation in the grey matter of the dominant hemisphere during memorization and reasoning, *Brain,* 96, 737, 1973.

4. **Ingvar, D. H. and Philipson, L.,** Distribution of cerebral blood flow in the dominant hemisphere during motor ideation and motor performance, *Ann. Neurol.,* 2, 230, 1977.

5. **Roland, P. E., Larsen, B., Lassen, N. A., and Skinhöj, E.,** Supplementary motor area and other cortical areas in organization of voluntary movements in man, *J. Neurophysiol.,* 43, 118, 1980.

6. **Roland, P. E.,** Somatotopical tuning of postcentral gyrus during focal attention in man. A regional cerebral blood flow study, *J. Neurophysiol.,* 46, 744, 1981.

7. **Risberg, J.,** Regional cerebral blood flow, in *Experimental Techniques in Human Neuropsychology,* Hannay, H. J., Ed., Oxford University Press, New York, 1986, chap. 15.

8. **Raichle, M. E.,** Circulatory and metabolic correlates of brain function in normal humans, in *Handbook of Physiology,* Mountcastle,V., Ed., Vol. 5, II, *Higher Functions of the Brain,* Plum, F., Ed., Williams & Wilkins, Baltimore, 1987, 643.

9. **Eriksson, L., Dahlbom, M., and Widén, L.,** Positron emission tomography—a new technique for studies of the central nervous system, *J. Microscopy,* 157, 305, 1990.

10. **Lassen, N. A., Vorstrup, S., and Mickey, B.,** Cerebral blood flow tomography using xenon-133 inhalation: methods and clinical applications, in *Positron Emission Tomography,* Reivich, M. and Alavi, A., Eds., Alan R. Liss, New York, 1985, 219.

11. **Pawlik, G. and Heiss, W.-D.,** Positron emission tomography and neuropsychological function, in *Neuropsychological Function and Brain Imaging,* Sigler, E. D., Ronald, A. Y., and Turkheimer, E., Eds., Plenum Press, New York, 1989, 65.

12. **Kety, S. S.,** Circulation and metabolism of the human brain in health and disease, *Am. J. Med.,* 8, 205, 1950.

13. **Reivich, M., Gur, R., and Alavi, A.,** Positron emission tomographic studies of sensory stimuli, cognitive processes and anxiety, *Hum. Neurobiol.,* 2, 25, 1983.

14. **Hagstadius, S.,** Brain Function and Dysfunction, Ph.D. thesis, Lund University, Lund, Sweden, 1989, 22.

15. **Johanson, A. M., Risberg, J., Silfverskiöld, P., and Smith, G.,** Regional changes in cerebral blood flow during increased anxiety in patients with anxiety neurosis, in *The Roots of Perception,* Hentschel, U., Smith, G, G., and Draguns, J. G., Eds., Elsevier, Amsterdam, 1986, 353.

16. **Roland, P. E. and Widén, L.,** Quantitative measurements of brain metabolism during physiological stimulation, in *Functional Brain Imaging,* Pfurtscheller, G. and Lopes da Silva, G. H., Eds., Huber, Toronto, 1988, 213.

17. **Petersen, S. E., Fox, P. T., Posner, M. I., Mintun, M., and Raichle, M. E.,** Positron emission tomographic studies of the cortical anatomy of single-word processing, *Nature,* 331, 585, 1988.

18. **Miyaoka, M., Shinohara, M., Batipps, M., Pettigrew, K. D., Kennedy, C. and Sokoloff, L.,** The relationship between the intensity of the stimulus and the metabolic response in the visual system of the rat, *Acta Neurol. Scand.,* 60 (Suppl. 70), 16, 1979.

19. **Toga, A., Horenstein, S., and Collins, R. C.,** The effect of stimulus rate and pattern in ^{14}C-deoxyglucose utilization in the visual system of the rat, *Soc. Neurosci. Abstr.,* 5, 811, 1979.

20. **Fox, P. T. and Raichle, M. E.,** Stimulus rate dependence of regional cerebral blood flow in human striate cortex, demonstrated by positron emission tomography, *J. Neurophysiol.,* 51, 1109, 1984.

21. **Roland, P. E., Eriksson, L., Widén, L., and Stone-Elander, S.,** Changes in regional cerebral oxidative metabolism induced by tactile learning and recognition in man, *Eur. J. Neurosci.,* 1, 3, 1989.

22. **Yarowsky, P. J. and Ingvar, D. H.,** Neuronal activity and energy metabolism, *Fed. Proc.,* 40, 2353, 1981.

23. **Roy, C. S. and Sherrington, C. S.,** On the regulation of the blood supply of the brain, *J. Physiol. London,* 11, 85, 1890.

24. **Sokoloff, L.,** Relationships among local functional activity, energy metabolism, and blood flow in the central nervous system, *Fed. Proc.,* 40, 2311, 1981.

25. **Sokoloff, L., Reivich, M., Kennedy, C., des Rosiers, M. H., Patlak, C. S., Pettigrew, K. D., Sakurada, O., and Shinohara, M.,** The [^{14}C] deoxyglucose method for the measurement of local cerebral glucose utilization: theory, procedure, and normal values in the conscious and anesthetized albino rat, *J. Neurochem.,* 28, 897, 1977.

26. **Baron, J. C., Lebrun-Grandie, P., Collard, P., Crouzel, C., Mestelan, G., and Bousser, M. G.,** Noninvasive measurement of blood flow, oxygen consumption, and glucose utilization in the same brain regions in man by positron emission tomography: concise communication, *J. Nucl. Med.,* 23, 391, 1982.

27. **Ginsberg, M. D., Dietrich, W. D., and Busto, R.,** Coupled forebrain increases of local cerebral glucose utilization and blood flow during physiologic stimulation of a somatosensory pathway in the rat, *Neurology,* 37, 11, 1987.

28. **Fox, P. T. and Raichle, M. E.,** Focal physiological uncoupling of cerebral blood flow and oxidative metabolism during somatosensory stimulation in human subjects, *Proc. Natl. Acad. Sci. U.S.A.,* 83, 1140, 1986.

29. **Fox, P. T., Raichle, M. E., Mintun, M. A. and Dence, C.,** Nonoxidative glucose consumption during focal physiologic neural activity, *Science,* 241, 462, 1988.

30. **Eriksson, L., Roland, P., Dahlbom, M., Blomqvist, G., and Widén, L.,** Coupling between regional cerebral metabolic rate of oxygen and regional cerebral blood flow during, *J. Cereb. Blood Flow Metab.,* 9(Suppl. 1), 578, 1989.

31. **Eriksson, L., Roland, P., Dahlbom, M., Blomqvist, G., Ingvar, M., and Widén, L.,** Coupling between regional cerebral oxidative metabolism and regional cerebral blood flow, *J. Cereb. Blood Flow Metab.,* submitted.

32. **Sokoloff, L.,** Circulation and energy metabolism of the brain, in *Brain Neurochemistry,* Siegel, G., Agranoff, B., Albers, R. W., and Molinoff, P., Eds., Raven Press, New York, 1989, 591.

33. **Blomqvist, G., Stone-Elander, S., Halldin, Ch., Roland, P., Widén, L., Lindqvist, M., Swahn, C.-G., Långström, B., and Wiesel, F. A.,** Positron emission tomographic measurements of cerebral glucose utilization using 1-[^{11}C]-D-glucose, *J. Cereb. Blood Flow Metab.,* 10, 467, 1990.

34. **Blomqvist, G., Halldin, C., Stone-Elander, S., Roland, P. E., Swahn, K. G., Haaparanta, M., Solin, O., Lindqvist, M., and Widén, L.,** A comparative PET study of CMRglu using [1-^{11}C]-D-glucose, [^{18}F] fluorodeoxyglucose, and the Fick principle, *J. Cereb. Blood Flow Metab.,* 9 (Suppl. 1), 123, 1989.

35. **Hossmann, K.-A. and Linn, F.,** Regional energy metabolism during functional activation of the brain, *J. Cereb. Blood Flow Metab.,* 7, S297, 1987.

36. **Prichard, J. W., Petroff, O. A. C., Ogino, T., and Shulman, R.,** Cerebral lactate evaluation by electroshock: a ^{1}H magnetic resonance study, *Ann. N.Y. Acad. Sci.,* 508, 54, 1987.

36A. **Lear, J. L. and Ackermann, R. F.,** Why the deoxyglucose method has proven so useful in cerebral activation studies: the unappreciated prevalence of stimulation-induced glycolysis, *J. Cereb. Blood Flow Metab.,* 9, 911, 1989.

37. **Hof, P. R., Pascale, E., and Magistretti, P. J.,** K$^+$ concentrations reached in the extracellular space during neuronal activity promotes a Ca^{++} dependent glycogen hydrolysis in mouse cerebral cortex, *J. Neurosci.,* 8, 1922, 1988.

38. **Reivich, M., Kuhl, D., Wolf, A., Greenberg, J., Phelps, M., Ido, T., Casella, V., Fowler, J., Gallagher, B., Hoffman, E., Alavi, A., and Sokoloff, L.,** Measurement of local cerebral glucose metabolism in man with [^{18}F]2-fluoro-2deoxy-D-glucose, *Acta Neurol. Scand.,* 56 (Suppl. 64), 190, 1977.

39. **Phelps, M. E., Huang, S. C., Hoffman, E. J., Selin, C., Sokoloff, L., and Kuhl, D. E.,** Tomographic measurement of local cerebral glucose metabolic rate in humans with [^{18}F]2-fluoro-2-deoxy-D-glucose: validation of method, *Ann. Neurol.,* 6, 371, 1979.

40. **Reivich, M., Alavi, A., Wolf, A., Greenberg, J. H., Fowler, J., Christman, D., MacGregor, R., Jones, S. C., London, J., Shive, C., and Yonecura, Y.,** Glucose metabolic rate kinetic model parameters determination in humans: the lumped constants and rate constants for [^{18}F] fluorodeoxyglucose and [^{11}C]deoxyglucose, *J. Cereb. Blood Flow Metab.,* 5, 179, 1985.

41. **Mintun, M. A., Raichle, M. E., Martin, W. R. W., and Herscovitch, P.,** Brain oxygen utilization measured with ^{15}O radiotracers and positron emission tomography, *J. Nucl. Med.,* 25, 177, 1984.

42. **Holden, J. E., Eriksson, L. A., Roland, P. E., Stone-Elander, S., and Widén, L.,** Direct comparison of single scan autoradiographic with multiple scan least-squares fitting approaches to PET CMRO$_2$ estimation, *J. Cereb. Blood Flow Metab.,* 8, 671, 1988.

43. **Paulson, O. B. and Newman, E. A.,** Does the release of potassium from astrocyte endfeet regulate cerebral blood flow?, *Science,* 237, 896, 1987.

44. **Lou, H. C., Edvinsson, L., and Mac Kenzie, E. T.** The concept of coupling blood flow to brain function: revision required?, *Ann. Neurol.,* 22, 289, 1987.

45. **Herscovitch, P., Markham, J., and Raichle, M. E.,** Brain blood flow measured with intravenous H$_2$^{15}O. I. Theory and error analysis. *J. Nucl. Med.,* 24, 782, 1983.

46. **Stone-Elander, S., Roland, P., Eriksson, L., Litton, J. E., Johnström, P., and Widén, L.,** The preparation of ^{11}C-labeled fluoromethane for the study of regional cerebral blood flow using positron emission tomography, *Eur. J. Nucl. Med.,* 12, 236, 1986.

47. **Roland, P. E., Eriksson, L., Stone-Elander, S., and Widén, L.** Does mental activity change the oxidative metabolism of the brain? *J. Neurosci.,* 7, 2373, 1987.

48. **Eichling, J. O., Raichle, M. E., Grubb, R. L., Jr., and Ter-Pogossian, M. M.,** Evidence of the limitations of water as a freely diffusible tracer in brain of the rhesus monkey, *Circ. Res.,* 35, 358, 1974.

49. **Herscovitch, P., Raichle, M. E., Kilbourne, M. R., and Welch, M. J.,** Positron emission tomographic measurement of cerebral blood flow and permeability surface area product of water using ^{15}O water and ^{11}C butanol, *J. Cereb. Blood Flow Metab.,* 7, 527, 1987.

50. **Berridge, M. S., Cassidy, E. H., and Miraldi, F. D.,** A routine synthesis of ^{15}O labeled butanol for positron tomography, *J. Nucl. Med.,* in press.

51. **Mata, M., Fink, D. J., Gainer, H., Smith, C. B., Davidsen, L., Savaki, H., Schwartz, W. J., and Sokoloff, L.,** Activity-dependent energy metabolism in rat posterior pituitary primarily reflects sodium pump activity, *J. Neurochem.,* 34, 213, 1980.

52. **Kadekaro, M., Crane, A. M., and Sokoloff, L.,** Differential effect of electrical stimulation of sciatic nerve on metabolic activity in spinal cord and dorsal root ganglion in the rat, *Proc. Natl. Acad. Sci. U.S.A.,* 82, 6010, 1985.

53. **Ackerman, R. F., Finch, D. M., Babb, T. L., and Engel, J.,** Increased glucose metabolism during long-duration recurrent inhibition of hippocampal pyramidal cells, *J. Neurosci.,* 4, 251, 1984.

54. **Popper, K. R. and Eccles, J. C.,** in *The Self and its Brain,* Springer-Verlag, New York, 1981, 243.

55. **Roland, P. E.,** Application of imaging of brain blood flow to behavioral neurophysiology: the cortical field activation hypothesis, in *Brain Imaging and Brain Function,* Sokoloff, L., Ed., Raven Press, New York, 1985, 87.

56. **Fox, P. T., Fox, J. M., Raichle, M. E., and Burde, R. M.,** The role of cerebral cortex in the generation of voluntary saccades: a positron emission tomographic study, *J. Neurophysiol.,* 54, 348, 1985.

57. **Seitz, R. J. and Roland, P. E.,** Errors in quantification of rCBF due to normalization, *J. Cereb. Blood Flow Metab.,* 9 (Suppl.1), 432, 1989.

58. **Bergström, M., Boethius, J., Eriksson, L., Greitz, T., Ribbe, T., and Widén, L.,** Head fixation device for reproducible positron alignment in transmission CT and positron emission tomography, *J. Comput. Assist. Tomogr.,* 5, 136, 1981.

59. **Greitz, T., Bohm, Ch., Eriksson, L., Mogard, J., Roland, P. E., Seitz, R. J., and Wiesel, F. A.,** The construction of a functional brain atlas: Elimination of bias from anatomical variations at PET by reformating three dimensional data into a standardized anatomy, in *Visualization of Brain Functions,* Ottoson, D. and Rostène W., Eds., Macmillan, London, 1989, 137.

60. **Fox, P. T., Burton, H., and Raichle, M. E.,** Mapping human visual cortex with positron emission tomography, *Nature,* 323, 806, 1986.

61. **Mintun, M. A., Fox, P. T., and Raichle, M. E.,** A highly accurate method of localizing regions of neuronal activation in the human brain with positron emission tomography, *J. Cereb. Blood Flow Metab.,* 9, 96, 1989.

62. **Roland, P. E. and Friberg, L.,** Localization of cortical areas activated by thinking, *J. Neurophysiol.,* 53, 1219, 1985.

63. **Sokoloff, L., Mangold, R. L., Wechsler, C., Kennedy, C., and Kety, S.,** The effect of mental activity on cerebral circulation and metabolism, *J. Clin. Invest.,* 34, 1101, 1955.

64. **Squire, L. R.,** *Memory and Brain,* Oxford University Press, New York, 1987.

65. **Maunsell, J. H. R. and Newsome, W. T.,** Visual processing in monkey extrastriate cortex, *Annu. Rev. Neurosci.,* 10, 363, 1987.

66. **Zeki, S. and Shipp, S.,** The functional logic of cortical connections, *Nature,* 335, 311, 1988.

67. **Risberg, J. and Ingvar, D. H.,** Patterns of activation in the gray matter of the dominant hemisphere during memorization and reasoning, *Brain,* 96, 737, 1973.

68. **Ingvar, D. H.,** "Memory of the future": an essay on the temporal organization of conscious awareness, *Hum. Neurobiol.,* 4, 127, 1985.

69. **Roland, P. E.,** Cortical organization of voluntary behavior in man, *Hum. Neurobiol.,* 4, 155, 1985.

70. **Goldman-Rakic, P. S.,** Topography of cognition: parallel distributed networks in primate association cortex, *Annu. Rev. Neurosci.,* 11, 137, 1988.

71. **Kosslyn, S. M.,** Aspects of a cognitive neuroscience of mental imagery, *Science,* 240, 1621, 1988.

72. **Tulving, E.,** Multiple memory systems and consciousness, *Hum. Neurobiol.,* 6, 67, 1987.

73. **Fox, P. T., Raichle, M. E., and Thach, W. T.,** Functional mapping of the human cerebellum with positron emission tomography, *Proc. Natl. Acad. Sci. U.S.A.,* 82, 7462, 1985.

74. **Seitz, R. J., Roland, P. E., Bohm, Ch., Greitz, T., Eriksson, L., and Stone-Elander, S.,** Functional mapping of motor learning in man, *J. Cereb. Blood Flow Metab.,* 9 (Suppl. 1), 577, 1989.

75. **Roland, P. E. and Seitz, R. J.,** Mapping of learning and memory functions in the human brain, in *Visualization of Brain Functions,* Ottoson, D. and Rostène, W., Macmillan, London, 1989, chap 14.

76. **Ito, M.,** *The cerebellum and neural control,* Raven Press, New York, 1984.

77. **Thompson, R. T.,** The neurobiology of learning and memory, *Science,* 233, 941, 1986.

78. **Gilbert, P. F. C. and Thach, W. T.,** Purkinje cell activity during motor learning, *Brain Res.,* 128, 309, 1977.

79. **Ryding, E., Brådvik, B., and Ingvar, D. H.,** Silent speech activates circumscribed speech centers selectively in the dominant hemisphere, *J. Cereb. Blood Flow Metab.,* 9 (Suppl. 1), 317, 1989.

80. **Ingvar, D. H.,** Serial aspects of speech, *Hum. Neurobiol.,* 2, 177, 1983.

81. **Snyder, A. Z., Peterson, S., Fox, P., and Raichle, M. E.,** PET studies of visual word recognition, *J. Cereb. Blood Flow Metab.,* 9 (Suppl. 1), 576, 1989.

82. **Mc Carthy, R. A. and Warrington, E. K.,** Evidence for modality-specific meaning system in the brain, *Nature,* 334, 428, 1988.

83. **Marshall, J. C.,** Sensation and semantics, *Nature,* 334, 378, 1988.

84. **Reiman, E. M., Raichle, M. E., Robins, E., Fusselman, M. J., Fox, P. T., Mintun, M. A., and Price, J. L.,** Involvement of temporal poles in pathological and normal forms of anxiety, *J. Cereb. Blood Flow Metab.,* 9 (Suppl. 1), 589, 1989.

85. **Reiman, E. M., Raichle, M. E., Robins, E., Butler, F. K., Herscovitch, P., Fox, P., and Perlmutter, J.,** The application of positron emission tomography to the study of panic disorder, *Am J. Psychiatry,* 143, 469, 1986.

Chapter 11

USE OF PET TO EVALUATE ACUTE STROKE AND OTHER CEREBROVASCULAR DISORDERS

K. Herholz and W.-D. Heiss

TABLE OF CONTENTS

I. INTRODUCTION

Use of positron emission tomography (PET) in cerebrovascular disease permits *in vivo* measurements of a variety of basic processes in the pathophysiology of ischemia and their neurological sequelae. Obviously, the most crucial point is the delivery of oxygen to brain tissue. If this process fails, neuronal loss or frank tissue infarction will occur. However, insufficient oxygen supply appears to be only the initial event triggering a complex cascade of metabolic derangements, often leading to final tissue necrosis in a process of infarct maturation even if blood flow and oxygen supply were restored after some time.[1-3] A large variety of tracers useful in the analysis of blood flow, oxygen extraction and consumption, glucose consumption, blood-brain barrier permeability, and pH are now available for clinical applications, providing unique possibilities to study the *in vivo* pathophysiology of cerebral ischemia in humans. Furthermore, new tracers for investigation of amino acid uptake and protein synthesis and receptor-binding ligands may also contribute to a more detailed understanding of infarct maturation in the future.

Under clinical aspects, it is of primary diagnostic and prognostic interest to check cerebral blood flow (CBF) and oxygen consumption in patients with suspected cerebrovascular disease. PET methods employing the ultrashort-living isotope ^{15}O (half-life of 2.1 min) are best suited for this task. It should be noted, and will be explained later in detail, that measurement of resting CBF alone will usually not be of much diagnostic value because primary changes caused hemodynamically cannot be distinguished from secondary alterations due to neuronal damage or dysfunction.

There has been only little progress in treatment of acute ischemic stroke during the last decades.[4] Even the effectiveness of commonly used concepts like isovolemic hemodilution is not well established. There is an urgent need to understand better which events after onset of ischemia are potentially reversible and which are not and to continue the search for effective measures to prevent secondary ischemic damage at an early stage. PET offers the possibility for testing hypotheses that have been derived from animal experiments in a noninvasive way in humans and to assess objectively the effects of therapeutic trials.

Current treatment of chronic cerebrovascular disease is mainly based on prevention of embolization of atheromatous material and control of risk factors for progression of cerebral atherosclerosis. In contrast, attempts to improve cerebral hemodynamics have been less successful.[5,6] Apparently, regulation of CBF in patients with cerebrovascular disease is not yet very well understood. PET studies have begun to attack this problem and may also help to clarify the role of ischemia or tissue hypoxia in slowly progressing senile dementia.

Most patients experience a variable degree of clinical recovery after a stroke. In part, recovery may be explained by resolution of brain edema and other less well-known factors causing transient impairment of brain function. Functional plasticity, i.e., the ability of some brain structures to adapt to and partially compensate for the dropout of infarcted tissue, may also play a role. Certain rehabilitation measures have been shown to be effective in facilitating these abilities. PET is a tool to elucidate the underlying neurophysiological processes since, apart from their alterations in acute ischemia, regional CBF and cerebral glucose metabolism (CMRGl) are also indicators of local neuronal activity in intact brain tissue.

The following review is an attempt to summarize what already has been achieved by the use of PET in cerebrovascular disease. First, a brief overview on the methods used clinically in the field will be given, followed by the results of their applications to acute stroke, chronic and reversible ischemia, multi-infarct syndrome, hemorrhages, functional remote effects of lesions, and recovery.

II. METHODS

In this paragraph a brief outline of the methods used clinically for investigation of cerebrovascular disease will be given. We will only address issues of direct relevance for clinical application; a more detailed review and evaluation has been compiled recently[7] and special aspects of modeling and chemistry are presented elsewhere in this book. A variety of tracers have been suggested to complement measurements of blood flow, oxygen, and glucose metabolism. Since they have not been applied systematically to cerebrovascular disease, they will only be mentioned briefly in connection with the particular clinical condition.

A. CEREBRAL BLOOD FLOW

Since the basic work of Kety and Schmidt,[8] most CBF measurements rely on diffusible tracers, whose wash-in and wash-out rate after local or systemic application depends on CBF. For calculation of CBF from measured tracer concentrations, the time course of arterial tracer input must be known, and knowledge of the tissue-blood distribution coefficient of the tracer is also required. Details of the calculation depend on the particular physical and biomedical constraints of each implementation, but the basic principles are rather the same for all CBF PET methods in current clinical use. Usually multiple arterial blood samples are collected from a radial artery during CBF measurement which with appropriate precautions — according to current experience — carries no significant risk of severe complications.[9]

Conceptually, the most straightforward implementation is the intravenous [^{15}O]-water bolus technique of Raichle et al.[10] The method has been validated against an intracarotid tracer injection technique. It relies essentially on the wash-in phase of the tracer in brain during the first 40 s after application. A correction for beginning wash-out during that period is applied, and arterial blood samples are needed for absolute quantitation. Its strength lies in its rapid repeatability, relatively low radiation exposure to the patient,[11] low demands with regard to patient cooperation, and a rather linear relation between measured tissue activity and CBF. Its limitations are incomplete first-pass extraction of [^{15}O]-water, resulting in underestimation of CBF in high-flow areas[12,13] and a high sensitivity of absolute CBF values to small timing errors in blood sampling, variable delay in vascular territories,[14,15] and dispersion effects at the radial artery and in the blood-sampling system.[16] The PET scanner must have the capability to handle the relatively high counting rates resulting from bolus injection of up to 1850 MBq [^{15}O]-water. A modification employing ^{11}C- or ^{15}O-labeled butanol, which has a complete first-pass extraction, was suggested,[12] but due to the more difficult chemical synthesis of the tracer, it has not been widely applied.

An alternative which is applicable to all inert diffusible tracers is dynamic PET scanning. The wash-in and wash-out phase may be recorded and CBF as well as the distribution coefficient may be determined for each pixel by nonlinear curve fitting using Kety's original model equation.[17] The full procedure is computationally demanding, but it has been implemented for clinical application of the freely diffusible gas [^{18}F]-fluoromethane,[13,18,19] which may also be labeled with ^{11}C,[20] [^{15}O]-water bolus injection,[13] and inhalation of [^{15}O]-carbon dioxide.[21] Rapid algorithms avoiding the time-consuming reconstruction of all sequential tomograms and nonlinear curve fitting have been described.[22-24]

The steady state method utilizes the flow dependence of the steady state between tracer supply to tissue and the rapid radioactive decay of ^{15}O. It requires continuous tracer administration, which is usually achieved by inhalation of [^{15}O]-carbon dioxide[25] which is converted in the lung to [^{15}O]-water. The method has mostly been applied in connection with steady state measurements of oxygen metabolism (see below). It is robust, but besides underestimation of high flow rates due to incomplete first-pass extraction, the highly nonlinear relation between the measured tissue activity and CBF as well may introduce additional underestimation of inhomogeneous tissue.

In the early days of PET, ^{13}N-labeled ammonia was also used as a flow tracer. However, extraction is low and tissue trapping may vary.[26] Therefore, the method has largely been abandoned.

Normal values depend to some extent on the particular method used. Besides limitations resulting from tracer properties, the spatial resolution of the tomograph and the strategy used for defining regions of interest affect regional normal values considerably. The reason lies in the large difference between actual gray and white matter CBF (approximately 80 vs. 20 ml/100 g/min) and the complex topographic relations between the thin cortical rim (thickness approximately 3 to 5 mm, below resolution of current PET scanners) and underlying white matter resulting in variable mixtures of the two components in every pixel and region of interest. Whole brain average blood flow values in young healthy adults should be close to 50 ml/100 g/min with any method.[27] As for most other parameters, such as cerebral blood flow (CBV), cerebral metabolic rate of O_2 (CMRO$_2$), or CMRGl, a decline of CBF with age has been reported by most investigators,[28-30] but has not always been seen in previous studies using invasive techniques. It is therefore still somewhat controversial, and the possible effect of variables like gender, brain size, and atrophy (partial volume effects) on regional values of CBF, CBV, CMRO$_2$, and CMRGl has not been fully elucidated. Typical coefficients of variation for most parameters range between 10 and 20% in samples of normal volunteers.

B. LOCAL BLOOD VOLUME

Cerebral blood volume can be determined by labeling of hemoglobin with [^{15}O]-CO or [^{11}C]-CO.[31-34] In principle, measurements yield the distribution volume of red blood cells which must be divided by the local hematocrit to obtain the local total blood volume. Alternatively, whole blood activity in a blood sample may be counted as a reference. Hematocrit is somewhat lower in capillaries than in large vessels due to plasma skimming, factors between 0.50 and 0.92 have been reported, and a value of 0.69 has been measured with PET.[35] Normal CBV values of 4.2 ml/100 g in whole brain and 2.4 and 5.9 ml/100 ml in white and gray matter, respectively, have been reported.[32,34]

Measurement of local blood volume has its own physiological interest, in particular with regard to compensatory vasodilatation in the presence of hemodynamically relevant stenoses of large arteries, but is also required for accurate measurements of oxygen metabolism.[34] Some early studies of CMRO$_2$ in acute brain infarcts[37,38] did not apply the correction and may therefore have overestimated OEF and CMRO$_2$ on average by approximately 10 to 28%,[34,38] with the largest effects in densely ischemic tissue.

C. OXYGEN METABOLISM

The steady state method for measurement of CMRO$_2$ is performed by continuous inhalation of ^{15}O-labeled oxygen gas, preceded by inhalation of ^{15}O-carbon dioxide for measurement of CBF and correction for circulating labeled metabolic water.[39] As mentioned already, a correction for nonextracted intravascular ^{15}O must also be applied by measurement of CBV. The method relies basically on the steady state between labeled water originating from the oxidation process in tissue, its wash-out depending on CBF, and the rapid nuclear decay of ^{15}O.

Oxygen extraction fraction (OEF) apparently does not differ much between gray and white matter (average normal values 0.37 and 0.41, respectively[34]), but CMRO$_2$ is higher in gray than in white matter (normal average 5.1 and 1.7 ml of O_2/100 ml/min, respectively, recalculated with correction for CBV from Frackowiak et al.[39]

Alternatively, a high dose of ^{15}O-labeled oxygen gas may be applied as a single breath.

CMRO$_2$ may be quantified including appropriate corrections for the effects of CBF and CBV on the basis of a single scan ("autoradiographic method"[40,41]) or sequential scanning.[42] In studies reported in this chapter from our laboratory, the autoradiographic method was used. The normal whole brain average with this technique is 2.9 ml/100 g/min.[40]

D. TISSUE ACID-BASE BALANCE

Weak acids that are able to cross the blood-brain barrier in undissociated but not in ionized form may be used to measure cerebral pH, since their equilibrium concentration in brain depends on the relation between plasma and tissue pH. Two tracers have been suggested for that purpose: ^{11}C-labeled 5,5-dimethyl-2,4-oxazolidinedione (^{11}C]-DMO)[43] and ^{11}C]-carbon dioxide.[44-46] Results depend on the total distribution space of the tracer in extracellular and intracellular compartments. Details of the theory and application of the measurement of brain pH are presented in Chapter 14.

E. CEREBRAL GLUCOSE METABOLISM

Until now, most PET measurements of CMRGl in cerebrovascular disease relied on some modification of the autoradiographic method with 2-fluoro-2-deoxy-D-glucose (FDG) developed by Sokoloff et al.[47] and adapted for PET by Reivich et al.[48] [18]F and ^{11}C may be used for labeling of the tracer. It should be noted that the method originally was not devised for application under pathological conditions, and therefore some limitations of quantitation accuracy exist in ischemic tissue.

The "autoradiographic" approach which requires only one single brain scan depends on estimates of the amount of unmetabolized FDG on the basis of the amount of activity delivered by blood plasma and the measured cerebral activity. A variety of formulas implying somewhat different assumptions have been suggested for that purpose.[49]

More accurate quantification of CMRGl under ischemic conditions may be achieved by sequential scanning and fitting of the rate constants for FDG transport and phosphorylation, as suggested by Hawkins et al.[50] Again, a large number of approaches, varying in mathematical detail, have been described,[17,49] but validation under ischemic conditions is lacking. We used the dynamic measurement protocol described by Heiss et al.[51] and quantification algorithms of Wienhard et al.[52] Normal whole brain average CMRGl is 30 ml/100 g/min,[53] with higher values in gray matter than in white matter (e.g., ranges 34 to 42 and 17 to 19 ml/100 g/min, respectively,[51]) depending on region placement and scanner resolution.

The main problem with the glucose analogue FDG under pathological conditions is that its affinity to the carrier and phosphorylation enzymes is different from that of native glucose. This is the price one has to pay for the convenient trapping that occurs after phosphorylation and permits recording of a stable tracer distribution in brain. A "lumped constant" was devised and validated in normal brain tissue by Sokoloff et al.[47] to take into account these special properties of FDG. It has been shown with PET that the lumped constant may be considerably increased in human brain infarcts,[54] resulting in unpredictable overestimation of CMRGl. Thus FDG PET is of great value for analysis of remote effects of focal infarcts in healthy tissue, but uptake in ischemic tissue does not fulfill the requirements for accurate quantification. Nevertheless, it can be used for clinical purposes as a very informative diagnostic tool even in acute ischemia, as will be demonstrated later.

Attempts have been made to overcome the limitations of the FDG method by using ^{11}C-labeled glucose,[55,56] but a significant amount of the metabolic product [^{11}C]-CO$_2$ is usually rapidly lost from the brain. The prospects may be better for glucose stereospecifically labeled with ^{11}C in position 6,[57,58] but the compound is currently not available for clinical applications.

III. APPLICATIONS IN CEREBROVASCULAR DISORDERS

A. ISCHEMIC STROKE

Per definition of the disease, CBF must be reduced at the onset of ischemic stroke. Usually, CBF will not be reduced to zero in most tissue parts even after complete thrombotic occlusion of the supplying artery, since leptomeningeal anastomoses may provide some residual supply, which may lead to partial tissue survival, at least at the borders of the affected area. In order to determine the severity of oligemia, blood flow measurement is well suited as a first approach.

Only few PET studies have been reported within the first 24 h after onset of symptoms.[59-61] The usual finding was a profound depression of CBF in the infarct core. Gross hyperfusion due to spontaneous reperfusion is apparently rare during that time (1 out of 11 in the series of Ackerman et al.[61]) but its frequency increases to about 33% within 48 h,[60] which is in our own experience the approximate frequency found over the next 2 weeks. Usually the spatial profile of blood flow alterations shows the lowest values (below 12 ml/100 g/min) in infarct core, gradually increasing at the borders to values above 18 ml/100 g/min, at which acute irreversible ischemic damage is unlikely to be induced.[62] In accordance with experimental data, a study by Baron et al.[63] in infarct patients 2 to 38 d after onset suggests that areas with CBF below 11 ml/100 g/min almost always proceed to tissue necrosis, whereas regions with higher values may have a variable outcome depending on metabolic parameters. Until now, mainly patients with infarcts involving gray and white matter in the territory of the middle cerebral artery have been studied in the acute stage and thresholds have not been attributed to specific infarct locations.

Although PET provides a better spatial resolution (typically 5 to 10 mm in-plane resolution with state-of-the-art equipment; up to 15 mm with older tomographs) than studies with nontomographic techniques or single photon emission computed tomography (SPECT), there is still considerable uncertainty about the significance of the borderzone in the "critical" blood flow range of 12 to 18 ml/100 g/min since its spatial extent is often within the resolution limits. Nevertheless, attempts have been made to analyze tissue outcome in this most interesting area, which is sometimes called the ischemic penumbra, although the full definition of the term including electrical silence but maintenance of the potassium ion membrane gradient[64] can usually not be checked *in vivo* in humans. It has been reported that postischemic hyperperfusion occurs earlier in this zone than in the infarct core.[61]

Apart from the extreme case of CBF values below approximately 10 ml/100 g/min, information on CBF alone is not sufficient to assess tissue viability but must be supplemented by information on oxygen consumption. Early applications of the ^{15}O inhalation techniques[36,65,66] indicated that CBF values well above 10 ml/100 g/min are frequently found in areas where severely decreased oxygen consumption ($CMRO_2$) indicates profound tissue damage. $CMRO_2$ thresholds between 1.25[36] and 1.7 ml of O_2/100 g/min[63] for irreversible damage have been reported. However, measurements had not been corrected for effects of CBV on $CMRO_2$. Subsequent studies of patients with subacute and chronic infarcts including appropriate corrections for the effects of CBV on $CMRO_2$ measurements indicates a threshold of 1.3 ml/100 g/min for tissue viability.[67]

The frequent situation of relative hyperperfusion in the presence of reduced $CMRO_2$ indicates that, albeit CBF is reduced below normal, it may already exceed the metabolic demands of the severely damaged tissue. Correspondingly, OEF is reduced. In that stage of infarct development, no benefit can be expected from increasing CBF. The situation may be called "luxury perfusion"[68] and has been recognized early in animal experiments by arterial blood appearing in veins draining an ischemic infarct. Only the extreme form of hyperemia with CBF values above those of surrounding intact tissue may also be diagnosed by CBF measurements alone.

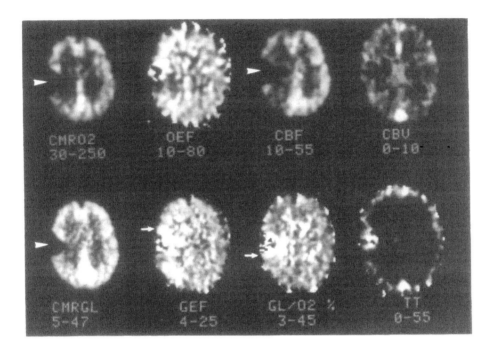

FIGURE 1.   Multitracer (^{15}O-water, ^{15}O-oxygen, ^{15}O-CO, FDG) PET study of an acute ischemic infarct in right MCA territory 24 h after onset of symptoms. Severely reduced CBF, CMRO$_2$, and CMRGl in infarct core (arrowheads). Increased glucose extraction fraction (GEF) and glucose to oxygen metabolic ratio (GL/O$_2$ %) at borders of infarct core (arrows). Increased oxygen extraction fraction (OEF) in the whole ipsilateral hemisphere due to critically reduced CBF caused by occlusion of the internal carotid artery.

The opposite situation, a more severe decrease of CBF than of CMRO$_2$, may be called "misery perfusion".[69] It indicates presence of actual ischemia (Figure 1). It is reflected by an increase in the OEF. The frequency of this condition has been reported to decline from 83% up to 12 h after onset, to 38% at 12 to 24 h, to 17% at later times.[59] Severe oligemia-ischemia with OEF over 80% in tissue with good outcome was seen by Baron et al.[70] in a few patients up to 4 d after onset at CBF values of 8 to 12 ml/100 g/min, whereas mild misery perfusion with significantly elevated OEF as compared with contralateral side was noted up to 12 d after onset of symptoms. The presumed penumbra zone has been studied in particular by Marchal et al.[71] Their data on a small group of seven patients indicate that usually metabolic deterioration occurs during infarct maturation in that area. Increased OEF was often found in the periphery of chronic minor ischemic infarcts.[72]

Glucose is the main substrate for energy production in the brain. It is transported across the blood-brain barrier by a carrier-mediated process. The extraction fraction is considerably lower (approximately 10%) than that of oxygen. Thus, under ischemic conditions, energy may also be produced, albeit much less efficiently, by anaerobic glycolysis (see Figure 1). With PET, the glucose analogue [^{18}F]-2-fluoro-2-deoxy-D-glucose can be used as a tracer for quantitation of glucose consumption. Some limitations, as discussed in the methods section, exist with regard to quantitation accuracy in pathological tissue.

In most cases of acute ischemic stroke, CMRGl is reduced in the affected tissue.[73-75] In only few cases — in our experience in less than 10% of acute and subacute infarcts — CMRGl may be increased in some parts of the infarcts to levels above those of normal gray matter (i.e., above approximately 40 µmol/100 g/min), probably indicating massive anaerobic glycolysis (Figure 2). It may lead to intracellular accumulation of high lactate levels that can induce additional brain damage.[76,77] Usually, the infarct appears inhomogeneous in these cases, also including areas of reduced CMRGl. In quantitative terms, CMRGl is often

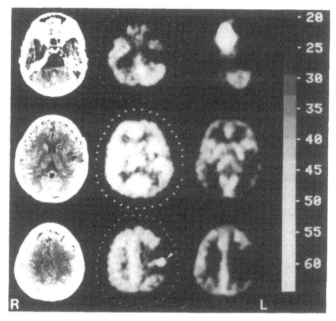

FIGURE 2A.

FIGURE 2.   (A) CMRGI and CBF measurement (three slices 14, 55, and 69 mm above the canthomeatal plane) in a patient with a 2-d-old ischemic MCA infarct. A focus of increased glucose consumption (presumably anaerobic glycolysis, marked by arrowhead) is shown within the underperfused area. Contralateral cerebellar deactivation is present in CMRGI and CBF image; (B) repeat study of the same patient 4 weeks later, showing a small demarcated infarct in CT, much more widespread reduction of CMRGI including the previously hypermetabolic focus, and recovery of CBF, indicating partial luxury perfusion. (From Wienhard, K., Wagner, R., and Heiss, W. D., *PET*, Springer-Verlag, Heidelberg, 1989. With permission.)

less reduced than $CMRO_2$[59] even if blood flow is already restored to some extent and the oxygen extraction fraction is reduced. In an investigation by Baron et al.,[78] the opposite condition with disproportionately low CMRGI relative to $CMRO_2$ was also noted in a few cases. Yet, the exact stochiometric relationships between oxygen and glucose consumption are still unclear because of methodological limitations.[79] Possibly, further elucidation of the complex issue may result from comparative studies with glucose labeled in the six position.[80]

Less reduction of FDG than of the semiquantitative CBF tracer [^{13}N]-ammonia soon after cerebravascular occlusion was already reported in 1980 by Kuhl et al.[73] and we found less reduction of CMRGI than CBF in 22.5% of infarcts within 14 d after onset.[81] These findings and the reports on relatively higher CMRGI than $CMRO_2$ correspond to the detection of lactate in experimental studies and by proton magnetic resonance spectroscopy in human ischemic infarcts.[82,83] Since it was observed not only during acute ischemia, but under conditions of luxury perfusion, it may indicate not only presence of anaerobic glycolysis but also production of lactate by aerobic glycolysis as a post-ischemic metabolic abnormality. In subacute infarcts, invasion by macrophages may possibly also contribute to an increase of CMRGI.

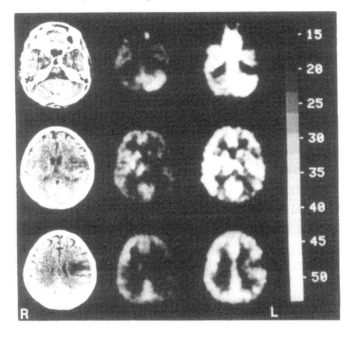

CT   CMRGl   CBF
µmol/100g/min ml/100g/min

FIGURE 2B.

Another aspect of infarct pathophysiology can be elucidated by measurements of tissue pH. Only studies of very few patients have been reported as yet. Tissue pH was found reduced under conditions of acute ischemia,[84] but in patients studied later than 5 d after onset of infarction, it was usually alkalotic.[43,44] Apparently, alkalosis is related to post-ischemic luxury perfusion[85] and may be due to increased washout of acidic CO_2 from tissue.

The damage of the blood-brain barrier in brain infarcts, which contributes mainly to the vasogenic type of edema, can also be assessed quantitatively by PET using ^{68}Ga-EDTA. Brain infarcts were visualized in most patients with acute and chronic brain infarcts.[86]

Amino acids and their analogues labeled with positron emitters, which until now have been mainly used in studies of brain tumors, may also be useful to study alterations of the blood-brain barrier and protein metabolism during and after ischemia. Experimental studies indicate that the latter may provide information on the probability of cell survival after ischemic injury.[87,88] Another tracer type of potential value in ischemia are ^{18}F-labeled nitroimidazoles, e.g., [^{18}F]-fluoromisonidazole, used clinically as hypoxic cell sensitizers, which may be able to mark hypoxic tissue selectively.[89]

Besides its value as a research tool in improving understanding of pathophysiology and treatment of cerebral ischemia, PET may also be used as a diagnostic instrument. In our experience with FDG, as reviewed recently,[90] it has a high sensitivity to detect ischemic lesions of supratentorial gray matter. Lesions appear usually larger than on CT,[73-75] and in some instances symptomatic lesions (usually with good clinical recovery) that were not visible in CT and magnetic resonance imaging (MRI) scans were detected (Figure 3). Extent and degree of hypometabolism appear to be related to the type of the clinical syndrome and the degree of eventual recovery.[75] Sensitivity is lower and clearly inferior to MRI for white matter and brainstem lesions due to lack of contrast and sufficient spatial resolution. Of course, FDG studies alone are rather unspecific with regard to the etiology of a lesion. One would expect that the diagnostic specificity might increase with multiparameter studies of

CT   MRT   CBF   CMRGl
                      ml/100g/min   μmol/100g/min

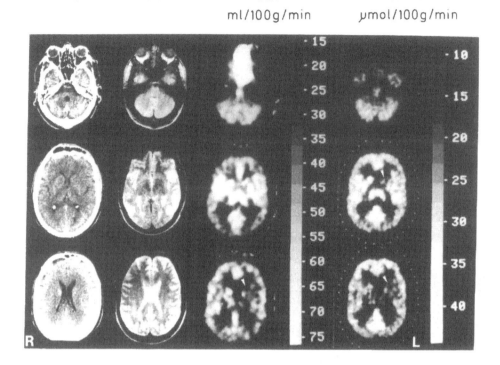

FIGURE 3. CT, MRI, and PET (CBF with ^{18}F-fluoromethane and CMRGl with FDG) of a patient after recovery from a transient ischemic attack. Only PET shows the damage to tissue function in the left striatum (arrowheads). (From Wienhard, K., Wagner, R., and Heiss, W. D., *PET*, Springer-Verlag, Heidelberg, 1989. With permission.)

CBF, CBV, and $CMRO_2$, but appropriate studies based on sufficiently large samples have not yet been published.

The *in vivo* measurement of CBF, $CMRO_2$, and CMRGl with PET may yield valuable information to evaluate the efficacy of therapeutic intervention which is difficult to obtain with clinical studies, since samples of patients with ischemic infarcts are often very inhomogeneous with respect to risk factors and additional complicating conditions. Attempts to increase the cerebral perfusion pressure in a few patients with misery perfusion have so far yielded variable results. Wise et al.[59] reported no success, whereas Ackerman et al.[61] presented preliminary evidence of improved survival of tissue with critical perfusion. Recently, better preservation or recovery of tissue metabolism was also seen under therapy with the calcium channel blocker nimodipine relative to placebo in patients with acute ischemic infarcts.[91,92] In one study,[91] less reduction of infarct core $CMRO_2$ was noted after 7 d of nimodipine treatment than with placebo; in the other study,[92] improvement of CMRGl in the noninfarcted tissue of the ipsilateral hemisphere was seen after 14 d of treatment.

B. MULTI-INFARCT DEMENTIA

Multiple ischemic infarcts may lead to development of multi-infarct dementia (MID). Although the frequency of MID is much lower than that of primary degenerative dementia such as dementia of Alzheimer type (DAT), little is known about the contribution of vascular factors to senile dementia. Decrease of CBF and $CMRO_2$ occurs in DAT and MID in correlation with disease severity.[93] It is probably due to the loss of neurons and normal neuronal communication in affected and functionally related areas. An increase of the oxygen extraction fraction indicating chronic misery perfusion has been found only in a minority of patients with MID (3 of 23)[94] who had bilateral internal carotid artery occlusion.

A different pattern of distribution of hypometabolic areas in MID vs. DAT was reported first by Benson et al.[95] with FDG PET. MID was characterized by multifocal reductions of CMRGl basically corresponding to the location of the infarcts without sparing of specific brain structures, whereas DAT was characterized by hypometabolism located specifically in the temporoparietal and frontal association cortex with relative sparing of primary visual and sensorimotor cortex, basal ganglia, and cerebellum. The DAT pattern also has been described by many other investigators. Instead of multiple cortical lesions, mild MID may also be associated with multiple, often lacunar infarcts of the basal ganglia and periventricular white matter. In that case, we found pronounced reductions of CMRGl in basal ganglia and cerebellum, but only a diffuse unspecific reduction of cortical CMRGl. The latter was also seen in Binswanger's subcortical hypertensive encephalopathy. Using a metabolic ratio between CMRGl in areas typically affected by DAT divided by CMRGl in typically unaffected areas, we found 84% correct classifications in the discrimination between predominantly presenile DAT patients and patients with cognitive deficits due to other conditions, including vascular dementia.[96]

C. REVERSIBLE AND CHRONIC ISCHEMIA

Transient ischemic attacks (TIAs) are the classical clinical sign of transient or chronic cerebral perfusion deficits. They are often associated with atherosclerotic lesions in large vessels supplying the respective vascular territories and may be caused by microemboli originating from these lesions. Only high-grade stenotic lesions of the carotid artery lead to a significant reduction of cerebral perfusion pressure at normal systemic blood pressure[97] and even chronic complete occlusion of the internal carotid artery has a relatively benign prognosis, probably due to the high collateralization capacity of the circle of Willis in most patients.[98] Yet, little is known about the combined effect of multiple stenotic lesions, especially under conditions of varying systemic blood pressure and cardiac output, as is common in patients with cerebrovascular disease. With PET, the hemodynamic situation can be studied in much more detail than has been possible with previous techniques.

The basic concept emerging from ^{15}O studies in patients with carotid artery occlusion or stenosis is that of a graded compensatory response in the cerebrovascular periphery. Gibbs et al.[99] and Powers et al.[100] described an increase of local CBV associated with a decrease of CBF in dependent MCA territories. The finding corresponds to that concept of peripheral vasodilation as a compensation for reduced perfusion pressure. Most conveniently, it can be analyzed in terms of CBF/CBV or CBV/CBF ratios, the latter corresponding formally to the microvascular transit time of an intravascular plasma tracer. In addition, an increase of OEF was noted in some patients, leading to less reduction of CMRO$_2$ than CBF, that can also be interpreted as a useful compensatory mechanism to maintain tissue oxygen supply in spite of reduced CBF (Figure 4). An inverse relation exists between CBF/CBV ratios and OEF[99] with a pronounced increase in OEF in only a few cases with very low CBF/CBV ratios (< approximately six per minute). Subsequently, Powers et al.[101] suggested two stages of reversibly compromised cerebral perfusion: the first stage with reduced perfusion pressure, but normal flow (CBF and OEF normal, but CBF/CBV reduced) and the second with reduced flow (CBF and CBF/CBV reduced, OEF increased). Leblanc et al.[102] found these effects most pronounced in the anterior vascular borderzone (between the anterior and middle cerebral artery). Observations by Powers et al.[103] and Itoh et al.[104] suggest that the relation between CBF/CBV ratios and OEF is rather variable. Thus, CBF/CBV ratios alone — which could also be estimated with single photon emission computed tomography (SPECT) — reveal only an incomplete picture of the degree of hemodynamic impairment.

Compensatory adaption of the cerebral circulation has been noted in two diseases with

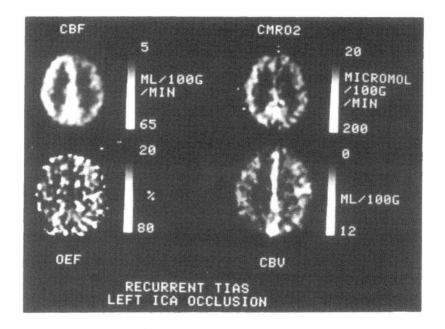

FIGURE 4.   PET measurement of CBF, CMRO$_2$, and CBV with ^{15}O-labeled tracers in a patient with recurrent TIAs due to left ICA occlusion. A more pronounced reduction of CBF than CMRO$_2$, corresponding to increased OEF (misery perfusion, stage 2 of hemodynamic decompensation), is shown in the left hemisphere (on viewer's right side), accompanied by increased CBV indicating peripheral vasodilatation.

increased risk of ischemic stroke: in sickle cell disease, an increase of CBF and CBV with normal OEF was found,[105] whereas in children with Moyamoya disease only CBV was increased, resulting in a decrease of CBF/CBV ratios.[106]

Another more widely available parameter of cerebral hemodynamics is the vascular responsiveness which can be assessed by comparing rCBF measurements at rest and under exposure to carbon dioxide or the carboanhydrase inhibitor diamox and is thus also accessible to the techniques like SPECT[107] and stable xenon-enhanced CT.[108] Kanno et al.[109] have shown variable (negative) correlations between OEF and vascular responsiveness. On average, a zero vasodilatory response, suggesting maximum vasodilation at rest, was observed at an OEF of 0.53. Reduced vascular responsiveness was found with [^{18}F]-fluoromethane PET in affected hemispheres (2.8% ± 1.9% per mmHg CO$_2$ change) and contralateral hemispheres (3.8% ± 1.3%) of patients with TIAs or minor stroke compared with age-matched normals (5.2% ± 0.8%.)[110] The reduction of vascular responsiveness was more pronounced in patients with a hemodynamic type of TIAs, characterized clinically by exertional, positional, orthostatic, or cardiac induction of TIAs.[111] Interestingly, the CBF response to functional activation is also reduced or globally altered in patients with severe carotid artery disease.[112]

The second stage of hemodynamic impairment with increased OEF, i.e., the condition of "misery perfusion", may be remedied by extracranial-intracranial arterial bypass surgery as demonstrated in one patient by Baron et al.[69] It may also be a sign of impending stroke.[113] Yet, apart from case reports, little is known about the prognosis of this condition which is apparently relatively rare even in patients with TIAs and significant carotid artery disease.[114] Attempts to overcome the disappointing results of extracranial-intracranial bypass surgery[5] by better patient selection using the described staging procedure are intriguing, but have so far not been successful. Powers et al.[115] reported a high incidence of perioperative and postoperative strokes (n = 9) during the first year in 29 patients with occlusive carotid artery disease, of whom 14 were in stage one and 10 were in stage two of hemodynamic impairment. In contrast, a similar patient group of 30 patients (16 in stage one, 5 in stage two) treated medically experienced only 2 strokes. The surgical procedure apparently does not improve $CMRO_2$ in these patients, although a preoperatively increased CBV may be normalized[117] or CBF may increase after surgery.[118]

Possibly, the prevailing uncertainty about proper patient selection may be due to the small patient samples studied by few research groups using different methods. Somewhat more encouraging results with respect to prognosis as dependent on hemodynamic impairment have been obtained by Rutigliano et al.[108] testing cerebrovascular responsiveness to carbon dioxide with xenon computed tomography (CT). Thus, in view of the high epidemiologic impact of chronic cerebrovascular disease, further research is urgently needed.

D. HEMORRHAGE

Only a few studies on intracerebral hemorrhage have been reported. With FDG, the findings have been compared with ischemic infarcts.[119] In the hematoma, low glucose metabolic rates of 10 ± 4.6 μmol/100 g/min were found, significantly below those of ischemic infarcts in comparable locations (15.6 ± 3.8 μmol/100 g/min), but remote effects were similar. Blood flow and oxygen metabolism are also largely reduced in intracerebral hematomas.[120] A study of cerebral hematomas in preterm newborn infants revealed profound widespread CBF reductions in the ipsilateral hemisphere, suggesting that the hematoma in those cases was only one part of a much larger ischemic lesion.[12]

Subarachnoid hemorrhage frequently causes arterial vasospasm and may thus lead to cerebral ischemia. The degree of ischemia, which is most important for tissue viability and prognosis, can be measured with ^{15}O PET. Martin et al.[122] report that patients with vasospasm and in poor clinical condition usually show marked decrease in $CRMO_2$, a somewhat less severe decrease of CBF, and increased CBV indicating peripheral vasodilatation to compensate for the spasm of the large arteries.

E. REMOTE EFFECTS

Even the earliest PET studies of ischemic forebrain lesions revealed reductions of metabolism and blood flow exceeding the extent of morphologically damaged tissue, a finding that had been noted also in previous studies employing nontomographic techniques. The most conspicuous effect was a reduction of CBF and metabolism in the contralateral cerebellum, sometimes called contralateral cerebellar diaschisis.[28,123] It was obviously due to some neuronally mediated effect, since a primary vascular cause could be excluded due to its location. Further remote effects were reductions of CBF and metabolism in ipsilateral cortex and basal ganglia.[173,124] Their cause was less clear since diffuse ischemic neuronal loss or inadequate blood supply could also contribute in these areas. Yet, similar effects have also been observed in nonischemic lesions such as brain tumors and intracerebral hematomas, and they seem therefore to be generally more closely related to the site than to the nature of the primary lesion. A study of the correlation between FDG PET and post-mortem findings in a patient with multiple brain infarcts concluded that degeneration of fiber tracts as well as microscopic infarcts not apparent on gross examination may contribute to metabolic inactivations.[125] Among cortical and subcortical lesions, infarcts of the parietal and frontal lobes most often cause significant reductions of CBF and metabolism in the contralateral cerebellum and ipsilateral basal ganglia.[126-128] Most likely this is explained by the damage of cortico-ponto-cerebellar pathways.

Infarcts of the basal ganglia may cause ipsilateral cortical as well as contralateral cerebellar deactivations. Among these, the effects of thalamic infarcts have been explored in several studies. Baron et al.[129] described mainly diffuse ipsilateral cortical effects. Significant contralateral cerebellar deactivations are apparently relatively rare; they were found in only

20% of patients with thalamic infarcts compared to over 50% in cortical infarcts.[130] Infarcts involving the medial thalamic nuclei apparently cause more widespread cortical metabolic reductions than those restricted to anterior, ventrolateral, and posterior nuclei.[131]

Only few studies on remote effects of white matter infarcts have been reported. White matter lesions are usually barely seen on PET scans because of spatial resolution limits and little contrast to low normal white matter blood flow and metabolism. Klinger et al.[132] found increased cerebellar metabolic rates in normals and patients with Alzheimer's disease who had periventricular white matter lesions predominantly in the frontal lobe. We have seen no significant effect of small white matter lesions on cortical blood flow, but only hemispheric and focal reductions of blood flow in cortical areas overlying large confluent white matter lesions that had been identified on T2-weighted MRI scans.[133] Similar cortical inactivations may also be caused by progressive multifocal leukoencephalopathy[134] and adrenoleukodystrophy.[135] Ischemic lesions of the optic radiation lead to inactivations of the occipital cortex corresponding to the type of the associated visual field defect.[136]

In our experience infarcts of the brainstem and cerebellum usually do not cause significant asymmetric inactivations of forebrain structures.

Brain infarcts in all locations often cause global reductions of cerebral glucose metabolism in addition to focal and asymmetric remote effects.[137] These global effects usually show changes corresponding to the clinical course of the patient after the acute ictus: in patients with good recovery, an increase of CMRGl was observed, whereas clinically deteriorating patients showed a further decline of global CMRGl.[138] In contrast to global alterations, asymmetric inactivations usually show relatively little changes at follow-up examinations.

Remote effects may explain clinical symptoms that are difficult to relate to the infarct proper. In early studies, Metter et al.[139-141] tried to relate various language functions with regional CMRGl and obtained complex correlations. In later studies, they found that Broca's aphasia was characterized by a most pronounced left-to-right metabolic asymmetry, whereas patients with Wernicke's aphasia showed less hemispheric asymmetry, and patients with conduction aphasia were metabolically symmetric.[142] Broca's aphasia was also always associated with contralateral cerebellar deactivation.[143] We found that receptive language function, as assessed by the Token test, could be explained to a high degree by temporoparietal metabolism, irrespective of infarct location.[144] For example, the degree of impairment of language understanding in patients with infarcts restricted to the basal ganglia was related to temporoparietal CMRGl, but not to basal ganglia metabolism (Figure 5). Other neuropsychological syndromes, such as hemianopia, Balint's syndrome, alexia, agraphia, and apraxia, have also been related to remote effects caused by ischemic lesions.[145]

F. FUNCTIONAL RECOVERY

The degree of functional recovery after an ischemic lesion of the brain is obviously somehow related to the activity of the remaining neurons. It is therefore intriguing to use PET to elucidate the mechanisms involved and to obtain prognostic clues. However, only few studies have attacked that issue as yet. Kushner et al.[75] observed a relation between the severity of the initial metabolic disorder, as judged by visual evaluation of PET scans, and functional outcome, but a preliminary quantitative analysis of another patient series did not confirm a correlation between CMRGl in the acute stage and clinical outcome.[138]

Functional activation studies may be suited to estimate the latent potential for recovery. We found in a preliminary analysis of 11 patients with various types of aphasia after a left hemispheric infarct that the increase of CMRGl in the ipsilateral cortex during functional activation by active speech production corresponded with later recovery from aphasia. In patients with poor outcome, metabolic activation was more or less restricted to the contralateral hemisphere and cerebellum (Figure 6).[146]

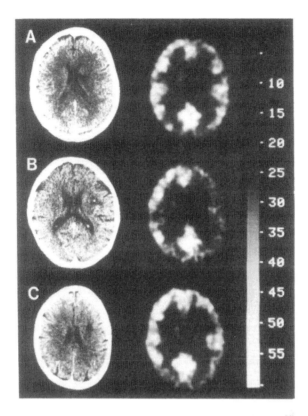

FIGURE 5. CT and FDG-PET scans of three right-handed patients with aphasia due to an ischemic subcortical lesion in the left hemisphere (on viewer's right). Only cortical CMRGl, but not size and location of the lesion, corresponded to aphasia severity: patient A with nearly symmetric metabolism had only minimal residual aphasia; patient B with gross hypometabolism had severe global aphasia; and patient C with moderate hypometabolism had mild Wernicke's aphasia. (From Karbe, H., Szelies, B., Herholz, K., and Heiss, W. D., *J. Neurol.*, 237, 19, 1990. With permission.

IV. SUMMARY

PET techniques for measurement of cerebral blood flow (CBF), cerebral blood volume (CBV), oxygen extraction (OEF), oxygen metabolism ($CMRO_2$), and glucose metabolism (CMRGl) have been used to explore the *in vivo* pathophysiology of ischemic brain disease. During acute ischemia the degree of the reduction of $CMRO_2$ appears to be of most importance for tissue viability. CBF is initially severely reduced. Recovery of CBF occurs to a variable degree within variable time, and luxury perfusion of irreversibly damaged tissue is a frequent finding. Therefore, isolated CBF measurements are only of very limited value in acute ischemic stroke. CMRGl is usually less reduced than $CMRO_2$ in acute ischemia. In a few cases, increased fluorodeoxyglucose (FDG) uptake indicates largely increased glycolysis and production of large amounts of lactate which may contribute to tissue damage.

PET has a very high sensitivity to detect gray matter damage and may therefore also be used as a diagnostic tool. The topographic pattern of metabolic alterations helps to differentiate vascular from degenerative dementia. With measurements of CBV, CBF, and $CMRO_2$, two stages of reversible perfusion impairment can be distinguished: first, peripheral vasodilatation (increased CBV); and second, increased oxygen extraction. The degree of ischemia

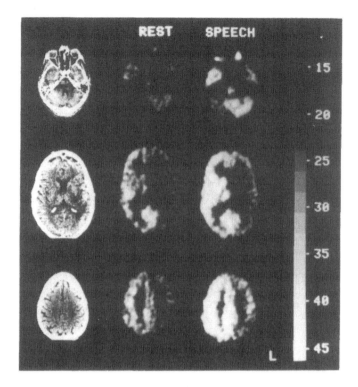

FIGURE 6A.

FIGURE 6.   (A) Three CT and corresponding PET slices of a patient with Broca's aphasia and eventually poor outcome. PET was performed at rest and during speech activation 3 weeks after onset of the left MCA infarct and showed severe metabolic impairment of the left hemisphere (on viewer's right) with little activation effect; (B) three CT, T-2-weighted MRI, and corresponding PET slices of a patient with Wernicke's aphasia, eventual good recovery, and good speech activation of areas adjacent to the 2-month-old left temporoparietal infarct. (From Heiss, W. D., Pawlik, G., Hebold, I., Beil, C., Herholz, K., Szelies, B., and Wienhard, K., *Cerebrovascular Disease,* Ginsberg, W. D. and Dietrich, W. D., Eds., Raven Press, New York, 1989, 345. With permission.)

due to vasospasm after subarachnoidal hemorrhage can also be assessed with these techniques.

Ischemic brain lesions often cause alterations of CBF and metabolism in remote brain areas. Apparently, they are mainly due to neuronal degeneration and deafferentiation. Hemipheric forebrain lesions typically cause asymmetric ipsilateral hemispheric and contralateral cerebellar deactivations. Most ischemic strokes in all locations also cause some degree of global metabolic impairment which tends to improve or deteriorate in parallel with the patient's general clinical condition. Remote deactivations may explain clinical symptoms; in particular they may elucidate the neuronal basis of neuropsychological deficits such as aphasia. Functional activation studies can be used to explore the potential for recovery from ischemic stroke by activation of compensatory mechanisms.

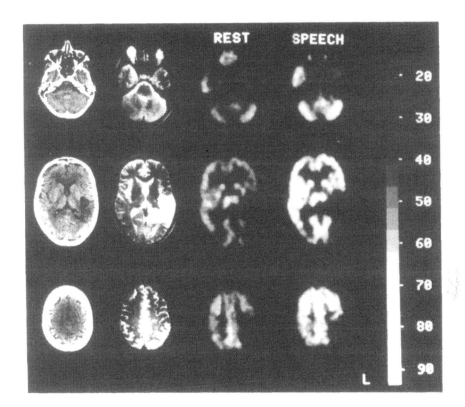

FIGURE 6B.

REFERENCES

1. **Raichle, M. E.**, The pathophysiology of brain ischemia, *Ann. Neurol.*, 13, 2, 1983.
2. **Plum, F.**, What causes infarction in ischemic brain?, The Robert Wartenberg Lecture, *Neurology*, 33, 222, 1983.
3. **Siesjo, B. K. and Bengtsson, F.**, Calcium fluxes, calcium-antagonists, and calcium-related pathology in brain ischemia, hypoglycemia, and spreading-depression. A unifying hypothesis, *J. Cereb. Blood Flow Metab.*, 9, 127, 1989.
4. **Heiss, W.-D.**, Medical management of cerebral infarction, in *Handbook of Clinical Neurology*, rev. ser., Vinken, P. J. et al., Eds., Elsevier Science, Amsterdam, 1988, 417.
5. **Barnett, H. J. M.**, Failure of extracranial-intracranial bypass to reduce risk of ischemic stroke. Results of an international randomized trial, *N. Engl. J. Med.*, 313, 1191, 1985.
6. Committee on Health Care Issues, Does carotid endarterectomy decrease stroke and death in patients with TIAs?, *Ann. Neurol.*, 22, 72, 1987.
7. **Baron, J. C., Frackowiak, R. S. J., Herholz, K., Jones, T., Lammertsma, A. A., Mazoyer, B., and Wienhard, K.**, Guidelines for the implementation and use of PET methods for measurement of cerebral energy metabolism and haemodynamics in cerebrovascular disease, *J. Cereb. Blood Flow Metab.*, 9, 723, 1989.
8. **Kety, S. S. and Schmidt, C. F.**, The nitrous oxide method for the quantitative determination of cerebral blood flow in man. Theory, procedure, and normal values, *J. Clin. Invest.*, 27, 476, 1948.
9. **Lockwood, A. H.**, Invasiveness of studies of brain function by positron emission tomography, *J. Cereb. Blood Flow Metab.*, 5, 487, 1985.
10. **Raichle, M. E., Herscovitch, P., Mintun, M. A., Markham, J., and Martin, W. R. W.**, Brain blood flow measured with intravenous $H_2{}^{15}O$. I. Implementation and validation, *J. Nucl. Med.*, 24, 790, 1983.
11. **Jones, S. C., Greenberg, J. H., Dann, R., Robinson, G. D., Kushner, M., Alavi, A., and Reivich, M.**, Cerebral blood flow with the continuous infusion of ^{15}O-water, *J. Cereb. Blood Flow Metab.*, 5, 566, 1985.

12. **Herscovitch, P., Raichle, M. E., Kilbourn, M. R., and Welch, M. J.**, Positron emission tomography measurement of cerebral blood flow and permeability surface-area-product of water using [^{15}O]water and [^{11}C]butanol, *J. Cereb. Blood Flow Metab.*, 7, 527, 1987.

13. **Herholz, K., Pietrzyk, U., Wienhard, K., Hebold, I., Pawlik, G., Wagner, R., Holthoff, V., Klinkhammer, P., and Heiss, W.-D.**, Regional cerebral blood flow measurement with intravenous [^{15}O]water bolus and [^{18}F]fluormethane inhalation, *Stroke*, 20, 1174, 1989.

14. **Koeppe, R. A., Hutchins, G. D., Rothley, J. M., and Hichwa, R. D.**, Examinations of assumptions for focal cerebral blood flow studies in PET, *J. Nucl. Med.*, 28, 1695, 1987.

15. **Iida, H., Higano, S., Tomura, N., Shishido, F., Kanno, I., Miura, S., Murakami, M., Takahashi, K., Sasaki, H., and Uemura, K.**, Evaluation of regional differences of tracer appearance time in cerebral tissues using [^{15}O]water and dynamic positron emission tomography, *J. Cereb. Blood Flow Metab.*, 8, 285, 1988.

16. **Iida, H., Kanno, I., Miura, S., Murakami, M., Takahashi, K., and Uemura, K.**, Error-analysis of a quantitative CBF measurement using H$_2$^{15}O autoradiography and PET. With respect to the dispersion of the input-function, *J. Cereb. Blood Flow Metab.*, 6, 536, 1986.

17. **Kety, S. S.**, The theory and applications of the exchange of inert gas at the lungs and tissues, *Pharamcol. Rev.*, 3, 1, 1951.

18. **Holden, J. E., Gatley, S. J., Hichwa, R. D., Shaugnessy, W. J., Nickels, R. J., and Polcyn, R. E.**, Cerebral blood flow using positron emission tomography measurements of fluoromethane kinetics, *J. Nucl. Med.*, 22, 1084, 1981.

19. **Koeppe, R. A., Holden, J. E., Polcyn, R. E., Nickles, R. J., Hutchins, G. D., and Reese, J. L.**, Quantitation of local CBF and partition-coefficient without arterial sampling. Theory and validation, *J. Cereb. Blood Flow Metab.*, 5, 214, 1985.

20. **Stone-Elander, S., Roland, P., Eriksson, L., Litton, J.-E., Johnström, P., and Widén, L.**, The preparation of ^{11}C-labeled fluoromethane for the study of regional cerebral blood flow using positron emission tomography, *Eur. J. Nucl. Med.*, 12, 236, 1986.

21. **Lammertsma, A. A., Frackowiak, R. S. J., Hoffman, J. M., Huang, S. C., Weinberg, I. N., Dahlbom, M., Macdonald, N. S., Hoffman, E. J., Mazziotta, J. C., Heather, J. D., Forse, G. R., and Phelps, M. E.**, The (CO$_2$)-^{15}O-buildup-technique to measure regional cerebral blood-flow and volume of distribution of water, *J. Cereb. Blood Flow Metab.*, 9, 461, 1989.

22. **Huang, S.-C., Carson, R. E., and Phelps, M. E.**, Measurement of local blood flow and distribution-volume with short-lived isotopes. A general input technique, *J. Cereb. Blood Flow Metab.*, 2, 99, 1982.

23. **Huang, S.-C., Carson, R. E., Hoffman, E. J., Carson, J., MacDonald, N., Barrio, J. R., and Phelps, M. E.**, Quantitative measurement of local cerebral blood flow in humans by positron computed tomography and ^{15}O-water, *J. Cereb. Blood Flow Metab.*, 3, 141, 1983.

24. **Alpert, N. M., Eriksson, L., Chang, J. Y., Bergstroem, M., Litton, J. E., Correia, J. A., Bohm, C., Ackerman, R. H., and Taveras, J. M.**, Strategy of the measurement of regional cerebral blood flow using short-lived tracers and emission tomography, *J. Cereb. Blood Flow Metab.*, 4, 28, 1984.

25. **Jones, T., Chesler, D. A., and Ter-Pogossian, M. M.**, The continuous-inhalation of oxygen-15 for assessing regional oxygen-extraction in the brain of man, *Br. J. Radiol.*, 49, 339, 1976.

26. **Phelps, M. E., Hoffman, E., and Raybaud, C.**, Factors which affect cerebral uptake and retention of ^{13}NH$_3$, *Stroke*, 8, 694, 1977.

27. **Lassen, N. A.**, Normal average value of cerebral blood flow in younger adults is 50 ml/100 g/min, *J. Cereb. Blood Flow Metab.*, 5, 347, 1985.

28. **Lenzi, G. L., Frackowiak, R. S. J., Jones, T., Heather, J. D., Lammertsma, A., Rhodes, C. G., and Pozzilli, C.**, CMRO2 and CBF by the oxygen-15 inhalation technique. Results in normal volunteers and cerebrovascular patients, *Eur. Neurol.*, 20, 285, 1981.

29. **Pantano, P., Baron, J.-C., Lebrun-Grandié, P., Duquesnoy, N., Bousser, M.-G., and Comar, D.**, Regional cerebral blood flow and oxygen consumption in human aging, *Stroke*, 15, 635, 1984.

30. **Leenders, K. L., Healy, M., Frackowiak, R., Buchingham, P., Lammertsma, A., and Jones, T.**, Effect of age on oxygen metabolism and CBF in healthy volunteers studied with positron emission tomography (PET), *J. Cereb. Blood Flow Metab.*, 7 (Suppl.), S398, 1987.

31. **Grubb, R. L., Raichle, M. E., Higgins, C. S., and Eichling, J. O.**, Measurement of regional blood-volume by emission tomography, *Ann. Neurol.*, 4, 322, 1978.

32. **Phelps, M. E., Huang, S. C., Hoffman, E. J., and Kuhl, D. E.**, Validation of tomographic measurement of cerebral blood volume with C-11-labeled carboxyhemoglobin, *J. Nucl. Med.*, 20, 328, 1979.

33. **Lammertsma, A. A. and Jones, T.**, Correction for the presence of intravascular O-15 in the steady-state technique for measuring regional oxygen extraction ratio in the brain. I. Description of the method, *J. Cereb. Blood Flow Metab.*, 3, 416, 1983.

34. **Lammertsma, A. A. and Jones, T.**, Correction for the presence of intravascular O-15 in the steady-state technique for measuring regional oxygen extraction ratio in the brain. II. Results in normal subjects and brain tumor and stroke patients, *J. Cereb. Blood Flow Metab.*, 3, 425, 1983.

35. **Lammertsma, A. A., Brooks, D. J., Beaney, R. P., Turton, D. R., Kensett, M. J., Healther, J. D., Marshall, J., and Jones, T.,** In vivo measurement of regional cerebral haematocrit using positron emission tomography, *J. Cereb. Blood Flow Metab.,* 4, 317, 1984.

36. **Lenzi, G. L., Frackowiak, R. S. J., and Jones, T.,** Cerebral oxygen metabolism and blood-flow in human cerebral ischemic infarction, *J. Cereb. Blood Flow Metab.,* 2, 321, 1982.

37. **Baron, J. C., Delattre, J. Y., Bories, J., Chiras, J., Cabanis, E. A., Blas, C., Bousser, M. G., and Comar, D.,** Comparison study of CT and positron emission tomography data in recent cerebral infarction, *A.J.N.R.,* 4, 536, 1983.

38. **Pantano, P., Baron, J. C., Crouzel, C., Collard, P., Sirou, P., and Samson, Y.,** The ^{15}O continuous-inhalation method. Correction for intravascular signal using C-^{15}O, *Eur. J. Nucl. Med.,* 10, 387, 1985.

39. **Frackowiak, R. S. J., Lenzi, G. L., Jones, T., and Heather, J. D.,** Quantitative measurement of regional cerebral blood flow and oxygen metabolism in man using ^{15}O and positron emission tomography. Theory, procedure, and normal values, *J. Comput. Assist. Tomogr.,* 4, 727, 1980.

40. **Mintun, M. A., Raichle, M. E., Martin, W. R. W., and Herscovitch, P.,** Brain oxygen utilization with O-15 radiotracers and positron emission tomography, *J. Nucl. Med.,* 25, 177, 1984.

41. **Herscovitch, P., Mintun, M. A., and Raichle, M. E.,** Brain oxygen utilization measured with ^{15}O radiotracers and positron emission tomography. Generation of metabolic images, *J. Nucl. Med.,* 26, 416, 1985.

42. **Huang, S. C., Feng, D. G., and Phelps, M. E.,** Model dependency and estimation reliability in measurement of cerebral oxygen utilization rate with oxygen-15 and dynamic positron emission tomography, *J. Cereb. Blood Flow Metab.,* 6, 105, 1986.

43. **Syrota, A., Casting, M., Rougemont, D., Berridge, M., Baron, J. C., Bousser, M. G., and Pocidalo, J. J.,** Tissue acid-base balance and oxygen metabolism in human cerebral infarction studied with positron emission tomography, *Ann. Neurol.,* 14, 419, 1983.

44. **Brooks, D. J., Lammertsma, A. A., Beaney, R. P., Leenders, K. L., Buckingham, P. D., Marshall, J., and Jones, T.,** Measurement of regional cerebral pH in human subjects using continuous inhalation of $^{11}CO_2$ and positron emission tomography, *J. Cereb. Blood Flow Metab.,* 4, 458, 1984.

45. **Buxton, R. B., Wechsler, L. R., Alpert, N. M., Ackerman, R. H., Elmaleh, D. R., and Correia, J. A.,** Measurement of brain pH using $^{11}CO_2$ and positron emission tomography, *J. Cereb. Blood Flow Metab.,* 4, 8, 1984.

46. **Buxton, R. B., Alpert, N. M., Babikian, V., Weise, S., Correia, J. A., and Ackerman, R. H.,** Evaluation of the $^{11}CO_2$ positron emission tomographic method for measuring brain pH. I. pH changes measured in states of altered pCO_2, *J. Cereb. Blood Flow Metab.,* 7, 709, 1987.

47. **Sokoloff, L., Reivich, M., Kennedy, C., DesRosiers, M. H., Patlak, C. S., Pettigrew, K. D., Sakurada, O., and Shinohara, M.,** The [^{14}C] deoxyglucose method for the measurement of local cerebral glucose utilization. Theory, procedure, and normal values in the conscious and anesthetized albino rat, *J. Neurochem.,* 28, 897, 1977.

48. **Reivich, M., Kuhl, D., Wolf, A., Greenberg, J., Phelps, M., Ido, T., Casella, V., Fowler, J., Hoffman, E., Alavi, A., Som, P., and Sokoloff, I.,** The ^{18}F-fluorodeoxyglucose method for the measurement of local cerebral glucose utilization in man, *Circ. Res.,* 44, 127, 1979.

49. **Lammertsma, A. A., Brooks, D. J., Frackowiak, R. S. J., Beaney, R. P., Herold, S., Heather, J. D., Palmer, A. J., and Jones, T.,** Measurement of glucose-utilisation with [^{18}F]2-fluoro-2-deoxy-D-glucose. A comparison of different analytical methods, *J. Cereb. Blood Flow Metab.,* 7, 161, 1987.

50. **Hawkins, R. A., Phelps, M. E., Huang, S.-C., and Kuhl, D. E.,** Effect of ischemia on quantification of local cerebral glucose metabolic rate in man, *J. Cereb. Blood Flow Metab.,* 1, 37, 1981.

51. **Heiss, W.-D., Pawlik, G., Herholz, K., Wagner, R., Goeldner, H., and Wienhard, K.,** Regional kinetic-constants and cerebral metabolic rate for glucose, in normal human volunteers determined by dynamic positron emission tomography of ^{18}F-2-fluoro-2-deoxy-D-glucose, *J. Cereb. Blood Flow Metab.,* 4, 212, 1984.

52. **Wienard, K., Pawlik, G., Herholz, K., Wagner, R., and Heiss, W.-D.,** Estimation of local cerebral glucose utilization by positron emission tomography of [^{18}F]2-fluoro-2-deoxy-D-glucose. A critical appraisal of optimization procedures, *J. Cereb. Blood Flow Metab.,* 5, 115, 1985.

53. **Phelps, M. E., Huang, S. C., Hoffman, E. J., Selin, C., Sokoloff, L., and Kuhl, D. E.,** Tomographic measurement of local cerebral glucose metabolic rate in humans with [F-18]2-fluor-2-deoxy-D-glucose. Validation of method, *Ann. Neurol.,* 6, 371, 1979.

54. **Gjedde, A., Wienhard, K., Heiss, W.-D., Kloster, G., Diemer, N. H., Herholz, K., and Pawlik, G.,** Comparative regional analysis of 2-fluorodeoxyglucose and methylglucose uptake in brain of four stroke patients. With special reference to the regional estimation of the lumped constant, *J. Cereb. Blood Flow Metab.,* 5, 163, 1985.

55. **Raichle, M. E., Larson, K. B., Phelps, M. E., Grubb, R. L., Welch, M. J., and Ter-Pogossian, M. M.,** In-vivo measurement of brain glucose transport and CMRGlu employing ^{11}C-glucose, *Am. J. Physiol.,* 228, 1936, 1975.

56. **Blomquist, G., Bergström, K., Bergström, M., Ehrin, E., Eriksson, L., Garmelius, B., Lindberg, B., Lilja, A., Litton, J.-E., Lundmark, L., Lundqvist, H., Malmborg, P., Moström, U., Nilsson, L., Stone-Elander, S., and Widén, L.,** Models for ^{11}C-glucose, in *The Metabolism of the Human Brain Studied with Positron Emission Tomography,* Greitz, T. et al., Eds., Raven Press, New York, 1985, 185.

57. **Hawkins, R. A., Mans, A. M., Davis, D. W., Vina, J. R., and Hibbard, L. S.,** Cerebral glucose use measured with ^{14}C-glucose labeled in the 1,2,or 6 position, *Am J. Physiol.,* 248, C170, 1985.

58. **Lear, J. L. and Ackermann, R. F.,** Comparison of cerebral glucose metabolic rates measured with fluorodeoxyglucose and glucose labeled in the 1-position, 2-position, 3-4-position, and 6-position using double label quantitative digital autoradiography, *J. Cereb. Blood Flow Metab.,* 8, 575, 1988.

59. **Wise, R. J. S., Bernardi, S., Frackowiak, R. S. J., Legg, N. J., and Jones, T.,** Serial observations on the pathophysiology of acute stroke the transient from ischemia to infarction as reflected in regional oxygen extraction, *Brain,* 106, 197, 1983.

60. **Hakim, A. M., Pokrupa, R. P., Villanueva, J., Diksic, M., Evans, A. C., Thompson, C. J., Meyer, E., Yamamoto, Y. L., and Feindel, W. H.,** The effect of spontaneous reperfusion on metabolic function in early human cerebral infarcts, *Ann. Neurol.,* 21, 279, 1987.

61. **Ackerman, R. H., Lev, M. H., Mackay, B. C., Katz, P. M., Babikian, V. L., Alpert, N. M., Correia, J. A., Panagos, P. D., and Senda, M.,** PET studies in acute stroke. Findings and relevance to therapy, *J. Cereb. Blood Flow Metab.,* 9 (Suppl. 1), S359, 1989.

62. **Heiss, W.-D.,** Flow threshold and morphological damage of brain tissue, *Stroke,* 14, 329, 1983.

63. **Baron, J.-C., Rougemont, D., Bousser, M. G., Lebrun-Grandie, P., Iba-Zizen, M. T., and Chiras, J.,** Local CBF, oxygen extraction fraction (OEF), and CMRO2. Prognostic value in recent supratentorial infarction in humans, *J. Cereb. Blood Flow Metab.,* 3 (Suppl. 1), S1, 1983.

64. **Astrup, J., Siesjoe, B. K., and Symon, L.,** Thresholds in cerebral ischemia. The ischemic-penumbra, *Stroke,* 12, 723, 1981.

65. **Ackerman, R. H., Correia, J. A., Alpert, N. M., Baron, J.-C., Gouliamos, A., Grotta, J. C., Brownell, G. L., and Taveras, J. M.,** Positron imaging in ischemic stroke disease using compounds labeled with oxygen 15. Initial results of clinicophysiologic correlations, *Arch. Neurol.,* 38, 537, 1981.

66. **Baron, J. C., Bousser, M. G., Comar, D., Soussaline, F., and Castiagne, P.,** Noninvasive tomographic study of cerebral blood flow and oxygen metabolism in-vivo. Potentials, limitations and clinical applications in cerebral ischemic disorders, *Eur. Neurol.,* 20, 273, 1981.

67. **Powers, W. J., Grubb, R. L., Darriet, D., and Raichle, M. E.,** Cerebral blood flow and cerebral metabolic rate of oxygen requirements for cerebral function and viability in humans, *J. Cereb. Blood Flow Metab.,* 5, 600, 1985.

68. **Lassen, N. A.,** The luxury-perfusion syndrome and its possible relation to acute metabolic acidosis localised within the brain, *Lancet,* 19, 1113, 1966.

69. **Baron, J. C., Bousser, M. G., Rey, A., Guillard, A., Comar, D., and Castaigne, P.,** Reversal of focal misery-perfusion syndrome by extra-intracranial arterial bypass in hemodynamic cerebral ischemia—a case study with ^{15}O positron emission tomography, *Stroke,* 12, 454, 1981.

70. **Baron, J. C., Samson, Y., Bousser, M. G., Derousné, C., D'Antona, R., Pantano, P., Pappata, S., Cambon, H., Rougemont, D., Lebrun-Grandié, P., Duquesnoy, N., Ottaviani, M., Sastre, J., Loc'h, C., Crouzel, C., Collard, P., Soussaline, F., Comar, D., and Castaigne, P.,** Measurement of regional CBF and oxygen metabolism in acute stroke with positron tomography: prognostic relevance and therapeutic implications, in *Acute Brain Ischemia. Medical and Surgical Therapy,* Battistini, N. et al., Eds., Raven Press, New York, 1986, 81.

71. **Marchal, G., Evans, A., Dagher, A., Meyer, E., and Hakim, A. M.,** The evolution of cerebral infarction with time. A PET study of the ischemic penumbra, *J. Cereb. Blood Flow Metab.,* 7(Suppl.1), S99, 1987.

72. **Pozzilli, C., Itoh, M., Matsuzawa, T., Fukuda, H., Abe, Y., Sato, T., Takeda, S., and Ido, T.,** Positron emission tomography in minor ischemic stroke using oxygen-15 steady-state technique, *J. Cereb. Blood Flow Metab.,* 7, 137, 1987.

73. **Kuhl, D. E., Phelps, M. E., Kowell, A. P., Metter, E. J., Selin, C., and Winter, J.,** Effects of stroke on local cerebral metabolism and perfusion. Mapping by emission computed tomography (PET) of [^{18}F]-FDG and ^{13}NH$_3$, *Ann. Neurol.,* 8, 47, 1980.

74. **Heiss, W.-D., Herholz, K., Boecher-Schwarz, H. G., Pawlik, G., Wienhard, K., Steinbrich, W., and Friedmann, G.,** PET, CT, and MR Imaging in Cerebrovascular Disease, *J. Comput. Assist. Tomogr.,* 10, 903, 1986.

75. **Kushner, M., Reivich, M., Fieschi, C., Siler, F., Chawluk, J., Rosen, M., Greenberg, J., Burke, A., and Alavi, A.,** Metabolic and clinical correlates of acute ischemic infarction, *Neurology,* 37, 1103, 1987.

76. **Rehncrona, S., Rosén, I., and Siesjö, B. K.,** Brain lactic acidosis and cell damage. I. Biochemistry and neurophysiology, *J. Cereb. Blood Flow Metab.,* 1, 297, 1981.

77. **Levine, S. R., Welch, K. M. A., Helpern, J. A., Chopp, M., Bruce, R., Selwa, J., and Smith, M. B.,** Prolonged deterioration of ischemic brain energy-metabolism and acidosis associated with hyperglycemia. Human cerebral infarction studied by serial P-31 NMR-spectroscopy, *Ann. Neurol.,* 23, 416, 1988.

78. **Baron, J. S., Rougemont, D., Soussaline, F., Bustany, P., Crouzel, C., Bousser, M. G., and Comar, D.,** Local interrelationships of cerebral oxygen consumption and glucose utilization in normal subjects and in ischemic stroke patients. A positron tomography study. *J. Cereb. Blood Flow Metab.,* 4, 140, 1984.

79. **Frackowiak, R. S. J., Herold, S., Petty, R. K. H., and Morganhughes, J. A.,** The cerebral metabolism of glucose and oxygen measured with positron tomography in patients with mitochondrial diseases, *Brain,* 111, 1009, 1988.

80. **Ackermann, R. F. and Lear, J. L.,** Simultaneous measurement of cerebral anaerobic and aerobic metabolism, *J. Cereb. Blood Flow Metab.,* 9 (Suppl.1), S238, 1989.

81. **Pawlik, G., Heiss, W.-D., Wienhard, K., Hebold, I. R., Ziffling, P., Staffen, W., Herholz, K., and Wagner, R.,** Brain glucose metabolism and blood flow in ischemic stroke, in *Clinical Efficacy of Positron Emission Tomography,* Heiss, W.-D., et al., Eds., Martinus Nijhoff, Boston, 1987, 37.

82. **Berkelbach van der Sprenkel, J. W. B., Luyten, P. R., Vanrijen, P. C., Tulleken, C. A. F., and DenHollander, J. A.,** Cerebral lactate detected by regional proton magnetic-resonance spectroscopy in a patient with cerebral infarction, *Stroke,* 19, 1556, 1988.

83. **Bruhn, H., Frahm, J., Gyngell, M. L., Merboldt, K. D., Hanicke, W., and Sauter, R.,** Cerebral metabolism in man after acute stroke. New observations using localized proton NMR-spectroscopy, *Magn. Reson. Med.,* 9, 126, 1989.

84. **Alpert, N. M., Senda, M., Buxton, R. B., Correia, J. A., Mackay, B., Weise, S., Ackerman, R. H., and Buonanno, F. S.,** Quantitative pH mapping in ischemic disease using $^{11}CO_2$ and PET, *J. Cereb. Blood Flow Metab.,* 9 (Suppl.1), S361, 1989.

85. **Yamamoto, Y. L., Hakim, A. M., Diksic, M., Pokrupa, R. P., Meyer, E., Tyler, J., Evans, A. C., Worsley, K., Thompson, C. J., and Feindel, W. H.,** Focal flow disturbances in acute strokes. Effects on regional metabolism and tissue pH, in *Functional Mapping of the Brain in Vascular Disorders,* Heiss, W. D., Ed., Springer-Verlag, Berlin, 1985, 85.

86. **Ericson, K., Bergstroem, M., Erikson, L., Hatam, A., Greitz, T., Soederstroem, C.-E., and Wilden, L.,** Positron emission tomography with ^{68}Ga-EDTA compared with transmission-CT in the evaluation of brain infarcts, *Acta Radiol.,* 22, 385, 1981.

87. **Xie, Y., Mies, G., and Hossmann, K. A.,** Ischemic threshold of brain protein-synthesis after unilateral carotid artery occlusion in gerbils, *Stroke,* 20, 620, 1989.

88. **Dwyer, B. E., Nishimura, R. N., Powell, C. L., and Mailheau, S. L.,** Focal protein synthesis inhibition in a model of neonatal hypoxic-ischemic brain injury, *Neurology,* 95, 277, 1987.

89. **Mathias, C. J., Welch, M. J., Kilbourn, M. R., Jerabek, P. A., Patrick, T. B., Raichle, M. E., Krohn, K. A., Rasey, J. S., and Shaw, D. W.,** Radiolabeled hypoxic cell sensitizers: tracers for assessment of ischemia, *Life Sci.,* 41, 199, 1987.

90. **Herholz, K. and Heiss, W.-D.,** Cerebrovascular disease. Use of fluordeoxyglucose, *Semin. Neurol.,* 9, 299, 1989.

91. **Hakim, A. M., Evans, A. C., Berger, L., Kuwabara, H., Worsley, K., Marchal, G., Biel, C., Pokrupa, R., Diksic, M., Meyer, E., Gjedde, A., and Marrett, S.,** The effect of nimodipine on the evolution of human cerebral infarction studied by pet, *J. Cereb. Blood Flow Metab.,* 9, 523, 1989.

92. **Heiss, W.-D., Holthoff, V., Pawlik, G., and Neveling, M.,** Effect of nimodipine on regional cerebral glucose metabolism in patients with acute ischemic stroke as measured by positron emission tomography, *J. Cereb. Blood Flow Metab.,* 10, 127, 1990.

93. **Frackowiak, R. S. J., Pozzilli, C., Legg, N. J., DuBoulay, G. H., Marshall, J., Lenzi, G. L., and Jones, T.,** Regional cerebral oxygen supply and utilization in dementia. A clinical and physiological study with oxygen-15 and positron tomography, *Brain,* 104, 753, 1981.

94. **Gibbs, J. M., Frackowiak, R. S. J., and Legg, N. J.,** Regional cerebral blood flow and oxygen metabolism in dementia due to vascular disease, *Gerontology,* 32 (Suppl. 1). 84, 1986.

95. **Benson, D. F., Kuhl, D. E., Hawkins, R. A., Phelps, M. E., Cummings, J. L., and Tsai, S. J.,** The fluorodeoxyglucose ^{18}F-scan in Alzheimer's disease and multi-infarct dementia, *Arch. Neurol.,* 40, 711, 1983.

96. **Herholz, K., Adams, R., Kessler, J., Szelies, B., and Heiss, W.-D.,** Specificity of FDG-PET findings in Alzheimer's disease, *Eur. J. Nucl. Med.,* 15, 432, 1989.

97. **Spetzler, R. F., Roskl, R. A., and Zabramski, J.,** Middle cerebral artery perfusion pressure in cerebrovascular occlusive disease, *Stroke,* 14, 552, 1983.

98. **Bornstein, N. M. and Norris, J. W.,** Benign outcome of carotid occlusion, *Neurology,* 39, 6, 1989.

99. **Gibbs, J. M., Wise, R. J. S., Leenders, K. L., and Jones, T.,** Evaluation of cerebral perfusion reserve in patients with carotid-artery occlusion, *Lancet,* 1, 310, 1984.

100. **Powers, W. J., Grubb, R. L., and Raichle, M. E.,** Physiological response to focal cerebral ischemia in humans, *Ann. Neurol.,* 16, 546, 1984.

101. **Powers, W. J., Press, G. A., Grubb, R. L., Gado, M., and Raichle, M. E.,** The effect of hemodynamically significant carotid artery disease on the hemodynamic status of the cerebral circulation, *Ann. Intern. Med.,* 106, 27, 1987.

102. **Leblanc, R., Yamamoto, Y. L., Tyler, J. L., and Hakim, A.,** Hemodynamic and metabolic effects of extracranial carotid disease, *Can. J. Neurol. Sci.,* 16, 51, 1989.

103. **Powers, W. J., Raichle, M. E., and Grubb, R. L.,** Positron emission tomography to assess cerebral perfusion, *Lancet,* 1, 102, 1985.

104. **Itoh, M., Hatazawa, J., Pozzilli, C., Fukuda, H., Abe, Y., Fujiwara, T., Kubota, K., Yamaguchi, K., Sato, T., Watanabe, H., Ido, T., and Matsuzawa, T.,** Haemodynamics and oxygen metabolism in patients with reversible ischaemic attacks or minor ischaemic stroke assessed with positron emission tomography, *Neuroradiology,* 29, 416, 1987.

105. **Herold, S., Brozovic, M., Gibbs, J., Lammertsma, A. A., Leenders, K. L., Carr, D., Fleming, J. S., and Jones, T.,** Measurement of regional cerebral blood flow, blood volume, and oxygen metabolism in patients with sickle-cell-disease using positron emission tomography, *Stroke,* 17, 692, 1986.

106. **Taki, W., Yonekawa, Y., Kobayashi, A., Ishikawa, M., Kikuchi, H., Nishizawa, S., Senda, M., Yonekawa, Y., Fukuyama, H., Harada, K., and Tanada, S.,** Cerebral circulation and oxygen metabolism in moyamoya disease of ischemic type in children, *Child's Nerv. Syst.,* 4, 259, 1988.

107. **Vorstrup, S.,** Tomographic cerebral blood flow measurements in patients with ischemic cerebrovascular disease and evaluation of the vasodilatory capacity by the acetazolamide test, *Acta Neurol. Scand.,* 77 (Suppl. 114), 3, 1988.

108. **Rutigliano, M. J., Yonas, H., and Johnson, D. W.,** Natural history of patients with compromised cerebral reserved, *J. Cereb. Blood Flow Metab.,* 9 (Suppl. 1), S609, 1989.

109. **Kanno, I., Uemura, K., Higano, S., Murakami, M., Iida, H., Miura, S., Shishido, F., Inugami, A., and Sayama, I.,** Oxygen extraction fraction at maximally vasodilated tissue in the ischemic brain estimated from the regional CO_2 responsiveness measured by positron emission tomography, *J. Cereb. Blood Flow Metab.,* 8, 227, 1988.

110. **Levine, R. L., Sunderland, J. J., Lagreze, H. L., Nickles, R. J., Rowe, B. R., and Tirski, P. A.,** Cerebral perfusion reserve indexes determined by fluoromethane positron emission scanning, *Stroke,* 19, 19, 1988.

111. **Levine, R. L., Lagreze, H. L., Dobkin, J. A., Hanson, J. M., Satter, M. R., Rowe, B. R., and Nickles, R. J.,** Cerebral vasocapacitance and TIAS, *Neurology,* 39, 25, 1989.

112. **Powers, W. J., Fox, P. T., and Raichle, M. E.,** The effect of carotid-artery disease on the cerebrovascular response to physiologic stimulation, *Neurology,* 38, 1475, 1988.

113. **Itoh, M., Hatazawa, J., Pozzilli, C., Matsuzawa, T., Abe, Y., Fukuda, H., Fujiwara, T., Watanuki, S., and Ido, T.,** Positron CT imaging of an impending stroke, *Neuroradiology,* 30, 276, 1988.

114. **Powers, W. J. and Raichle, M. E.,** Positron emission tomography and its application to the study of cerebrovascular disease in man, *Stroke,* 16, 361, 1985.

115. **Powers, W. J., Grubb, R. L., and Raichle, M. E.,** Clinical results of extracranial-intracranial bypass-surgery in patients with hemodynamic cerebrovascular disease, *J. Neurosurg.,* 70, 61, 1989.

116. **Powers, W. J., Tempel, L. W., and Grubb, R. L.,** Influence of cerebral treated patients, *Ann. Neurol.,* 25, 325, 1989.

117. **Gibbs, J. M., Wise, R. J. S., Thomas, D. J., Mansfield, A. O., and Ross Russel, R. W.,** Cerebral haemodynamic changes after extracranial-intracranial bypass surgery, *J. Neurol. Neurosurg. Psychiatry,* 50, 140, 1987.

118. **Powers, W. J., Martin, W. R. W., Herscovitch, P., Raichle, M. E., and Grubb, R. L.,** Extracranial-intracranial bypass surgery. Hemodynamic and metabolic effects, *Neurology,* 34, 1168, 1984.

119. **Heiss, W.-D., Pawlik, G., Herholz, K., Szelies, B., Wagner, R., and Wienhard, K.,** Nontraumatic intracerebral hematoma versus ischemic stroke: Differences in regional pattern of glucose metabolism, *J. Cereb. Blood Flow Metab.,* 5 (Suppl. 1), S5, 1985.

120. **Ackerman, R. H., et al.,** Positron imaging of cerebral blood flow and metabolism in *Hypertensive Intracerebral Hemorrhage,* Mizukami, M., Ed., Raven Press, New York, 1983, 165.

121. **Volpe, J. J., Herscovitch, P., Perlman, J. M., and Raichle, M. E.,** Positron emission tomography in the newborn. Extensive regional cerebral blood flow with intracerebral involvement, *Pediatrics,* 72, 589, 1983.

122. **Martin, W. R. W., Baker, R. P., Grubb, R. L., and Raichle, M. E.,** Cerebral blood volume, blood flow, and oxygen metabolism in cerebral ischemia and subarachnoid haemorrhage. An *in vivo* study using positron emission tomographys, *Acta Neurochir.,* 70, 3, 1984.

123. **Baron, J. C., Bousser, M. G., Comar, D., Duquesnoy, N., Sastre, J., and Castaigne, P.,** Crossed cerebellar diaschisis in human supratentorial brain infarction, *Trans. Am. Neurol. Assoc.,* 105, 459, 1980.

124. **Heiss, W.-D., Ilsen, H. W., Wagner, R., Pawlik, G., and Wienhard, K.,** Remote functional depression of glucose metabolism in stroke and its alteration by activating drugs, in *Positron Emission Tomography of the Brain* , Heiss, W.-D. and Phelps, M. E., Eds., Springer-Verlag, Berlin, 1983, 162.

125. **Metter, E. J., Mazziotta, J. C., Itabashi, H. H., Mankovich, N. J., Phelps, M. E., and Kuhl, D. E.,** Comparison of glucose metabolism, X-ray-CT, and postmortem data in a patient with multiple cerebral infarcts, *Neurology,* 35, 1695, 1985.

126. **Heiss, W.-D., Pawlik, G., Wagner, R., Ilsen, H. W., Herholz, K., and Wienhard, K.,** Functional hypometabolism of noninfarcted brain regions in ischemic stroke, in *J. Cereb. Blood Flow Metab.,* 3 (Suppl. 1), 582, 1983.

127. **Martin, W. R. W. and Raichle, M. E.,** Cerebellar blood flow and metabolism in cerebral hemisphere infarction O-15 positron emission tomography, *Ann. Neurol.,* 14, 168, 1983.

128. **Kushner, M., Alavi, A., Reivich, M., Dann, R., Burke, A., and Robinson, G.,** Contralateral cerebellar hypometabolism following cerebral insult. A positron emission tomography study, *Ann. Neurol.,* 15, 425, 1984.

129. **Baron, J. C., D'Antona, R., Pantano, P., Serdaru, M., Samson, Y., and Bousser, M. G.,** Effects of cerebral cortex. A positron tomography study in man, *Brain,* 109, 1243, 1986.

130. **Pawlik, G., Beil, C., Herholz, K., Szelies, B., Wienhard, K., and Heiss, W.-D.,** Comparative dynamic FDG-PET study of functional deactivation in thalamic versus extrathalamic focal ischemic brain lesions, *J. Cereb. Blood Flow Metab.,* 5 (Suppl. 1), S9.

131. **Szelies, B., Herholz, K., Pawlik, G., Karbe, H., Hebold, I., and Heiss, W.-D.,** Patterns of regional deactivation associated with pure thalamic infarction, *Arch. Neurol.,* in press.

132. **Klinger, A., Deleon, M. J., George, A. E., Miller, J. D., and Wolf, A. P.,** Elevated cerebellar glucose-metabolism in disease. Normal aging and Alzheimer's disease, *J. Cereb. Blood Flow Metab.,* 8, 433, 1988.

133. **Herholz, K., Heindel, W., Rackl, A., Neubauer, I., Steinbrich, W., Pietrzyk, U., Erasmi-Körber, H., and Heiss, W.-D.,** Regional cerebral blood flow in patients with leukoaraiosis and atherosclerotic carotid disease, *Arch. Neurol.,* 47, 392, 1990.

134. **Kiyosawa, M., Bosley, T. M., Alavi, A., Gupta, N., Rhodes, C. H., Chawluk, J., Kushner, M., Savino, P. J., Sergott, R. C., Schatz, N. J. et al.,** Positron emission tomography in a patient with progressive leukoencephalopathy, *Neurology,* 38, 1864, 1988.

135. **Volkow, N. D., Patchell, L., Kulkarni, M. V., Reed, K., and Simmons, M.,** Adrenoleukodystrophy: imaging with CT, MRI, PET, *J. Nucl. Med.,* 28, 524, 1987.

136. **Bosley, T. M. Rosenquist, A. C., Kushner, M., Burke, A., Stein, A., Dann, R., Cobbs, W., Savino, P. J., Schatz, N. J., Alavi, A., and Reivich, M.,** Ischemic lesions of the occipital cortex and tomography, *Neurology,* 35, 470, 1985.

137. **Pawlik, G., Herholz, K., Beil, C., Wagner, R., and Heiss, W.-D.,** Remote effects of focal lesions on cerebral flow and metabolism, in *Functional Mapping of the Brain in Vascular Disorders,* Heiss, W.-D., Ed., Springer-Verlag, Berlin, 1985, 59.

138. **Beil, C., Hebold, I., Pawlik, G., Wienhard, K., and Heiss, W.-D.,** Correlative clinico-metabolic short-term follow-up in ischemic stroke studied by positron emission tomography of 2(^{18}F)-fluoro-deoxyglucose and standardized clinical ratings, *J. Cereb. Blood Flow Metab.,* 7 (Suppl.1), S20, 1987.

139. **Metter, E. J., Wasterlain, C. G., Kuhl, D. E., Hanson, W. R., and Phelps, M. E.,** [^{18}F]FDG positron emission computed tomography in a study of aphasia, *Ann. Neurol.,* 10, 173, 1981.

140. **Metter, E. J., Riege, W. H., Hanson, W. R., Kuhl, D. E., Phelps., M. E., Squire, L. R., Wasterlain, C. G., and Benson, D. F.,** Comparison of metabolic rates, language, and memory in subcortical aphasia, *Brain Language,* 21, 187, 1984.

141. **Metter, E. J., Riege, W. H., Hanson, W. R., Camras, L. R., Phelps, M. E., and Kuhl, D. E.,** Correlations of glucose metabolism and structural damage to language function in aphasia, *Brain Language,* 21, 187, 1984.

142. **Metter, E. J., Kempler, D., Jackson, C., Hanson, W. R., Mazziotta, J. C., and Phelps, M. E.,** Cerebral glucose metabolism in Wernicke's, Broca's, and conduction-aphasia, *Arch. Neurol.,* 46, 27, 1989.

143. **Metter, E. J., Kempler, D., Jackson, C. A., Hanson, W. R., Riege, W. H., Camras, L. R., Mazziotta, J. C., and Phelps, M. E.,** Cerebellar glucose metabolism in chronic aphasia, *Neurology,* 37, 1599, 1987.

144. **Karbe, H., Herholz, K., Szelies, B., Pawlik, G., Wienhard, K., and Heiss, W.-D.,** Regional metabolic correlates of Token test results in cortical and subcortical left hemispheric infarction, *Neurology,* 39, 1083, 1989.

145. **Pawlik, G. and Heiss, W. D.,** Positron-emission-tomography and function in *Neuropsychological function and brain imaging,* Bigler, E. D. et al., Eds., Plenum Press, New York, 1989, 65.

146. **Heiss, W.-D., Pawlik, G., Hebold, I., Beil, C., Herholz, K., Szelies, B., and Wienhard, K.,** Can positron emission tomography be used to gauge the ischemic stroke? (A European perspective), in *Cerebrovascular Diseases,* Ginsberg, M. D. and Dietrich, W. D., Eds., Raven Press, New York, 1989, 345.

Chapter 12

EVALUATION OF CEREBRAL HEMODYNAMICS WITH POSITRON EMISSION TOMOGRAPHY

William J. Powers

TABLE OF CONTENTS

I. INTRODUCTION

The role of hemodynamic factors in the pathogenesis and treatment of ischemic cerebrovascular disease remains unclear. Although hemodynamic or embolic etiologies for cerebral infarction can sometimes be clearly identified (e.g., border zone infarction following cardiac arrest or combined systemic embolism and stroke in young patients with rheumatic mitral stenosis), these are a minority. For the more typical patient with focal ischemic symptoms and atherosclerosis of the carotid or vertebrobasilar systems, the relative impact of embolic and hemodynamic factors on the subsequent occurrence of stroke has been difficult to determine. Observations of platelet thrombi both in retinal vessels and in surgical specimens removed at carotid endarterectomy have lent credence to the idea that platelet emboli are of primary importance in patients with minimal or minor degrees of carotid stenosis. However, when carotid stenosis becomes more severe or when carotid occlusion occurs, the relative importance of hemodynamic and embolic factors is uncertain. This distinction is of more than just academic interest since treatment with antithrombotic drugs such as aspirin is unlikely to prevent hemodynamic stroke. Indeed, although aspirin is of value in preventing subsequent stroke in symptomatic patients, the effect is incomplete. Stroke still occurs at a rate of 3 to 5%/year.[1,2] The explanation for these "medical failures" is not known, but they may, in part, be hemodynamically mediated. Surgical revascularization procedures have the potential to improve regional cerebral perfusion pressure (rCPP) and cerebral blood flow (rCBF) to such areas and to prevent hemodynamic infarction.

A better understanding of the relationship between hemodynamics and cerebral infarction is critical for the proper design of studies to evaluate both medical and surgical therapies to prevent stroke. Therapeutic trials are most effective when restricted to those patients who will benefit. This approach improves statistical power, decreases the likelihood that benefits in small subgroup will be overlooked, and permits more specific applications of the results to general clinical practice. Thus, trials of platelet inhibitory drugs typically exclude patients with possible etiologies for stroke believed not to involve platelet aggregation, e.g., cardiac thrombi. Exclusion, if possible, of those with hemodynamically mediated cerebral ischemia would further improve the selectivity of such trials in the future.

Consideration of hemodynamic factors is, perhaps, even more important in the proper design of trials of surgical therapy for cerebrovascular disease. Superficial temporal artery-middle cerebral artery (STA-MCA) bypass surgery was developed to improve CBF in patients with complete carotid occlusion or intracranial carotid stenosis not amenable to conventional endarterectomy. Since this surgery is unlikely to provide protection from embolic stroke, its efficacy in preventing stroke should be greatest in those patients in whom hemodynamic factors are important in the pathogenesis of cerebral infarction. A large prospective randomized trial has demonstrated no value for this surgery in preventing subsequent stroke.[3] This trial has been criticized for failing to identify and separately analyze the subgroup of patients with reduction in perfusion pressure in whom surgery might be more beneficial.[4] In addition, STA-MCA bypass has been criticized for not providing adequate augmentation of flow to restore hemodynamics to normal. This has led to the development of new surgical revascularization strategies based on the premise that hemodynamic factors are important.[5] Thus, a better understanding of the importance of cerebral hemodynamics is also critical in determining future research directions into improved surgical methods for stroke prevention.

In order to determine what role hemodynamic factors play in the pathogenesis, prognosis, and choice of treatment for patients with ischemic cerebrovascular disease, a method for determining the hemodynamic status of the cerebral circulation, accurately and in awake subjects under normal conditions, must be available. It is obviously impractical to measure directly the perfusion pressure in the distal arterial bed of the cerebral circulation. It has been necessary, therefore, to rely on indirect assessments.

Measurements in the cervical carotid artery have demonstrated that reductions in pressure and flow distal to a stenosis occur only when the lumen diameter is reduced by more than 60 to 65% producing a residual lumen of less than 1 to 2 mm in diameter.[6-8] The presence of such a hemodynamically significant carotid artery lesion is generally regarded as an indication that the hemodynamic status of the cerebral circulation is impaired and is often used as an indication of the need for carotid endarterectomy in patients with both symptomatic and asymptomatic disease. Studies investigating the relation between the hemodynamic significance of a carotid artery lesion and subsequent stroke have produced conflicting results. In asymptomatic patients, some studies have shown an increased risk for stroke with increasing stenosis,[9-11] whereas others have not.[12,13] These studies have largely failed to account for such confounding prognostic variables as age, hypertension, and different treatment regimens. Their results are therefore difficult to interpret and the question remains unresolved. For symptomatic patients with transient ischemic attacks or minor stroke in the carotid artery distribution, the data relating severity of stenosis to stroke risk are much clearer and the results are somewhat surprising. The best available evidence indicates that the risk for ipsilateral stroke in medically treated patients with symptomatic carotid artery disease is not influenced by the degree of stenosis.[14] However, the number of patients in these studies with severe stenosis is small; the power to detect a difference is therefore low. Such an emphasis on the hemodynamic significance of a carotid artery lesion has thus failed to provide a definitive answer about hemodynamic factors and stroke risk. This is not surprising since this approach ignores the contribution of collateral circulatory pathways in maintaining cerebral perfusion. The importance of collateral channels in maintaining CBF distal to an occluded or stenotic carotid artery is well recognized, but its relationship to the subsequent risk of stroke is less clear.

To gain a better idea of the effect of proximal atherosclerotic lesions and angiographic collateral circulatory patterns on cerebral hemodynamics, a variety of indirect methods for evaluating the status of the distal cerebral vasculature have been developed. The rationale for these methods is based upon the compensatory responses made by the brain to progressive reductions in CPP.

II. COMPENSATORY RESPONSES TO REDUCED CEREBRAL PERFUSION PRESSURE

When CPP is normal (stage 0), CBF is regulated by changes in arterial diameter. There is a direct correlation between CBF and the intravascular cerebral blood volume (CBV); the ratio CBV/CBF is therefore independent of CBF.[15-17] Under these conditions, CBF is also closely matched to the resting metabolic rate of the tissue. Gray matter areas with higher metabolic rates have higher CBF and white matter areas with lower metabolic rates have lower CBF. A fairly uniform value for the ratio between CBF and metabolism exists in all areas of the brain. As a consequence of this resting balance between flow and metabolism, the fractional extraction of oxygen from the blood shows little regional variation. The actual value for oxygen extraction fraction (OEF) will vary from person to person and from measurement to measurement within the same person but it is approximately one-third for normal individuals.[18]

Changes in CPP over a wide range have little effect on CBF.[19,20] This phenomenon is known as autoregulation. A rise in mean arterial pressure produces vasoconstriction of pial arterioles and a fall produces vasodilation.[21,22] Similar changes in the intravascular volume of cerebral vessels have been detected by *in vivo* measurements of CBV in primates over a wide range of CPP.[23,24]

The ratio CBV/CBF continues to increase as CPP falls.[25] This ratio is mathematically equivalent to the mean vascular transit time for red blood cells through the cerebral vessels;

an increase in this ratio indicates slowing of the cerebral circulation.[26] It is believed to be a more sensitive index of reduced CPP than CBV alone since increased ratios may occur when the CBV is still within the upper range of normal.[27,28] Autoregulatory vasodilation impairs the response of cerebral blood vessels to other vasoactive stimuli. At normal CPP, increases in P_aCO_2 produce marked vasodilation and increases in CBF. This response is attenuated as CPP is reduced and eventually is lost altogether.[20]

When the capacity for compensatory vasodilation has been exceeded, autoregulation fails and CBF begins to decline. Measurements of arteriovenous oxygen differences have demonstrated the brain's capacity to increase OEF when oxygen supply is diminished due to decreasing CBF.[29-31] As the perfusion pressure of the brain falls further, CBF progressively declines until the increase in OEF is no longer adequate to supply the energy needs of the brain. Clinical evidence of brain dysfunction now appears. Persistent or further declines in CPP and CBF can lead to permanent tissue damage.

It is immediately apparent that the measurement of multiple physiologic variables is necessary to assess accurately cerebral hemodynamics. Measurements of CBF alone are inadequate for this purpose since they cannot detect reductions in CPP when CBF is maintained by compensatory vasodilation. Furthermore, they cannot differentiate decreased CBF caused by reduced blood supply from that caused by reduced metabolic demands. While these two causes of reduced CBF may not be difficult to distinguish in areas of known infarction, there can be real problems in the interpretation of CBF measurements when the reduction in metabolic demand occurs in structurally normal brain tissue. Destruction of afferent or efferent fibers pathways by a cerebral lesion may cause reductions in both CBF and metabolism in regions which show no abnormality by X-ray CT or MRI. Such areas of reduced blood flow and metabolism may be seen overlying subcortical lesions or in distant structures such as contralateral cerebellum.[18,32] Determination of the hemodynamic status of such areas of preserved brain when supplied by a stenotic or occluded vessel is not possible with simple measurements of CBF.

The development of PET has made it practical to perform a variety of accurate, quantitative measurements of cerebral physiology *in vivo*. PET measurements of rCBF, rCBV, and rOEF can be combined with known cerebrovascular physiology as described above to detect the presence and severity of reductions in local CPP. Two basic strategies have emerged. The first involves measurement in the resting brain of rCBV/rCBF and rOEF.[27,28] Based on these measurements, regional cerebral hemodynamics can be separated into three stages: when CPP is normal (Stage 0), rCBV/rCBF and rOEF are also normal. When CPP is reduced and cerebral vessels dilate to maintain flow (Stage 1), rCBV/rCBF will increase, but rOEF will remain normal. When the capacity for compensatory vasodilation is overwhelmed and rCBF begins to fall (Stage 2), both CBV/CBF and rOEF will increase.

The second strategy involves the use of two rCBF measurements, one performed at rest and the second during the application of some vasodilatory stimulus. The stimulus may be CO_2 inhalation[33-35] or a physiological task involving movement or sensory stimulation.[36,37] An impairment in the normal CBF increase is taken as evidence of preexisting autoregulatory vasodilation due to reduced CPP.

Kanno et al.[35] and Herold et al.[38] have investigated the relationship between these two strategies, comparing CO_2 reactivity and rOEF. Both found a significant negative linear relationship between CO_2 reactivity and rOEF, the highest OEF values occurring in brain regions without CO_2 reactivity. Herold et al. also studied the relationship between CBV/CBF and CO_2 reactivity and found a significant relationship as well.[38] (These investigators actually reported CBF/CBV ratios. For simplicity, all results will be given here in terms of CBV/CBF regardless of what was originally reported.) While these two strategies are derived from the same physiological bases and appeared to yield the same information in the above studies, there is some reason to believe that they might not be completely interchangeable. The CO_2 response may be impaired by atherosclerosis itself whereas autoregulation (and by extension CBV/CBF) is apparently unaffected.[39] A state of dissociated vasoparalysis with absent CO_2 responsivity and preserved autoregulation has also been demonstrated in following dogs with severe ischemia.[40]

Powers et al. have examined the relationship between the rCBF response to physiological stimulation and resting hemodynamics measurements in 11 patients with severe carotid artery disease, but no evidence of functional or structural brain damage.[36] Three of the four patients with normal hemodynamics at rest had normal rCBF responses to bilateral hand stimulation and six of seven patients with abnormal resting hemodynamics had reduced or absent rCBF responses. All three patients with increased OEF had such abnormal CBF responses. This association between an abnormal rCBF response to physiologic stimulation and evidence of poor perfusion pressure at rest (increased CBV/CBF ratio or increased OEF) suggests that poor vascular responsiveness due to preexisting vasodilation may be responsible in most cases. Improvement following cerebral revascularization in two of three patients supports this view. The hypothesis of reduced vascular responsiveness due to preexisting vasodilation does not, however, adequately explain the reduced rCBF responses in the patient with normal resting hemodynamics. The mechanism involved here remains unknown.

These two strategies for assessment of intracranial hemodynamics are appropriate only for uninfarcted tissue at a time remote from an ischemic event. The changes in CBF, CBV, OEF, and vascular reactivity that occur in newly ischemic or infarcted tissue are so complex and poorly understood that they cannot be used to infer the hemodynamic effect of large vessel occlusive disease.[18,41-43]

III. APPLICATION OF PET TO THE STUDY OF CEREBRAL HEMODYNAMICS IN PATIENTS WITH CAROTID ARTERY DISEASE

A. METHODOLOGICAL CONSIDERATIONS

Discussion of the methodological considerations necessary for the proper application of PET to the accurate measurement of the physiological variables of interest here is beyond the scope of this chapter. Many of these issues are dealt with in other chapters of this book and in recent reviews.[44-46] The interested reader should also consult the references cited in this chapter, many of which contain excellent discussion of methodological issues.

Two issues of data analysis directly bear in the interpretation of studies to be discussed here. These have to do with sample size and correction for the effect of multiple comparisons on the statistical significance of observed differences. Most PET studies of cerebral hemodynamics have employed relatively few (5 to 20) normal controls. Conclusions about patient populations based on small deviations from the range of values or calculated 99% confidence limits from such small normal groups must be viewed with caution. It is only when consistent results occur across studies from different centers that some confidence in the findings can be justified.

The problem of multiple comparisons is a difficult one. When multiple independent physiological measurements are taken from two groups of subjects, the chances of finding a significant difference at least one measurement with a probability of $p = \alpha$ is increased. Various methods to deal with this problem are available. The most common employs the Bonferroni inequality. With this method the desired study probability (conventionally $\alpha = 0.05$) is divided by the number of measurements to provide the appropriately reduced value to apply to each comparison. Since many PET studies of cerebral hemodynamics involve the measurement of multiple physiological variables (CBF, CBV, CBV/CBF, OEF, and $CMRO_2$), some correction for the effect of multiple comparisons must be made. Since these variables are not truly independent, however, the Bonferroni correction is overly conservative.[47,48] However, no other method more appropriately suited to this situation has been forthcoming and the use of the Bonferroni correction permits conclusions to be drawn with little chance of committing a Type I error.

B. PREVALENCE OF HEMODYNAMIC ABNORMALITIES

Several large studies of patients with occlusive disease of the carotid arterial system have provided data on the prevalence of hemodynamic abnormalities in this population. All of these studies have relied on PET measurements taken from uninfarcted tissue in the middle cerebral artery territory distal to the involved carotid or middle cerebral artery. Gibbs et al. described 32 patients with internal carotid artery (ICA) occlusion in whom PET measurements were carried out.[27] They found small, but not significant, decreases in both CBF and $CMRO_2$ distal to the occlusion with significant increases in both CBV and OEF. A total of 22 of 24 patients with unilateral occlusion and 8 of 8 patients with bilateral occlusion (90% of total) had CBV/CBF above the normal range. Six patients (19%) had elevations in regional OEF. In a group of 51 patients with ICA occlusion, Samson and Baron[49] found absolutely high values of rOEF in 12%, but significant side-to-side asymmetries of rOEF in 41%. The frequency of OEF asymmetries was approximately the same in patients with unilateral and bilateral ICA obstruction. They concluded that asymmetries are a more sensitive index than absolute values. Powers et al.[50,51] described 61 patients with greater than 50% stenosis or occlusion of the carotid or middle cerebral arteries. Significant side-to-side asymmetries in CBV/CBF were observed in 47 patients (77%) and asymmetries of OEF were observed in 16 (26%). Levine et al.[34] found an impaired CBF response to 5% CO_2 inhalation in 9 of 32 patients with carotid transient ischemic attacks (TIAs).

C. RELATIONSHIP OF ANGIOGRAPHY TO PET

In an attempt to understand further the factors associated with PET findings of impaired cerebral hemodynamics, several studies have looked at the relationship between angiographic findings and PET measurements. Gibbs et al. noted that values for both CBV/CBF and OEF were higher in patients with bilateral carotid occlusion as compared to those with unilateral occlusion.[27] Samson and Baron did not observe this in their series of 51 patients.[49] Powers et al.[28] assessed the relationship between the PET measurements of hemodynamics and both the degree of carotid stenosis and the pattern of collateral circulation in 19 patients with no clinical or CT evidence of cerebral infarction who underwent selective bilateral biplane carotid arteriography for clinical indications. All had at least one carotid artery with a "hemodynamically significant" lesion of 66% or more diameter reduction. Five patients also underwent selective vertebral arteriography for evaluation of the posterior circulation. In the middle cerebral artery territory distal to the "hemodynamically significant" carotid artery stenosis, seven patients had normal cerebral hemodynamics (Stage 0), eight had increased CBV/CBF only (Stage 1), and four had both increased CBV/CBF and increased OEF flow (Stage 2). There was no significant relationship between the hemodynamic status of the cerebral circulation as measured by PET and either the percent carotid stenosis or the residual carotid lumen diameter in millimeters. There was, however, a highly significant relationship between the three PET categories of cerebral hemodynamics and the arteriographic collateral circulation. Leptomeningeal collaterals from anterior cerebral artery to middle cerebral artery were present only in patients with reduced CBF (3 of 4 patients in Stage 2) and not in patients with normal CBF (0 of 15 patients in Stages 0 and 1, $p < 0.005$). Ophthalmic collaterals were observed only in patients with reduced cerebral perfusion pressure (7 of 12 patients in Stages 1 and 2) and not in patients with normal hemodynamics (0 of 7 patients in Stage 0; $p < 0.02$). Of five patients who underwent vertebrobasilar arteriography, only one demonstrated collateral flow via the posterior communicating artery to the middle cerebral artery territory ipsilateral to the severe carotid lesion. This patient had normal cerebral hemodynamics by PET. Four other patients showed no perfusion of the middle cerebral artery branches via the posterior communicating artery. All four had abnormal cerebral hemodynamics in that hemisphere with one classified as Stage 1 and three as Stage 2. Three of these four patients demonstrated collateral circulation via leptomeningeal vessels from the posterior cerebral artery to the middle cerebral artery. Samson et al.[52] studied 27 patients with four-vessel angiography and PET. An OEF increase distal to carotid occlusion correlated with either ipsilateral ophthalmic artery collaterals or collateral circulation through the circle of Willis from a stenosed artery. Levine et al.[34] reported the relationship between CO_2 reactivity and degree of carotid stenosis in 32 patients with TIA. Distal to the symptomatic vessel, CO_2 response was reduced in 0 of 7 with less than 50% stenosis, 4 of 15 with 50 to 99% stenosis, and 5 of 10 with carotid occlusion. In those with carotid occlusion, abnormal hemodynamics were associated with ipsilateral external carotid artery collaterals (n = 3) or collateral circulation across the circle of Willis from a contralateral carotid artery with greater than 50% stenosis (n = 2).

These studies indicate that the concept of the "hemodynamically significant" carotid artery stenosis needs to be reconsidered. While a stenosis greater than 50% appears to be a necessary condition for hemodynamic compromise, it is not sufficient. Of those with greater than 50%, the pressure distally ranged from normal to severely reduced with no relationship to the degree of stenosis. Thus, neither the presence nor severity of a unilateral, "hemodynamically significant" carotid artery lesion is a reliable indicator of the hemodynamic status of the cerebral circulation in the ipsilateral hemisphere. The primary determinant of cerebral hemodynamics in patients with severe carotid artery stenosis or occlusion is the adequacy of collateral circulatory pathways.

Leblanc and co-workers have recently challenged these conclusions.[53] They rightly pointed out that the PET measurements were made in the middle cerebral artery territory and that the borderzone regions between middle cerebral artery and anterior cerebral artery (ABZ) and between the middle cerebral and posterior cerebral artery (PBZ) might be more sensitive to the effects of carotid occlusive disease. They studied seven patients with 80% or greater carotid stenosis and compared them with six controls. They report a significant increase in CBV/CBF in the ipsilateral ABZ ($p < 0.05$) with 6/7 outside the normal range. In the ipsilateral PBZ, OEF was outside the normal range in four, but not significantly different from normal; CBV/CBF was not elevated. In this analysis, no correction of the effect of 40 separate pair-wise comparisons on the statistical significance level was carried out. Due to the small numbers of subjects and high likelihood of Type I error in this study, Carpenter et al. recently have attempted to confirm these findings in a larger series of 17 controls and 35 patients with internal carotid artery or middle cerebral artery stenosis, all of whom had normal neurological examinations and normal head CT scans.[57] A total of 12 patients had stenosis of 50 to 79%, 6 had 80 to 99% stenosis, and 17 had complete occlusion. ABZ and PBZ were located using stereotactic coordinates based on neuropathological studies. Ratios of each borderzone value to the ipsilateral core MCA territory were calculated to increase sensitivity for detecting selective borderzone hemodynamic abnormalities even in the presence of hemodynamic abnormalities in the MCA territory. There was no significant difference from control in patients with >75% stenosis, occlusion, or in those with contralateral stenosis of >50%. Patients with abnormal hemodynamics in the MCA territory (increased rCBV/rCBF) did not have any evidence of selectively worse hemodynamic impairment in border zones. No patient had any ratio that exceeded the normal range. This statistical analyses was done with a Bonferroni correction for multiple comparisons. Even with no such correction, however, no evidence for selective hemodynamic compromise in the borderzones was demonstrated for any single comparison at a significance level of $p < 0.05$.

D. RELATIONSHIP BETWEEN CLINICAL SYMPTOMS AND PET

Gibbs et al.[27] noted that CBV/CBF was higher in 11 patients who had clinical features suggesting hemodynamic perfusion failure (borderzone infarction, symptoms precipitated by sudden standing, limb shaking, or prolonged fluctuating deficit) than in 18 patients without such features. Levine et al.[34] found that CO_2 reactivity was reduced in the symptomatic carotid territory of 6 of 8 patients with hemodynamic TIAs (induced by postural change, exertion, orthostasis, cardiac perfusion abnormality, or combinations thereof), but in only 2 of 24 without such features.

Powers et al.[55] investigated the relationship of PET measurements of cerebral hemodynamics to clinical presentation in 27 patients with predominantly unilateral carotid artery disease of greater than 67% diameter reduction. Thirteen patients had complete carotid occlusion. These patients had no clinical or CT evidence of cerebral infarction: 7 were asymptomatic and 20 had had 1 or more carotid territory transient ischemic attacks. Seven of these had had transient monocular blindness alone, ten had had hemispheric TIAs alone, and three had had transient monocular blindness and hemispheric TIAs. Of these 27, 13 had normal cerebral hemodynamics by PET and 14 had reduced cerebral perfusion pressure (Stage 1 or 2). Four of seven asymptomatic patients had normal hemodynamics and three had abnormal hemodynamics. In the 20 patients with carotid territory transient ischemic attacks, the following clinical characteristics of the TIA were not significantly different between patients with normal and abnormal hemodynamics: number, stereotypy, duration, and tempo at onset. Neither transient monocular blindness, hemispheric symptoms, or symptoms restricted to a single limb were significantly more common in patients with hemodynamic disease. TIA with limb shaking occurred only in patients with hemodynamic cerebrovascular disease (5 of 11) as did posturally precipitated TIA (2 of 11), but these differences were not statistically significant when corrected for multiple comparisons.

These data indicate that such clinical manifestations as the number, stereotypy, duration and anatomic location of TIA are not reliable indicators of the presence of hemodynamic cerebrovascular disease. There are some uncommon characteristics of TIA such as limb shaking and posturally precipitated attacks that may be specific, but not sensitive, indicators of abnormal cerebral hemodynamics. It must be remembered, however, that this type of study cannot determine the actual cause of the TIAs in these patients. Although hemodynamic TIA is unlikely to occur in patients who have normal cerebral hemodynamics at rest (unless there is a substantial fall in systemic arterial pressure), embolic TIA can certainly occur in patients with hemodynamic cerebrovascular compromise. We have recently observed a patient with carotid occlusion who had limb shaking TIAs precipitated by standing. PET showed increased OEF in the MCA territory distal to the occluded vessel — 3 d later she developed a large frontal infarct. Repeat angiography the next day demonstrated several new occlusions of MCA branches most consistent with multiple emboli.

E. THE EFFECT OF EXTRACRANIAL-INTRACRANIAL BYPASS SURGERY ON CEREBRAL HEMODYNAMICS

The use of PET to determine regional cerebral hemodynamic status has been applied to patients undergoing surgical revascularization of the cerebral circulation for two purposes: (1) to monitor the circulatory and metabolic effects of the surgery and (2) to try to define subgroups within an otherwise homogeneous population who might respond differently to treatment.

Baron et al. reported results of STA-MCA bypass surgery in five patients: one with TIA, two with cerebral infarction and TIA, and two with stable cerebral infarction.[56,57] Preoperatively, all had uninfarcted regions with decreased CBF and two had concomitant increases in OEF. Following bypass surgery, CBF improved in four and OEF returned to normal in the two patients where it had been elevated. One of these also showed a slight increase in $CMRO_2$. Powers et al. have studied with PET 21 patients undergoing STA-MCA bypass anastomosis.[58] Twelve had TIA only, three had cerebral infarction with subsequent TIA, and six had stable cerebral infarction. Following surgery, there was a significant improvement in CBF to the operated hemisphere in six patients. The preoperative PET studies in these six patients showed large areas of uninfarcted brain with decreased CBF. In five of the six, hemispheric $CMRO_2$ was also decreased. Hemispheric OEF was increased in three and hemispheric CBV was increased in five. Postoperatively, along with the increase in hemispheric CBF, these six patients showed a significant decrease in hemispheric OEF implying that an increase in the local fractional extraction of oxygen had served as a compensatory mechanism preoperatively. There was no significant postoperative change in $CMRO_2$ indicating that the compensation had been adequate. There was also no postoperative change in CBV suggesting that, although the bypass improved blood flow, perfusion pressure was still below the autoregulatory limit and maximal vasodilation was still present. Both of these studies were based on the assessment of ratios between the operated and nonoperated hemisphere, not on quantitative values in the operated hemisphere. Samson et al. reported measurements of CBF, OEF, and $CMRO_2$ on patients undergoing STA-MCA bypass for carotid occlusion (n = 11) or middle cerebral artery occlusion (n = 1).[59] Contrary to the results of the previous two studies, they found that an asymmetry in OEF was not a reliable predictor of a relative increase in CBF postoperatively. Interestingly, when they analyzed quantitative values, they found a bilateral postoperative increase in $CMRO_2$ restricted to those seven patients with either severe contralateral carotid artery disease (n = 6) or MCA occlusion (n = 1). This increase in $CMRO_2$ occurred in concert with a rise in CBF, but no consistent change in OEF. Postoperative improvement in neurological deficit occurred in three of the seven patients with and in one of five patients without the $CMRO_2$ increase. The authors postulated reversal of chronic ischemic suppression of oxygen metabolism to explain these observations and noted that the functional benefit, if any, associated with the increase in $CMRO_2$ remains to be established. Gibbs et al. reported CBF, CBV, $CMRO_2$, and OEF measurement on 12 patients with carotid occlusion before and after STA-MCA bypass.[60] The principal finding of this study was a significant postoperative reduction in both CBV and CBV/CBF in the hemisphere ipsilateral to the bypass. They found no significant changes in CBF, $CMRO_2$, or OEF measurements in the operated hemisphere. OEF was elevated above normal preoperatively in four patients. It returned to normal postoperatively in all four, but in only one was this due to an increase in rCBF. In the other three patients, OEF decreased due to a reduction in $CMRO_2$; two of these suffered perioperative infarcts. Two patients (one with unilateral and one with bilateral severe carotid occlusive disease) showed bilateral increases in $CMRO_2$ with "no detectable alteration in the patient's clinical state." Leblanc et al. have recently reported a small study of five patients.[61] They also found postoperative reductions in both CBV and CBV/CBF with no consistent changes in CBF, $CMRO_2$, or OEF. Three of their subjects underwent neuropsychological testing before and after surgery. One of these had bilateral carotid occlusion and was suffering from progressive dementia preoperatively. He showed dramatic improvement in higher mental function postoperatively in association with a marked bilateral increase in $CMRO_2$ (88% ipsilateral; 44% contralateral). A second subject showed improvements in attention and memory associated with a 17% ipsilateral increase and 16% contralateral decrease in $CMRO_2$. The third showed no change in neuropsychological function with 22 and 26% reductions in hemispheric $CMRO_2$.

The results of these studies are somewhat difficult to reconcile with the concepts of cerebrovascular responses to reduced perfusion pressure presented earlier. It is true that reductions in the CBV/CBF ratio have been consistent findings postoperatively as would be expected from an improvement on cerebral perfusion pressure by revascularization. However, OEF has not been shown to be a consistent predictor of postoperative improvement in CBF as originally thought. Finally, the explanation for and importance of the bilateral increases in $CMRO_2$ observed in some patients remains unclear. Samson et al. studied this phenomenon further in ten patients with internal carotid artery occlusion, five with and five without significant contralateral disease.[62] After a baseline study, mean arterial blood pressure was raised 30 ± 7 mmHg by intravenous infusion of angiotensin II. All brain regions with asymmetries in rOEF at baseline showed increased rCBF and reduced rOEF during induced hypertension. In five patients with bilateral ICA obstruction, infusion of angiotensin II resulted in a significant increase in both CBF ($21 \pm 16\%$, $p < 0.01$) and $CMRO_2$ ($16 \pm 7\%$, $p < 0.001$) in both hemispheres. No significant change occurred in either hemisphere of the five patients with unilateral disease. While these data allowed the authors to confirm the hemodynamic basis for the findings in their STA-MCA bypass study, they bring us no closer to understanding the underlying pathophysiology.

F. ROLE OF HEMODYNAMIC FACTORS IN PROGNOSIS AND CHOICE OF THERAPY

Accurate knowledge of cerebral hemodynamics should permit the identification of a subgroup of patients at high risk for subsequent stroke on hemodynamic grounds (either spontaneously or secondary to hypotensive episodes occurring during major cardiovascular surgery) and who need surgical revascularization by STA-MCA bypass or carotid endarterectomy to reduce this risk. Following publication of the results of the EC/IC bypass trial in 1985,[3] Powers et al. undertook to test this hypothesis at Washington University Medical Center with a pilot study to investigate the relationship between cerebral hemodynamics as determined by PET and the subsequent risk of stroke in medically treated patients with symptomatic occlusion or greater than 75% intracranial stenosis of the carotid arterial system.[51] During 1986, 13 patients were enrolled prospectively. Another 19 patients were added by retrospective review of records back to 1981, primarily those who had been evaluated for STA-MCA bypass, but not operated. Follow-up to 1 year was obtained for all 32 patients. The incidence at 1 year for all stroke was 1 of 9 for patients with normal hemodynamics and 1 of 23 for patients with abnormal hemodynamics (increased rCBV/rCBF ratio with or without increased rOEF). The 1-year incidence for ipsilateral ischemic stroke was 1 of 9 for hemodynamically normal patients and 0 of 23 in the abnormal group. Twenty-one patients with PET evidence of abnormal cerebral hemodynamics had had symptoms within 3 months prior to PET and thus met clinical and arteriographic entry criteria for the EC/IC Bypass Trial.[3] The 1-year occurrences of both total and ipsilateral ischemic stroke in this sample of 21 patients were compared with all 714 medically treated patients in the Bypass Trial. Subsequent medical treatment with aspirin was similar in both groups. Both the total stroke occurrence (1 of 21 = 0.048) and the ipsilateral ischemic stroke occurrence (0 of 21) were lower than the comparable values from the EC/IC Trial of 0.127 and 0.109. Powers et al. were able to reject with better than 75% certainty ($p = 0.23$) the hypothesis that their sample came from a population with a total stroke rate greater than that in the trial and reject with better than 90% certainty ($p = 0.09$) that it came from a population with a greater ipsilateral ischemic stroke rate. Thus, it is quite unlikely that patients with abnormal cerebral hemodynamics defined by the rCBV/rCBF ratio constitute a subgroup of symptomatic patients with carotid system occlusion or intracranial stenosis who have a high 1-year risk for ipsilateral ischemic stroke.

In a related study, clinical follow-up data were obtained on all patients studied with PET measurements of CBF, CBV, and OEF at Washington University Medical Center who had subsequently undergone STA-MCA bypass.[50] Of 29 patients with symptomatic occlusion or intracranial stenosis of the carotid arterial system prior to undergoing STA-MCA bypass surgery, 24 patients had evidence of reduced rCPP (increased rCBV/rCBF ratio) distal to the arterial lesion. Of 21 patients who survived surgery without stroke, 3 suffered ipsilateral

TABLE 1
Two Year Follow-Up of 56 Patients at High Risk for Stroke Studied With Positron Emission Tomography[63]

Group	Medical			Surgical		
Stage	0	1	2	0	1	2
n	9	15	5	5	13	9
All stroke	1	2	2	1	2	2
Ipsilateral ischemic stroke	1	1	1	1	2	2
All death	2	3*	0	0	1	0

Note: Asterisk (*) indicates one fatal stroke.

ischemic strokes during the first postoperative year. These results were compared to the 23 medically treated patients with similar clinical, epidemiologic, arteriographic, and PET characteristics. As noted above, no ipsilateral ischemic strokes occurred in the medically treated group during the first year following PET. Based on these results in 44 patients, the probability that successful surgery reduces the occurrence of ipsilateral ischemic stroke one year later was calculated. This probability ranged from 0.045 for a 50% reduction to 0.168 for a 10% reduction. Thus, there was little evidence to suggest that abnormal cerebral hemodynamics defined by CBV/CBF ratio can identify a group of patients who would benefit from STA-MCA bypass surgery.

Follow-up to 2 years has been continued for all 56 patients who had symptoms within 3 months of initial PET.[63] One surgical patient was lost to follow-up at 16 months. One medical patient who underwent contralateral endarterectomy at 22 months was dropped from the study. The remaining medical (n = 29) and surgical (n = 27) groups were comparable with regard to major epidemiological risk factors (age, sex, hypertension, and cigarette smoking), but differed somewhat in that the surgical group had more patients with TIA (81 vs. 52%). Aspirin use was the same in both groups (56 vs. 55%). Results are listed in Table 1. Included in these results for the surgical group are two perioperative ipsilateral ischemic strokes (one each in stages 0 and 1) and one fatal perioperative myocardial infarction in a Stage 1 patient.

In the medically treated group, although the incidence of all strokes was higher in patients with abnormal cerebral hemodynamics (Stages 1 and 2) than in those with normal hemodynamics (4 of 20 vs. 1 of 9), this difference was not statistically significant. Only two of these strokes were ipsilateral. At 2 years, there was still no evidence that STA-MCA bypass could prevent either ipsilateral or all stroke in the patients with hemodynamic abnormalities (Stages 1 and 2). A total of 4 of 20 medically treated patients and 4 of 22 surgically treated patients had had strokes.

This analysis has limitations. It is based on a small number of patients followed for only 2 years. The criterion of an abnormal rCBV/rCBF ratio that was used to define abnormal cerebral hemodynamics may not be restrictive enough. In our experience, one half of patients with >50% carotid stenosis and three quarters of patients with carotid occlusion will demonstrate this finding by PET. A smaller subgroup of patients with even more severe hemodynamic abnormality, such as those with increased OEF may be at increased risk for stroke. For all 14 medical and surgical patients with increased OEF from this study, the risk of stroke at 2 years was 4 of 14 = 28% vs. 6 of 42 = 14% for the other two groups (Stages 1 and 0). Although this difference is not statistically significant, it suggests that patients with increased OEF may be at higher risk for stroke.

Any study of the influence of hemodynamic factors on the risk of stroke must take into account the possible effects of coexisting angiographic, epidemiologic and clinical factors in a multivariate analysis. It is possible that hemodynamic factors may be an important prognostic indicator simply because they occur more often in patients at high risk for other reasons and thus would not be expected to retain significance in a multivariate analysis. Alternatively, the effect of other risk factors may be mediated via local hemodynamic compromise and they would no longer be significant predictors in a multivariate analysis. The fact that STA-MCA bypass does not eliminate subsequent stroke may indicate that the increase in OEF is a predictive marker for other than hemodynamic reasons. A larger study of medically treated patients studied by PET with appropriate multivariate analysis of potentially confounding risk factors is necessary to resolve these issues.

IV. CONCLUSIONS

Positron emission tomography (PET) provides previously unavailable information about regional CBF, CBV, OEF, and $CMRO_2$ in patients with vascular disease of the large arteries supplying the brain. This information can be used to infer the regional hemodynamic status of uninfarcted brain distal to a carotid lesion. Such studies have demonstrated that the severity of the carotid stenosis is a poor predictor of cerebral hemodynamics (collateral circulation is much more important) and have provided insight into TIA symptoms which are likely to be hemodynamic in origin. Several studies of the effect of STA-MCA bypass have demonstrated evidence of improved hemodynamics (reduced CBV/CBF) postoperatively. Changes in OEF and $CMRO_2$ have been variable and difficult to interpret. One study of 56 patients followed for 2 years produced no evidence that PET determination of autoregulatory vasodilation can identify patients at high risk for stroke if treated medically or can identify a subgroup of patients who will benefit from STA-MCA bypass. This study is not definitive due to its nonrandomized design and small number of patients with severe hemodynamic compromise manifested by increased OEF. Further studies to elucidate the pathophysiology of changes in OEF and $CMRO_2$ in patients with severe carotid occlusive disease and to determine the relationship between these PET measurements of cerebral hemodynamics and subsequent stroke risk need to be carried out.

In order for PET to be clinically valuable for making prognostic or therapeutic decisions in cerebral ischemia, the data it provides must be unobtainable by other cheaper or less invasive means. PET must be compared with other diagnostic tests, such as arteriography, Doppler, ultrasound, oculoplethysmography, and SPECT (single photon emission computed tomography), as well as with careful clinical evaluation of the patient to determine empirically which are the most accurate, safest, and least expensive means for predicting stroke risk and choosing therapy. This research is only just beginning.

ACKNOWLEDGMENT

This work was supported by National Institutes of Health (NIH) Grants NS00647, NS06833, HL13851, and AG03991, the McDonnell Center for Studies of Higher Brain Function, and the Lillian Strauss Institute for Neurosciences of the Jewish Hospital at Washington University Medical Center, St. Louis, Missouri.

REFERENCES

1. Canadian Cooperative Study Group, A randomized trial of aspirin and sulfinpyrazone in threatened stroke, *N. Engl. J. Med.*, 299, 53, 1978.
2. **Bousser, M. G., Eschwege, E., Haguenau, M. et al.,** "AICLA" controlled trial of aspirin and dipyridamole in the secondary prevention of athero-thrombotic cerebral ischemia, *Stroke*, 14, 5, 1983.

3. The EC-IC Bypass Study group, The international cooperative study of extracranial-intracranial arterial anastomosis (EC-IC bypass study): methodology and entry characteristics, *Stroke*, 16, 397, 1985.

4. Editorial: Extracranial to intracranial bypass and the prevention of stroke, *Lancet*, 2, 1401, 1985.

5. **Diaz, F. G., Umansky, F., Mehta, B., Montoya, S., Dujovny, M., Ausman, J. I., and Cabezudo, J.,** Cerebral revascularization to a main limb of the middle cerebral artery in the Sylvian fissure: an alternative approach to conventional anastomosis, *J. Neurosurg.*, 63, 21, 1985.

6. **Archie, J. P., Jr. and Feldtman, R. W.,** Critical stenosis of the internal carotid artery, *Surgery*, 89, 67, 1981.

7. **Brice, J. G., Dowsett, D. J., and Lowe, R. D.,** Haemodynamic effects of carotid artery stenosis, *Br. Med. J.*, 2, 1363, 1964.

8. **DeWeese, J. A., May, A. G., Lipchik, E. O., and Rob, C. G.,** Anatomic and hemodynamic correlations in carotid artery stenosis, *Stroke*, 1, 149, 1970.

9. **Durward, Q. J., Ferguson, G. G., and Barr, H. W. K.,** The natural history of asymptomatic carotid bifurcation plaques, *Stroke*, 13, 459, 1982.

10. **Roederer, G. O., Langlois, Y. E., Jager, K. A. et al.,** The natural history of carotid arterial disease in asymptomatic patients with cervical bruits, *Stroke*, 15, 605, 1984.

11. **Meissner, I., Wiebers, D. O., Whisnant, J. P., and O'Fallon, W. M.,** The natural history of asymptomatic carotid artery occlusive lesions, *JAMA*, 258, 2704, 1987.

12. **Grotta, J. C., Bigelow, R. H., Hu, H., Hankins, L., and Fields, W. S.,** The significance of carotid stenosis or ulceration. *Neurology*, 34, 437, 1984.

13. **Roederer, G. O., Langlois, Y. E., Lusiani, L. et al.,** Natural history of carotid artery disease on the side contralateral to endarterectomy, *J. Vasc. Surg.*, 1, 62, 1984.

14. **Powers, W. J.,** What is the significance of hemodynamically significant carotid stenosis in symptomatic patients?, in *Cerebrovascular Diseases*, 15th Res. (Princeton) Conf., Powers, W. J. and Raichle, M. E., Eds., Raven Press, New York, 1987, 215.

15. **Risberg, J., Ancri, D., and Ingvar, D.,** Correlation between cerebral blood volume and cerebral blood flow in the cat, *Exp. Brain Res.*, 8, 321, 1969.

16. **Grubb, R. L., Raichle, M. E., Eichling, J. O., and Ter-Pogossian, M. M.,** The effects of changes in P_aCO_2 on cerebral blood volume, blood flow and mean transit time, *Stroke*, 5, 630, 1974.

17. **Herold, S., Brozovic, M., Gibbs, J. et al.,** Measurement of regional cerebral blood flow, blood volume and oxygen metabolism in patients with sickle cell disease using positron emission tomography, *Stroke*, 17, 692, 1986.

18. **Powers, W. J. and Raichle, M. E.,** Positron emission tomography and its application to the study of cerebrovascular disease in man, *Stroke*, 16, 361, 1985.

19. **Rapela, C. E. and Green, H. A.,** Autoregulation of canine cerebral blood flow, *Circ. Res.*, 15 (Suppl. 1), 205 1964.

20. **Harper, A. M. and Glass, H. I.,** Effect of alterations in the arterial carbon dioxide tension on the blood flow through cerebral cortex at normal and low arterial blood pressures, *J. Neurol. Neurosurg. Psychiatry*, 28, 449, 1965.

21. **Fog, M.,** Cerebral circulation. The reaction of pial arteries to a fall in blood pressure, *Arch. Neurol. Psychiatry*, 37, 351, 1937.

22. **Fog, M.,** Cerebral circulation. II. Reaction of pial arteries increase in blood pressure, *Arch. Neurol. Psychiatry*, 41, 260, 1939.

23. **Grubb, R. L. and Phelps, M. E.,** The effects of arterial blood pressure on the regional cerebral blood volume by X-ray fluorescence, *Stroke*, 4, 390, 1973.

24. **Grubb, R. L., Jr., Raichle, M. E., Phelps, M. E., and Ratcheson, R. A.,** Effects of increased intracranial pressure on cerebral blood volume, blood flow and oxygen utilization in monkeys, *J. Neurosurg.*, 43, 385, 1975.

25. **Ferrari, M., Wilson, D. A., Hartmann, J. F., Rogers, M. C., and Traystman, R. J.,** Effects of graded hemorrhagic hypotension on cerebral blood flow, blood volume and transit time in the dog, *Anesthesiology*, 67, A85, 1987.

26. **Powers, W. J., Grubb, R. L., Jr., and Raichle, M. E.,** Physiologic responses to focal cerebral ischemia in humans, *Ann. Neurol.*, 16, 546, 1984.

27. **Gibbs, J. M., Wise, R. J. S., Leenders, K. L., and Jones, T.,** Evaluation of cerebral perfusion reserve in patients with carotid-artery occlusion, *Lancet*, 1, 310, 1984.

28. **Powers, W. J., Press, G. A., Grubb, R. L., Jr., Gado, M., and Raichle, M. E.,** The effect of hemodynamically significant carotid artery disease on the hemodynamic status of the cerebral circulation, *Ann. Intern. Med.*, 106, 27, 1987.

29. **Lennox, W. G., Gibbs, F. A., and Gibbs, E. L.,** Relationship of unconsciousness to cerebral blood flow and to anoxemia, *Arch. Neurol. Psychiatry*, 34, 1001, 1935.

30. **Shenkin, H. A., Cabieses, F., van den Noordt, G., Sayers, P., and Copperman, R.,** The hemodynamic effect of unilateral carotid ligation on the cerebral circulation of men, *J. Neurosurg.*, 8, 38, 1951.

31. **Kety, S. L., King, B. D., Horvath, S. M., Jeffers, W. A., and Hafkenschiel, J. H.,** The effects of an acute reduction in blood pressure by means of differential spinal sympathetic block on the cerebral circulation of hypertensive patients, *J. Clin. Invest.,* 29, 402, 1950.

32. **Feeney, D. M. and Baron, J. C.,** Diaschisis, *Stroke,* 17, 817, 1986.

33. **Levine, R. L., Sunderland, J. J., Lagreye, H. L. et al.,** Cerebral perfusion reserve indexes determined by fluoromethane positron emission scanning, *Stroke,* 19, 19, 1988.

34. **Levine, R. L., Lagreye, H. L., Dobkin, J. A. et al.,** Cerebral vasocapacitance and TIAs, *Neurology,* 39, 25, 1989.

35. **Kanno, I., Uemura, K., Higano, S. et al.,** Oxygen extraction fraction at maximally vasodilated tissue in ischemic brain estimated from regional CO_2 responsiveness measured by positron emission tomography, *J. Cereb. Blood Flow Metab.,* 8, 227, 1988.

36. **Powers, W. J., Fox, P. T., and Raichle, M. E.,** The effect of carotid artery disease on the cerebrovascular response to physiologic stimulation, *Neurology,* 38, 1475, 1988.

37. **Chang, J. Y., Kelley, R. E., Ginsberg, M. D. et al.,** Assessment of "cerebral reserve" in patients with occlusive cerebrovascular disease by a somatosensory activation strategy and positron emission tomography, *J. Cereb. Blood Flow Metab.,* 9 (Suppl. 1), S358, 1989.

38. **Herold, S., Brown, M. M., Frackowiak, R. S. J., Mansfield, A. O., Thomas, D. J., and Marshall, J.,** Assessment of cerebral hemodynamic reserve: correlation between PET parameters and CO_2 reactivity measured by intravenous [133] xenon injection technique, *J. Neurol. Neurosurg. Psychiatry,* 51, 1045, 1988.

39. **Heistad, D. D., Marcus, M. L., Piegors, D. J., and Armstrong, M. L.,** Regulation of cerebral blood flow in atherosclerotic monkeys, *Am J. Physiol.,* 239, H539, 1980.

40. **Nemoto, E. M., Snyder, J. F., Carroll, R. G., and Marita, H.,** Global ischemia in dogs: cerebrovascular CO_2 reactivity and autoregulation, *Stroke,* 6, 429, 1975.

41. **Baron, J. C.,** Positron tomography in cerebral ischemia, *Neuroradiology,* 27, 509, 1985.

42. **Frachowiak, R. S. J.,** The pathophysiology of human cerebral ischemia: a new perspective obtained with positron tomography, *Q. J. Med.,* 57, 713, 1985.

43. **Hakim, A. M., Pokrupa, R. P., Villaneuva, J. et al.,** The effect of spontaneous reperfusion on metabolic function in early human cerebral infarcts, *Ann. Neurol.,* 21, 279, 1987.

44. **Baron, J. C., Frackowiak, R. S. J., Herholz, K. et al.,** Use of PET methods for measurement of cerebral energy metabolism and hemodynamics in cerebrovascular disease, *J. Cereb. Blood Flow Metab.,* 9, 723, 1989.

45. **Phelps, M., Mazziotta, J., and Schelbert, H., Eds.,** *Positron Emission Tomography and Autoradiography: Principles and Applications for the Brain and Heart,* Raven Press, New York, 1986.

46. **Reivich, M. and Alavi, A.,** *Positron Emission Tomography,* Alan R. Liss, New York, 1985.

47. **Wallenstein, S., Zucker, C. L., and Fleiss, J. L.,** Some statistical methods useful in circulation research, *Circ. Res.,* 47, 1, 1980.

48. **O'Brien, P. C., and Shampo, M. A.,** Statistical considerations for performing multiple tests in a single experiment. V. Comparing two therapies with respect to several endpoints, *Mayo Clinic Proc.,* 63, 1140, 1988.

49. **Samson, Y. and Baron, J. C.,** PET studies in internal carotid occlusion patients: base-line hemodynamic and metabolic alterations; effects of surgical and revascularization and induced arterial hypertension, in *Clinical Efficacy of Positron Emission Tomography,* Heiss, W.-D., Pawlik, G., Herholz, K., and Weinhard, K., Eds., Martinus Nijhoff, Boston, 1987, 55.

50. **Powers, W. J., Grubb, R. L., Jr., and Raichle, M. E.,** Clinical results of extracranial-intracranial bypass surgery in patients with hemodynamic cerebrovascular disease, *J. Neurosurg.,* 70, 61, 1989.

51. **Powers, W. J., Tempel, L. W., and Grubb, R. L., Jr.,** Influence of cerebral hemodynamics on stroke risk: one-year follow-up of 30 medically treated patients, *Ann. Neurol.,* 25, 325, 1989.

52. **Samson, Y., Baron, J. C., and Bousser, M. G.,** Cerebral hemodynamic and metabolic changes in carotid artery occlusion, A PET study, in *Cerebral Vascular Disease,* Vol. 5, Meyer, J. S., Lechner, H., Reivich, M., and Ott, E. O., Eds., Excerpta Medica, Amsterdam, 1985, 128.

53. **Leblanc, R., Yamamoto, Y. L., Tyler, J. L., Diksic, M., and Hakim, A.,** Borderzone ischemia, *Ann. Neurol.,* 22, 707, 1987.

54. **Carpenter, D. A., Grubb, R. L., Jr., and Powers, W. J.,** Borderzone hemodynamics in cerebrovascular disease, *Neurology,* 39 (Suppl. 1), 251, 1989.

55. **Powers, W. J., Tempel, L. W., Grubb, R. L., Jr. et al.,** Clinical correlates of cerebral hemodynamics, *Stroke,* 18, A284, 1987.

56. **Baron, J. C., Bousser, M. G., Rey, A., Guillard, A., Comar, D., and Castaigne, P.,** Reversal of focal "misery-perfusion syndrome" by extra-intracranial arterial bypass in hemodynamic cerebral ischemia, *Stroke,* 12, 454, 1981.

57. **Baron, J. C., Rey, A., Guillard, A., Bousser, M. G., Comar, D., and Castaigne, P.,** Non-invasive tomographic imaging of cerebral blood flow (CBF) and oxygen extraction fraction (OEF) in superficial temporal artery to middle cerebral artery (STA-MCA) anastomosis, in *Cerebral Vascular Disease,* Vol. 3, Meyer, J. S., Lechner, H., Reivich, M., and Ott, E. O., Eds., Excerpta Medica, Amsterdam, 1981, 58.

58. **Powers, W. J., Martin, W. R. W., Herscovitch, P., Raichle, M. E., and Grubb, R. L., Jr.,** Extracranial-intracranial bypass surgery: hemodynamic and metabolic results, *Neurology,* 34, 1168, 1984.

59. **Samson, Y., Baron, J. C., Bousser, M. G. et al.,** Effects of extra-intracranial arterial bypass on cerebral blood flow and oxygen metabolism in humans, *Stroke,* 16, 609, 1985.

60. **Gibbs, J. M., Wise, R. J. S., Thomas, D. J. et al.,** Cerebral haemodynamic changes after extracranial-intracranial bypass surgery, J. Neurol. *Neurosurg. Psychiatry,* 50, 140, 1987.

61. **Leblanc, R., Tyler, J. L., Mohr, G. et al.,** Hemodynamic and metabolic effects of cerebral revascularization, *J. Neurosurg.,* 66, 529, 1987.

62. **Samson, Y., Baron, J. C., Pappata, S. et al.,** Angiotensin II infusion improves perfusion and oxygen consumption in both cerebral hemispheres in patients with bilateral carotid artery obstruction, *J. Cereb. Blood Flow Metab.,* 7 (Suppl. 1), S177, 1987.

63. **Powers, W. J., Grubb, R. L., Jr., Tempel, L. W., Carpenter, D. A., and Raichle, M. E.,** Cerebral hemodynamics and stroke risk, *J. Cereb. Blood Flow Metab.,* 9 (Suppl. 1), S356, 1989.

Chapter 13

THE USE OF SPECT IN FOCAL EPILEPSY

Samuel F. Berkovic and Christopher C. Rowe

TABLE OF CONTENTS

I. INTRODUCTION

Single photon emission computed tomography (SPECT) studies of patients with epilepsy have been reported since the early 1980s. The initial studies were of low resolution, the correlations with electrically defined seizure foci were imprecise, and the technique was of no clinical use. In the last few years, however, there have been a number of major technical developments, both in instrumentation and in the availability of new radiopharmaceuticals. SPECT can now be regarded as a useful clinical tool in the localization of the epileptic focus in patients with focal epilepsy.

To date, clinical SPECT studies have been largely confined to the imaging of cerebral blood flow (CBF), although there is certainly potential to investigate a variety of neuro-transmitter systems with this technique. Here we review CBF changes in human focal epilepsy and show how current SPECT techniques can be used clinically to exploit these changes.

II. HISTORICAL BACKGROUND

In the latter part of the 19th century, it was generally believed that cerebral vasocon-striction with resultant cortical anemia (hypoperfusion) was the cause of epileptic seizures. In 1892 Sir Victor Horsely vigorously attacked this notion by marshalling evidence from direct visual observation of the cerebral cortex during seizures in human subjects and ex-perimental animals. Horsely reported hyperemia (increased blood flow) at the cortical focus during seizures induced by both electrical stimulation and by the chemical convulsant ab-sinthe.[1] Horsely's view was generally disregarded. Treatment of epilepsy by cervical sym-pathectomy, to increase cerebral blood flow, was still advocated around the turn of the century.[2]

In 1934, following preliminary animal experiments, Gibbs and colleagues in Boston documented increased cerebral blood flow in human epileptic seizures by inserting a ther-mocouple into the internal jugular vein. They demonstrated a marked increase in blood flow during seizures with increased blood flow persisting for a short time in the postictal phase.[3-5]

Contemporaneously with Gibbs' work, Wilder Penfield and colleagues at the Montreal Neurological Institute began a careful study of cerebrovascular changes during epileptic seizures in humans and experimental animals. In 1933 Penfield reported observations of the human cerebral cortex of 30 patients where he electrically induced seizures. He clearly was still attracted to the idea that cerebral vasoconstriction was related to seizures, but he only observed constriction of pial arteries in 6 of his 30 cases. Pallor of the cerebral cortex was sometimes noted following a seizure and did not occur during the ictus. Cessation of visible pulsation of the arteries, darkening of the vessels, and a gyral flush were usually observed at the site of stimulation.[6] By 1938 Penfield's human studies and animal experiments clearly showed that a localized increase in blood flow occurred at the region of the epileptic focus; pial vasoconstriction was only rarely noted and was regarded as an artifact of stimulation or as a secondary effect.[7,8] These observations finally refuted the earlier concept of cerebral vasoconstriction causing seizures, but were of little practical value in the clinical management of epilepsy.

During the 1960s and 1970s, a small number of studies confirmed Gibbs' findings, of a global increase in CBF during and immediately after seizures in man, by the Kety-Schmidt modification of the Fick principle of arteriovenous differences using nitrous oxide[9,10] or xenon[11] and by using blood flow probes in the major neck vessels.[12,13] Measurements of global interictal blood flow by these methods did not reveal any marked deviations from normal.[14,15] Penfield's finding, of a localized ictal increase in blood flow at the seizure focus, was elegantly confirmed by Dymond and Crandall in 1976 by the use of thermocouples attached to depth electrodes during the study of temporal lobe seizures. Again, increased blood flow during the seizures, persisting into the postictal state was observed.[16] These observations of increased CBF during and immediately after seizures in man have been repeatedly confirmed by animal studies, as reviewed by Chapman.[17]

The early methods were not of value for routine clinical use. Global measurements of CBF by the Kety-Schmidt technique do not provide localizing information, and methods for measuring local increases in blood flow were too invasive. Noninvasive measurements of regional CBF were required and this became possible with planar and tomographic imaging using radioactive CBF tracers.

III. REGIONAL CBF IN EPILEPSY

Early planar imaging and SPECT studies of regional CBF in epilepsy yielded inconsistent results. Positron emission tomography (PET) provides quantitative regional data with good spatial resolution. Although many of the PET studies in epilepsy postdated the early planar imaging and SPECT work, PET studies in focal epilepsy are briefly discussed first, as they provide the "gold standard" to judge other regional techniques.

A. POSITRON EMISSION TOMOGRAPHY

The majority of studies with PET in epilepsy have measured glucose metabolism with ^{18}F-fluorodeoxyglucose (FDG). Studies of CBF using ^{13}N-ammonia and ^{15}O-CO_2 tracers have yielded similar results.[18]

1. Ictal Studies with PET

Focal epileptic seizures are usually unpredictable in their time of occurrence and typically last only 1 to 2 min. PET is therefore poorly suited for the systematic study of ictal CBF and metabolism, and most reports have reported serendipitous observations when a patient had a seizure during a planned interictal study. Increased blood flow and metabolism has been found in these isolated cases, in agreement with the earlier studies reviewed above.[18-21] The rare entity of *epilepsia partialis continua* comprises focal motor seizures lasting for hours and PET studies have shown localized regions of increased CBF and metabolism, sometimes with surrounding areas of depressed activity.[19,22,23]

2. Interictal Studies with PET

The major contribution of PET in epilepsy to date has been the discovery that approximately 70% of patients with intractable temporal lobe epilepsy have an area of decreased metabolism and CBF on the side of the epileptic focus in the interictal state.[18-28] There was no reliable information on regional CBF and metabolism in the interictal state prior to PET. Indeed, the finding of decreased CBF and metabolism was intuitively somewhat of a surprise, as the epileptic focus was generally conceived of as a region of increased, albeit abnormal, neuronal activity.

Interictal hypoperfusion and hypometabolism determined by PET is quite a reliable guide to the lateralization of the epileptic focus in temporal lobe epilepsy. PET is now used, where available, in the preoperative assessment of patients for therapeutic temporal lobectomy. The area of hypometabolism and hypoperfusion does not correlate with the activity of interictal spiking, nor with the extent of morphological abnormalities in resected temporal lobes and its biological significance remains to be determined.[24-26] PET is less often useful in extra-temporal focal epilepsies but, in some cases, diagnostically significant areas of hypometabolism and hypoperfusion are found.[29]

Perfusion and metabolic defects can also be found in some patients with nonfocal epilepsies such as the Lennox-Gastaut syndrome.[30,31] Therefore, the significance of focal hypometabolism and hypoperfusion must always be interpreted in the light of the electro-clinical data of each patient.

B. PLANAR IMAGING STUDIES

Planar regional imaging using intraarterial ^{133}Xe and multiple detectors was developed by Lassen and Ingvar in the 1960s. In 1973 and 1976 they reported, as predicted from earlier work, markedly increased blood flow at the site of the epileptic focus in five patients with on-going focal seizures.[32,33]

Surprisingly, these workers,[33] and another group[34] using the xenon inhalation washout technique, reported regional *increased* blood flow at the focus in the interictal state. These latter observations are difficult to explain in view of the overwhelming evidence against this from subsequent PET and SPECT studies. Moreover, other contemporaneous[35] and subsequent studies using xenon inhalation washout[36] found the expected decrease in regional CBF around the area of the focus in many patients with focal epilepsy. Recently, Valmier et al.,[37] using xenon, reported decreased regional CBF in the region of the focus, with a paradoxical increase following intermittent photic stimulation. This observation, if confirmed, might partially explain the apparently anomalous interictal results initially reported with the xenon method.

Planar imaging with inhaled xenon has major limitations, particularly as regards spatial resolution. Furthermore, it is very insensitive to abnormalities in the mesial temporal region which is the usual site of origin of temporal lobe seizures and thus it has not proven useful for routine clinical purposes.

IV. SPECT STUDIES IN FOCAL EPILEPSY

SPECT instruments were developed in the late 1970s[38,39] and the first studies of regional CBF in epilepsy were reported in 1983. SPECT studies can be divided into two types. "Dynamic" studies refer to tomographic images performed during the xenon washout technique. "Static" studies use tracers given by bolus intravenous injection, and the images reflect regional CBF at the time of injection.

A. DYNAMIC SPECT STUDIES

Dynamic SPECT studies with ^{133}Xe were first reported by Bonte and colleagues in 1983.[40] They found decreased regional perfusion in 12 of 18 patients with focal epilepsy, but the correlations with the EEG focus were inexact. A single ictal study showed increased perfusion at the putative focus. A brief communication in the same year extended their series to 25 patients with similar results[41] and this group has also correlated hypoperfusion on dynamic SPECT with focal neuropsychological deficits.[42] Other workers have also found interictal regional hypoperfusion with dynamic SPECT.[43] More recently, Bonte's group reported, in abstract form, that 15 of 100 subjects with focal epilepsy had regional areas of *increased* perfusion in the resting state in areas generally unrelated to the focus.[44] This finding is difficult to accept in the light of previous knowledge and will need careful evaluation when published in detail.

Although dynamic SPECT has advantages of being quantitative, and studies can be rapidly repeated in the one subject, the spatial resolution with xenon is poor and current interest now focuses on static SPECT methods.

B. STATIC SPECT STUDIES

The ^{123}I-iodoamines and ^{99m}Tc tracers developed for static SPECT behave like "chemical microspheres".[45,46] They have a high first-pass extraction rate, their distribution reflects regional CBF and they remain fixed for hours after injection. These properties offer great advantages for imaging seizures as the patient can be injected during a seizure; the scan can then be performed, when convenient, during the next few hours. In addition, the image resolution with these tracers is far superior to that of ^{133}Xe. Disadvantages are few, but include difficulties with quantitation, which although possible, is not routinely performed, and uncertainties regarding the nonlinear relationship between count rate and true regional CBF (see Chapter 3).

1. Studies with Iodoamines

The introduction of the iodoamine compounds promised to increase the resolution of SPECT because of the higher energy characteristics of ^{123}I.[45] In 1983 Uren et al. reported the use of ^{123}I-isopropyl-iodoamphetamine (IMP) and found hyperperfusion during seizures in four cases with focal epilepsy and interictal hypoperfusion in seven patients, with the most marked hypoperfusion being seen in the immediate postictal state.[47] Using IMP, Sanabria and colleagues reported interictal hypoperfusion at the region of the EEG focus in 22 of 32 patients with focal epilepsy.[48]

An important series of studies has been reported by Lee and colleagues using the isotope trimethyl-(hydroxy-methyl-^{123}I-iodobenzyl)-propanediamine (HIPDM). They reported correct localization of the seizure focus using ictal studies in 13 of 16 patients with temporal lobe epilepsy.[49,50] They also found that ictal SPECT still showed a focal hypermetabolic zone when partial seizures secondarily generalized.[51] Unfortunately, the epileptic focus was not clearly defined by EEG in all of their cases. More recently, they have reported interictal hypoperfusion at the site of the seizure focus in 16 of 40 patients, with hypoperfusion in other regions in 8 cases, highlighting the lack of specificity of interictal SPECT with HIPDM.[52]

2. Initial Studies with HMPAO

^{99m}Tc hexamethyl-propyleneamineoxime (HMPAO) has several advantages over the previously used isotopes in SPECT. Images of cerebral blood flow are of much higher resolution than seen with xenon, and indeed are superior to those with the iodoamines as well. Moreover, because the compound can be labeled with technetium, it is readily available in all nuclear medicine departments, and is not dependent on a cyclotron as is the case with the iodoamines.[53-55]

A number of groups have reported findings with HMPAO in focal epilepsy.[56-62] Unfortunately, many of the studies have incorporated patients with poorly characterized epilepsies and the relationship of the findings with HMPAO to the seizure focus cannot be reliably determined.

Biersack et al. first reported focal hypoperfusion in 4 of 10 cases with apparent hyperperfusion in another 2 patients and, in a later report, abnormalities were noted in HMPAO images in 32 of 40 cases.[56,57] The electroclinical details of these 40 cases were not described, nor was the relationship of the site of HMPAO abnormalities to the ictal focus discussed. In a detailed separate report of ten patients, however, this group found significant interictal abnormalities at the side of the focus in only five of the ten cases.[58] SPECT with HMPAO was inferior to PET with ^{18}F-FDG in providing useful localizing information in this series.[58]

Podreka et al. claimed focal hypoperfusion was present in 12 of 14 patients (86%) with complex partial seizures, and in 7 of 9 patients with generalized seizures, but the electroclinical details were sparse.[59] These findings are difficult to reconcile with those from the more thoroughly evaluated patient series studied with both SPECT and PET. This group also reported ictal hyperperfusion in three patients and, based on four serial scans in one case, suggested that hyperperfusion can persist for up to 14 d. This patient did not have chronic epilepsy and his history was more compatible with an acute focal encephalitis, which is a more likely cause of the blood flow changes described.[60] Two other groups have reported ictal hyperperfusion in isolated cases studied with HMPAO,[61,62] and one of these centers claimed focal interictal hypoperfusion was present in 12 of 14 cases.[62]

3. Studies with HMPAO at Austin Hospital, Melbourne

We have completed a study of interictal and immediate postictal SPECT using HMPAO in 51 patients with carefully characterized temporal lobe epilepsy. All the patients had intractable temporal lobe epilepsy and were being considered for surgical treatment. Seizures were localized by ictal EEG recordings in all cases, including depth electrode exploration in 33 patients. A total of 45 cases were found to have a unilateral temporal lobe focus, whereas 6 had bitemporal epilepsy. We also studied a group of ten healthy young male volunteers. Scans were interpreted blindly by a panel of three observers and were analyzed quantitatively using right to left asymmetries. This work forms part of a doctoral dissertation[63] and has been published in part.[64-67]

a. Interictal Findings

Visual analysis of the interictal scans of the 45 unilateral temporal cases by two blinded observers revealed focal hypoperfusion in the epileptogenic temporal lobe in 38%, with incorrect localization in 7% and 56% being regarded as inconclusive.[63] A separate blind analysis by the third reader resulted in 49% correct localization, 16% incorrect, and 36% inconclusive.[65] Our experience with visual analysis suggests that as one's criteria for hypoperfusion becomes looser, the error rate increases.

Visual assessment was confirmed quantitatively by a right/left asymmetry index using an antero-mesial temporal region of interest. With the criterion of normality being within 1.96 standard deviations (SD) from the mean of the controls, 47% of cases were correctly localized, 7% were incorrectly localized, and 47% were normal. Using less strict criteria, the sensitivity increased, but with an unacceptable rate of incorrectly localized cases.

It is clear that, at present, interictal studies with SPECT in temporal lobe epilepsy yield useful diagnostic information in no more than half the patients. Indeed, even with conservative interpretation of the scans there is a significant false positive rate, thus clouding the clinical value of this technique. Our unfavorable assessment of the value of interictal SPECT is supported by previous studies using rigorous criteria for seizure localization.[52,58] Studies claiming diagnostic accuracy in 80% or more of cases were flawed by a lack of blinded interpretation and by imprecise EEG seizure localization.[56,57,59,62]

Currently, interictal SPECT, with diagnostically useful scans in about 40% of cases, is inferior to interictal PET scanning with FDG, where approximately 70% of patients have an appropriate area of hypometabolism. It is unclear whether this relates wholly to the poorer resolution of SPECT or to a fundamental difference in the frequency of abnormalities of blood flow vs. glucose metabolism in temporal lobe epilepsy. The latter factor may be important as the rate of detection of abnormalities in FDG scans does not appear to have increased with improved resolution of PET cameras. Moreover, small-scale parallel studies of glucose metabolism and blood flow using PET have suggested the superiority of glucose imaging.[28,69]

b. Postictal SPECT Studies

We performed 78 postictal studies in 51 patients with temporal lobe epilepsy, the isotope being injected 4.3 ± 4.5 min after seizure onset. A characteristic pattern of blood flow change was found that had great diagnostic significance. This consisted of a focal area of relatively increased perfusion, usually in the mesial and anterior parts of the temporal lobe, with decreased perfusion in the adjacent lateral temporal cortex (Plate 1*). This lateral temporal hypoperfusion often extended into the parietal and frontal regions as well. Where this combination of mesial hyperperfusion and lateral hypoperfusion was seen, it was found to be a highly reliable diagnostic aid to the site of the focus.[63-67]

---

*   Plate 1 follows page 198.

In all, postictal SPECT resulted in reliable localization in 70% of our 45 unilateral temporal cases. Initially, one case was incorrectly localized. This was the first scan of our series, performed 20 min after the complex partial seizure and, in retrospect, the persistent hypoperfusion was misinterpreted as hyperperfusion on the other side.[64,64] With increasing experience of the patterns of postictal blood flow change, we have not made any further errors. Indeed, in a subsequent blinded review, both observers correctly interpreted this scan as nondiagnostic.[68] Of the six bitemporal cases, five showed bilateral changes, but one had a strongly lateralized postictal SPECT study. It is unclear whether this patient was misdiagnosed on depth EEG as bitemporal because of inadequate electrode placement, or whether the SPECT was misleading.

These studies have shown that there is a sequence of evolution of blood flow changes after temporal lobe seizures. During seizures blood flow appears to be markedly increased at and around the site of the focus. This has been shown by Lee et al. with HIPDM,[50] and recently by our group in 12 consecutive ictal studies using HMPAO.[70] The area of hyperperfusion rapidly shrinks after the seizure and characteristically is seen as a small crescent in the mesial and anterior parts of the epileptogenic temporal lobe. In our experience, this mesial hyperperfusion rarely persists longer than 6 to 8 min after the seizure and may even disappear in 1 or 2 min. However, the associated lateral cortical hypoperfusion often persists for a longer period (up to 20 min), and then the scan returns to its usual interictal state.[63-68] It is only with an understanding of this sequence that postictal blood flow images can be properly interpreted.

Approximately 30% of patients do not have diagnostic lateralized information on their postictal SPECT studies. There are at least two reasons for this. First, in a small number of patients, the pattern of evolution described above may occur within 2 or 3 min, rather than 15 or 20 min. Unless such patients are given a very early injection, the diagnostic changes will not be seen. Second, some patients show prominent bilateral postictal changes, either bilateral mesial hyperperfusion with lateral hypoperfusion, or just bilateral hypoperfusion. Some of these cases appear to have bitemporal epilepsy, but the majority have a unilateral EEG onset with rapid involvement of the contralateral temporal lobe. It is clear from our postictal studies, that there is a higher diagnostic yield of lateralized changes with early injection of HMPAO.[63,64,68] This has been confirmed by our recent experience with ictal studies, yielding correct lateralized information in all 12 consecutive cases of temporal lobe epilepsy studied.[70]

V. CONCLUSION: CURRENT CLINICAL ROLE OF SPECT IN FOCAL EPILEPSY

Functional imaging with both PET and SPECT has an established place as an adjunctive clinical tool in the localization of epileptic foci, particularly in the preoperative assessment for the surgical treatment of focal epilepsy. In centers where PET is available, some 70% of patients with temporal lobe epilepsy have an interictal abnormality in the appropriate temporal lobe and this information is extremely useful in reaching a clinical decision.

SPECT has certain advantages and disadvantages compared to PET. Its disadvantages include a lower spatial resolution, the present inability to do metabolic as opposed to blood flow studies, and the lack of readily available quantitative techniques. However, it has major advantages including its ready availability, its relative low cost, the fact that it is far less draining on manpower than PET, the availability of isotopes that do not require a cyclotron, and the ability to do routine ictal or postictal studies which are impossible with PET.

It is clear that interictal SPECT is unreliable for the localization of epileptic foci. At best interictal SPECT leads to a correct localization in only 40 to 50% of patients, with some false positives. However, ictal and postictal SPECT are far more reliable with a diagnostic yield of 70% or more. In our center, the use of postictal SPECT has led to a marked reduction in the frequency with which depth electrode exploration is required to localize the epileptic foci.

In the future, the role of SPECT in the localization of extratemporal epilepsies needs to be explored. The possibility of using ligands directed at the benzodiazepine, GABA, and other receptors may further enhance the role of SPECT in the clinical management of patients with focal epilepsy.

ACKNOWLEDGMENTS

We are grateful to our colleagues at the Austin Hospital in the Departments of Nuclear Medicine (Mr. Mark Austin, Dr. W. John McKay, and Dr. S. T. Benjamin Sia) and Neurology (Dr. Peter F. Bladin and Dr. Mark R. Newton) whose skilled help was essential in the performance of our studies. This work was supported by the National Health and Medical Research Council of Australia.

REFERENCES

1. **Horsely, V.**, An address on the origin and seat of epileptic disturbance, *Br. Med. J.*, 1, 693, 1892.
2. **Alexander, W.**, *The Treatment of Epilepsy*, Y. J. Pentland, Edinburgh, 1889.
3. **Gibbs, F. A.**, Cerebral blood flow preceding and accompanying experimental convulsions, *Arch. Neurol. Psychiatry*, 30, 1003, 1933.
4. **Gibbs, F. A., Lennox, W. G., and Gibbs, E. L.**, Cerebral blood flow preceding and accompanying epileptic seizures in man, *Arch. Neurol. Psychiatry*, 32, 257, 1934.
5. **Lennox, W. G.**, The physiological pathogenesis of epilepsy, *Brain*, 59, 113, 1936.
6. **Penfield, W.**, The evidence for a cerebral vascular mechanism in epilepsy, *Ann. Intern. Med.*, 7, 303, 1933.
7. **Penfield, W.**, The circulation of the epileptic brain, *Assoc. Res. Nerv. Ment. Dis.*, 18, 605, 1938.
8. **Penfield, W., von Sántha, K., and Cipriani, A.**, Cerebral blood flow during induced epileptiform seizures in animals and man, *J. Neurophysiol.*, 2, 257, 1939.
9. **White, P. T., Grant, P., Mosier, J., and Craig, A.**, Changes in cerebral dynamics associated with seizures, *Neurology*, 11, 354, 1961.
10. **Posner, J. B., Plum, F., and van Poznak, A.**, Cerebral metabolism during electrically induced seizures in man, *Arch. Neurol.*, 20, 388, 1969.
11. **Brodersen, P., Paulson, O. B., Bolwig, T. G., Rogon, E., Rafaelson, O. J., and Lassen, N. A.**, Cerebral hyperemia in electrically induced epileptic seizures, *Arch. Neurol.*, 28, 334, 1973.
12. **Meyer, J. S., Gotoh, F., and Favale, E.**, Cerebral metabolism during epileptic seizures in man, *Electroencephalogr. Clin. Neurophysiol.*, 21, 10, 1966.
13. **Magnaes, B. and Nornes, H.**, Circulatory and respiratory changes in spontaneous epileptic seizures in man, *Eur. Neurol.*, 12, 1974, 104.
14. **Grant, F. C., Spitz, E. B., Shenkin, H. A., Schmidt, C. F., and Kety, S. S.**, The cerebral blood flow and metabolism in idiopathic epilepsy, *Trans. Am. Neurol. Assoc.*, 72, 82, 1947.
15. **Kennedy, C., Anderson, W. B., and Sokoloff, L.**, Cerebral blood flow in epileptic children during the interseizure period, *Neurology*, 8, 100, 1958.
16. **Dymond, A. M., and Crandall, P. H.**, Oxygen availability and blood flow in the temporal lobes during spontaneous epileptic seizures in man, *Brain Res.*, 102, 191, 1976.
17. **Chapman, A. G.**, Cerebral energy metabolism and seizures, in *Recent Advances in Epilepsy 2*, Pedley, T. A. and Meldrum, B. S., Eds., Churchill Livingstone, Edinburgh, 1985, 19.
18. **Engel, J., Jr.**, The use of PET scanning in epilepsy, *Ann. Neurol.*, 15 (Suppl.), s180, 1984.
19. **Engel, J., Kuhl, D. E., Phelps, M. E., Rausch, R., and Nuwer, M.**, Local cerebral metabolism during partial seizures, *Neurology*, 33, 400, 1983.
20. **Theodore, W. H., Newmark, M. E., Sato, S. et al.**, [^{18}F]Fluorodeoxyglucose positron emission tomography in refractory complex partial seizures, *Ann. Neurol.*, 14, 429, 1983.
21. **Abou-Khalil, B. W., Siegel, G. J., Sackellares, J. C. et al.**, Positron emission tomography studies of cerebral glucose metabolism in chronic partial epilepsy, *Ann. Neurol.*, 22, 48, 1987.

22. **Kuhl, D. E., Engel, J., Phelps, M. E., and Selin, C.,** Epileptic patterns of local cerebral metabolism and perfusion in humans determined by emission computed tomography of ^{18}FDG and ^{13}NH$_2$, *Ann. Neurol.,* 8, 348, 1980.

23. **Franck, G., Sadzot, B., Salmon, E. et al.,** Regional cerebral blood flow and metabolic rates in human focal epilepsy and status epilepticus, *Adv. Neurol.,* 44, 935, 1986.

24. **Engel, J., Jr., Kuhl, D. E., Phelps, M. E., and Crandall, P. H.,** Comparative localization of epileptic foci in partial epilepsy by PCT and EEG, *Ann. Neurol.,* 12, 529, 1982.

25. **Engel, J., Kuhl, D. E., Phelps, M. E., and Mazziotta, J. C.,** Interictal cerebral glucose metabolism in partial epilepsy and its relation to EEG changes, *Ann. Neurol.,* 12, 510, 1982.

26. **Engel, J., Brown, W. J., Kuhl, D. E. et al.,** Pathological findings underlying focal temporal hypometabolism in partial epilepsy, *Ann. Neurol.,* 12, 518, 1982.

27. **Bernardi, S., Trimble, M. R., Frackowiak, R. S. J., Wise, R. J. S., and Jones, T.,** An interictal study of partial epilepsy using positron emission tomography and the oxygen-15 inhalation technique, *J. Neurol. Neurosurg. Psychiatry,* 46, 473, 1983.

28. **Yamamoto, Y. L., Ochs, R., Gloor, P. et al.,** Patterns of rCBF and focal energy metabolic changes in relation to electroencephalographic abnormality in the interictal phase of partial epilepsy, in *Current Problems in Epilepsy I,* Baldy-Moulinier, M., Ingvar, D. H., and Meldrum, B. S., Eds., John Libbey, London, 1983, 51.

29. **Henry, T. R., Engel, J., Jr., and Phelps, M. E.,** Extratemporal cortical hypometabolism on interictal PET in complex partial seizures, *Epilepsia,* 29 (Abstr.), 676, 1988.

30. **Gur, R., Sussman, N., Alavi, A. et al.,** Positron emission tomography in two cases of childhood epileptic encephalopathy (Lennox-Gastaut syndrome), *Neurology,* 32, 1191, 1984.

31. **Chugani, H. T., Mazziotta, J. C., Engel, J., and Phelps, M. E.,** The Lennox-Gastaut syndrome: metabolic subtypes determined by 2-deoxy-2[^{18}F]fluoro-D-glucose positron emission tomography, *Ann. Neurol.,* 21, 4, 1987.

32. **Ingvar, D. H.,** Regional cerebral blood flow in focal cortical epilepsy, *Stroke,* 4, 359, 1973.

33. **Hougaard, K., Oikara, T., Sveinsdottir, E., Skinhog, E., Ingvar, D. H., and Lassen, N. A.,** Regional cerebral blood flow in focal cortical epilepsy, *Arch. Neurol.,* 33, 527, 1976.

34. **Sakai, F., Meyer, J. S., Naritomi, H., and Hsu, M. C.,** Regional cerebral blood flow and EEG in patients with epilepsy, *Arch. Neurol.,* 35, 648, 1978.

35. **Lavy, S., Melamed, E., Portnoy, Z. et al.,** Interictal regional cerebral blood flow in patients with partial seizures, *Neurology,* 26, 418, 1978.

36. **Valmier, J., Touchon, J., Daures, P., Zanca, M., and Baldy-Moulinier, M.,** Correlations between cerebral blood flow variations and clinical parameters in temporal lobe epilepsy: an interictal study, *J. Neurol. Neurosurg. Psychiatry,* 50, 1306, 1987.

37. **Valmier, J., Touchon, J., and Baldy-Moulinier, M.,** Interictal regional cerebral blood flow during nonspecific activation test in partial epilepsy, *J. Neurol. Neurosurg. Psychiatry,* 52, 364, 1989.

38. **Kuhl, D. E., Edwards, R. Q., Ricci, A. R., Yacob, R. J., Mich, T. J., and Alavi, A.,** The mark 4 system for radionuclide computed tomography of the brain, *Radiology,* 121, 405, 1976.

39. **Stokely, E. M., Sveinsdottir, E., Lassen, N. A., and Rommer, P.,** A single photon dynamic computer assisted tomograph (DCAT) for imaging brain function in multiple cross-sections, *J. Comput. Assist. Tomogr.,* 4, 230, 1980.

40. **Bonte, F. J., Stokely, E. M., Devous, M. D., and Homan, R. W.,** Single-photon tomographic study of regional cerebral blood flow in epilepsy: a preliminary report, *Arch. Neurol.,* 40, 267, 1983.

41. **Bonte, F. J., Devous, M. D., Stokely, E. M., and Homan, R. W.,** Single-photon tomographic determination of regional cerebral blood flow in epilepsy, *Am. J. Neuroradiol.,* 4, 544, 1983.

42. **Homan, R. W., Paulman, R. G., Devous, M. D., Walker, P., Jennings, L. W., and Bonte, F. J.,** Cognitive function and regional cerebral blood flow in partial seizures, *Arch. Neurol.,* 46, 964, 1989.

43. **Inoue, Y., Wolf, P., Hedde, J. P., and Meencke, H. J.,** Single-photon emission computed tomography with xenon-133 in the diagnosis of temporal lobe epilepsy, in *Advances in Epileptology,* Vol. 16, Wolf, P., Dam, M., Janz, D., and Dreifuss, F. E., Eds., Raven Press, New York, 1987, 291.

44. **Homan, R. W., Devous, M. D., LeRoy, R. F., and Bonte, F. J.,** Interictal focal cerebral blood flow elevations in partial seizures, *Neurology,* 39 (Suppl.1, Abstr.), 300, 1989.

45. **Winchell, H. S., Baldwin, R. M., and Lin, T. H.,** Development of I-123 labelled amines for brain studies: localization of I-123 iodophenyl-alkyl amines in rat brain, *J. Nucl. Med.,* 21, 940, 1980.

46. **Leonard, J. P., Nowotnik, D. P., and Neirinckx, R. D.,** Technetium-99m-d, 1-HM-PAO: a new radiopharmaceutical for imaging regional cerebral brain perfusion using SPECT — a comparison with iodine-123 HIPDM, *J. Nucl. Med.,* 27, 1819, 1986.

47. **Uren, R. F., Magistretti, P. L., Royal, H. D., Parker, J. A. et al.,** Single-photon emission computed tomography: a method of measuring cerebral blood flow in three dimensions (preliminary results of studies in patients with epilepsy and stroke), *Med. J. Aust.,* 1, 411, 1983.

48. **Sanabria, E., Chauvel, P., Askienazy, J. P. et al.,** Single photon emission computed tomography (SPECT) using ^{123}I-isopropyl-iodo-amphetamine (IAMP) in partial epilepsy, in *Current Problems in Epilepsy I,* Baldy-Moulinier, M., Ingvar, D. H., and Meldrum, B. S., Eds., John Libbey, London, 1983, 82.

49. **Lee, B. I., Markand, O. N., Siddiqui, A. R. et al.,** Single photon emission computed tomography (SPECT) brain imaging using HIDPM: intractable complex partial seizures, *Neurology,* 36, 1471, 1986.

50. **Lee, B. I., Markand, O. N., Wellman, H. N. et al.,** HIPDM-SPECT in patients with medically intractable complex partial seizures: ictal study, *Arch. Neurol.,* 45, 397, 1988.

51. **Lee, B. I., Markand, O. N., Wellman, H. N. et al.,** HIPDM single photon emission computed tomography brain imaging in partial onset secondarily generalized tonic-clonic seizures, *Epilepsia,* 28, 305, 1987.

52. **Lee, B. I., Park, H. M., Siddiqui, A. R. et al.,** Interictal HIPDM-SPECT in patients with complex partial seizures, *Neurology,* 38 (Suppl.1, Abstr.), 406, 1988.

53. **Costa, D. C., Ell, P. J., Cullum, I. D., and Jarritt, P. H.,** The *in vivo* distribution of ^{99m}Tc-HMPAO in normal man, *Nucl. Med. Commun.,* 7, 647, 1986.

54. **Leonard, J. P., Nowotnik, D. P., and Neirinckx, R. D.,** Technetium-99-m-d,1-HM-PAO: a new radiopharmaceutical for imaging regional brain perfusion using SPECT — a comparison with Iodine-123-HIPDM, *J. Nucl. Med.,* 27, 1819, 1986.

55. **Anderson, A. R., Friberg, H., Lassen, N. A., Kristensen, K., and Neirinckx, R. D.,** Serial studies of cerebral blood flow using ^{99m}Tc-HMPAO: a comparison with ^{133}Xe, *Nucl. Med. Commun.,* 8, 549, 1987.

56. **Biersack, H. J., Reichmann, K., Winkler, C. et al.,** ^{99m}Tc-labelled hexamethylpropyleneamineoxime photon emission scans in epilepsy, *Lancet,* 2, 1436, 1985.

57. **Biersack, H. J., Stefan, H., Reichmann, K., Linke, D. et al.,** HM-PAO brain SPECT and epilepsy, *Nucl. Med. Commun.,* 8, 513, 1987.

58. **Stefan, H., Pawlik, G., Bocher-Schwarz, H. G. et al.,** Functional and morphological abnormalities in temporal lobe epilepsy: a comparison of interictal and ictal EEG, CT, MRI, SPECT and PET, *J. Neurol.,* 234, 377, 1987.

59. **Podreka, I., Suess, E., Goldenberg, G., Steiner, M. et al.,** Initial experience with Technetium-99m HMPAO brain SPECT, *J. Nucl. Med.,* 28, 1657, 1987.

60. **Lang, W., Podreka, I., Suess, E., Muller, C., Zeitlhofer, J., and Deecke, L.,** Single photon emission computerized tomography during and between seizures, *J. Neurol.,* 235, 277, 1988.

61. **Duncan, R., Patterson, J., Bone, I., and Wyper, D. J.,** Reversible cerebellar diaschisis in focal epilepsy, *Lancet,* 2, 625, 1987.

62. **Ryding, E., Rosen, I., Elmqvist, D., and Ingvar, D. H.,** SPECT measurements with ^{99m}Tc-HMPAO in focal epilepsy, *J. Cereb. Blood Flow Metab.,* 8 (Suppl. 1), S95, 1988.

63. **Rowe, C. C.,** Interictal and Postictal Cerebral Blood Flow in Temporal Lobe Epilepsy, thesis for Doctor of Medicine, University of Melbourne, Australia, 1989.

64. **Rowe, C. C., Berkovic, S. F., Sia, S. T. B., Austin, M., McKay, W. J., Kalnins, R. M., and Bladin, P. F.,** Localization of epileptic foci with postictal single photon emission computed tomography, *Ann. Neurol.,* 26, 660, 1989.

65. **Rowe, C. C., Berkovic, S. F., Austin, M., McKay, W. J., and Bladin, P. F.,** Postictal SPECT in epilepsy, *Lancet,* 1, 389, 1989.

66. **Rowe, C. C., Berkovic, S. F., Austin, M., McKay, W. J., and Bladin, P. F.,** Patterns of postictal blood flow in focal epilepsy determined by single photon computerized tomography (SPECT), *Epilepsia,* (Abstr.), 29, 677, 1988.

67. **Rowe, C. C., Berkovic, S. F., Sia, B., Austin, M., Bladin, P. F., and McKay, W. J.,** Postictal cerebral blood flow patterns: qualitative and quantitative analysis in temporal lobe epilepsy, *Aust. N.Z. J. Med.,* 19 (Suppl.1) (Abstr.), 612, 1989.

68. **Rowe, C. C., Berkovic, S. F., Austin, M., McKay, W. J., and Bladin, P. F.,** Patterns of postictal cerebral blood flow in temporal lobe epilepsy: qualitative and quantitative analysis, *Neurology,* in press.

69. **Leiderman, D., Balish, M., Bromfield, E., Sato, S., and Theodore, W. H.,** Comparison of interictal FDG and ^{15}O-H$_2$O PET scanning in patients with uncontrolled complex partial seizures, *Neurology,* 39 (Suppl.1, Abstr.), 301, 1989.

70. **Newton, M. R. and Berkovic, S. F.,** unpublished data.

Chapter 14

MAPPING OF LOCAL CEREBRAL pH WITH POSITRON EMISSION TOMOGRAPHY

N. M. Alpert, M. Senda, and J. A. Correia

TABLE OF CONTENTS

I. INTRODUCTION

Ordinary brain pH is tightly regulated by physicochemical buffering, metabolic and catabolic processes, and by active transport of H^+, but, this regulation can be impaired by systemic processes and by local tissue pathology. The available data suggest that measurement of local brain pH is an important indicator of tissue status, particularly in stroke and ischemic injury.

Animal studies have demonstrated disturbances in acid/base balance resulting from experimental ischemic injury.[1-13] Disruptions in brain pH were found extending beyond the boundaries of corresponding disruptions in NADH and ATP.[14] Impairment of tissue pH regulation was found in severe hypoglycemia.[15] Rehncrona[16] suggested that in complete ischemia, tissue lactate and pH are approximately linearly correlated; and this idea was amplified by Pulsinelli and Duffy[17] whose finding of selective neuronal destruction in a normoglycemic state vs. necrosis of all tissue elements during equivalent ischemia plus hyperglycemia led to the postulation of a $[H^+]$ threshold, beyond which all cells are destroyed. Kraig et al. extended these observations, using pH-sensitive microelectrodes to measure brain interstitial pH, pH_o, in anesthetized rats during complete ischemia and found:

1. pH_o dropped immediately in response to a fall in blood pressure.[18]
2. pH_o changes abruptly (threshold) as a function of lactic acid accumulation.[18]
3. Depletion of intracellular $[HCO^-]_e$ limited the extracellular increase in $[H^+]$.[18]
4. The increase in $[H^+]_o$ during complete ischemia suggests that brain infarction develops after the plasma membrane in brain cells can no longer transport ions to regulate $[H^+]$.[10]
5. Direct injection of sodium lactate solution (pH 4.5) caused neuronal necrosis as soon as 1 h postinjection. Necrosis of astrocytes occurred later, at 3 to 6 h.[19] Furthermore, another line of investigation has suggested that treating brain acidosis reduces mortality in experimental brain injury.[20]

There is also a growing body of evidence from human studies documenting the importance of alterations in acid-base balance: studies, such as the one performed by DeSalles et al. on subjects with severe head injury, showed that subjects with a poor outcome had a higher ventricular CSF lactate level than did those with moderate disabilities or a good outcome.[21] In three head-injured patients, microelectrode measurements of extracellular brain pH showed brain tissue acidosis in areas of contusion or compression by mass lesion.[22] Alterations in acid-base balance have also been found in human subjects with PET. For example, regions of tissue alkalosis were observed 10 to 19 d after the onset of symptoms in patients with stroke[23] and with brain tumors;[24] more recently, regions of tissue acidosis were observed within 48 h of the onset of symptoms in stroke patients.[25,26]

Local alterations in brain pH may also arise in association with neoplastic tissue. Increased levels of anaerobic glycolysis have been found in malignant cells,[27] suggesting increased production of lactate and a lowered intracellular pH. Alterations in local pH may also affect the delivery of chemotherapeutic agents either directly, by inhibition of enzyme-catalyzed transport mechanisms, or indirectly, by action at the level of the microvasculature effecting CBF. The pH of the tissue environment may also play a role in radiation therapy and hyperthermia, with evidence that tumor cells may be less radiation sensitive[28] but more heat sensitive[29] at low pH. Studies of tumor pH in animal models have typically indicated tissue pH which is either not more acidic than normal or alkalotic; but work by Arnold and colleagues,[30] using ^{14}C-DMO in rats with RG-2 gliomas, interpreted their alkalotic findings to represent an increased extravascular water content and a plasma-like extracellular pH.

The standard methods for measuring brain pH in animals cannot be used in man because they are too invasive, requiring either tissue samples or placement of microelectrodes. Investigations have recently employed nuclear magnetic resonance (NMR) or positron emission tomography (PET), modalities which are much less invasive and more suitable for human studies. NMR spectroscopy of ^{31}P using surface coils has been used to measure intracellular pH; but, at present, this technique can only provide pH measurements averaged over large tissue volumes.[31-33] PET techniques offer the possibility for measurement of local brain pH in human subjects with volume resolutions of less than 1 cc, but they require arterial blood sampling. Two PET methods, the ^{11}C-CO_2 method and the ^{11}C-DMO method, have been developed, validated, and applied to clinical research. This chapter reviews the scientific basis of these methods. Results from clinical research are presented along with a brief review of the relevant data from animal studies. An effort is made, when possible, to elucidate the basic and clinical research questions of current interest.

II. THEORY OF pH MEASUREMENT

A. SIMPLIFIED THEORY

All of the PET measurements of brain pH proposed to date are based on the classical methods involving the distribution of a weak acid or base.[34] The theoretical basis of pH measurements with PET can be illustrated most simply with a closed two-compartment system in which blood plasma and tissue are separated by a membrane permeable to neutral molecules, but impermeable to charged molecules. Consider the distribution of a weak acid, labeled with a positron emitter, which is allowed to come to equilibrium with the system. The distribution of label between the two compartments can be deduced theoretically from the conservation of tracer mass, with the simultaneous imposition of chemical and difussion equilibria.

Following standard chemical notation, we denote the acid by AH and observe that it will be present in solution in each compartment in two chemical forms, as indicated by the ionization reaction:

$$AH \rightleftharpoons A^- + H^+ \tag{1}$$

Within each compartment, chemical equilibrium is determined by the pH, according to the Henderson-Hasselbalch equation:

$$\frac{[A^-]}{[AH]} = 10^{pH - pK} \tag{2}$$

where pK is the dissociation constant for the acid.

Since the neutral form can diffuse between the two compartments, diffusion equilibrium requires that

$$[AH]_p = [AH]_t \tag{3}$$

where subscripts p and t denote blood plasma and tissue compartments, respectively.

The total concentration of labeled acid in the plasma compartment, $[A]_p$, is $[A]_p = [A^-]_p + [AH]_p$ and in the tissue compartment, $[A]_t$, is $[A]_t = [A^-]_t + [AH]_t$. Using the Henderson-Hasselbalch equation to eliminate $[A^-]$,

$$[A]_p = [AH]_p(1 + 10^{pH_p - pK})$$
$$[A]_t = [AH]_t(1 + 10^{pH_t - pK})$$

The concentration [AH] can be eliminated to give the tissue-to-blood partition coefficient, L, in the form:

$$L = \frac{1 + 10^{pH_t - pK}}{1 + 10^{pH_p - pK}} \tag{4}$$

or, equivalently, "tissue pH" may be expressed as

$$pH_t = pK + \log[L (1 + 10^{pH_p - pK}) - 1] \tag{5}$$

showing explicitly that in a closed, two-compartment system, knowledge of the equilibrium tissue-to-blood partition coefficient and the blood pH are sufficient to determine tissue pH. We will see below that the equilibrium tissue-to-plasma partition coefficient plays a similar role in more realistic situations.

B. EFFECTIVE TISSUE pH

In order to interpret pH measurements, it is necessary to consider a more realistic tissue model and the fact that PET imaging devices measure radioactivity concentrations averaged over the microscopic constituents of tomographically defined volume elements. A tomographic volume element can be viewed as a system composed of several spaces and cell types (e.g., vascular and interstitial spaces, neurons and glia), so measurements with PET provide an average "tissue pH," rather than estimates of intracellular or extracellular pH. The tracer is assumed to have access to a distribution volume, V_d, where V_d is the tissue volume in which the acid is distributed per volume of tissue. V_d is usually taken to be the tissue water content. In a tissue element with n compartments and/or spaces, the individual pH's are denoted by pH_i, where $i = 1, 2 \ldots , n$. Each compartment or space, i, occupies a fraction X_i of the distribution volume, with the constraint that $\sum_i^n X_i = 1$. The organization of these compartments need not be specified, but at least one must be in contact with the blood plasma and free diffusion of the neutral molecule [AH] between compartments must be maintained.

The concentration of labeled acid in the ith compartment is given by

$$[A]_i = [AH]_i(1 + 10^{(pH_i - pK)}) \tag{6}$$

Since $[AH]_i = [AH]_p$ for all i the observed tissue-to-plasma partition coefficient, L, can be expressed as

$$L = V_d \frac{\sum_i^n X_i(1 + 10^{(pH_i - pK)})}{1 + 10^{(pH_p - pK)}} \tag{7}$$

The quantity which can be directly measured in PET is L. Hence, it is useful to define an effective tissue pH, pH_t. The operational equation for pH measurement is taken to be Equation 5, but pH_t is interpreted by equating Equations 4 and 7.

$$pH_t = pK + \log\left[V_d \sum_i (1 + V_i 10^{pH_i - pK}) - 1)\right] \tag{8}$$

An alternative formulation is to define the average tissue pH, pH_{av}[23,34,35] as

$$pH_{av} = \log\left(V_d \sum_i X_i 10^{pH_i}\right) \tag{9}$$

In this formulation, L can be expressed as

$$L = V_d \frac{1 + 10^{pH_{av} - pK}}{1 + 10^{pH_p - pK}} \qquad (10)$$

As pointed out by Buxton et al.,[35] Equation 4 is more convenient for PET measurement when the local tissue water content is unknown.

The magnitude of pH_t is largely determined by two compartments, extracellular fluid and intracellular cytoplasm, and thus pH_t can be approximated by

$$pH_t \simeq \log[V_d (X_E 10^{pH_E} + (1 - X_E)10^{pH_I})] \qquad (11)$$

where X_E is the fraction of tissue water in the intracellular space and pH_E and pH_I are the values of pH in the extracellular fluid and intracellular cytoplasm, respectively. These results show explicitly that the effective tissue pH is a logarithmically weighted average of reciprocal H^+-ion concentration, depending on the volume of distribution of the tracer, the fractional volume occupied by each compartment, and the pH of the individual compartments. Variations in tissue water content or the relative size of the extracellular space can potentially cloud the interpretation of measurements of tissue pH and measurements to determine the size of the extracellular space may be desirable in some situations. Senda et al.[26] have used Equation 11 to graph pH_t vs. pH_I for different values of X_E and V_d. This equation predicts that $pH_t = 7.4$, $V_d = 0.77$ and $X_E = 0.2$; and as we shall see, this value is close to the results found in experiment. For these normal values of pH_e, V_d, and X_E, pH_t is sensitive to changes in the intracellular pH, exhibiting an approximately linear response over the range $6.6 < pH_I < 7.4$. Increasing either X_E or V_d increases the effective tissue pH. A variation in tissue water content has the effect of displacing the curve by a (nearly) constant amount for all values of intracellular pH. However, as X_E increases from its normal value, the response of pH_t to changes in pH_I diminishes; and, when X_E reaches about 0.5, the theory indicates that pH_t would not reflect even profound tissue acidosis. Instead, the method would likely indicate a tissue alkalosis because the weighing of the extracellular space becomes dominant. As we shall see, pH studies in stroke and tumor patients exhibit results consistent with the theory discussed above.

C. THE $^{11}CO_2$ METHOD

As is well known, CO_2 behaves as a weak acid, dissociating *in vivo* into bicarbonate and H^+ ions. Raichle and colleagues suggested the use ^{11}C-CO_2 in pH measurements[36] with PET, but several objections were raised to their approach, including the need to account for incorporation of the ^{11}C label into compounds other than CO_2 and bicarbonate (i.e., fixation)[37] and the assumed need to estimate the local tissue P_aCO_2. Another difficulty with the original suggestion is the fact that ^{11}C-CO_2 does not reach an equilibrium distribution in the body, but is continuously expelled from the lungs. Even though the tissue-to-blood concentration ratio may approach a constant, the total ^{11}C-CO_2 concentration is decreasing, and this constant value will be higher than the equilibrium ratio. To overcome these objections, Buxton et al.[38] developed a three-compartment kinetic model and a PET measurement strategy employing continuous inhalation rather than bolus injection. The kinetic model of Buxton et al.[38] relates the time-dependent tissue concentration during the continuous inhalation of ^{11}C-CO_2 and the clearance of the label to the local values of L, the blood-to-tissue transport rate for CO_2 and the rate of CO_2 fixation in other compounds. By using continuous inhalation instead of bolus administration, the buildup of labeled compounds other than CO_2 (fixation) was made negligible during the period of administration. The model parameters were determined by least squares fitting. Buxton and colleagues[39] evaluated their pH method by

TABLE 1
Dosimetry-^{11}C-CO$_2$
Method

Organ	Dose (mrad/mCi)
Whole body	11
Lungs	30
Brain	12
Kidney	8

performing PET studies on dogs during a hypocapnic and a hypercapnic state. Direct measurement of ^{11}C-CO$_2$ in blood samples verified that CO$_2$ fixation in blood was negligible throughout the inhalation period. The kinetic model fit the experimental data well throughout the inhalation period and during the washout of label. Tissue pH calculated by the method was in good agreement with previously reported measurements of brain pH, both in absolute value and in variation with P$_a$CO$_2$.

We have estimated the radiation dose from the measurement procedure. PET studies performed in our laboratory on dogs have shown that ^{11}C-CO$_2$ is distributed more or less uniformly in the whole body, with the exception of the lung and airways which form the route of administration. Doses were estimated, assuming clearance is by physical decay only. This procedure overestimates the dose, since ^{11}C-CO$_2$ clears also through respiration, and thus provides an upper limit. Table 1 lists the dose per millicurie of ^{11}C-CO$_2$ administered for several organs. Assuming administration of 25 mCi, these data lead to an absorbed dose of about 750 mrad to the lung and a whole body dose of about 275 mrad.

Senda et al.[26] have extended Buxton's method to allow for more efficient mapping of local tissue pH by standardizing the CO$_2$ fixation rate and by substituting an analytic parameter estimation method in place of least squares fitting. Their results also showed (1) that the blood-to-tissue transport rate of CO$_2$ was positively correlated with regional cerebral blood flow (rCBF) and (2) that the ratio of CO$_2$ transport rate to rCBF was negatively correlated with CBF, a finding they interpreted to mean that CO$_2$ extraction varies inversely with flow.

D. THE ^{11}C-DMO METHOD

One of the classic methods for measurement of tissue pH employs the weak acid, 5,5-dimethyl-2-4-oxazolidine-dione (DMO), an agent which is essentially nontoxic, nonmetabolizable in mammalian tissue, and which does not bind to human plasma proteins. Following intravenous injection, DMO diffuses slowly from blood plasma into the tissue, over times longer than the 20-min half-life of ^{11}C, reaching an equilibrium distribution which depends on values of the intra- and extracellular pH's, the size of the local extracellular space, and the local tissue water content. Data in the literature from measurements in animals suggested the equilibration times for DMO were greater than 1 h,[40-42] but DMO studies with PET in human subjects indicate the times are much shorter, 20 to 60 min.[23,25,43] In practice, sequential PET measurements can be performed to gauge the time when equilibrium is achieved. Sampling of arterial plasma and PET measurement of the tissue concentration of ^{11}C-DMO at equilibrium determines the tissue-to-blood concentration ratio, L, thereby allowing the calculation of an effective tissue pH. However, because DMO equilibrates slowly, pH measurements with ^{11}C-DMO[44,45] may suffer from relatively poor statistical precision.

Kearfott et al.[46] have estimated the absorbed dose for pH measurements with ^{11}C-DMO from the biodistribution of ^{14}C-DMO in rats. Their study assayed the dose to a wide variety of tissues, with the dose per millicurie administered ranging from 8 to 23 mrad. The tissues with the highest dose were small intestine, heart wall and uterus. In a study with an injected dose of 25 mCi, the critical organ receives about 575 mrad and the whole body dose is 275 mrad.

III. MEASUREMENT OF BRAIN pH

A. NORMAL pH

The ^{11}C-CO$_2$ method was first applied to human studies by Brooks et al., who reported examination of four normal subjects[47] and five normal subjects, ages 30 to 43 years, and 12 patients with cerebral tumors.[24] It is not clear if the five normal subjects in the 1986 study included the four reported in 1984. In the five normal subjects, a mean tissue pH of 7.02 ± 0.03 (SD) was found for measurements in cerebral cortex thought to be composed largely of gray matter, while a mean tissue pH of 6.98 ± 0.05 was found for regions composed largely of white matter. Hakim and co-workers,[25] using ^{11}C-DMO, have evaluated five stroke-age controls with no medical or neurological illness who had normal CT scan, finding pH$_t$ was 7.12 ± 0.02. Senda et al.[26] reported values of pH$_t$, measured with the ^{11}C-CO$_2$ method, for three normal males (ages 42 to 43 years); the average pH$_t$ in gray matter was 7.04; the average pH$_t$ in white matter was 6.91.

B. pH IN TUMORS

pH studies have also been performed on patients with primary and metastatic brain tumors to assess acid-base status. Rottenberg et al.[48] studied nine patients with ^{11}C-DMO. Their data indicate an alkalosis rather than acidosis; but a definitive interpretation is complicated by the possibility of changes in tissue water content, the size of the extracellular space, and by the likelihood of blood-brain-barrier (BBB) disruptions. Calculations taking into account such effects suggest that the intracellular pH must be at least 7.0 to be consistent with the measured data. Tyler et al.[49] evaluated the metabolic and hemodynamic characteristics of untreated cerebral gliomas in 16 patients; pH$_t$ was also measured in seven gliomas, one grade III and six grade IV, showing an increased tumor pH relative to contralateral tissue in all cases. Plate 1 (A, C, and D)* shows examples of work performed by Tyler and colleagues at the Montreal Neurological Institute (MNI) illustrating the increased uptake of ^{11}C-DMO (consistent with tissue alkalosis) in malignant gliomas with corresponding FDG-hypometabolism. In similar pH studies with ^{11}C-CO$_2$, Brooks et al. reported on 12 patients with cerebral tumors.[24] Six patients with cerebral gliomas who showed no evidence of BBB disruptions had a mean tissue pH of 7.09 ± 0.05, while homologous contralateral brain had mean pH$_t$ = 7.01 ± 0.06. Six patients with tumors showing enhancement on CT scan (glioma IV, glioma [n = 2], adenocarcinoma, melanoma, and oligodendroglioma) had mean pH$_t$ = 6.98 ± 0.04 as compared to mean pH$_t$ in homologous contralateral tissue of 6.99 ± 0.02. The authors concluded that in tumors with intact BBB, there was a consistent finding of tissue alkalosis.

C. pH IN STROKE

Syrota and colleagues were the first to perform PET studies with DMO in stroke patients.[23,50] Their initial studies, performed 10 to 34 d after onset, demonstrated increased DMO concentration in the infarcted region. Explanations for this increase included increases in the size of the extracellular space, the tissue water content, and possibly intracellular pH. Later studies, performed on ten patients during the period 10 to 19 d following the ictus, evaluated pH, using ^{11}C-DMO and the extracellular water fraction. Studies with ^{76}Br showed that extracellular water in infarcted regions was elevated. Even so, the increase in the extracellular space could not fully account for the increases in the equilibrium partition

---

* Plate 1 follows page 198.

observed with ^{11}C-DMO; it was concluded that intracellular alkalosis was present, at least in some studies.

Hakim and co-workers evaluated 12 stroke patients[25] within 48 h of the onset of symptoms to determine tissue pH (^{11}C-DMO), CBF, CBV, CMRO$_2$, and CMRGlu. Tissue in the cortical rim was considered injured when rCMRO$_2 \leqq 67 \mu$mol/100 g/min (1.5 ml/100 g/min). Based on their rCBF measurements, injured and/or infarcted tissues were classified as either hypoperfused (Group I, n = 8) or hyperperfused (Group II, n = 4). Group II patients had, on average, significantly ($p <0.05$) lower OEF, higher glucose utilization, and a higher CMRGlu to CMRO$_2$ ratio. The hypoperfused regions (Group I) had, on average, significantly lower pH$_t$ than either control regions or regions which were hyperperfused, a finding illustrated in Plate 2B.*

Because it can be assumed that all subjects had regions with severe hypoperfusion at the onset of symptoms, the increase in pH$_t$ found in the Group II subjects was associated with reperfusion. The authors observed that in stroke-injured tissue with hyperperfusion, anaerobic glycolysis is enhanced, but acidosis is not apparent, while in stroke-injured tissue with hypoperfusion, anaerobic glycolysis is not as intense and a tendency to acidosis is evident.

Senda et al. reported pH studies in 12 subjects with cerebral ischemia or stroke using the ^{11}C-CO$_2$ method.[26] Their study also included measurement of local CBF, OEF, and CMRO$_2$ performed during the same scanning session. In five stroke patients, measured within 48 h of the ictus, injured cortex had lower CBF (20.6 ml/min/100 g), higher OEF (78.1%), and lower pH$_t$ (6.96) than homologous tissue in the contralateral hemisphere (CBF = 41.4 ml/min/100 g, OEF = 53.3%, pH$_t$ = 7.00). Their findings are illustrated by measurement of local brain pH, cerebral blood flow, and oxygen consumption in acute stroke (11 h postictus) in Plate 2. In three stroke patients measured 5 to 8 d after the ictus, the injured cortex had higher CBF (60.9 ml/ min/100 g), lower OEF (32.0%), and higher pH$_t$ (7.12) than tissue in homologous regions of contralateral cortex (CBF = 45.3 ml/min/100 g, OEF = 58.0%, pH$_t$ = 7.06). Based on their data and arguments similar to those presented above, the authors hypothesized that early in the course of ischemic injuries, such as stroke, tissue acidosis may be observed arising from injured cells with intact cellular membranes, but, later in the course of irreversible injury, cell membrane integrity is compromised and the extracellular space increases, leading to the observations of a relative tissue alkalosis.

IV. DISCUSSION

We have reviewed the status of pH measurements in the brain with PET. The PET methods are based on the distribution of a weak acid, an adaptation of a classical approach, in use in laboratory measurement for decades.[34] Since the last review by Buxton and colleagues[35] in 1985, the strengths and weaknesses of the methodologies have become more evident and several research applications have been reported. There are now two established methods for measurement of brain pH with PET, the ^{11}C-CO$_2$ method and the ^{11}C-DMO method. The ^{11}C-CO$_2$ method may be preferable on practical grounds: ^{11}C-CO$_2$ is easier to prepare than ^{11}C-DMO, and the statistical precision attainable with the DMO method is limited by the 20.3-min physical half-life of the ^{11}C-label and the long equilibration period required before measurement can be performed. Theoretical considerations suggest that interpretation of the pH$_t$ data may be complicated by changes in water content and changes in the relative fraction of extracellular and intracellular space. These changes in water compartmentation can be assessed in separate studies, if necessary.[23] Both pH methods may also suffer from limitations in the delivery of the label to tissues with very low flow and thus interpretation

---

* Plate 2 follows page 198.

of pH_t should be guided by measurement of CBF in the same tissue. Finally, pH_t data should be interpreted cautiously where partial volume effects may average regions with no or low flow and those with normal flow.

Normal pH_t in human subjects, is not well established; less than 20 subjects have been studied. The two studies using $^{11}C\text{-}CO_2$ are in reasonable agreement with one another, but are about 0.1 pH unit lower than the DMO study reported by Hakim et al. in five stroke-age controls. Given the present understanding of the methodologies, their accuracy, and precision, it is not possible to confidently ascribe the difference between the CO_2 and DMO results to technical factors; physiological mechanisms must also be considered. Based on the differences in tissue water content, one would expect to observe a gray-white pH_t difference of about 0.068 pH units. No significant gray-white pH difference was found by Brooks et al.,[24] but Senda et al.,[26] using a higher-resolution tomograph, reported about a 0.1 pH unit gray-white difference in their 3 normal subjects and in 12 patients with ischemic disease.

The role of pH measurements in evaluating cerebral tumors is not yet clear. PET studies of CBF and $CMRO_2$ in tumors indicate low flow and oxygen extraction fraction and most, but not all, PET studies indicate increased glycolysis. Some studies of glioma metabolism, with ^{18}FDG have shown a positive correlation with glioma grade and glycolytic activity.[51-53] supporting the earlier suggestion that the glycolysis rate increases with increasing malignancy.[27] One might expect increased production of lactate, but in fact, all PET studies of tissue pH indicate a relative alkalosis, as do studies with ^{31}P magnetic resonance spectroscopy. The predictive ability of pH data has not been established.

There is abundant evidence that the sequence of events referred to as stroke is triggered by an interruption or severe decrease in blood flow to tissue in the brain. However, the underlying mechanism initiating the event and/or the ensuing pattern of changes may vary from patient to patient; nevertheless, a few general patterns can be assumed. In the first few minutes to hours following the acute event, one would expect to observe anaerobic metabolism, with production of lactate and acidosis. Later, one expects irreversible damage to the cell membrane, leading first to loss of neurons and then glia. Later, but still in the acute phase of the injury, one expects cell swelling, leading finally to increases in extracellular water and disruption of the BBB. The studies of pH_t with PET appear to be in accord with these general patterns. Senda et al.[26] reported five studies performed within the first 48 h of onset; all had evidence of low CBF and increased OEF; most had evidence of tissue acidosis; one subject, measured 36 h after onset, had evidence of reperfusion with normal CBF and $CMRO_2$ but alkalotic pH_t. When Senda et al.[26] considered data from all subjects (n = 12) in their study, 89% of the variance could be explained as a linear combination (principal component analysis) of CBF, OEF, and pH_t, with acute stroke patients exhibiting lower CBF, increased OEF, and lower pH_t than subjects with established stroke (5 d to 3 weeks) who exhibited increased CBF, low OEF, and alkalotic pH_t. The data of Senda et al. are largely in accord with those of Hakim et al.,[25] who found that stroke patients exhibiting low CBF flow had tissue acidosis, while those with increased flow (i.e., reperfusion), but low $CMRO_2$, had tissue pH that did not differ from control. In chronic stroke, pH_t has been found to be increased, with increased extracellular fluid fraction.[50] Taken together, these studies have led to a hypothesis on the predictive ability of pH studies. If pH_t of injured tissue is lower than the contralateral tissue, the cell is ischemic with anaerobic metabolism, the cell membrane is intact, and tissue damage may be reversible, but if pH_t is increased, cell destruction has already occurred and the tissue is irreversibly injured.

REFERENCES

1. **Tenny, R. T., Sharbrough, R. W., Anderson, R. E., and Sundt, T. M., Jr.** Correlation of intracellular redox states and pH with blood flow in primary and secondary seizure foci, *Ann. Neurol.*, 8, 564, 1981.
2. **Paschen, W., Djuricic, B., Mies, G., Schmidt-Kastner, R., and Linn, F.,** Lactate and pH in the brain: association and dissociation in different pathophysiological states, *J. Neurochem.*, 48, 154, 1987.
3. **Hakim, A. M.,** Cerebral acidosis in focal ischemia. II. Nimodipine and verapoamil normalize cerebral pH following middle cerebral artery occlusion in the rat, *J. Cereb. Blood Flow Metab.*, 6, 676, 1986.
4. **Meyer, F. B., Anderson, R. E., Sundt, T. M., Jr., and Yaksh, T. L.,** Treatment of experimental focal cerebral ischemia with mannitol. Assessment by intracellular brain pH, cortical blood flow, and electroencephalography, *J. Neurosurg.*, 66, 109, 1987.
5. **Marsh, W. R., Anderson, R. E., and Sundt, T. M., Jr.,** Effect of hyperglycemia on brain pH levels in areas of focal incomplete ischemia in monkeys, *J. Neurosurg.*, 65, 693, 1986.
6. **Smith, M. L., von Hanwehr, R., and Siesjo, B. K.,** Changes in extra- and intracellular pH in the brain during and following ischemia in hyperglycemic and in moderately hypoglycemic rats, *J. Cereb. Blood Flow Metab.*, 6, 574, 1986.
7. **Meyer, F. B., Anderson, R. E., Sundt, T. M., Jr., and Yaksh, T. L.,** Intracellular brain pH, indicator tissue perfusion, electroencephalography, and histology in severe and moderate focal cortical ischemia in the rabbit, *J. Cereb. Blood Flow Metab.*, 6, 71, 1986.
8. **Meyer, F. B., Anderson, R. E., Yaksh, T. L., and Sundt, T. M., Jr.,** Effect of nimodipine on intracellular brain pH, cortical blood flow and EEG in experimental focal cerebral ischemia, *J. Neurosurg.*, 64, 617, 1986.
9. **von Hansher, R., Smith, M. L., and Siesjo, B. K.,** Extra- and intracellular pH during near-complete forebrain ischemia in the rat, *J. Neurochem.*, 46, 331, 1986.
10. **Kraig, R. P., Pulsinelli, W. A., and Plum, F.,** Hydrogen ion buffering during complete brain ischemia, *Brain Res.*, 342, 281, 1985.
11. **Anderson, R. E. and Sundt, T. M., Jr.,** Brain pH in focal cerebral ischemia and the protective effects of barbiturate anesthesia, *J. Cereb. Blood Flow Metab.*, 3, 493, 1983.
12. **Mabe, H., Blomqvist, P., and Siesjo, B. K.,** Intracellular pH in the brain following transient ischemia, *J. Cereb. Blood Flow Metab.*, 3, 109, 1983.
13. **Yoshida, S., Busto, R., Martinez, E., Scheinberg, P., and Ginsberg, M. D.,** Regional brain energy metabolism after complete versus incomplete ischemia in the rat in the absence of severe lactic acidosis, *J. Cereb. Blood Flow Metab.*, 5, 490, 1985.
14. **Kima, S., Handa, H., Ishikawa, M., Hirai, O., Yoshida, S., and Imadaka, K.,** Brain tissue acidosis and changes of energy metabolism in mild incomplete ischemia-topographical study, *J. Cereb. Blood Flow Metab.*, 5, 432, 1985.
15. **Pelligrino, D. A. and Siesjo, B. K.,** Regulation of extra- and intracellular pH in the brain in severe hypoglycemia, *J. Cereb. Blood Flow Metab.*, 1, 85, 1981.
16. **Rehncrona, S. and Kagstrom, E.,** Tissue lactic acidosis and ischemic brain damage. *Am. J. Emerg. Med.*, 2, 168, 1983.
17. **Pulsinelli, W. A. and Duffy, T. E.,** Regional energy balance in rat brain after transient forebrain ischemia, *J. Neurochem.*, 40, 1500, 1983.
18. **Kraig, R. P., Pulsinelli, W. A., and Plum, F.,** Heterogeneous distribution of hydrogen and bicarbonate ions during complete brain ischemia, *Prog. Brain Res.*, 63, 155, 1985.
19. **Petito, C. K., Kraig, R. P., and Pulsinelli, W. A.,** Light and electron microscope evaluation of hydrogen ion-induced brain necrosis, *J. Cereb. Blood Flow Metab.*, 7, 625, 1987.
20. **Rosner, M. J. and Becker, D. P.,** Experimental brain injury: successful therapy with the weak base tromethamine, with an overview of CNS acidosis, *J. Neurosurg.*, 60, 961, 1984.
21. **DeSalles, A. A. F., Kontos, H. A., Becker, D. P., Yang, M. S., Ward, J. D., Moulton, R., Gruemer, H. D., Lutz, H., Maset, A. L., Jenkins, L., Marmarou, A. and Muizelaar, P.,** Prognostic significance of ventricular csf lactic acidosis in severe head injury, *J. Neurosurg.*, 65, 615, 1986.
22. **DeSalles, A. A. F., Kontos, H. A., Ward, J. D., Marmarou, A., and Becker, D. P.,** Brain tissue pH in severely head-injured patients: a report of three cases, *Neurosurgery*, 20, 297, 1987.
23. **Syrota, A., Castaing, M., Rougemont, D., Merridge, M., Bazierre, B., Baron, J. C., Bousser, M. G., and Pocidalo, J. J.,** Regional tissue pH and oxygen metabolism in human cerebral infarction studied with PET, *Ann. Neurol.*, 14, 419, 1983.
24. **Brooks, D. J., Beaney, R. P., Thomas, D. G. T., Marshall, J., and Jones, T.,** Studies on regional cerebral pH in patients with cerebral tumors using continuous inhalation of $^{11}CO_2$ and positron emission tomography, *J. Cereb. Blood Flow Metab.*, 6, 529, 1986.
25. **Hakim, A. M., Pokrupa, R. P., Villanueva, M. D., Diksic, M., Evans, A. C., Thompson, C. J., Meyer, E., Yamamoto, Y. L., and Feindel, W. H.,** The effect of spontaneous reperfusion on metabolic function in early human cerebral infarcts, *Ann. Neurol.*, 21, 279, 1987.

26. **Senda, M., Buxton, R. B., Alpert, N. M., Correia, J. A., Mackay, B. C., Weise, S. B., and Ackerman, R. H.,** The [15]O steady state method: correction for variation in arterial concentration, *J. Cereb. Blood Flow Metab.,* 8, 681, 1988.

27. **Warburg, O.,** *The Metabolism of Tumors,* Vol., A. Constable, New York, 1930.

28. **Haveman, J.,** The influence of pH on the survival after X-irradiation of cultured malignant cells: effects of carbonyl-cyanide-3-chlorophenylhydrazone, *Int. J. Radiat. Biol.,* 37, 210, 1980.

29. **Gerweck, L. E. and Richards, B.,** Influence on pH on the thermal sensitivity of cultured human glioblastoma cells, *Cancer Res.,* 41, 845, 1981.

30. **Arnold, J. B., Junck, L. and Rottenberg, D. A.,** *In vivo* measurement of regional brain and tumor pH using [14C]dimethyloxazolidinedione and quatitative autoradiography, *J. Cereb. Blood Flow Metab.,* 5, 369, 1985.

31. **Petroff, O. A. C., Prichard, J. W., Behar, K. L., Rothman, D. L., Alger, J. R., and Shulman, R. G.,** Cerebral metabolism in hyper- and hypocarbia: [31]P and ^{1}H nuclear magnetic resonance studies, *Neurology,* 35, 1681, 1985.

32. **Welch, K. M. A., Levine, S. R., Helpern, S. A., Ewing, J. R., Bruce, R., and Smith, M. B.,** Acute human cerebral infarction studied serially by ^{31}P-NMR spectroscopy, *J. Cereb. Blood Flow Metab.,* 7 (Abstr.), S131, 1987.

33. **Levine, S. R., Welch, K. M. A., Helpern, J. A., Chopp, M., Bruce, R., Selwa, J., and Smith, M. B.,** Prolonged deterioration of ischemic brain energy metabolism and acidosis associated with hyperglycemia: human cerebral infarction studied by serial ^{31}P NMR spectroscopy, *Ann. Neurol.,* 23, 416, 1988.

34. **Roos, A. and Boron, W. F.,** Intracellular pH, *Physiol. Rev.,* 61, 296, 1981.

35. **Buxton, R. B., Alpert, N. M., Ackerman, R. H., Wechsler, L. R., Elmaleh, D. R., and Correia, J. A.,** Measurement of brain pH with positron emission tomography, in *Positron Emission Tomography,* Reivich, M., Ed., Alan R. Liss, New York, 1985, 451.

36. **Raichle, M. E. and Grubb, R. L., Jr.,** Measurement of brain tissue carbon dioxide content *in vivo* by emission tomography, *Brain Res.,* 166, 413, 1979.

37. **Lockwood, A. H. and Finn, R. D.,** ^{11}C-carbon dioxide fixation and equilibration in rat brain: effects on acid base measurements, *Neurology,* 32, 451, 1982.

38. **Buxton, R. B., Wechsler, L. R., Alpert, N. M., Ackerman, R. H., Elmaleh, D. R., and Correia, J. A.,** Measurement of brain pH using [11]CO_2 and positron emission tomography, *J. Cereb. Blood Flow Metab.,* 4, 8, 1984.

39. **Buxton, R. B., Alpert, N. M., Babikian, V., Weise, S., Correia, J. A., and Ackerman, R. H.,** An evaluation of the [11]CO_2/PET method for measuring brain pH. 1. pH changes measured in states of altered PCO_2, *J. Cereb. Blood Flow Metab.,* 7, 709, 1987.

40. **Roos, A.,** Intracellular pH and intracellular buffering power of the cat brain, *Am J. Physiol.,* 209, 1233, 1965.

41. **Arieff, A. I., Kerian, A., Massry, S. G., and DeLima, J.,** Intracellular pH of brain: alterations in acute respiratory acidosis and alkalosis, *Am J. Physiol.,* 230, 804, 1976.

42. **Pelligrino, D. A., Musch, T. I., and Dempsey, J. A.,** Interregional differences in brain intracellular pH and water compartmentation during acute normoxic and hypoxic hypocapnea in the anesthetized dog, *Brain Res.,* 214, 387, 1981.

43. **Rottenberg, D. A., Ginos, J. Z., Kearfott, K. J., Junck, L. R., and Dhawan, V.,** Determination of regional cerebral acid-base status using ^{11}C-dimethyloxazolidinedione and dynamic positron emission tomography, *J. Cereb. Blood Flow Metab.,* 3 (Suppl. 1), S150, 1983.

44. **Berridge, M., Comar, D., Roeda, D., and Syrota, A.,** Synthesis and *in vivo* characteristics of (2-^{11}C) 5,5-dimethyloxazolidine-2,4-dione (DMO), *Int. J. Appl. Radiat. Isot.,* 33, 647, 1982.

45. **Ginos, J. Z., Tilbury, R. S., Haber, M. T., and Rottenberg, D. A.,** Synthesis of (2-^{11}C) 5,5-dimethyl-2,4-oxazolidinedone for studies with positron tomography, *J. Nucl. Med.,* 23, 255, 1982.

46. **Kearfott, K. J., Junck, L., and Rottenberg, D. A.,** C-11 dimethyloxazolidinedione (DMO): biodistribution, radiation absorbed dose, and potential for PET measurement of regional brain pH: concise communication, *J. Nucl. Med.,* 24, 805, 1983.

47. **Brooks, D. J., Lammertsma, A. A., Beaney, R. P., Leenders, K. L., Buckingham, P. D., Marshall, J., and Jones, T.,** Measurement of regional cerebral pH in human subjects using continuous inhalation of [11]CO_2 and positron emission tomography, *J. Cereb. Blood Flow Metab.,* 4, 458, 1984.

48. **Rottenberg, D. A., Ginos, J. Z., Kearfott, K. J., Junck, L., and Dhawan, Y.,** *In vivo* measurement of brain tumor pH using [11C]DMO and positron emission tomography. *Ann. Neurol.,* 17, 70, 1985.

49. **Tyler, J. L., Diksic, M., Villemure, J. G., Evans, A. C., Meyer, E., Yamamoto, Y. L., and Feindel, W.,** Metabolic and hemodynamic evaluation of gliomas using positron emission tomography, *J. Nucl. Med.,* 28, 1123, 1987.

50. **Syrota, A., Samson, Y., Boullais, P., Boullais, C., Wajnberg, P., Loch, C., Crouzel, C., Mazière, B., Soussaline, F., and Baron, J. C.,** Tomographic mapping of brain intracellular pH and extracellular water space in stroke patients. *J. Cereb. Blood Flow Metab.,* 5, 358, 1985.

51. **Patronas, N. J., DiChiro, G., Kufta, C. et al.**, Prediction of survival in glioma patients by means of positron emission tomography, *J. Neurosurg.*, 62, 816, 1985.
52. **DiChiro, G.**, Positron emission tomography using (^{18}F) fluordeoxyglucose in brain tumors, *Invest. Radiol.*, 22, 360, 1987.
53. **DiChiro, G. and Brooks, R. A.**, PET-FDG of untreated and treated cerebral gliomas, *J. Nucl. Med.*, 29, 421, 1988.

Chapter 15

REGIONAL CEREBRAL METABOLIC AND STRUCTURAL CHANGES IN NORMAL AGING AND DEMENTIA AS DETECTED BY PET AND MRI

David W. Weiss, Elaine Souder, and Abass Alavi

TABLE OF CONTENTS

I. INTRODUCTION

The rapidly enlarging population of aged individuals adds particular importance to the evaluation of normal aging and dementia. The average life expectancy in the U.S. in 1900 was 47 years. By 1980, it had risen to 71 years for males and to 78 years for females, with a continued increase expected.[1] With improved prevention and treatment of cancer and cardiovascular disease, dementia has become the fourth most common cause of death, with over 120,000 deaths attributed to severe dementia each year. While 10% of those over 65 have some form of mental impairment, involvement may be as high as 48% in those over the age of 85 years.[2] As the present population of approximately 1,000,000 demented individuals increases, a blossoming health care burden is expected.

The DSM-III defines dementia as a progressive decline in cognitive, intellectual, and memory functions secondary to disease processes of the central nervous system.[3] No single etiology for dementia exists, although Alzheimer's disease contributes to approximately 55% of all cases. Multi-infarct dementia alone accounts for 15% of cases, and a combination of Alzheimer's disease and multiple infarcts occurs in 15% of the cases.[4-9] The remaining 15% of cases of dementia have a variety of etiologies, some of which are potentially reversible,[6,10-15] including post-encephalitic/meningitic dementia, nutritional deficiencies, intracranial mass lesions, normal pressure hydrocephalus, endocrine and metabolic disorders, and dementia associated with AIDS.

The clinical presentation of Alzheimer's disease is often insidious, with early signs of forgetfulness, untidiness, errors in judgment, transient confusion, and periods of restlessness and lethargy. Dysphasia, dyspraxia, and agnosia follow with progression of involvement to include slowing of gait, frontal release signs, and mild extrapyramidal signs. Remote memory is often intact until late in the patient's course. The typical course of the disease lasts 5 to 10 years with the final stages represented by marked mental emptiness, loss of control of bodily functions, and loss of motor functions, leaving the patient bed-ridden or in a wheelchair.

II. STRUCTURAL CHANGES IN AGING AND DEMENTIA

A. POST-MORTEM FINDINGS

1. Normal Aging

Determination of the cellular changes in the brain due to normal aging have yielded variable results. Early assessments of total neuronal counts using manual counting methods found extensive global neuronal loss with increased age, with the area of greatest decrement in the temporal lobes.[16] Neuronal loss in specimens from the frontal lobe has been estimated to be 0.8 to 1.6%/year, with a 10 to 12% loss between the ages of 70 to 85.[16-18] However, more recent automated computerized methods of neuronal counting suggest a more stable total number of neurons, but with a definite loss in neuronal size.[19,20] The total number of glial cells does increase with aging, however, causing a decrease in the neuron to glial cell ratio.

Gross structural changes in the aging brain include a loss of both weight and volume with subsequent increase in size of cortical sulci. Increasing ventricular size is also noted with normal aging.[20-24] Although general agreement exists that a global loss of brain weight does occur with age, the relative regional contributions are not as clear. Tomlinson studied 28 patients who were without reported neurologic symptoms and found cortical atrophy which was most prominent in frontal and parasagittal parietal regions.[25] In an examination of 51 whole brains from carefully screened normal patients, Terry and colleagues found a significant age related decrease in cortical thickness with greatest involvement in frontal and temporal lobes.[20] Others have described regional atrophy involving superior frontal and temporal lobes, the precentral gyrus, the hippocampus, and area striata.[16,18]

2. Dementia

Pathologic changes usually associated with dementia, but often seen to a lesser extent with normal aging include intracytoplasmic lipofuscin granules, granulovacuolar organelles, neuritic or senile plaques, neurofibrillary tangles, and loss of dendrites. Although not specific for Alzheimer's disease, afflicted patients often have these findings in greater abundance with preferential involvement of the cortex when compared to aged patients without clinical evidence of disease.[26] Price found a selective loss of cholinergic neurons in the nucleus basalis of Meynert in patients with Alzheimer's disease.[27] However, more recent reports suggest that the neuronal count in the nucleus basalis is stable and that the apparent neuronal loss may indeed be due to cell shrinkage rather than an absolute loss in neurons.[28,29] Terry has reported a 40% reduction in the number of larger neurons in the frontal cortex and a 46% reduction of the same cell types in the temporal lobes in patients with Alzheimer's disease.[30] However, the advent of accurate *in vivo* anatomic assessment by computed tomography (CT) and magnetic resonance imaging (MRI) has opened a new chapter in the assessment of changes in normal aging and dementia.

B. *IN VIVO* FINDINGS

1. General Considerations

The *in vivo* anatomic assessment of the structural changes associated with aging or dementia can be grouped into two major categories: (1) enlargement of the cerebrospinal fluid spaces, in particular the ventricular system and the cortical sulci, and (2) white matter and to a lesser extent gray matter abnormalities. Each method of evaluation offers advantages and disadvantages for the assessment of these findings.

For the demonstration of ventricular size, computed tomography (CT) provides a fast and accurate method of assessment. The study can be performed quickly and is acquired in single-slice increments, allowing for repetition of slices if technically inadequate due to patient movement. Cost and availability also favor CT over MRI for brain assessment. However, due to attenuation of the less energetic portion of the X-ray beam spectrum by the dense bony calvarium, adjacent structures deep to the calvarium are penetrated by the remaining more energetic X-ray beam. Artifactually low CT number assignments are therefore made to areas penetrated by the more energetic X-rays and make CT a less accurate indicator of the degree of sulcal abnormality than MRI.[31]

MRI allows for acquisition and display of images in any plane allowing for more accurate assessment of the high convexity and temporal lobes than by CT scanning. Because of the lack of beam hardening artifact, more accurate assessment of the posterior fossa and the overall sulcal pattern can be determined with MRI. Ventricular size and configuration are also readily available from MR imaging. The greatest advantage of MRI over CT, however, is in the detection of white matter disease. Although the acquisition of a volume of information can allow for multiplanar reconstruction, patient motion can severely hamper a study and may require repeating the entire study.

2. Normal Aging

Computed tomography has shown in numerous studies that the ventricular system and cortical sulci both progressively increase in size with age.[32-37] Both pathologic and imaging studies suggest that the cerebrospinal fluid volume is static in the normal individual until the fifth to seventh decade (Figures 1B and 1C). Subsequent aging results in mild to moderate dilatation of the ventricular system and cortical sulci[30,38-42] not ascribable to other specific processes and apparently due to the process of aging (Figure 2B).

Findings in the brain due to aging and those due to other processes overlap to some

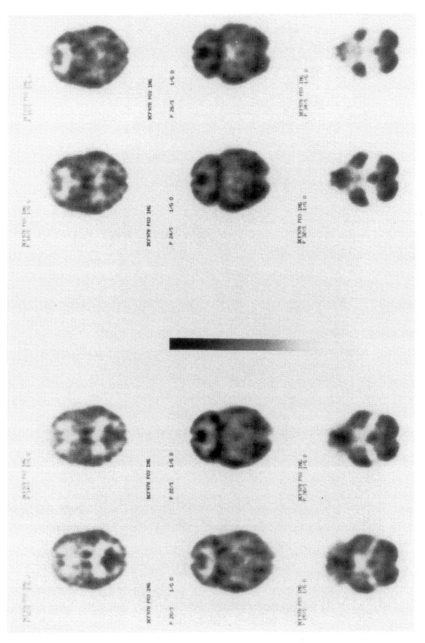

FIGURE 1A.

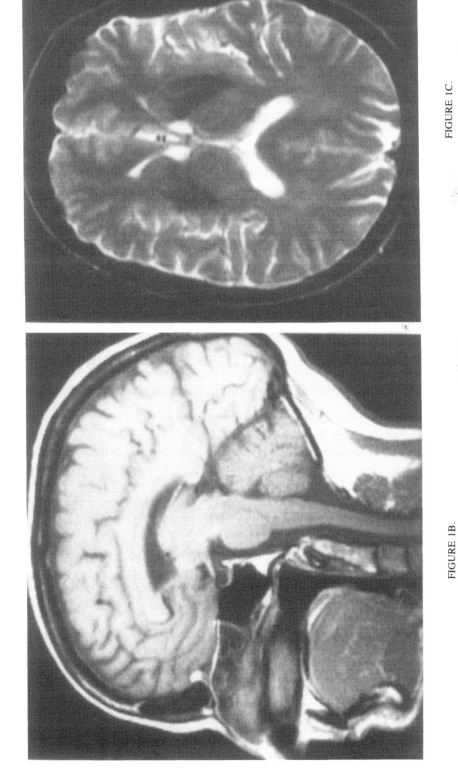

FIGURE 1B.

FIGURE 1C.

FIGURE 1. (A) Normal distribution of F-18-FDG by positron emission tomography; (B) structurally normal brain. Sagittal T1 weighted MRI near the mid-sagittal plane. (C) Structurally normal brain. Axial T2 weighted magnetic resonance image slice at the level of the foramenae of Monro.

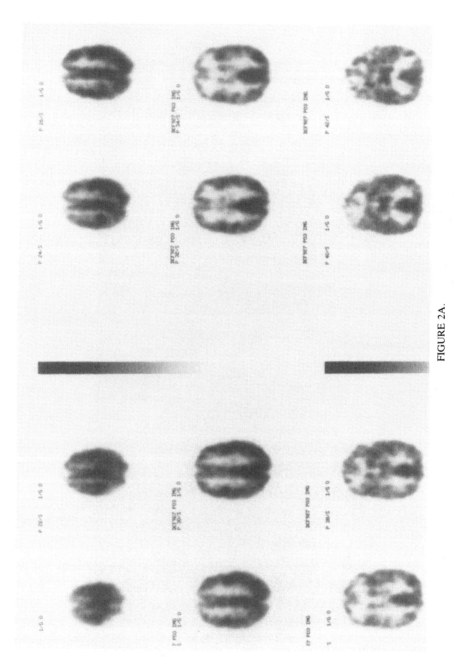

FIGURE 2A.

FIGURE 2.   Metabolic and structural findings of a normal aged patient. (A) Mild bilateral decrease in frontal glucose metabolism by PET. The other cortical areas and subcortical structures retain normal glucose metabolism; (B) T2 weighted axial MRI demonstrates a normal sized ventricular system and mild enlargement of the sulcal spaces.

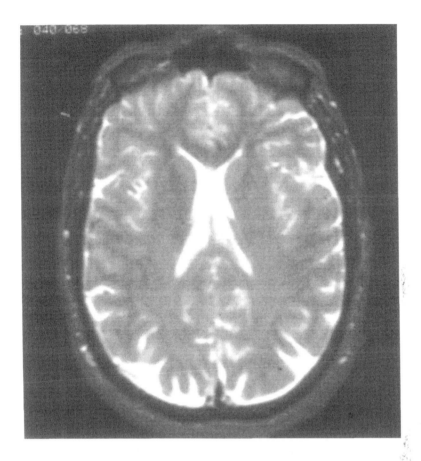

FIGURE 2B.

extent, but many investigators have attempted to isolate changes that are specific for aging. de Leon et al. suggested that cortical atrophy defined by the degree of sulcal prominence (and separate from ventricular size) is a better indicator of normal aging.[37] However, an increasing linear width of the third ventricle has also been shown to correlate well with age.[37]

The ubiquitous white matter signal abnormalities found on T2 weighted MR images have been extensively studied and compared with areas of white matter decreased density by CT scanning.[43-56] As many as 30% of elderly individuals without known neurologic abnormalities will have areas of white matter signal abnormalities by MRI. A typical pattern of involvement would include focal or even confluent areas of abnormality deep in the white matter in the frontal and parieto-occipital areas, usually in a periventricular distribution. While the scan pattern is common, it is very nonspecific and can also be seen in a number of entities, such as infarction, Binswanger's encephalopathy, demyelinating processes, hydrocephalus, and hypoperfusion states.[57-64] A generic term proposed by Hachinski[48] to describe white matter disease by MRI or CT is "leuko-araiosis", which means a diminution in white matter density. In studies which correlate the MRI findings with both gross and microscopic examinations of the brain, it can be concluded that most of the white matter signal abnormalities detected by MRI in elderly patients represent the result of chronic vascular insufficiency or infarction. Other authors have suggested that these lesions may in part be due to hypoperfusion due to extracranial factors such as cardiac disease or hypotension.[65] Although the specific etiology of white matter disease detected by MRI or CT is variable, it is clear that MRI provides a more sensitive means than CT of detecting this process.

Enlargement of the perivascular spaces may explain a distinctly different MRI finding from the white matter signal abnormalities above. Lesions are typically bright on T2 weighted sequences and are located in the base of the brain. In aging individuals, the enlargement of the perivascular or Virchow-Robin spaces has been shown to account for these signal abnormalities.[66] Scan findings therefore reflect the presence of cerebrospinal fluid around penetrating vessels due to the loss of volume in the surrounding brain. Pathologic examination suggests that these findings are associated with demyelination and/or gliosis in the brain parenchyma surrounding the vessels.

3. Dementia

A significant overlap exists between the structural changes present in the aged normal brain and those noted in the brains of demented individuals. Since most forms of dementia occur in the elderly population, the utility of using structural changes as an indicator of early dementia is significantly impaired. It is not surprising, therefore, that the results of many studies undertaken to identify structural changes of the demented brain have been conflicting. Since most studies have reported findings present in Alzheimer's disease (AD) and multi-infarct dementia (MID), the following discussion will focus on these two major forms of dementia.

Studies of ventricular and sulcal changes in dementia by MRI and CT have failed to demonstrate a clear distinction between the presence of disease and findings noted in normal aging (Figures 3B, 4B, and 5B). Brinkman et al. found no significant difference in ventricular or sulcal measurements between a group of patients with clinical evidence of dementia and aged matched controls.[67] Other investigators have suggested that more severe atrophy seems to be associated with dementia.[68,69] Of course, the changes of atrophy are not specific for dementia, since they are found in unrelated disorders, such as alcoholism, trauma, inflammation, radiation therapy, chemotherapy, demyelinating disorders, infarction, and normal aging.[70] The evaluation of regional atrophy may provide a more accurate assessment of the presence of dementia. George et al. and others have reported focal atrophy of the temporal lobes may be a more accurate predictor of AD.[71,72] Other findings associated with AD include marked enlargement of the Sylvian fissures, preferential frontal lobe atrophy, and enlargement of the choroid/hippocampal fissure.[73] The latter finding is of particular interest due to the presence of histopathologic changes found in the hippocampal region.[74]

The spectrum of white matter patterns associated with AD has run the gamut from the total lack of white matter disease in some series to the detection of white matter disease which is nearly as prominent as patterns found in patients with MID. No clear relationship has yet been established between white matter abnormalities and AD. However, categorization of white matter into three patterns, as proposed by Fazekas, may allow the separation of white matter abnormalities into groups related to the underlying entiology. The three groups are as follows: (1) periventricular hyperintensity (PVH), (2) deep white matter hyperintensities (DWMH), and (3) miscellaneous patterns which include other abnormalities, such as basal ganglion lesions, infarcts, and cortical hyperintensities. When classified in such a manner, patients with AD were found to have a halo of PVH compared to controls which tended to have pencil-thin PVH, caps of PVH, or a complete absence of PVH. PVH in MID patients tends to be irregular with deep extension into the white matter. DWMH does seem to differ in MID and AD patients, in that a confluent pattern of DWMH is most often associated with MID (Figure 6B). No specific pattern of cortical disease is present in AD, with a significant overlap in these areas between AD and MID. Basal ganglion lesions are more common in MID than in AD. Fazekas concluded that the finding of a halo of PVH may be specific for AD. The presence of mild degrees of DWMH is not specific and is found in normal aging and aging-related disorders. Patterns of involvement which suggest MID include classic infarcts, basal ganglion lesions, and irregular DWMH that extends into the deep white matter.[75]

Anatomic changes in dementia due to other etiologies have also been described. Preferential and symmetric frontotemporal atrophy has been described in Pick's disease.[76] Patients with Huntington's disease often show preferential atrophy of the basal ganglia in advanced cases.[77] The dementia associated with acquired immune deficiency syndrome (AIDS) appears to be due to a subacute encephalitis caused by the AIDS virus. Although early scans may be normal, atrophy or diffuse white matter abnormalities are common.[78] Creutzfeldt-Jacob disease and progressive multifocal leukoencephalopathy (PML), both believed to be due to viral infection, cause clinical signs of dementia. However, PML preferentially affects the white matter, whereas Creutzfeldt-Jacob disease causes a nonspecific generalized and progressive pattern of atrophy with gradual enlargement of both sulci and ventricles.[79]

III. FUNCTIONAL CHANGES IN NORMAL AGING

The first method described for the assessment of cerebral metabolism was by Kety and Schmidt in 1948.[80] By measuring the levels of nitrous oxide, an inert gas, in the arterial supply and venous drainage of the brain, global cerebral blood flow (CBF) was calculated. Additional calculations could then be performed to assess both the global cerebral metabolic rate of oxygen (CMRO$_2$) and the global cerebral metabolic rate for glucose (CMRgl). Freyhan used this method and reported a fall in global CMRO$_2$ and CMRgl in senile dementia.[81] The original series of subjects evaluated by Kety and Schmidt in 1956 showed a significant decrease in CBF and CMRO$_2$ with age. However, their subjects were selected from a hospital population and their general medical health and educational background, which might influence results, were not considered. Schieve and Wilson et al. had previously reported a lack of significant difference in global CBF and CMRO$_2$ in healthy people with mean ages of 29, 40, and 60 years.[82] Lassen et al. confirmed the work done earlier by Freyhan, although in a later series (1960) of demented and normally aged patients, he was able to show only a modest fall in global CMRO$_2$ with age.[83-84] Perhaps with foresight for the need for regional assessment of cerebral metabolism, Lassen suggested that the small decrease in CMRO$_2$ in whole brain measurements may actually be a reflection of a more significant fall in the focal areas of greatest involvement.

Dastur et al. in 1963 evaluated aging men (mean 71 years) considered normal and compared them to young controls with a mean age of 21 years. A small group of elderly subjects with hypertension, arteriosclerosis, or senile psychosis was also studied. Four global parameters were assessed:

1. Cerebral blood flow (CBF)
2. Cerebral vascular resistance (CVR)
3. Cerebral metabolic rate for oxygen (CMRO$_2$)
4. Cerebral metabolic rate for glucose utilization (CMRgl)

They found no significant difference in CBF or CMRO$_2$ between normal aged subjects and the group of young subjects. However, a significant reduction in CBF (approximately 16%) was documented in the elderly with health problems when compared to young controls. No similar significant difference existed between healthy aged patients and young controls. However, a statistically significant reduction of approximately 26% in CMRGl was found, suggesting that aging produces a subtle decrease in cerebral glucose metabolism that results in a lowering of CMRgl before a reduction in CMRO$_2$ or CBF becomes apparent.[85] Gottstein et al. later confirmed the findings of Dastur of a greater fall in CMRgl compared to CMRO$_2$ with normal aging.[86]

A milestone in the assessment of regional cerebral glucose metabolism occurred when Sokoloff et al. described a mathematical model of cerebral glucose metabolism based on a

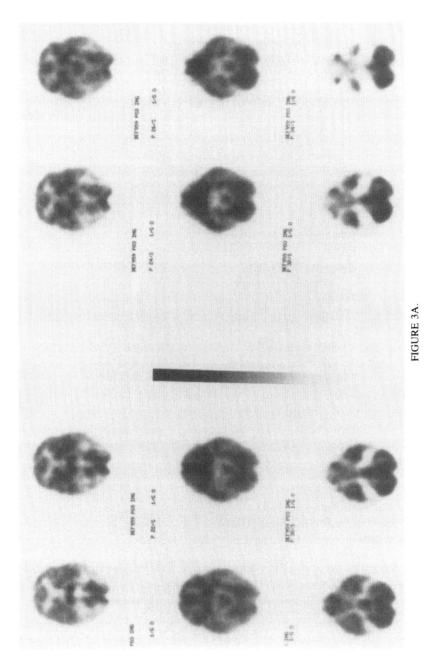

FIGURE 3A.

FIGURE 3.   Metabolic and structural findings in mild Alzheimer's dementia. (A) Bilateral decrease in parietal glucose metabolism by PET. The remaining cortex and subcortical gray matter have normal glucose metabolism; (B) T2 weighted axial MRI demonstrates moderate enlargement of the lateral ventricles and areas of very mild periventricular increased signal in the white matter, which in this case are primarily located in the frontal white matter.

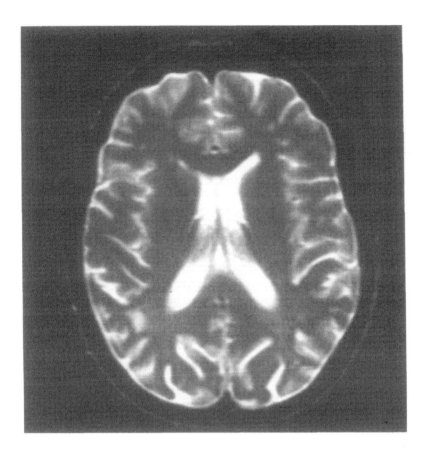

FIGURE 3B.

^{14}C-labeled deoxyglucose.[87] Once ^{14}C-deoxyglucose is transported within the neuron, it is phosphorylated, which traps the molecule within the cell. However, it does not progress further along the path of glycolysis, remaining intact within the neuron, making it an ideal tracer for localization of intracellular glucose. When labeled with the positron emitter fluorine-18, fluoro-deoxyglucose (F-18-FDG) provided a physiologic tracer which can be accurately localized by positron emission tomography[88] (see Figure 1A).

An F-18-FDG model has been applied to the evaluation of normal aging using positron emission tomography (see Figure 2A). Kuhl et al. studied 40 normal volunteers aged 18 to 78 years with F-18-FDG and found a gradual decline in mean cerebral regional metabolism of glucose (CMRgl) with age.[89] The areas which declined most rapidly with age were the superior frontal and posterior inferior frontal regions. Alavi et al. used F-18-FDG to study four normal elderly subjects (mean age, 72 years; range, 60 to 86 years) and nine young controls (mean age, 22 years; range 19 to 26 years). While no significant difference in regional metabolic rates was found between the young and the elderly controls, a general decrease in metabolism in the frontal area was found in the elderly.[90] Duara et al. studied a group of healthy men aged 21 to 83 years with F-18-FDG and found no correlation between mean hemispheric CMRgl or mean hemispheric gray matter CMRgl with age.[91] He also scanned 31 midline and bilateral structures and found no significant change in regional CMRgl with age. Rapoport et al. were able to demonstrate that regional functional activities of the frontal and parietal lobes did correlate with each other, but seemed to be independent of the activity in the temporal and occipital lobes and in the subcortical nuclei. This was interpreted to indicate a loss of integration among these areas with aging.[92]

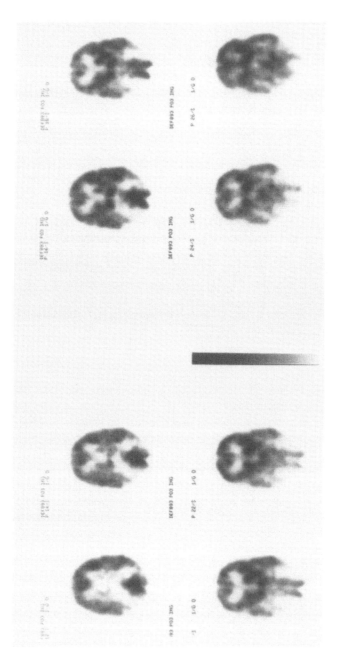

FIGURE 4A.

FIGURE 4.   Metabolic and structural findings with moderate clinical impairment due to Alzheimer's dementia. (A) Bilateral decrease in parietal and posterior temporal glucose metabolism, more severe than in Figure 3A. The remaining cortex and subcortical structures retain normal glucose metabolism. Mild splaying of the subcortical activity suggests enlargement of the ventricles. (From Jolles, P. R., Chapman, P. R., and Alavi, A., *J. Nucl. Med.*, 30, 1589, 1989. With permission.) (B) T2 weighted axial MRI with clear enlargement of the lateral ventricles and sulcal spaces. Extensive punctate areas of periventricular increased signal are present. A small periventricular cap of increased signal intensity is evident about the frontal horn of the lateral ventricle on the right, a nonspecific pattern of white matter disease.

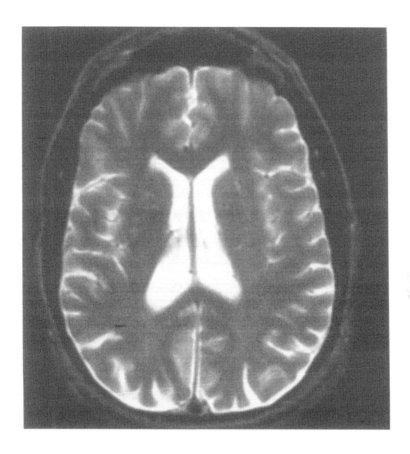

FIGURE 4B.

de Leon et al. studied 15 young and 22 elderly normal subjects with PET and found no significant change in CMRgl, prompting the suggestion that structural changes may be a more sensitive method to assess the effects of aging.[93] However, with a larger series of patients (n = 81), deLeon et al. did find relative frontal lobe hypometabolism with increasing age.[94] Chawluk found a 17% reduction in absolute frontal metabolic rate when comparing essentially healthy elderly patients with a mean age of 63 years to a group of healthy young controls with a mean age of 27 years. Of the group of elderly patients, 17 of 23 had minor health problems not shared by the younger controls. In the 21 young patients, an average CMRgl of 4.24 mg/100 g/min was found compared to 3.54 mg/100 g/min for the aged group. When the same groups were evaluated with normalization of values to the calcarine cortex, additional areas of hypometabolism were found in the inferior parietal, left superior temporal, and primary sensorimotor areas. When the elderly group was subdivided into three groups based on the presence of medical diseases, no significant difference in CMRgl was found between those with no medical problems, those with minor noncardiovascular disease, and those with cardiovascular disease. This suggests that the decline in CMRgl with age is independent of the presence of cardiovascular and systemic health problems.[95] In a subsequent report, Alavi compared regional metabolic rates in 21 normal young adults (mean age 27 years) with 23 older healthy adults (mean age 65 years). The young controls had a frontal metabolic rate of 4.02 compared to 3.19 (corrected for atrophy) in the older controls, a 21% difference.[96]

Yoshii et al. evaluated 76 subjects with FDG PET and found higher frontal, temporal, and parietal CMRgl in younger patients compared to the elderly patients.[97] Cerebrovascular

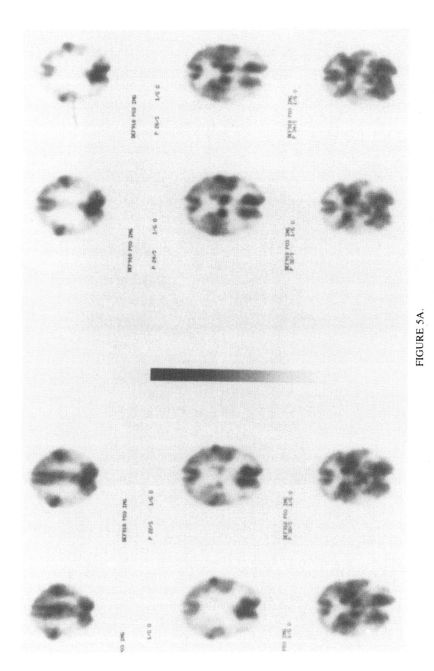

FIGURE 5A.

FIGURE 5. Metabolic and structural findings in a patient with severe clinical impairment due to Alzheimer's dementia. (A) Marked bilaterally symmetric decrease in metabolism involving the parietal lobes, frontal lobes, and posterior temporal lobes. Relative sparing of the somato-sensory strip, subcortical gray matter, and the occipital lobes is present. Definite splaying of the activity in the caudate and lentiform nuclei is consistent with ventriculomegaly. (From Jolles, P. R., Chapman, P. R., and Alavi, A., *J. Nucl. Med.*, 30, 1589, 1989. With permission.) (B) T2-weighted axial MRI reveals generalized ventricular and sulcal enlargement and periventricular increased signal in the white matter. Unlike the pattern of abnormal metabolism, the findings are without a definite lobal predominance.

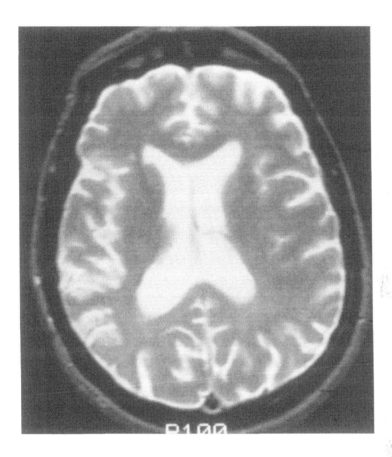

FIGURE 5B.

risk factors did not affect glucose metabolism, supporting the earlier findings by Chawluk. Mean gray matter CMRgl was higher in women and in younger patients. An expected strong relationship was found between age and atrophy. However, by using covariate analysis to exclude the effects of differences in brain volume and atrophy, age and gender no longer had a significant influence on mean gray matter CMRgl. The authors indicated, however, that 80% of the variance in CMRgl was unaccounted for statistically. Possible explanations include variance due to methodological factors or perhaps some physiological factor such as state of arousal.

A recent analysis of F-18-FDG PET data was performed in our laboratory on 85 individuals ranging from third to ninth decade in age who had no significant medical problems. Absolute whole brain metabolism was significantly decreased in the elderly compared to young controls. In addition, a significant decrease in lobar glucose metabolism was found in the frontal, parietal, and sensorimotor strips bilaterally with age.

^{15}O PET offers another method to study CBF and $CMRO_2$ in the aging brain. Frackowiak et al. initially found a decrease in CBF and $CMRO_2$ with increasing age for gray matter, but no similar change for white matter.[98] However, further study by the same investigator failed to confirm the drop in $CMRO_2$ with age although the relationship of decreasing CBF with age was reproduced.[99] Frackowiak et al. showed that an increase in the regional oxygen extraction ratio (rOER) allowed maintenance of a normal $CMRO_2$ in the face of decreasing CBF.[100]

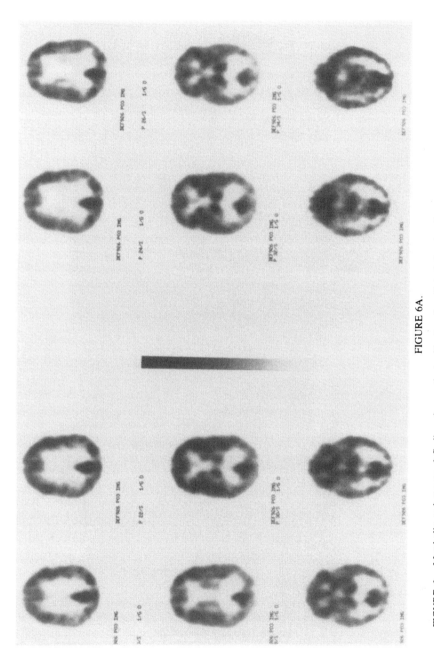

FIGURE 6A.

FIGURE 6. Metabolic and structural findings in a patient with stepwise clinical deterioration suggestive of multi-infarct dementia. (A) Generalized decrease in glucose metabolism throughout the white matter without clear cut cortical abnormalities; (B) T2-weighted axial MRI with extensive periventricular white matter disease which in some areas appear confluent. The areas of abnormal signal are clearly larger than those in AD shown in Figure 4B. Marked ventricular enlargement is also present.

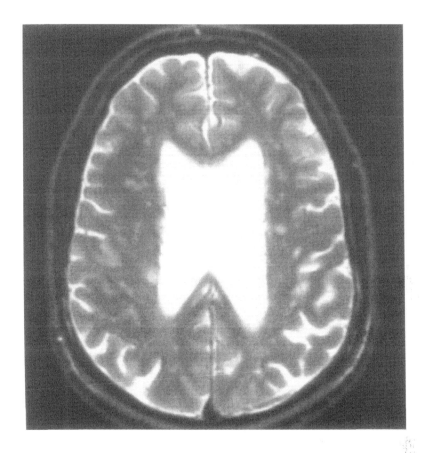

FIGURE 6B.

IV. REGIONAL METABOLIC ABNORMALITIES IN DEMENTIA

A. ALZHEIMER'S DISEASE

The early evaluation of AD patients by Alavi et al. demonstrated a 20 to 30% drop in metabolic rates in demented patients compared to elderly controls.[90] Duara evaluated 21 patients with AD who were subdivided into three groups based on severity. In the mildly demented patients, the parietal region was decreased compared to the somatosensory strip (see Figure 3A). In the moderately demented group, additional areas of involvement were found in the midfrontal and superior temporal regions. The most severely demented patients had hypometabolism in the left midfrontal, bilateral inferior and superior parietal, and left superior temporal regions.[101] Glucose metabolic defects were also correlated to nonmemory cognitive defects by the same group of investigators. Mildly and moderately affected patients with AD were examined with both neuropsychological testing and regional glucose metabolic determination. Decreased glucose metabolism was found in the mild AD group prior to the development of neuropsychologic deficits, suggesting that metabolic assessment may allow earlier detection of AD when compared to neuropsychological testing.[102]

Other investigators have not found abnormal glucose metabolism in the mild AD patients. Cutler found no significant absolute metabolic abnormalities in patients with mild AD.[103] However, when the data was normalized to whole brain values to reduce intersubject variability, decreases were demonstrated in the parietal and temporal lobes. Jamieson also reported no significant difference in CMRgl in mild AD compared to elderly controls, but did find hypometabolism in the temporal and parietal regions as well as in the frontal regions in moderate AD. The findings in severe AD followed a similar, but more pronounced, pattern to that noted in moderate AD (see Figures 4A and 5A). Of note is the fact that the pattern of involvement correlates well with the loss of cognitive function noted clinically. An example is the language deficits found in the more severely affected patients and the left inferior parietal lobe hypometabolism. No metabolic decrement was found in the cerebellum, brainstem, or calcarine cortex.[104]

FDG-PET has been used to document the progression of disease over time in patients with AD. Jagust et al. studied six deteriorating AD patients with two scans over a period of 11 to 22 months (mean 15.5 months).[105] Controls were scanned only once, but were age matched. Using a frontal to parietal percent difference calculation, the AD group was found to be significantly different from the age matched controls after the first scan. In five to six AD patients the frontal to parietal percent difference increased with the second study, indicating a more severe decrement in parietal compared to frontal metabolism. In one patient, progressive decrease in glucose metabolism in the parietal region was documented in three separate studies performed 8 months apart.

Comparison of PET, MRI, and CT scanning for the detection of abnormalities in aging and dementia was evaluated in blinded readings by Alavi et al.[106] In the review of 30 patients with AD and 25 elderly controls, a highly significant difference between AD patients and controls was found for reader ratings of cortical atrophy by MRI and CT and metabolic abnormalities by PET. However, FDG PET scanning seemed to be a more reliable method of separating the AD patients from controls. Cortical hypometabolism by PET did not correlate with the extent or location of white matter lesions by CT or MRI. Areas of focal PET abnormalities did not always correlate with readings of focal atrophy by anatomic imaging.

B. OTHER FORMS OF DEMENTIA

Attempts have been made to determine if characteristic patterns of metabolism occur in the less common forms of dementia. Multi-infarct dementia is characterized clinically by a typical stepwise deterioration, indicating sequential focal insults. The scan pattern is therefore one of focal or asymmetric areas of hypometabolism, usually distinct from the pattern of involvement in AD. MRI appears to be quite sensitive in detecting the white matter disease found frequently in patients with MID. In some patients, MRI may be the only study to clearly delineate the underlying cause of dementia (see Figure 6A). Pick's disease causes a dementia which is clinically similar to AD; however, anatomic findings are those of frontal atrophy with occasional temporal involvement.[107] Not surprisingly, metabolic abnormalities in Pick's disease are much more likely to involve the frontal regions than the temporal and parietal distribution found in AD patients. Huntington's disease (HD) is an autosomal dominant disorder with characteristic neuronal loss in the caudate nuclei and putamina. PET studies have shown decreased glucose metabolism in the basal ganglia in early HD, prior to the demonstration of loss of volume by CT scanning.[108] The degree of hypometabolism by PET correlates well with the degree of impairment in verbal learning.[109] AIDS dementia complex (ADC) includes impaired memory and concentration, psychomotor slowing, and motor deficits. However, Rottenberg et al. found no statistically significant difference in global metabolic rates or in regional metabolic rates (absolute or normalized) for a group of patients with clinical ADC when compared to normal controls.[110]

V. PET DATA ANALYSIS

Cerebral atrophy correction — The resolution limitations of PET imaging make separation of cerebrospinal fluid (CSF) spaces from white matter difficult. As previously noted, both aging and dementia are associated with loss of brain volume and subsequent expansion of the CSF spaces. When assignments of regions of interest (ROIs) are made in order to quantitate regional glucose metabolism, the inclusion of abnormally enlarged CSF spaces in the brain ROIs may cause an incorrectly low measurement of parenchymal glucose metabolism.[111] Initial attempts to correct for atrophy were performed on global metabolic values using the fraction of intracranial volume occupied by brain tissue on CT scan.[112] Chawluk et al. reported atrophy correction of regional metabolic values using CT scans on four elderly controls and four subjects with AD. Frontal hypometabolism was found with aging of a magnitude not explained by focal atrophy, which correlates well with the effect of aging noted by other investigators. However, errors in the regional metabolic rates due to atrophy ranged from 30 to 133%, with the greatest magnitude of error in the parietal lobes.[31] Although performed in a small number of subjects, the wide range of potential error underscores the need for accurate correction for the effect of atrophy on regional glucose metabolic values. Yoshii et al. scored MRI scans according to the degree of atrophy for normal subjects undergoing PET. In their analysis, brain atrophy associated with aging was second only to brain volume as a contributor to the variance noted in CMRgl.[97] At present, however, data on the regional metabolism of glucose by the brain are generally presented without atrophy correction. Methods for precise atrophy correction are presently being developed in our lab utilizing volumetric analysis and differences in MRI signal characteristics between brain and CSF.

VI. CONCLUSIONS

Recent advances in neuroimaging methodologies have permitted study of structural and physiological effects of the aging process on the brain. CT and MRI both reveal ventricular and cortical sulcal enlargement with aging. In addition, MRI has proven to be sensitive in the identification of white matter lesions. While findings have not been consistent, understanding of metabolic changes with aging has been advanced with PET technology. Alzheimer's disease and multi-infarct dementias have been associated with characteristic hypometabolic patterns, and there is ongoing research to identify patterns in dementias of other etiologies. Atrophy correction represents a method to further improve the accuracy of assessment of regional metabolic values in aging and dementia and in other processes with disordered brain glucose metabolism.

ACKNOWLEDGEMENT

This work was supported by National Institutes of Health (NIH) Grants AG 03934-08 and NS 14867.

REFERENCES

1. **Cote, L.,** Aging of the brain and dementia, in *Principles of Neural Science,* 2nd ed., Kandel, E. R. and Schwartz, J. H., Eds., Elsevier, New York, 1985, 784.
2. **Evans, D., Funkenstein, H., Albert, M. S., Scherr, S., Coo, N., Chown, M. J., Heaber, L. E., Hennekens, C. H., and Taylor, J. O.,** Prevalence of Alzheimer's disease in a community population of older persons, *JAMA,* 262(18), 2551, 1989.
3. American Psychiatric Association, *Diagnostic and Statistical Manual III,* American Psychiatric Association Press, Washington, D.C., 1980.
4. **Beck, J. C., Benson, D. F., Scheibel, A. B., Spar, J. E., and Rubenstein, L. Z.,** Dementia in the elderly: the silent epidemic, *Ann. Intern. Med.,* 97, 231, 1982.
5. **Smith, J. S. and Kiloh, L. G.,** The investigation of dementia: results in 200 consecutive admissions, *Lancet,* 1, 824, 1981.

6. **Tomlinson, B. E., Blessed, G., and Roth, M.,** Observations on the brains of demented old people, *J. Neurol. Sci.,* 11, 205, 1970.

7. **Terry, R. D. and Katzman, R.,** Senile dementia of the Alzheimer type, *Ann. Neurol.,* 14, 497, 1983.

8. **Tomlinson, B. E. and Henderson, G.,** Some quantitative cerebral findings in normal and demented old people, in *Aging,* Vol. 3, *Neurobiology of Aging,* Terry, R. D. and Gershon, S., Eds., Raven Press, New York, 1976, 183.

9. **Jellinger, K.,** Neuropathological aspects of dementia resulting from abnormal blood and cerebrospinal fluid dynamics, *Acta Neurol. Belg.,* 76, 83, 1976.

10. **Todorov, A. B., Go, R. C. P., Constantinidis, J., and Elston, R. C.,** Specificity of the clinical diagnosis of dementia, *J. Neurol. Sci.,* 26, 81, 1975.

11. **Brant-Zawadski, M., Fein, G., Van Dyke, D., Kiernan, R., Davenport, L., and de Groot, J.,** MR imaging of the aging brain: patchy white-matter lesions and dementia, *Am. J. Neuroradiol.,* 6, 675, 1985.

12. **Ekholm, S. and Simon, J. H.,** Magnetic resonance imaging and the acquired immunodeficiency syndrome dementia complex, *Acta Radiol.,* 29, 227, 1988.

13. **Koenig, S., Gendelman, H. E., and Orestein, J. M. et al.,** Detection of AIDS virus in macrophages in brain tissue from AIDS patients with encephalopathy, *Science,* 233, 1089, 1986.

14. **Gabuzuda, D. H., Ho, D. D., and De La Monte, S. M. et al.,** Immunohistochemical identification of HTLV-III antigen in brains of patients with AIDS, *Ann. Neurol.,* 20, 289, 1986.

15. **Haase, G. R.,** Disease presenting as dementia, in *Dementia,* 2nd ed., Wells, C. E., Ed., F. A. Davis, Philadelphia, 1977, 27.

16. **Brody, H.,** Organization of the cerebral cortex. III. A study of aging in the human cerebral cortex, *J. Comp. Neurol.,* 102, 511, 1955.

17. **Henderson, G., Tomlinson, B. E., and Gibson, P. H.,** Cell counts in human cerebral cortex in normal adults throughout life using an image analyzing computer, *J. Neurol. Sci.,* 46, 113, 1980.

18. **Anderson, J. M., Hubbard, B. M., Coghill, G. R. et al.,** The effect advanced old age on the neurone content of the cerebral cortex: observations with an automatic image analyser point counting method, *J. Neurol. Sci.,* 58, 235, 1983.

19. **Haug, H., Kuhl, S., Mecke, E. et al.,** The significance of morphometric procedures in the investigation of age changes in cytoarchitectonic structures of human brain, *J. Hirnforsch.,* 25, 353, 1984.

20. **Terry, R. D., DeTeresa, R., and Hansen, L. A.,** Neocortical cell counts in normal human adult aging, *Ann. Neurol.,* 21, 530, 1987.

21. **Blinkov, S. M. and Glezer, I. I.,** *The Human Brain in Figures and Tables: A Quantitative Handbook,* Plenum Press, New York, 1968.

22. **Dekaban, A. S. and Sadowsky, D.,** Changes in human brain weights during the span of human life: relation of brain weights to body heights and body weights, *Ann. Neurol.,* 4, 345, 1978.

23. **Creasy, H. and Rappoport, S. I.,** The aging human brain, *Ann. Neurol.,* 17, 2, 1985.

24. **Riederer, P. and Jellinger, K.,** Morphological and biochemical changes in the aging brain: pathophysiological and possible therapeutic consequences, in *The Aging Brain,* Hoyer, S., Ed., Springer-Verlag, Berlin, 1982, 158.

25. **Tomlinson, B. E., Blessed, G., and Roth, M.,** Observations on the brains of nondemented old people, *J. Neurol. Sci.,* 7, 331, 1968.

26. **Robbins, S. L. and Cotran, R. S.,** *Pathologic Basis of Disease,* W. B. Saunders, Philadelphia, 1979, 1582.

27. **Price, D. L., Whitehouse, P. J., Sruble, R. G., Clark, A. W., Coyle, J. T., DeLong, M. R., and Hedreen, J. C.,** Basal forebrain cholinergic systems in Alzheimer's disease and related dementias, *Neurosci. Comment,* 1(2), 84, 1982.

28. **Chui, H. C., Bondareff, W., Zarow, C., and Slager, U.,** Stability of neuronal number in the human nucleus basalis of Meynert with age, *Neurobiol. Aging,* 5, 83, 1983.

29. **Whitehouse, P. J., Parhad, I. M., Hedreen, J. C. et al.,** Integrity of the nucleus basalis of Meynert in normal aging, *Neurology,* 33 (Suppl. 2), 159, 1983.

30. **Terry, R. D., Pick, A., DeTeresa, R., Fisman, M., Lau, C., and Merskey, H.,** Some morphometric aspects of the brain in senile dementia of the Alzheimer type, *Ann. Neurol.,* 10, 184, 1981.

31. **Chawluk, J. B., Alavi, A., Dann, R., Hurting, H. I., Bais, S., Kushner, M. J., Zimmerman, R. A., and Reivich, M.,** Positron emission tomography in aging and dementia: effect of cerebral atrophy, *J. Nucl. Med.,* 28, 431, 1987.

32. **Baron, S. A., Hacobs, L., and Kinkel, W.,** Changes in size of normal lateral ventricles during aging determined by computerized tomography, *Neurology,* 26, 1011, 1976.

33. **Glydensted, C. and Kosteljanetz, M.,** Measurements of the normal ventricular system with computer tomography, *Neuroradiology,* 10, 205, 1976.

34. **Hahn, F. J. Y. and Rim, K.,** Frontal ventricular dimensions on normal computed tomography, *Am. J. Roentgenol.,* 126, 593, 1976.

35. **Haug, G.,** Age and sex dependence of the size of normal ventricles on computed tomography, *Neuroradiology,* 14, 201, 1977.
36. **Zatz, L. M., Jernigan, T. L., and Ahumada, A. J., Jr.,** Changes on computed cranial tomography with aging: intracranial fluid volume, *Am. J. Neuroradiol.,* 3, 1, 1982.
37. **de Leon, M. J., George, A. E., Ferris, S. H., Christman, D. R., Fowler, J. S., Gentes, C. I., Brodie, J., Reisberg, B., and Wolf, A. P.,** Positron emission tomography and computed tomography assessment of the aging human brain, *J. Comput. Assist. Tomogr.,* 8, 88, 1984.
38. **Nagata, K., Basugi, N., Fukushima, T., Tango, T., Suzuki, I., Kaminuma, T., and Kurashina, S.,** A quantitative study of physiological cerebral atrophy with aging: a statistical analysis of the normal range, *Neuroradiology,* 29, 327, 1987.
39. **Yamamura, H., Ito, M., Kubota, K., and Matsuzawa, T.,** Brain atrophy during aging: a quantitative study with computed tomography, *J. Gerontol. Aging,* 35, 492, 1980.
40. **Schwartz, J., Creasy, H., Grady, C. L., De Leo, J. M., Frederickson, H. A., Cutler, N. R., and Rapoport, S. I.,** Computed tomographic analysis of brain morphometrics in 30 healthy men, aged 21 to 81 years, *Ann. Neurol.,* 17(2), 146, 1985.
41. **Hubbard, B. M. and Anderson, J. M.,** Age, senile dementia, and ventricular enlargement, *J. Neurol. Neurosurg. Psychiatry,* 44, 631, 1985.
42. **Morel, J. and Wildi, E.,** General and cellular pathochemistry of senile and presenile alterations of the brain, in *Proc. 1st Int. Congr. Neuropathology,* Vol. 2, Rosenberg and Sellier, Torino, 1952, 347.
43. **Brant-Zawadski, M., Fein, G., Van Dyke, K., Kiernan, R., Davenport, L., and de Groot, J.,** MR imaging of the aging brain: patchy white-matter lesions and dementia, *Am J. Neuroradiol.,* 6, 675, 1985.
44. **George, A. E., de Leon, M. J., Gentes, C. I., Miller, J., London, E., Budzilovich, G. N., Ferris, S., and Chase, N.,** Leukoencephalopathy in normal and pathologic aging. I. CT of brain lucencies, *Am. J. Neuroradiol.,* 7, 561, 1986.
45. **George, A. E., de Leon, M. J., Kalnin, A., Rosner, L., Goodgold, A., and Chase, N.,** Leukoencephalopathy in normal and pathologic aging. II. MRI of brain lucencies, *Am. J. Neuroradiol.,* 7, 567, 1986.
46. **Kertesz, A., Black, S. E., Tokar, G., Benke, T., Carr, T., and Nicholson, L.,** Periventricular and subcortical hyperinstensities on magnetic resonance imaging, *Arch. Neurol.,* 45, 404, 1988.
47. **Erkinjuntti, T., Ketonen, L., Sulkava, R., Sipponen, J., Vuorialho, M., and Iivanainen, M.,** Do white matter changes on MRI and CI differentiate vascular dementia from Alzheimer's disease?, *J. Neurol. Neurosurg. Psychiatry,* 50, 37, 1987.
48. **Hachinski, V. C., Potter, P., and Merskey, H.,** Leuko-araiosis, *Arch. Neurol.,* 44, 21, 1987.
49. **Inzitari, D., Diaz, F., Fox, A., Hachinski, V. C., Steingart, A., Lau, G., Donald, A., Wade, J., Mulic, H., and Merskey, H.,** Vascular risk factors and leuko-araiosis, *Arch. Neurol.,* 44, 42, 1987.
50. **Steingart, A., Hachinski, V. C., Lau, C., Fox, A. J., Diaz, F., Cape, R., Lee, D., Inzitari, D., and Mersky, H.,** Cognitive and neurologic findings in subjects with diffuse white matter leucines on computed tomographic scan (leuko-araiosis), *Arch. Neurol.,* 44, 32, 1987.
51. **Fazekas, F., Chawluk, J. B., Alavi, A., Hurtig, H. I., and Zimmerman, R. A.,** MR signal abnormalities at 1.5 T in Alzheimer's dementia and normal aging, *Am. J. Neuroradiol.,* 8, 421, 1987.
52. **Zatz, L. M., Jernigan, T. L., and Ahumada, A. J., Jr.,** White matter changes in cerebral computed tomography related to aging, *J. Comput. Assist. Tomogr.,* 6, 19, 1982.
53. **Goto, K., Ishii, N., and Fakasawa, H.,** Diffuse white-matter disease in the geriatric population, *Radiology,* 141, 687, 1981.
54. **Gerard, G. and Weisberg, L.,** MRI periventricular lesions in adults, *Neurology,* 36, 998, 1986.
55. **Zimmerman, R. D., Fleming, C. A., Lee, B. C. P., Saint-Louis, L. A., and Deck, M. D.,** Periventricular hyperintensity as seen by magnetic resonance: prevalence and significance, *Am J. Roentgenol.,* 146, 443, 1986.
56. **Kirkpatrick, J. and Hayman, L.,** White-matter lesions in MR imaging of clinically healthy brains of elderly subjects: possible pathologic basis, *Radiology,* 162, 509, 1987.
57. **Erkinjuntti, T., Sipponen, J. T., Iivanainen, M., Ketonen, L., Suklava, R., and Sepponen, R. E.,** Cerebral NMR and CT imaging in dementia, *J. Comput. Assist. Tomogr.,* 8, 614, 1984.
58. **Bydder, G. M., Steiner, R. E., Young, I. R., Hall, A. S., Thomas, D. J., Marshall, J., Pallis, C. A., and Legg, N. J.,** Clinical NMR imaging of the brain: 140 cases, *Am J. Roentgenol.,* 139, 215, 1982.
59. **Rosenberg, G. A., Kornfield, M., Stovring, J., and Bicknell, J. M.,** Subcortical arteriosclerotic encephalopathy (Binswager): computerized tomography, *Neurology,* 29, 1102, 1979.
60. **Woodarz, R.,** Watershed infarctions and computed tomography. A topographical study in cases with occlusion of the carotic artery, *Neuroradiology,* 19, 245, 1980.
61. **De Witt, L. D., Buonanno, F. S., Kistler, J. P., Brady, T. J., Pykett, I. L., Goldman, M. R., and Davis, K. R.,** Nuclear magnetic resonance imaging in evaluation of clinical stroke syndromes, *Ann. Neurol.,* 16, 535, 1984.
62. **Jack, C. R., Mokri, B., Laws, E. R., Jr., Houser, O. W., Baker, H. L., Jr., and Peterson, R. C.,** MR findings in normal-pressure hydrocephalus: significance and comparison with other forms of dementia, *J. Comput. Assist. Tomogr.,* 11(6), 923, 1987.

63. **Braffman, B. H., Zimmerman, R. A., Trajanowski, J. Q., Gonatas, N. K., Hickey, W. F., and Schlaegfer, W. W.,** Brain MR: pathologic correlation with gross and histopathology. I. Lacunar infarction and Virchow-Robin spaces, *Am. J. Roentgenol.,* 151, 551, 1988.

64. **Braffman, B. H., Zimmerman, R. A., Trajanowski, J. Q., Gonatas, N. K., Hickey, W. F., and Schlaegfer, W. W.,** Brain MR: pathologic correlation with gross and histopathology. II. Hyperintense white matter foci in the elderly, *Am. J. Roentgenol.,* 151, 559, 1988.

65. **Amano, T., Meyer, J. S., Okabe, T., Shaw, T., and Mortel, K. F.,** Stable Xenon CT cerebral blood flow measurements computed by a single compartment-double integration model in normal aging and dementia, *J. Comput. Assist. Tomogr.,* 6, 923, 1982.

66. **Awad, I. A., Johnson, P. C., Spetzler, R. F., and Hodak, J. A.,** Incidental subcortical lesions identified on magnetic resonance imaging in the elderly. II. Postmortem pathologic correlations, *Stroke,* 17, 1090, 1986.

67. **Brinkman, S. D., Sarwar, M., Levin, H. S., and Morris, H. H.,** Quantitative indexes of computed tomography in dementia or normal aging, *Neuroradiology,* 138, 89, 1981.

68. **Gado, M. H., Hughes, C. P., and Danziger, W. et al.,** Volumetric measurements of the cerebrospinal fluid spaces in subjects with dementia and in controls, *Radiology,* 144, 535, 1982.

69. **Albert, M., Naeser, M. A., and Levine, H. L. et al.,** Ventricular size in patients with presenile dementia of the Alzheimer type, *Arch. Neurol.,* 41, 1258, 1984.

70. **TerBrugge, K. G., Rao, K. C. V. G., and Lee, S. H.,** Hydrocephalus and atrophy, in *Cranial Computed Tomography and MRI,* Lee, S. H. and Rao, K. C. V. G., Eds., McGraw-Hill, New York, 1987, 231.

71. **George, A. E., Sylopoulos, L. A., de Leon, M. J., Klinger, A., Kluger, A., and Miller, J. D.,** Temporal lobe CT diagnostic features of Alzheimer's disease, *Am. J. Neuroradiol.,* 8 (Abstr.), 931, 1987.

72. **Kido, D. K., Caine, E. D., Booth, H. A., and Ekholm, S. E.,** Temporal lobe atrophy in patients with Alzheimer's disease, *Am. J. Neuroradiol.,* 8 (Abstr.), 931, 1987.

73. **George, A. E. and de Leon, M. J.,** Computed tomography and positron emission tomography in aging and dementia, in *Computed Tomography of the Head, Neck, and Spine,* Latchaw, R. E., Ed., Grune & Stratton, New York, 1985, 872.

74. **Hyman, B. T., Van Hoesen, G. W., Damasio, A. R., and Barnes, C. L.,** Alzheimer's disease: cell specific pathology isolates the hippocampal formation, *Science,* 22(5), 1168, 1984.

75. **Fazekas, F., Chawluk, J. B., Alavi, A., Hurtig, H. I., and Zimmerman, R. A.,** MR signal abnormalities at 1.5 T in Alzheimer's dementia and normal aging, *Am. J. Neuroradiol.,* 8, 421, 1987.

76. **Davis, D. O. and Pressman, B. D.,** Computerized tomography of the brain, *Radiol. Clin. North Am.,* 12, 297, 1974.

77. **Anderson, R. E. and Jarcho, L. W.,** CT scan in Huntington's chorea, paper presented at the 19th Annu. Meet. of the American Society of Neuroradiology, Chicago, 1981 (Abstr.).

78. **Snider, W. D., Simpson, D. M., and Nielsen, S. et al.,** Neurological complications of acquired immune deficiency syndrome: analysis of 50 patients, *Ann. Neurol.,* 14, 403, 1983.

79. **Schneck, S. A. and Hedley-White, E. T.,** Case records of the Massachusetts General Hospital: Case 49, *N. Engl. J. Med.,* 309, 1440, 1983.

80. **Kety, S. S. and Schmidt, C. F.,** The nitrous oxide method for the quantitative determination of cerebral blood flow in man: theory, procedure, and normal values, *J. Clin. Invest.,* 27, 476, 1948.

81. **Freyhan, F. A., Woodford, R. B., and Kety, S. S.,** Cerebral blood flow and metabolism in psychoses of senility, *J. Nerv. Ment. Dis.,* 113, 449, 1951.

82. **Scheive, J. F. and Wilson, W. P.,** The influence of age, anesthesia and cerebral arteriosclerosis on cerebral vascular activity to CO_2, *Am J. Med.,* 15, 171, 1953.

83. **Lassen, N. A., Munck, O., and Tottey, E. R.,** Mental function and cerebral oxygen consumption in organic dementia, *Arch. Neurol. Psychiatry,* 77, 126, 1957.

84. **Lassen, N. A., Feinberg, J., and Lane, M. H.,** Bilateral studies of cerebral oxygen uptake in young and aged normal subjects and in patients with organic dementia, *J. Clin. Invest.,* 39, 495, 1960.

85. **Dastur, D. K., Lane, M. H., Hansen, D. B., Kety, S. S., Butler, R. N., Perlin, S., and Sokoloff, L.,** Effects of aging on cerebral circulation and metabolism in man, in Human Aging—A Biological and Behavioral Study, DHEW Publ. No. 986, Birrin, J. E., Butler, R. N., Greenhouse, S. W., Sokoloff, L., and Yarrow, M. R., Eds., U.S. Department of Health, Education, and Welfare, National Institute of Mental Health, Bethesda, MD, 1963, 59.

86. **Gottstein, U., Held, K., Moller, W., and Berghoff, W.,** Utilisation of ketone bodies by the human brain, in *Research on Cerebral Circulation,* Meyer, J. S., Reivich, M., and Lechner, H., Eds., Charles C Thomas, Springfield, IL, 1970, 137.

87. **Sokoloff, L., Reivich, M., Kennedy, C., Des Rosiers, M. H., Patlak, C. S., Pettigrew, K. D., Kakurada, D., and Sinohara, M.,** The [^{14}C] deoxyglucose method for the measurement of local cerebral glucose utilization: theory; procedure and normal values in the conscious and anesthetized albino rat, *J. Neurochem.,* 28, 897, 1977.

88. **Reivich, M., Kuhl, D., Wolf, A., Greenberg, J., Phelps, M., Ido, T., Casella, V., Fowler, J., Gallagher, B., Hoffman, E., Alavi, A., and Sokoloff, L.,** Measurement of local cerebral glucose metabolism in man with ^{18}F-2-fluoro-2-deoxy-glucose, *Acta Neurol. Scand.,* 56 (Suppl. 64), 190, 1977.

89. **Kuhl, D. E., Metter, E. J., Reige, W., II, and Phelps, M. E.,** Effect of human aging on patterns of local cerebral glucose utilization determined by the [^{18}F]fluorodeoxyglucose method, *J. Cereb. Blood Flow Metab.,* 2, 163, 1982.

90. **Alavi, A., Reivich, M., Ferris, S., Christman, D., Fowler, J., MacGregor, R., Farkas, T., Greenberg, J., Dann, R., and Wolf, A.,** Regional cerebral glucose metabolism in aging and senile dementia as determined by [^{18}F]-deoxyglucose and positron emission tomography, in *The Aging Brain—Physiological and Pathophysiological Aspects,* Hoyer, S., Ed., Springer-Verlag, Berlin, 1982, 187.

91. **Duara, R., Margolin, R. A., Robertson-Tchabo, E. A., London, E. D., Schwartz, M., Renfrew, J. W., Koziarz, B. J., Sundaram, M., Grady, C., Moore, A. M., Ingvar, D. H., Sokoloff, L., Weingartner, H., Kessler, R. M., Manning, R. G., Channing, M. A., Cutler, N. R., and Rapoport, S. I.,** Cerebral glucose utilisation, as measured with positron emission tomography in 21 resting healthy men between the ages of 21 and 83 years, *Brain,* 106, 761, 1983.

92. **Rapoport, S. I., Duara, R., Horwitz, B., Kessler, R. M., Sokoloff, L., Ingvar, D. H., Grady, C., and Cutler, N.,** Brain aging in 40 healthy men: rCMRglc and correlated functional activity in various brain regions in the resting state, *J. Cereb. Blood Flow Metab.,* 3 (Suppl. 1), S484, 1983.

93. **deLeon, M. J., Ferris, S. H., George, A. E. et al.,** Positron emission tomography studies of normal aging and Alzheimer's disease, *Am J. Neuroradiol.,* 4, 568, 1983.

94. **deLeon, M. J., George, A. E., and Tomanelli, J. et al.,** Positron emission tomography studies of normal aging, a replication of PET III and ^{18}FDG using PET VI and ^{11}CDG, *Neurobiol. Aging,* 8(4), 319, 1987.

95. **Chawluk, J. B., Alavi, A., and Jamieson, D. G. et al.,** Changes in local cerebral glucose metabolism with normal aging, the effects of cardiovascular health and systemic health factors, *J. Cereb. Blood Flow Metab.,* 7 (Suppl. 1), S411, 1987.

96. **Alavi, A.,** The aging brain, *J. Neuropsychiatry,* 1 (Suppl. 1), S51, 1989.

97. **Yoshii, F., Barker, W. W., and Chang, J. Y. et al.,** Sensitivity of cerebral glucose metabolism to age, gender, brain volume, brain atrophy, and cerebrovascular risk factors, *J. Cereb. Blood Flow Metab.,* 8, 654, 1988.

98. **Frackowiak, R. S. J., Lenzi, G. L., Jones, T., and Heather, J. D.,** Quantitative measurements of regional cerebral blood flow and oxygen metabolism in man using ^{15}O and positron emission tomography: theory, procedure, and normal values, *J. Comput. Assist. Tomogr.,* 4, 727, 1980.

99. **Frackowiak, R. S. J., Wise, R. J. S., and Gibbs, J. M. et al.,** Positron emission tomograhic studies in aging and cerebrovascular disease at Hammersmith Hospital, *Ann. Neurol.,* Suppl. 15, S112, 1984.

100. **Frackowiak, R. S. J., Wise, R. J. S., Gibbs, J. M., and Jones, T.,** Oxygen extraction in the aging brain, *Eur. Neurol.,* 22 (Suppl. 2), 24, 1983.

101. **Duara, R., Grady, C., and Haxby, M. et al.,** Positron emission tomography in Alzheimer's disease, *Neurology,* 36, 879, 1986.

102. **Haxby, J. V., Grady, C. L., and Duara, R. et al.,** Neocortical metabolic abnormalities precede non-memory cognitive defects in early Alzheimer's-type dementia, *Arch. Neurol.,* 43, 882, 1986.

103. **Cutler, N. R., Haxby, J. V., Duara, R., Grady, C. L., Kay, A. D., Kessler, R. M., Sundaram, M., and Rapoport, S. I.,** Clinical history, brain metabolism, and neuropsychological function in Alzheimer's disease, *Ann. Neurol.,* 18, 298, 1985.

104. **Jamieson, D. G., Chawluk, J. B., and Alavi, A. et al.,** The effect of disease severity on local cerebral glucose metabolism in Alzheimer's disease, *J. Cereb. Blood Flow Metab.,* 7 (Suppl. 1), S410, 1987.

105. **Jagust, W. J., Friedland, R. P., Budinger, T. F., Koss, E., and Ober, B.,** Longitudinal studies of regional cerebral metabolism in Alzheimer's disease, *Neurology,* 38, 909, 1988.

106. **Alavi, A., Fazekas, F., Chawluk, J. C., Zimmerman, R. A., Hackney, D., Bulaniuk, L., Rosen, M., Alves, W. M., Hurtig, H. I., Jamieson, D. G., Kushner, M. J., and Reivich, M.,** A comparison of CT, MR and PET in Alzheimer's dementia and normal aging, *J. Nucl. Med.,* 29 (Suppl.)(Abstr.), 852, 1988.

107. **Kamo, H., McGeer, P. L., and Harrop, R. et al.,** Positron emission tomography and histopathology in Pick's disease, *Neurology,* 37, 439, 1987.

108. **Mazziotta, J. C., Phelps, M. E., and Pahl, J. J.,** Reduced glucose metabolism in asymptomatic subjects at risk for Huntington's disease, *N. Engl. J. Med.,* 316, 357, 1987.

109. **Berent, S., Giordani, B., and Lehtinen, S. et al.,** Positron emission tomographic scan investigations of Huntington's disease: cerebral metabolic correlates of cognitive function, *Ann. Neurol.,* 23, 541, 1988.

110. **Rottenberg, D. A., Moeller, R. R., and Strother, S. G. et al.,** The metabolic pathology of the AIDS dementia complex, *Ann. Neurol.,* 22, 700, 1987.

111. **Alavi, A., Leonard, J. C., and Chawluk, J.,** Correlative studies of the brain with positron emission tomography, nuclear magnetic resonance, and X-ray computed tomography, in *Cerebral Blood Flow and Metabolism Measurement,* Hartmann, A. and Hoyer, S., Eds., Springer-Verlag, Berlin, 1985, 524.
112. **Herscovitch, P., Auchus, A. P., and Gado, M. et al.,** Correction of positron emission tomography data for cerebral atrophy, *J. Cereb. Blood Flow Metab.,* 6, 120, 1986.

Chapter 16

EVALUATION OF BRAIN TUMORS BY PET

G.-J. Meyer and O. Schober

TABLE OF CONTENTS

I. INTRODUCTION

The most terrifying feature of tumor cells is their potential uncontrolled proliferation. As long ago as the late 1950s, this pathophysiological behavior became a target for metabolically oriented research with radiotracers. Much research was pointed toward the analyses of proliferation mechanisms with radioiodinated nucleosides, such as iodo-deoxyuridine. However, except for valuable basic biochemical information, little diagnostic or therapeutic advancement was achieved with these compounds.[1,2]

Although tumor research with positron emitting compounds was started even before the development of PET-scanners,[3] it must be admitted that meaningful data for the understanding of tumor physiology and results which had direct impact on the treatment of cancer have been obtained only since the development of positron emission tomography (PET). However, unlike today the main focus of earlier research was not directed toward the brain, but to neoplasms of other organs, e.g., after the development of synthetic pathways for ^{11}C-labeled amino acids, the pancreas became the most well-studied organ. Groups from Oak Ridge,[4] Orsay,[5,6] Hannover,[7] and Chicago[8,9] studied the possible differentiation of pancreatic cancer from other dysfunctions, mostly pancreatitis.

Subsequently, protein catabolism of many other tumor types was investigated mainly with the amino acids from Oak Ridge, including ^{11}C-labeled tryptophan, valine, ACPC, ACBC, ACHC, and proposed others.[8-11] ^{15}O-labeled compounds like water and oxygen were used by the Hammersmith group, who studied blood-flow and oxygen metabolism in breast carcinoma.[12] They found increased blood-flow with decreased OER and MRO_2 in this tumor type.

In various proven neoplasms including lymphoma, pancreas, lung, and breast carcinoma, the overall sensitivity for detection of the tumor with amino acids was reported to be 85 to 90%.[9,10] Consequently, it followed from these studies that brain tumors were also analyzed;[13] however, results were poor, probably because of the limitations of the scanning device used in these studies, and these investigations were abandoned.

It was some years later, that brain tumors again became a focus for PET investigators. Now, at some institutions PET investigations on neoplasms in the brain present a major fraction of all PET studies.

Since PET relies on the observation that the biological function of any organism or part of organism, independent of its physiological or pathophysiological state, is directly related to the sum of all biochemical processes in its organs, medical functional diagnostics must concentrate on the analysis of the biochemical functions. The spectrum of physiological functions which can be analyzed with positron emission tomography and which have been applied to cancer research is quite large already, as shown in Table 1.

Furthermore, it is now clear that the spectrum of physiological processes that can be investigated by PET and the specificity of the analyses will further increase by the development of more sophisticated tracer compounds and analytical methods for the interpretation of the data.

Because data analysis and the interpretation of PET signals have been shown to be the "rate-limiting" step in progress of the method itself, it seems appropriate to discuss the methodological background of the various analytical methods applied for tumor investigations.

II. BLOOD BRAIN BARRIER INTEGRITY

The first PET studies on intracranial tumors were directed at measurements of the integrity of the blood-brain barrier (BBB) by Yamamoto et al. using ^{68}Ga-EDTA.[14] This tracer is distributed within the extracellular volume which includes the plasma volume and penetrates into tumor tissue because of damage of the BBB.[15] Subsequent studies with this tracer have

TABLE 1
Brain Tumor Studies with PET

Analyzed function	Tracer	Investigators
Glucose metabolism	[18F]-2-fluoro-2-deoxyglucose	Bethesda, Köln, Montreal, Jülich, Tohoku, Hannover, Michigan, Kyoto
	[11C]-glucose	Stockholm
Glucose transport	[11C]-6-methylglucose	Jülich, Hammersmith
	3-[18F]-2-deoxyglucose	Jülich
Oxygen metabolism	[15O]-O₂,[15O]-water	Hammersmith, Montreal, Tohoku
Blood flow	[15O]-water	Hammersmith, Montreal, Tohoku, Hannover
Blood volume	[11C]-/[15O]-CO labeled erythrocytes	Hammersmith, Montreal, Tohoku
BBB damage	[68Ga]-EDTA,[82Rb]	Hammersmith, Montreal, Tohoku, Stockholm
Tissue pH	[11C]-DMO	Montreal
	[11C]-CO₂	Sloan Kettering
Amino acid uptake	[11C]-Methionine	Stockholm, Baltimore, Hannover, Tohoku, Orsay
	[11C]-ACPC, etc.	Oak Ridge
AA analog uptake	[11C]-putrescine	Brookhaven
Polyamine synthesis	[11C]-ornitine	Brookhaven
Proliferation	[18F]-desoxyuridine	Tohoku

Note: BBB, blood-brain barrier; AA, amino acid.

been reported by Kessler et al.,[16,17] Hawkins et al.,[18] Blasberg et al.,[19] Bergström et al.,[20] Wapenski et al.,[21] Wong et al.,[22] Herholz et al.,[23] Ericson et al.,[24] and Mosskin et al.[25] (Figure 1).

In general, a region with a disrupted barrier is outlined similarly by the findings in contrast enhanced CT or Gd-EDTA MRI. Comparison of the three methods usually reveals a better outline of peritumerous edematous tissue with Gd-EDTA enhanced MRI.

Attempts for a quantification of the BBB damage with 68Ga-EDTA PET have been reported by Kessler et al.,[16,17] Hawkins et al.,[18] and Schlageter et al.[26] In all approaches, the method relies on a steady-state concentration measurement of the tracer in the regions of interest (ROI) and a quantification of the input function.

Another tracer which has been proposed for the measurement of BBB integrity is 82Rb.[27-32] The direct measurements of 82Rb are likely to be contaminated by a fraction present in the blood pool. This has to be corrected for in a second measurement of the blood pool itself with labeled erythrocytes. A model for the quantification of BBB damage with 82Rb measurements which requires an independent blood flow measurement has been developed by Lammertsma et al.[29] The model takes into account that the first pass uptake signal after injection of a short-lived tracer is perfusion limited. A low perfusion does not necessarily lead to a high extraction, however, since the influx is also dependent on the vascularization of the tissue, which is quite variable, especially in the case of tumors (also see Schober and Meyer.[33])

A third PET method for quantitative measurement of BBB damage which has been proposed by Hara et al.[34] refers to a determination of the inulin space, using 11C-labeled inulin. It has, however, not yet been applied to the measurement of tumors.

According to the pathological situation, the well-perfused meningiomas with missing BBB usually exhibit a high uptake of 68Ga-EDTA and 82Rb. While relatively benign tumors such as low-grade astrocytomas usually show no uptake of these tracers, the influx increases

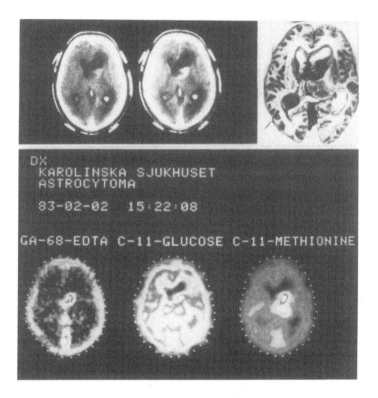

FIGURE 1.   High grade astrocytoma as analyzed by CT with and without contrast medium and by PET with Ga EDTA (left), glucose (middle), and methionine (right). A postmortem histological slice delineates the location of the tumor *"in situ"*. Contrast-enhanced CT and Ga EDTA-PET both delineate the tumor as being much smaller than the methionine PET, which describes the histological situation best (From Bergström, M., Collins, V. P., Ehrin, E., Ericson, K., Eriksson, L., Greitz, T., Halldin, C., Holst, H. V., Langstrom, B., Lilja, A., Lundquist, H., and Nagreu, K., *J. Comput. Assist. Tomogr.*, 7, 1062, 1983. With permission.)

with the malignancy of the neoplasms, while metastases exhibit a quite variable pattern.[23-25,28,35] It seems to be of some diagnostic importance that contrast enhanced CT, Gd-EDTA MRI, and ^{68}Ga-EDTA or [82]Rb PET are of complementary value. This allows an assessment of BBB damage with PET after therapeutic intervention by radiation or surgery, when necrotic changes and scar impede the evaluation of morphologically oriented images.[36]

III. pH MEASUREMENTS

Regional tissue pH measurements allow an evaluation of functional conditions, which often cannot be described by simple physiological and biochemical processes, e.g., changes in tumor tissue pH have been shown to be indicators for cell thermo-sensitivity, radiation sensitivity, and proliferation.[37]

Two methods have been proposed for the calculation of tissue pH with PET. Both rely on the following equations, which describe the dissociation of an acid according to the law of mass action (Equation 1) and to the definition of pH according to Hasselbach-Henderson (Equation 2):

$$AH = A^- + H^+ \tag{1}$$

$$pH = pK_d + \log[A^-]/[AH] \tag{2}$$

The first proposal to measure tissue pH with these equations was made by Raichle et al.,[35] using $^{11}CO_2$ according to the reaction:

$$^{11}CO_2 + H_2O = H^{11}CO_3^- + H^+ \tag{3}$$

This model assumes that $^{11}CO_2$ is a freely diffusible gas so that its intra- and extracellular concentrations are equal under steady-state conditions. In both compartments $^{11}CO_2$ dissociates according to Equation 3. Carbonate ions may, however, not pass the BBB. Measuring the intravascular $^{11}CO_2$ concentration and the $H^{11}CO_3^-$ concentration in the tissue then allows the calculation of intracellular pH according to Equation 2.[38]

Since $^{11}CO_2$ is not inert and may enter other metabolic pathways as mentioned by Lockwood and Finn,[39] the model was expanded to include the fixation of $^{11}CO_2$ by Buxton et al.[40] and Brooks et al.[41]

The second approach for pH determination with PET uses ^{11}C-labeled dimethyloxazolidendion (DMO). DMO is metabolically neutral, passes the BBB in its protonated form by free diffusion and dissociates according to Equation 4.

$$HDMO = DMO^- + H^+ \tag{4}$$

The method is based on experiments described by Waddell and Butler[42] and has been validated using ^{14}C-labeled DMO by Junck et al.[43] An overview of intracellular pH measurements and the associated pitfalls is given by Roos and Boron.[44] An implementation of the DMO method for analysis of tissue pH in brain tumors by PET with ^{11}C-DMO was described by Kearfott et al.[45] and Rottenberg et al.[46,47] Unfortunately, the method becomes less accurate at lower pH levels due to statistical imprecision and the need for blood volume corrections. Rottenberg et al.[47] state a low sensitivity for discrimination of tumor tissue from surrounding gray and white matter despite an accuracy of ± 0.1 pH at levels >6.5 pH. The importance of pH measurements for factors like tumor description, therapy planning, and clinical patient handling remains to be demonstrated (also see Chapter 14 in this volume).

IV. BLOOD VOLUME

Presently blood volume has not been considered to be of major importance in tumor description alone, but has been measured as a correction factor in various protocols, especially in oxygen extraction measurements.

The method is based on the isotope dilution technique under steady-state conditions as developed by Meyer and Zierler.[48] After application of a tracer which remains intravascular and is homogeneously distributed in the circulating blood, the regional blood volume can be measured directly by determination of the distribution volume of the tracer. The only possible variability in the methodological approach is determined by the selection of the tracer, which can either be a blood particle like red cells or plasma constituents like a protein. The former can be labeled *in vivo* by ^{15}O- or ^{11}C-carbon monoxide inhalation[49-54] or by *in vitro* gas exchange of autologous erythrocytes.[50,55,56] The biological half-life of this label is >180 min as shown by Weinreich et al.[57] Plasma proteins have been labeled by ^{11}C-methylation[58] and ^{68}Ga-DTPA.[59,60]

The quantitative determination is simple, when compared with other functional measurements, nevertheless some methodological considerations have to be made. Using either ^{15}O- or ^{11}C-carbon monoxide labeled erythrocytes, the regional changes of hematocrit in tissue must be corrected for, e.g., by comparison with labeled plasma constituents, as ^{68}Ga-

transferrin.[61] The difference may reach up to 20% in brain[62,64] when compared with systemic values. In the case of the short lived ^{15}O-labeled erythrocytes, concentration gradients in the tissue under investigation which result from nonnegligible transit times must be considered.[64]

As mentioned above, blood volume has not been reported to be of general importance for the description of brain tumors. However, in order to analyze the hemodynamic parameters of a tumor adequately, it must be measured in combination with perfusion, oxygen supply, and oxygen extraction. Although a combination of these parameters has not yet led to significant impact on the clinical management of tumors, it can be expected to be of value for the surgical management of brain tumors.

V. PERFUSION

Beyond doubt, perfusion is the most essential physiological function for any organ, since it is the basic requirement for nutritional supply. Because of the importance and simplicity of its measurement which does not require corrections for metabolic interference, perfusion measurements by PET have reached quite a state of excellence since its initiation by Ter-Pogossian.[65] A large number of reviews on its theoretical basis,[33,66-70] as well as on its clinical applications in cerebral blood flow have since been published.[33,71-73] Since perfusion in the brain is dealt with in special chapters of this book, general remarks on the basic protocols will be kept short.

The "gold standard" for blood flow measurements in any tissue are labeled microspheres, which stick by microembolization in the capillary bed of the microvasculature. Its application in PET is troublesome, however, since it requires arterial application and allows the determination of flow in only one hemisphere at a time. Therefore, all commonly used techniques make use of freely diffusible tracers, as ^{15}O-labeled water, [15]O- or ^{11}C-labeled butanol, or ^{18}F-labeled methylfluoride.

While ^{15}O-labeled water can be used with various protocols such as steady-state inhalation of [11]CO$_2$, continuous infusion, bolus injection, and wash-in techniques, the others all require bolus injection techniques with a dynamic data analysis. Because of their short half-life ^{15}O-labeled tracers allow repeated measurements or combined investigations of blood flow with metabolic functions. For most practical applications, the limited accuracy of some protocols, which mainly stems from the limited partition of ^{15}O-labeled water between blood and tissue,[74] is of minor importance.

Perfusion measurements of brain tumors have been reported by several groups. In all cases, these measurements have been performed in combination with the determination of oxygen extraction and oxygen utilization.[76-78] Some studies in which perfusion has been related to regional glucose utilization will be discussed in the next paragraphs. All these studies have shown that perfusion studies by themselves do not allow a discrimination of tumor tissue from normal structures. The variability of perfusion in tumors is quite large and averages 37 ± 27 ml/min/100 g in gliomas,[77] which falls between average grey and white matter flow values. The only type of brain tumor which has consistently been reported to exhibit significantly high flow values is meningioma. Metastases of other tumors in the brain are very variable in flow also, depending not only on their primary tumor source, but also on their developmental stage and individual location. General rules for their perfusion status have not yet been established.

VI. OXYGEN EXTRACTION AND OXYGEN UTILIZATION

Unlike the measurements of distribution volumes and flow, the analysis of functions which include metabolic steps require the determination of the metabolic fate of the tracer and a differentiation of the signal stemming from the metabolites by either changes with time or by a subtraction method. For oxygen extraction and oxygen utilization steady state and dynamic measurement protocols have been established. First attempts have been reported by Ter-Pogossian et al.[79] The development of the steady-state technique was first described by Jones et al.[53] and Subramanyam et al.[54] Further improvements and validation were reported by Frackowiak et al.,[80,81] Jones et al.,[82] Bigler et al.,[83,84] and Lammertsma et al.[85] Dynamic PET measurements, which basically are derived from the Kety-Schmidt washout methodology,[86] have been implemented and improved by the St. Louis group.[87,88]

Briefly, the St. Louis two compartment model method requires the single breath inhalation of $^{15}O_2$, determination of the arterial time courses of $^{15}O_2$ and the only metabolite, $H_2^{15}O$, measurement of arterial O_2 content, and separate determinations of rCBF($H_2^{15}O$) and rCBV($C^{15}O$). CMRO$_2$ is then computed as the product of the arterial O_2 content, rCBF and rOE. The model is simple and all model parameters are known because assumptions are few and are unequivocal. The test may be rapidly and easily repeated. A detailed description of the compartment model can be found in Chapter 9 of this book and in recent reviews.[33,66]

All investigators who have measured oxygen metabolism in brain tumors agree that this metabolic activity is markedly decreased.[73,75,77,85,89] Furthermore, there is no flow correlation with metabolic activity or grade of malignancy. These finding are in agreement with the observation that the flow variability in tumors is large and that it cannot be used as a predictive indicator for metabolic activity.

The decreased oxygen extraction and utilization are usually interpreted to be the result of the increased fraction of anaerobic metabolism in tumors. It is interesting to note that oxygen extraction deficits are usually limited to the area of the tumor tissue, whereas oxygen utilization is often decreased in peritumerous areas also, an effect which has been observed for some other functions also, as will be discussed later in the section on multiparameter studies. Up to now, the absolute amount of decrease of either function has not yet been related to tumor types and their grades, but it seems likely that the effect can be related to other metabolic factors and a general nonspecific measure of malignancy.

VII. GLUCOSE METABOLISM

Beside perfusion measurements the quantitative determination of regional cerebral metabolic rate of glucose (rCMRGl) has become the main driving force for the development of PET. Based on Sokoloff's deoxyglucose model,[90,91] the development of 2-[^{18}F]-2'-deoxyglucose (FDG)[92] by the Brookhaven group in 1978 led to the implementation of a method for in vivo rCMRGl measurements.[93,94] For a comprehensive review on the methodology, see References 33, 66, and 95 and other chapters in this book.

Besides some recent investigations on experimental tumor models in animals,[96-102] an impressive number of glucose utilization measurements on patient groups with brain tumors have been reported by authors from nearly all PET groups, especially, however, from Bethesda,[103-116] London,[117,118] Stockholm,[119] Montreal,[120] Akita,[89,121] Cologne,[122] Hannover,[123] and Berkeley.[124]

The rationale for investigations of glucose utilization in tumors is based on Warburg's hypothesis[125] of an increased anaerobic glycolysis in tumorous tissue. Experimental investigations corroborated this hypothesis by demonstration of an increased hexokinase activity in tumors.[126,127] Finally, the hypothesis was strongly confirmed by the above-discussed PET findings of a generally decreased oxygen utilization in tumors, especially when compared with glucose utilization.[117] DiChiro et al.[103,104] were the first to demonstrate a correlation of tumor grade (malignancy) and glucose utilization of brain tumors. In a further series of investigations,[105-116] this was corroborated not only for gliomas and other primary brain tumors, but also for metastases and cerebral lymphomas[128] (Figure 2).

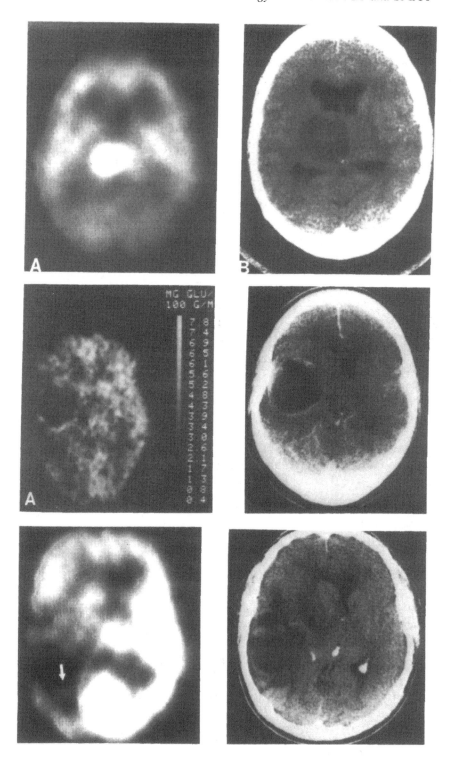

FIGURE 2.   Three gliomas analyzed with (A) FDG (left) and (B) CT (right). All tumor systems are high grade astrocytomas. Case 1 (upper) exhibits uniform higher uptake than gray matter. Case 2 (middle) shows uptake in a small rim comparable to the uptake in gray matter and no uptake in the large tumor volume. Case 3 (lower) shows nearly no FDG uptake, except for a thin tumoral rim with an intensity which is comparable or even less than in white matter (From Di Chiro, G., Brooks, R. A., Bairamian, D., Patronas, N. J., Komblith, P. L., Smith, B. H., and Mansi, L., in *Positron Emission Tomography*, Reivich, M. and Alavi, A., Eds., Alan R. Liss, New York, 1985, 291. With permission.)

Other investigators reported similar findings, although their correlation was less pronounced.[118-123] Several studies state a good differentiation of viable tumor vs. necrotic tissue and edema, which allows a judgment on the success of radiation and/or chemotherapy after several months[110-112,124] or one month[89] post-treatment. In the case of good trapping of [18]FDG by the tumor, semiquantitative analytical approaches are suitable for a fast determination of therapy response[129] which is otherwise difficult to judge.

It must be mentioned that the results of the various studies remain somewhat inconsistent, however. The NIH group claims a strong correlation of the glucose utilization rate and glioma malignancy,[104-108] although some cautions regarding the data interpretation are important.

Since most brain tumor tissue consists of structural, i.e., white matter rather than grey matter, DiChiro and co-workers point out that for any comparison of tumor tissue with normal tissue, whether qualitative or quantitative, one has to use white matter as reference.[130,131] Using this approach, they outlined a good reciprocal correlation of glucose utilization and survival time in recurrent gliomas.[109,114] Other investigators, however, found an inconsistent uptake in the same type of tumors, even if they were histologically similar.[119-123]

Some explanations for the different observations can be derived from the literature itself. Besides the difficulties in selecting patients with the same type of tumor, the quantification problems associated with altered transport kinetics in extremely abnormal tissue seem to limit generalization of a quantitative approach. Furthermore, the glucose utilization pattern in brain is highly structured in normal brain, which limits a simple outline of tumorous structures. These difficulties seem to be negligible only if the tumor location and its outline are well known and delineated by other imaging modalities, or as in the case of a judgment of recurrence, its histology and grading is most probably known beforehand.

Furthermore, it seems that a general approach for rating of tumor activity by glucose utilization is hampered by the necessity to evaluate the questionable structures in comparison with white-matter structures.[130,131] Although a careful comparison with CT and NMR will usually allow a correlation of structures, this may become difficult in the case of grade III gliomas which seem to vary in their glucose utilization rate to such an extent, that, depending on their location, they may not be distinguishable from surrounding structures. As has been shown by papers on the differentiation of posttherapeutic necrotic and scar tissue from recurrences,[89,110-112,124] the latter can be easily outlined as hypermetabolic regions in cases of highly malignant systems, whereas a delineation seems to be more difficult for medium- and low-grade tumors.

A possibility has been outlined by the Cologne group to overcome or at least reduce the problems associated with quantitative analysis of pathological tissue. They analyzed transport rate and phosphorylation rate with a dynamic aquisition protocol and reported a mismatch and decoupling of these parameters in tumors vs. reference normal tissue.[122] Their findings underline that despite limitations and the warning not to overemphasize quantitative interpretation of PET data,[130] dynamic quantification adds not only valuable, but essential information in PET. Similar findings of a decoupling of several parameters have been corroborated by Koeppe et al.[132] Besides these recent findings, little attention has yet been paid to the possibility of measuring the glucose transport function of tumorous tissue independently from the phosphorylation-glycolytic steps, by using the labeled glucose derivatives 3-^{18}F]-fluoro-2-deoxyglucose or 6-^{11}C]-methyl-glucose.[133]

VIII. AMINO ACID MEASUREMENTS

A. INTRODUCTION

Tumor growth is associated with the increase of tissue mass, a large fraction of which is made from protein-containing structures. Since these are built up from amino acids, an assumption is that these building blocks are accumulated in tumorous tissue to a larger extent than in the surrounding tissue. As indicated above, early results on pancreatic cancer remained ambiguous because of the extended use of amino acids under physiological conditions and the low resolution of the PET instruments. However, if the surrounding tissue has a low amino acid turnover rate, as in the case of brain, the approach proved to be extremely successful. Now amino acids seem to be the most promising agents to study the metabolic activity of brain tumors. As shown in a study by the Karolinska group, the amino acid image delineates the metabolically active tumor tissue over a quite larger area than would have been expected from all other modalities, and more exactly when compared with stereotactic biopsies.[24]

Unlike in the case of glucose utilization, there was no quantitative model for the metabolism of amino acids available that could have helped in the interpretation of the PET data obtained from the first applications of amino acids in tumors. Despite a broad knowledge on the basic mechanisms of metabolic pathways of amino acids and protein synthesis,[134,135] specific questions arising from the PET data under pathological conditions could only be answered by further basic investigations in animal experiments, especially designed to increase the knowledge about the mechanisms of tumor uptake and to establish a quantitative model for protein synthesis in normal brain.

There is still an ongoing discussion in the current literature about which amino acid is most suitable for a given problem. Since any essential amino acid should lead to the same protein synthesis rate, several amino acids are under investigation, especially in animal experiments. The most suitable amino acid for protein synthesis rate determinations should be the one with the least secondary metabolic pathways, which usually interfere with PET quantification. The degree of interference is dependent on the fate of the label, which may change with its position in the molecule. However, since all amino acids turn out to exhibit a series of interfering metabolic pathways, practical reasons, like ease of synthesis and purification of the amino acid, begin to play an important role in the use of these compounds for clinical applications in tumor diagnosis and follow-up of treatment. Furthermore, it is not clear yet whether a single parameter like protein synthesis or a less exactly defined mixture of metabolic processes are best suited to describe the metabolic activity of a tumor. Anyhow, transport phenomena have to be considered of predominant importance and especially their distortion under pathophysiological conditions.[136]

B. BASIC STUDIES

Based on first pass extraction measurements of many amino acids, Oldendorf et al. concluded that the transfer of amino acids from blood into brain tissue is an enzyme catalyzed transport, following Michaelis/Menten kinetics.[137,138] They found that there are at least three groups of amino acids. Within the groups the amino acids compete with each other for the same carrier.[139,140] Further experiments showed an individual stereospecificity for various amino acids, which ranged from high values for DOPA, valine, histidine, lysine, and tryptophan to medium values for isoleucine and tyrosine to amino acids with low stereospecificity like phenylalanine, leucine, and methionine.[141,142]

With respect to PET applications of amino acids as tools for the investigation of brain tumors, the mechanisms of uptake and the metabolic interpretation of the accumulation has been studied by many groups. Although basic studies and modeling approaches cannot be separated, since the former are necessary for the validation of the latter, the animal experiments and basic studies can be grouped by the different amino acids. ^{14}C- and/or ^{11}C-leucine, which can be regarded as the most classic model amino acid with relatively simple metabolic pathways, has been used by the groups from NIH[143-148] and UCLA.[149-152] Most of their effort has been in finding quantitative models for a description of protein synthesis, with little attention to application in brain tumors, however.

On the other hand, ^{11}C-methionine has been used most widely with a direct aim of tumor analysis, despite obvious difficulties in establishing quantitative models for the measurement of protein synthesis. The problems with methionine stem from the considerable fraction of metabolites which are formed in trans-methylation reactions. Although these difficulties have been underestimated in first reports by Bustany et al.[153,154] and Lestage et al.,[155] there is an agreement on the usefulness of methionine for an analysis of the metabolic activity of brain tumors in general, regardless of the limitations that are associated with the interpretation of methionine data in terms of protein synthesis.[156-161] Some of these problems have been addressed by comparison of carboxylic- and methyl-labeled methionine, although both cannot be considered especially suitable for protein synthesis measurements.[162]

With respect to its lower abundance in brain protein and the decreased possibility of cross-contamination of the precursor pool from protein degradation, ^{11}C-labeled valine has been used by the group in Montreal.[102,163] [14]C-/^{11}C-labeled phenylalanine has been used by several investigators[139,164-167] because it can be regarded as a model compound for aromatic amino acids and has been well studied in its ^{14}C-labeled form.[168] Furthermore, its own metabolic spectrum is simple, as long as the spectrum of its main metabolite tyrosine can be neglected. Finally, the ^{11}C-label can be put in different positions in phenylalanine more easily than in other amino acids, thereby changing and simplifying the metabolite spectrum further.[169]

The protein incorporation of ^{11}C-tyrosine has been reported by the group from Groningen,[170] mainly in animal experiments which suggest a high protein incorporation rate for this amino acid, however, the analysis disregarded the potentially large spectrum of metabolites. Furthermore, ^{11}C-glycine[156,171] and the common amino acid metabolite ^{11}C-pyruvate[172] have been used in preliminary studies for brain tumor analysis.

Several groups have analyzed the uptake patterns of D-amino acids.[173-180] Whereas the results of early investigators who reported a positive contrast of D-amino acid uptake in some nonbrain tumors[10,11] remained ambiguous, it is quite surprising that in brain tumors the uptake of the unphysiological enantiomer was in all investigations nearly as high as for the L-enantiomer or even better in some cases. Animal experiments as well as patient investigations have shown that the similar uptake occurs despite strongly decreased protein incorporation.[177-179] This phenomenon has been used to attribute the uptake to diffusion processes following BBB damage. It was shown, however, that low-grade tumors with intact BBB exhibit a similar uptake pattern, thereby relating the accumulation of the amino acids to active transport phenomena.

Few ^{18}F-labeled amino acids have been investigated, among them ^{18}F-p- and ^{18}F-o-fluorophenylalanine[181-184] and ^{18}F-m-/^{18}F-o-fluorotyrosine.[183,185] Only the latter one has been shown to be incorporated into brain protein to the same degree as natural amino acids. For this amino acid analogue, only few and negligible amounts of metabolites have been identified. Therefore, it seems to be suitable for the measurement of protein synthesis rates. Preliminary studies have been performed in dynamic mode which allowed a kinetic analysis according to a three-compartment model as discussed by Coenen et al.[185] The k-map analysis showed that most of the PET signal information can be attributed to the transport of the amino acid analogue into tumor tissue, whereas the metabolic rate map added only little additional information.[186]

It can be concluded from the various animal experiments and basic patient investigations that the uptake is surprisingly similar for various amino acids, including D-enantiomers, especially of methionine, in normal brain tissue as well as in brain tumors. In contrast to the uptake, the protein incorporation is a slow process which is markedly different for each of the investigated amino acids. The plasma clearance curves for all investigated amino acids are relatively similar, but show differences with respect to the fractions of protein bound activity and low molecular weight components. Several of these results have been corroborated in dynamic PET investigations in patients,[204,208,209,211] e.g., it is clear by now that the uptake is a fast process, which in tumors seems to be triggered by demand.

This hypothesis is further supported by the observation that iodinated amino acids, like *p*-iodophenylalanine[187,188] and iodo-α-methyl-tyrosine[189-191] which have been developed for SPECT investigations of brain tumors following the promising PET results discussed above, show a very similar accumulation in brain tumors, despite a total lack of incorporation into brain proteins.

C. MODELING APPROACHES

Early modeling approaches in order to interpret brain amino acid uptake data in terms of protein synthesis rate (PSR) have been suggested by Bustany et al.[192-194] and Lestage et al.[155] for methionine and by Dienel et al.,[195] Dweyer et al.,[196] Mies et al.,[197] Smith et al.,[143] and Phelps et al.[149] for leucine. While it has become clear that a three-compartment model for methionine cannot be validated by experimental results,[198] efforts have been intensified to validate a model for leucine.[150,151] The only major degradation pathway for leucine labeled in the carbon-1 position leads to a decarboxylation with loss of the label as CO_2, which then rapidly clears via the lungs. Therefore, fast dynamic measurements in the early phase after uptake allow a calculation of the CO_2 loss in regions of interest.

A kinetic model for tyrosine has been discussed briefly.[199] Despite high uptake values in tumors, the relatively complicated metabolic pathways, which this amino acid enters and which potentially interfere with a conclusive model, have discouraged many PET researchers from further studies.

In principle, this holds true for phenylalanine[164-168] as well. However, a smaller fraction of the amino acid undergoes metabolic pathways other than protein incorporation or degradation, initiated by decarboxylation. This small fraction then is almost totally converted to tyrosine and may enter the various possible pathways for this amino acid.[200,201]

An approach for the measurement of brain protein synthesis by a simplified three-compartment model with ^{14}C-valine has been reported by Kirikae et al.[102,163] It has, however, not yet been applied to PET investigations in humans.

In a study by Anders,[198] three- and four-compartment models have been compared on two sets of experimental data from leucine and methionine.[177] The deterministic analysis was generalized by describing the concentration changes in the compartments by matrices, which simplified the calculations upon variation of the compartment models. The results of this approach have shown that a three-compartment model is definitely insufficient to fit the data, while the generally accepted four-compartment model can fit the data sufficiently for leucine as well as for methionine despite their different metabolic pathways. Since in the case of methionine the metabolic compartment cannot be analyzed by PET, however, the set of differential equations remains unsolved for PSR with this amino acid.

To summarize, it can be stated that for the calculation of PSR with ^{11}C-carboxyl labeled amino acids the loss of labeled [11]CO_2 is the major complicating factor, which according to recent results can possibly be corrected sufficiently in actual PET experiments, provided this is the only major competing metabolic pathway. The contamination of the plasma curve by a large fraction of protein bound activity must also be corrected.

For methyl-labeled ^{11}C-methionine, the loss of activity is nearly negligible, but the transmethylated products, which seem to be eliminated to a small extent and which cannot be differentiated from protein bound material in the PET signal, make it impossible to calculate a protein synthesis rate. The plasma curve represents the input function better than in the case of many carboxylic labeled amino acids, however, it was shown that contaminants contribute to the signal in the case of methionine also, requiring a chemical analysis of plasma samples for good fitting.[202,203] Despite these difficulties and limitations, kinetic analytical approaches on methionine data by the simplified three-compartment model have revealed some insight into the accumulation process. As indicated above for ^{18}F-tyrosine,[186] Ericson et al.[204] have shown that most of the information which can be obtained from a kinetic analytical approach is in the transport parameters.

Although the images presented by Ericson et al.[204] as shown in Figure 3 underscore the impact of the transport function, because of some oversimplification in the applied model, their data lead them to the following statement: "Our results are indicative of a specific physiological process in the tumors without a disruption of the BBB, where demand and transport are closely related" (Figure 3).

Under such physiological conditions, a Gjedde-Patlak analytical approach[205-207] seems to be the method of choice. It describes the unidirectional transport into the irreversible compartment, disregarding whether this compartment is split or not, as in the case of methionine, where the transmethylated components cannot be distinguished from the protein bound fraction. This approach has been evaluated by Anders,[198] Hatazawa et al.,[203] and Bergström et al.[180]

The belief that an increase of the transport into tumor tissue is more relevant than increased protein synthesis rates is strengthened by a finding of Meyer et al.[208] In this study, measurements of protein incorporation of methionine in human brain tumor samples indicate that the protein synthesis rate is not significantly changed in brain tumors of different malignancy, but that the transport is highly increased.

It is reasonable, however, to postulate a regulation mechanism which itself is governed by demand. This interpretation is corroborated by Swedish investigators,[209] who like researchers from Baltimore,[161,210] have found that the uptake can be blocked by saturation of the carrier in the BBB with other amino acids.

D. PATIENT RESULTS

Clinical results on the usefulness of amino acid uptake in brain tumors for analysis of metabolic activity are as numerous as those reported on glucose metabolism. Most of them have been reported by Swedish groups. Their main focus was first to establish a comparison of tumor extension by morphologically oriented methods like CT and MRI with functional PET imaging.[20,25,119,157,211-214] Most of their results have been corroborated by investigators from France,[153,154] Germany,[123,178,179,215-219] the U.S.,[161,210] and Japan.[159,160,220]

The comparison of CT and MRI with PET has revealed that the active tumor tissue is most accurately delineated by the ^{11}C-methionine PET. This has been verified by relating the PET data with multiple stereotactic biopsy examinations.[157,213,214] Although the spatial resolution of MRI is not reached by PET, the PET images allow a clear differentiation of perioedematous regions from active tumor tissue, which is most often difficult by MRI. Besides these visual aids for therapy planning, the quantitative or even semi-quantitative analysis bears more advantages for the PET method. In high-grade tumors (grade IV) which are usually well characterized by CT and MRI, PET offers the advantage of demonstrating the most metabolically active areas better than the former methods. The growth direction can be determined by one ^{11}C-methionine PET investigation, whereas this may require sequential images over a longer time span with other methods. This information can influence therapeutic approaches significantly.

While high-grade tumors are easy to identify by all radiological methods, medium- and low-grade systems are often difficult to specify. Investigations which tried to grade the tumors by quantifying the uptake of amino acids in tumors have shown that despite a good correlation of the uptake with the histological grade, there is some overlap between the histological grades II and III as well as between grades III and IV.[160,215-219] The relatively large variability especially in grade III tumors suggests that histological grading is not the ultimate answer on the metabolic state of these tumors. The amino acid PET data suggest that within the histological class III at least two subgroups can be defined according to

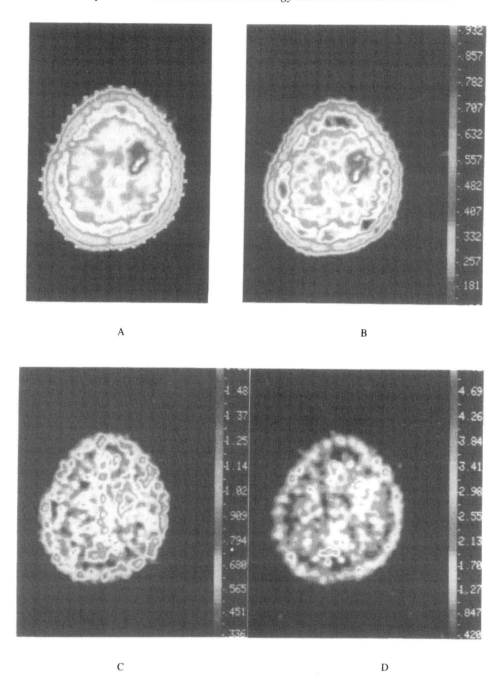

FIGURE 3.   Dynamic analysis of methionine accumulation in an oligo astrocytoma. (A) Steady-state distribution of methionine; (B) accumulation rate; (C) influx; (D) partition coefficient. For comparison, CT image is given also (E). (From Ericson, K., Blomquist, G., Bergström, M., Eriksson, L., and Stone-Elander, S., *Acta Radiol.*, 28, 505, 1987. With permission.)

their metabolic activity. Until now, however, too few PET measurements have been performed in follow-up studies, in order to correlate these findings with clinical data like survival times, treatment response, and prognosis. It should be mentioned here that recent findings in lung cancer investigations with ^{11}C-L-methionine indicate a good correlation of differential uptake ratios with the histological type of tumor.[221]

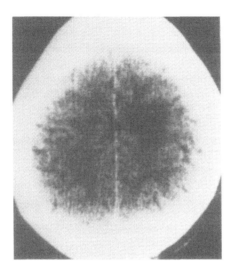

FIGURE 3E.

Some cases of low-grade malignancies have been reported in which amino acid uptake measurement could differentiate tumors from other malformations or could identify areas of increased metabolic activity despite the absence of delineated morphological signs.[33,220]

Two examples of investigations with amino acids are shown in Figure 4 and Figure 5. Figure 4 refers to a small glioma which is rarely visible in the CT (Figure 4, upper left). As can be seen in Figure 4 (upper right), the tumor is well delineated in MR. The PET scan with L-methionine (Figure 4, lower left) reveals the tumor as a central structure with a medium range uptake of the amino acid. The T/NT ratio is 1.7 which is usually found in grade III malignancies. This was later corroborated by histology, which classified the tumor as a grade III glioma. The HMPAO-SPECT image (Figure 4, lower right) indicates a significant flow reduction in a large area of the right cortex indicating the presence of secondary effects.

Figure 5 shows a glioma grade IV as analyzed by methionine PET, CT, and MRI. In this example, the advantages of PET in deliniating the metabolically active parts of the tumor are demonstrated, whereas the capability of MRI in outlining the edematous zones is highlighted.

Futhermore, amino acid uptake in tumors can be used as an excellent tool for therapy modification.[212] All therapeutic interventions, surgery, radiation, and chemotherapy, induce necrosis and morphological changes at the tumor location which complicate the interpretation of CT and MRI images. ^{11}C-methionine PET has shown to be suitable to differentiate necrotic changes and scar from residual tumor tissue and recurrences.[222] The possibility to evaluate metabolic activity on a quantitative basis also allows direct measurement of the response of the tumor tissue to radiation and chemotherapy. Whereas the morphologically oriented methods allow a judgment of the tissue response only after several weeks or months, PET measurements can give quantitative information on the metabolic response directly after the treatment. A special example has been demonstrated by Bergström and Muhr[223] who reported on the effects of bromocryptine therapy on a prolactinoma with ^{11}C-L-methionine.

As shown in Plate 1*, the therapeutic response was already detectable 3 h after treatment, whereas tumor regression was not detectable with CT and MRI until several weeks later.

---

* Plate 1 follows page 198.

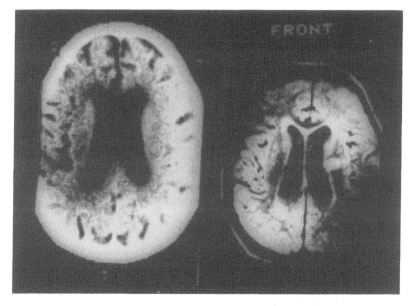

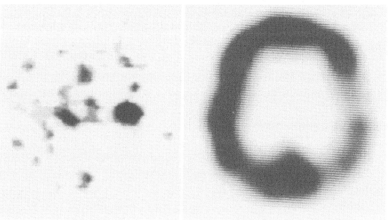

FIGURE 4.   Small glioma grade III as analyzed by CT and MRI (upper row) and methionine-PET and HMPAO-SPECT (lower row). The tumor is rarely visible in CT and clearly seen in MRI and methionine-PET. The HMPAO-SPECT clearly indicates peripheral deactivation of the ipsilateral cortex (Hannover).

IX. AMINO ACID ANALOGUES

Preliminary data are available for some amino acid analogues such as putrescine and ornithine. Both plus methylputrescine have been labeled with ^{11}C,[224-228] and a putrescine derivative with ^{18}F,[229] in order to evaluate their potential for tumor localization and determination of tumor metabolism. Putrescine and ornithine enter pathways for polyamine synthesis, which has long been correlated with cell growth and proliferation.[230-232] Especially noteworthy is the high accumulation of these compounds in prostate tissue.

Besides results in animal tumor models,[224-227] first results from investigations with ^{11}C-putrescine in patients with brain tumors have been reported by Hiesiger et al.[233] They showed that the putrescine accumulation correlates quite well with the histological grade, whereas ^{11}C-deoxyglucose metabolism was quite variable. The putrescine uptake was negligible, however, in several medium grade tumors.

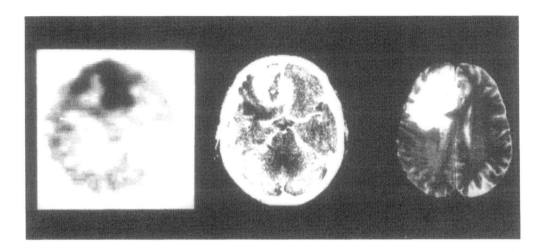

FIGURE 5. Glioma grade IV, analyzed with methionine PET (left), CT (middle), and MRI (right). Metabolically active part of the tumor can be delineated with PET accurately. MRI outlines edematous areas. These do not accumulate methionine (Hannover).

X. PROLIFERATION MEASUREMENTS

Attempts to determine the proliferation of tumor cells by measuring the incorporation of radiolabeled nucleosides have been numerous. It is beyond the scope of this article to refer to the many reports on radioiodine and radiobromine labeled nucleosides and their applications for diagnosis and therapy; for a recent review, see Adelstein et al.[234]

Long before the development of PET scanners [methyl-^{11}C]-thymidine had been proposed as an agent for the *in vivo* determination of DNA synthesis by Christman et al.[235] Although the chemical preparation was subsequently improved[236,237] and preliminary results have been encouraging,[238] the data obtained with this tracer with respect to an interpretation in terms of proliferation remained ambiguous, because of the considerable and rapid metabolism especially when the label is in the methyl position.[239] Therefore, it seems to be desirable to label thymidine in a different position, preferably in the ring. Appropriate syntheses are under development at various PET centers.[239]

[18]F-fluorinated nucleosides and nucleotides have been prepared for the analysis of tumor metabolism by several groups.[240-242] While ^{18}F-uracil[243-246] has mainly been used to predict the efficacy of FU-chemotherapy, ^{18}F-deoxyuridine[247-249] has been used to investigate the metabolic state of tumors. It has been shown that high grade malignancies accumulate ^{18}F-deoxyuridine and that the outline of the accumulation corresponds to the morphological extent as seen by CT and MRI, as well as to grades of malignancy as found in biopsies.[250,251] Little is known, however, about the differentiation of transport phenomena and metabolic processes with this compound, especially in brain and brain tumors. Discussions continue about these issues and whether a major fraction of the accumulation in brain tumors must be attributed to diffuse pathways, opened by BBB disruption. It is still unclear if true proliferation rates can be measured with ^{18}F-deoxyuridine.[239]

XI. CYTOSTATIC DRUGS

Cytostatic drugs like several nitroso urea derivatives have been labeled with different positron emitters such as ^{11}C, ^{13}N, and ^{18}F,[252-254] in order to analyze the pharmacokinetics of these therapeutic drugs in general and to predict their accumulation for therapy planning.[255-257] The ^{11}C-labeled derivatives have yielded the deepest insight in the metabolism of these compounds, however, clinical studies applying this research have not yet been reported.

XII. MULTIPARAMETER ANALYSES

The physiological status of an organ cannot be described by any single parameter. Usually a set of parameters including perfusion, energy consumption, and some organ specific metabolic function need to be analyzed. The same holds true for brain tumors. Most of the relevant parameters for a description of tumor physiological status have been referred to in the above sections of this chapter. Because of the large interpatient variability of tumors, even within one histological class, multiparameter studies on one individual tumor are of special interest for investigations of a correlation of these parameters. Several studies for the analysis of such possible correlations have been undertaken.

Some of these investigations, especially those measuring hemodynamic parameters in combination with glucose utilization, have been referred to already in earlier sections. The results from these can be summarized by the statements that the generally decreased oxygen consumption of brain tumors is accompanied by a variable flow pattern and that glucose utilization is often related to the perfusion characteristics although the correlation coefficient is weak.

There have been few studies which compare amino acid uptake and metabolism with other parameters. Experimental three-parameter studies on brain tumors which compared blood flow, glucose utilization, and amino acid uptake have been reported by Mies et al.[197] Kirikae et al.[102] compared glucose utilization and valine uptake in a rat brain tumor model and Abe et al.[258] measured blood flow and methionine uptake in a nonbrain rat tumor model. All studies agree that amino acid uptake is a sensitive and reliable parameter for increased metabolic activity, as analyzed by histological examination of tumor slices. Kirikae et al. identified valine synthesis as being more useful than glucose utilization to assess the effectiveness of cytotoxic drugs and their toxicity to normal brain tissue.[102]

Several multiparameter investigations in patients[20,119,123,259] have shown that glucose utilization may be inferior to amino acid uptake in identifying the extent and the metabolic activity of brain tumors.

In a detailed study from Hannover,[123] the physiological function of individual brain tumors was analyzed by the three parameters: blood flow (^{15}O-water), glucose utilization (^{18}F-FDG), and amino acid uptake (^{11}C-methionine). All three parameters were recorded by a steady-state protocol. The data were analyzed in four relative ways using four methods: (1) tumor region over contralateral region ratio, (2) tumor region over average slice minus tumor region, (3) tumor region over average white matter and, (4) tumor region over average grey matter ratio as indicated in Figure 6.

A total of 112 patients were analyzed with ^{11}C-methionine. All measurements were carried out prior to surgery in 82 patients, thus leading to the same number of histologically proven samples; 31 patients were studied with 2 parameters; 15 were studied with all 3 modalities.

The results can be summarized as follows. All tumors showed ratios $\geqslant 1$ in terms of amino acid uptake by all analytic methods. The majority of tumors of grade III showed a ratio < 1 for glucose utilization by methods 1, 2, and 4. This was matched in all cases by a decreased blood flow as well. Using method 3 about half of these tumors showed a ratio < 1 for glucose utilization. This again was matched individually by the blood flow ratios obtained with the corresponding method. Only some high grade gliomas of grade IV exhibited glucose utilization ratios > 1 by all analytical methods. These cases showed matched high blood flow ratios as well. An example of this approach is given in Figure 7.

Figure 7 shows an astrocytoma grade III in the right parietooccipital cortex which is

Data Analysis

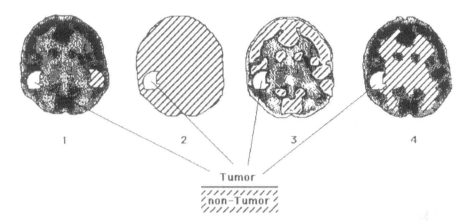

FIGURE 6. Four ways of relative data analysis to charac erize tumor tissue from surrounding brain tissue. (1) Tumor region over nontumor region of same size on the contralateral side (T/NT); (2) tumor region over whole slice minus tumor region (contrast); (3) tumor region over average gray matter region (T/GM); (4) tumor region over average white matter region (T/WM). (Courtesy of Dr. Mats Bergström, Uppsala, Sweden.)

characterized by a high methionine uptake, but low perfusion and glucose utilization. Calculation of the various tumor over reference region ratios gives the results shown in Table 2.

As demonstrated in Figure 8, even recurrencies of high grade astrocytoma (IV) may fail to exhibit a significant glucose uptake, but can clearly be delineated with an amino acid.

Methionine yields a high contrast number with all analytical methods, with tumor over white matter ratio giving the highest value. The contrast is much smaller with FDG. In this case, the visual inspection would clearly rate the tumor as having a decreased glucose utilization. However, by comparing the average white matter CMRGl with tumor-MRGl, a slight increase can be seen. The ratios for flow are quite similar to those for FDG. These relationships were present in almost all cases, including those tumors with high glucose utilization. Only meningioma exhibited a mismatch of flow and glucose utilization. In meningiomas, however, flow and amino acid uptake correlated significantly. A comparison of the results of all investigated tumors is given in Figure 9.

The methodological limitation of this analytical approach is, of course, highly dependent on the accuracy of region of interest identification. Especially in the case of low contrast of the tumor region over adjacent brain structures, as in the case of those FDG images where the tumor is embedded in gray matter structures or within the borders of grey and white matter, errors in the definition of tumor region and clean white matter regions may influence the ratio of tumor over reference region dramatically. The least error occurs from ratios of tumor regions over nontumor regions of the same size on the contralateral side or on tumor regions over average slice minus tumor regions (methods 1 and 2). Since the metabolically active tumor region is delineated most accurately by the amino acid uptake image, this seems to be the most sensitive (least error) approach for a comparative analysis of brain tumors. Figure 10 summarizes the multiparameter analyses in various lesions of the brain.

It can be seen that the amino acid uptake is the only parameter which gives a positive contrast in all types of brain tumors. The small fraction of low grade tumors with slightly decreased uptake results from two calcified astrocytomas II which were in this study. Flow and glucose utilization were very variable, but mostly matched (Figure 9). Figure 11 shows the methionine uptake ratios for various lesions.

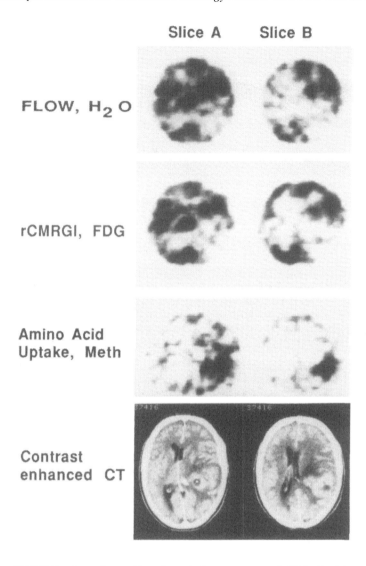

FIGURE 7.   Two adjacent slices of a typical astrocytoma grade III (in the right parietooccipital cortex). These tumors most often exhibit a low perfusion, low glucose utilization, and high methionine uptake. Contrast-enhanced CT is shown for comparison (Hannover).

TABLE 2
Calculated Tumor over Nontumor Ratios According to Four Different Methods as Indicated by Figure 6

	Contr.	T/NT	T/WM	T/GM
Meth.	2.0	1.8	2.4	1.6
FDG	0.9	1.1	1.1	0.8
H_2O	1.0	1.0	1.1	0.6

Note: Contr., contrast; Meth., methionine; T, tumor; NT, nontumor; WM, white matter; GM, gray matter.

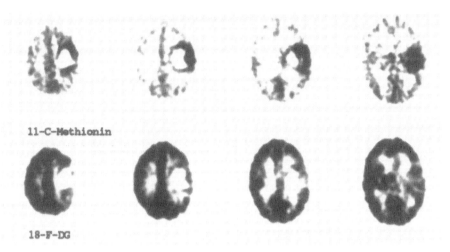

11-C-Methionin

18-F-DG

FIGURE 8. Four consecutive slices of a recurrency of an astrocytoma IV as seen with ^{11}C-L-methionine (upper row) and [18]FDG (lower row). Post-operatively a cyst has formed which caused increasing symptoms due to pressure build-up because of tumor growth at the rim. The tumor growth can clearly by seen by the amino acid uptake, whereas glucose utilization is still negligible.

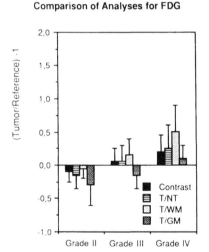

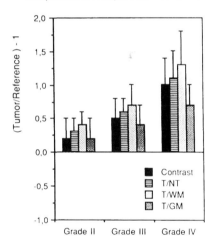

FIGURE 9. Comparison of four analytical methods of data evaluation for methionine and FDG from Figure 6 (Hannover).

It is concluded from this study that amino acid uptake is the method of choice for clinically oriented routine analysis of brain tumors, which usually does not allow a multi-parameter approach for each patient.

The PET investigations in brain tumors can be summarized as follows. Several multi-parameter studies have shown that in gliomas oxygen extraction and oxygen metabolism is generally reduced. There is less agreement on the hemodynamic parameters blood volume and blood flow, which are reported to be quite variable. Glucose utilization has been reported frequently to be a useful parameter for the judgment of metabolic activity, but there remains a not yet fully understood variability in medium grade tumors. Amino acid uptake has been compared with all other functional parameters, except for pH measurements.

The interpretation of the uptake data in terms of quantitative functional parameters as,

Tumor / non-Tumor ratios

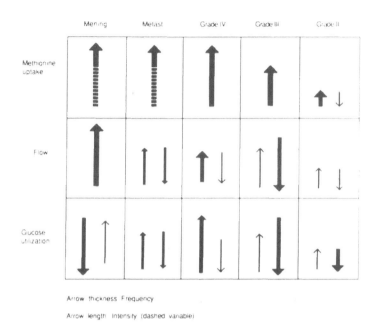

FIGURE 10.   Graphic display of contrast ratios obtained with amino acid uptake, flow, and glucose utilization in various tumors of the brain (Hannover).

Methionine uptake in various lesions

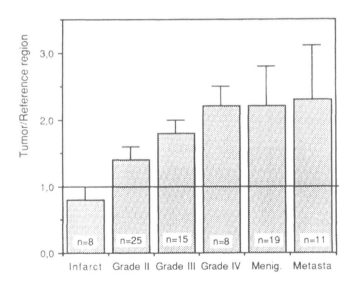

FIGURE 11.   Correlation of methionine uptake ratios with various lesions of the brain indicating the possible use of methionine for grading and differentiation in brain tumors and its limitations (Hannover).

e.g., protein synthesis rate seems possible, but remains difficult. Nevertheless, amino acid uptake has been shown to be a reliable measure of metabolic activity.

Since broad multiparameter studies are far beyond the scope of a routine clinical application for positron emission tomography, amino acid uptake could well become the standard application for diagnostic brain tumor studies with PET.

REFERENCES

1. **Prusoff, W. H., Jaffe, J. J., and Gunther, H.**, Studies in the mouse of the pharmacology of 5-iodo-deoxyuridine, an analogue of thymidine, *Biochem. Pharmacol.*, 3, 110, 1960.
2. **Ertl, H. H., Feinendegen, L. E., and Heiniger, H. J.**, Iodine-125, a tracer in cell biology: physical properties and biological aspects, *Phys. Med. Biol.*, 15, 447, 1970.
3. **Matthews, C. M. E.**, Comparison of coincidence counting and focussing collimators with various isotopes in brain tumor detection, *Br. J. Radiol.*, 37, 351, 1964.
4. **Hübner, K. F., Andrews, G. A., Buonocore, E., Hayes, R. L., Washburn, L. C., Collman, I. R., and Gibbs, W. D.**, Carbon-11-labelled amino acids for the rectilinear and positron tomographic imaging of the human pancreas, *J. Nucl. Med.*, 20, 507, 1979.
5. **Syrota, A., Comar, D., Cerf, M., Plummer, D., Maziere, M., and Kellershohn, C.**, (¹¹C)-Methionine pancreatic scanning with positron emission computed tomography, *J. Nucl. Med.*, 20, 778, 1979.
6. **Syrota, A., Duquesnoy, N., Paraf, A., and Kellershohn, C.**, The role of positron emission tomography in the detection of pancreatic disease, *Radiology*, 143, 249, 1982.
7. **Meyer, G.-J., Schober, O., Gielow, P., and Hundeshagen, H.**, Functional imaging of the pancreas by positron emission tomography: routine production of ¹¹C-L-methionine, quality control and methodology, in *Nuclear Medicine and Biology*, Raynaud, C., Ed., Pergamon Press, Paris, 1980, 1977.
8. **Kirchner, P. T., Ryan, J., Zalutsky, M., and Harper, P. V.**, Positron emission tomography for the evaluation of pancreatic disease, *Sem. Nucl. Med.*, 10, 374, 1980.
9. **Cooper, M. D. and Harper, P. V.**, Radionuclide scintigraphy of the pancreas: perspectives on its role in the diagnosis of pancreatic neoplasms, in *Tumors of the Pancreas*, Mossa, A. R., Ed., Williams & Wilkins, Baltimore, 1980, 245.
10. **Hübner, K. F., King, P., Gibbs, W. D., Partain, L. C., Washburn, L. C., Hayes, R. L., and Holloway, E.**, Clinical investigations with carbon-11-labelled amino acids using positron emission computerized tomography in patients with neoplastic diseases, in *Medical Radionuclide Imaging*, IAEA, Vienna, 1980, 515.
11. **Hübner, K. F., Krauss, S., Washburn, L. C., Gibbs, W. D., and Holloway, E. C.**, Tumor detection with 1-aminocyclopentane and 1-aminocyclobutane C-11-carboxylic acid using positron emission computerized tomography, *Clin. Nucl. Med.*, 6, 249, 1981.
12. **Beaney, R. P., Lammertsma, A. A., Jones, T., McKenzie, C. G., and Halnan, K. E.**, Positron emission tomography for *in vivo* measurement of regional blood flow, oxygen utilization, and blood volume in patients with breast carcinoma, *Lancet*, 1, January 21, 131, 1984.
13. **Hübner, K. F., Purvis, J. T., Mahaley, S. M., Robertson, J. T., Rogers, S., Gibbs, W. D., King, P., and Partain, C. L.**, Brain tumor imaging by positron emission computed tomography using 11-C-labeled amino acids, *J. Comput. Assist. Tomogr.*, 6, 544, 1982.
14. **Yamamoto, Y. L., Thompson, C. J., Meyer, E., Robertson, S. J., and Feindel, W.**, Dynamic positron emission tomography for study of cerebral hemodynamics in a cross section of the head using positron-emitting ⁶⁸Ga-EDTA and ⁷⁷Kr, *J. Comp. Assist. Tomogr.*, 1, 43, 1977.
15. **Moerlein, S. M. and Welch, M. J.**, The chemistry of gallium and indium as related to radiopharmaceutical production, *Int. J. Nucl. Med. Biol.*, 8, 277, 1981.
16. **Kessler, R. M., Goble, J. C., Barranger, J. A., Bird, J. H., and Rapoport, S. I.**, Quantitative measurement of blood-brain barrier permeability following osmotic opening, *J. Nucl. Med.*, 24, 107, 1983.
17. **Kessler, R. M., Goble, J. C., Bird, J. H., Girton, M. E., Doppman, J. L., Rapoport, S. I., and Baranger, J. A.**, Measurement of blood-brain barrier permeability with positron emission tomography and [⁶⁸Ga]EDTA, *J. Cereb. Blood Flow Metab.*, 4, 323, 1984.
18. **Hawkins, R. A., Phelps, M. E., Huang, S. C., Wapenski, J. A., Grimm, P. D., Parker, R. G., Juillard, G., and Greenberg, P.**, A kinetic evaluation of blood-brain barrier permeability in human brain tumors with [⁶⁸Ga]EDTA and positron computed tomography, *J. Cereb. Blood Flow Metab.*, 4, 507, 1984.

19. **Blasberg, R. G., Wright, D. C., Patlak, C. S., Brooks, R. A., Carson, R. E., Groothuis, D. R., and Di Chiro, G.,** Determination of regional blood tissue transfer constants and initial (plasma) volume in brain tumors using ^{68}Ga-EDTA and dynamic positron emission tomography, *J. Nucl. Med.,* 25, P51, 1984.

20. **Bergström, M., Collins, V. P., Ehrin, E., Ericson, K., Eriksson, L., Greitz, T., Halldin, C., Holst, H. V., Langström, B., Lilja, A., Lundquist, H., and Nagren, K.,** Discrepancies in brain tumor extent as shown by computed tomography and positron emission tomography using ^{68}Ga-EDTA, ^{11}C-Glucose, and ^{11}C-Methionine, *J. Comput. Assist. Tomogr.,* 7, 1062, 1983.

21. **Wapenski, J. A., Hawkins, R. A., Mazziotta, J. C., Phelps, M. E., Huang, S.-C., and Verity, A.,** Relationship of blood brain barrier (BBB) permeability by [^{68}Ga]EDTA positron tomography with characteristics of human brain tumors, *Neurology,* 35(Suppl. 1), 288, 1985.

22. **Wong, D. F., Inoue, Y., Rosenbloom, S., Wharam, M., Carson, B., and Wagner, H. N., Jr.,** Gallium-68 ethylene diamine tetra acetic acid (EDTA) imaging of brain tumors by positron tomography, *J. Nucl. Med.,* 25, P50, 1984.

23. **Herholz, K., Wienhard, K., Pawlik, G., Seldon, L., Beil, C., and Heiss, W. D.,** Functional imaging of reversible and irreversible ^{68}Ga-EDTA uptake in brain tumors, *Acta Neurol. Scand.,* 72, 105, 1985.

24. **Ericson, K., Bergström, M., Eriksson, L., Hatam, A., Greitz, T., Söderström, C. E., and Widén, L.,** Positron emission tomography with ^{68}Ga-EDTA compared with transmission computed tomography in the evaluation of brain infarcts, *Acta Radiol.,* 22, 385, 1981.

25. **Mosskin, M., von Holst, H., Ericson, K., and Noren, G.,** The blood tumor barrier in intracranial tumors studied with X-ray computed tomography and positron emission tomography using 68-Ga EDTA, *Neuroradiology,* 28, 259, 1986.

26. **Schlageter, N. L., Carson, R. E., and Rapoport, S. I.,** Examination of blood-brain barrier permeability in dementia of the Alzheimer type with [^{68}Ga]EDTA and positron emission tomography, *J. Cereb. Blood Flow Metab.,* 7, 1, 1987.

27. **Yen, C.-K. and Budinger, T. F.,** Evaluation of blood brain barrier permeability changes in rhesus monkeys and man using Rb-82 and positron emission tomography, *J. Comput. Assist. Tomogr.,* 5, 792, 1981.

28. **Yen, C.-K., Yano, Y., Budinger, T. F., Friedland, R. P., Derenzo, S. E., Huesman, R. H., and O'Brien, H. A.,** Brain tumor evaluation using Rb-82 and positron emission tomography, *J. Nucl. Med.,* 23, 532, 1982.

29. **Lammertsma, A. A., Brooks, D. J., Frackowiak, R. S. J., Heather, J. D., and Jones, T.,** A method to quantitate the fractional extraction of rubidium-82 across the blood-brain barrier using positron emission tomography, *J. Cereb. Blood Flow Metab.,* 4, 523, 1984.

30. **Brooks, D. J., Beaney, R. P., Lammertsma, A. A., Leenders, K. L., Horlock, P. L., Kensett, M. J., Marshall, J., Thomas, D. G., and Jones, T.,** Quantitative measurement of blood-brain barrier permeability using [82]Rb and positron emission tomography, *J. Cereb. Blood Flow Metab.,* 4, 535, 1984.

31. **Jarden, J. O., Dhawan, V., Kerfott, J. K., and Rottenberg, D. A.,** Measurement of brain/tumor capillary permeability using [82]Rb and positron emission tomography, *Ann. Neurol.,* 16, 131, 1984.

32. **Jarden, J. O., Dhawan, V., Poltorak, A., Posner, J. B., and Rottenberg, D. A.,** Positron emission tomographic measurement of blood-to-brain and blood-to-tumor transport of [82]Rb: the effect of dexamethasone and whole-brain radiation therapy, *Ann. Neurol.,* 18, 636, 1985.

33. **Schober, O. and Meyer, G.-J.,** Klinische Anwendung der Positronen Emissions Tomographie, in *Handbuch der Medizinischen Radiologie Band XV/1B,* Diethelm, L., Heuck, F., Olsson, O., Strnad, F., Vieten, H., Zuppinger, A., Eds., Springer-Verlag, Berlin, 1988, 315.

34. **Hara, T., Iio, M., Tsukiyama, T., and Yokoi, F.,** Measurement of human blood barrier integrity using ^{11}C-inulin and positron emission tomography, *Eur. J. Nucl. Med.,* 14, 173, 1988.

35. **Ilsen, H. W., Sato, M., Pawlik, G., Herholz, K., Wienhard, K., and Heiss, W.-D.,** [^{68}Ga]-EDTA positron emission tomography in the diagnosis of brain tumors, *Neuroradiology,* 26, 393, 1984.

36. **LaFrance, N. D., Links, J., Williams, J., Holcomb, H., Dannals, R., Ravert, H., Wilson, A., Drew, H., Herda, S., Wong, D., Brem, H., Long, D., and Wagner, H. N.,** ^{11}C-Methionine and ^{18}F-Deoxyglucose (FDG) in the postoperative management of patients with brain tumors with positron emission tomography, *J. Nucl. Med.,* 27, P890, 1986.

37. **Gerson, D. F., Kiefer, H., and Eufe, W.,** Intracellular pH mitogen-stimulated lymphocytes, *Science,* 216, 1009, 1982.

38. **Raichle, M. E., Grubb, R. L., and Higgins, C. S.,** Measurement of brain tissue carbon dioxide content *in vivo* by emission tomography, *Brain Res.,* 166, 413, 1979.

39. **Lockwood, A. H. and Finn, R. D.,** ^{11}C-carbon dioxide fixation and equilibration in rat brain: effects on acid base measurements, *Neurology,* 32, 451, 1982.

40. **Buxton, R. B., Wechsler, L. R., Alpert, N. M., Ackerman, R. H., Elmaleh, D. R., and Correia, J. A.,** Measurement of brain pH using [11]CO_2 and positron emission tomography, *J. Cereb. Blood Flow Metab.,* 4, 8, 1984.

41. **Brooks, D. J., Lammertsma, A. A., Beaney, R. P., Leenders, K. L., Buckingham, P. D., Marshall, J., and Jones, T.**, Measurement of regional cerebral pH in human subjects using continuous inhalation of $^{11}CO_2$ and positron emission tomography, *J. Cereb. Blood Flow Metab.*, 4, 458, 1984.

42. **Waddell, W. J. and Butler, T. C.**, Calculation of intracellular pH from the distribution of 5,5-dimethyl-2,4-oxazolidinedione (DMO): application to skeletal muscle of the dog, *J. Clin. Invest.*, 38, 720, 1959.

43. **Junck, L., Blasberg, R., and Rottenberg, D. A.**, Brain and tumor pH in experimental leptomenigeal carcinomatosis, *Trans. Am. Neurol. Assoc.*, 106, 298, 1981.

44. **Roos, A. and Boron, W. F.**, Intracellular pH, *Physiol. Rev.*, 61, 296, 1981.

45. **Keafott, K. J., Junck, L., and Rottenberg, D. A.**, ^{11}C-dimethyloxozolidinedione (DMO): biodistribution, estimates of radiation absorbed dose, and potential for positron emission tomographic (PET) measurements of regional brain tissue pH, *J. Nucl. Med.*, 24, 805, 1983.

46. **Rottenberg, D. A., Ginoz, J. Z., and Kearfott, K. J.**, Determination of regional cerebral acid-base status using ^{11}C-dimethyloxazolidinedione and dynamic positron emission tomography, *J. Cereb. Blood Flow Metab.*, 3, S150, 1983.

47. **Rottenberg, D. A., Ginos, J. Z., Kearfott, K. J., Junck, L., Dhawan, V., and Jarden, J. O.**, *In vivo* measurement of brain tumor pH using [^{11}C]DMO and positron emission tomography, *Ann. Neurol.*, 17, 70, 1985.

48. **Meier, P. and Zierler, K. L.**, On the theory of the indicator dilution method for measurement of blood flow and volume, *J. Appl. Physiol.*, 12, 731, 1954.

49. **Glass, H., Brant, A., and Clark, J. C.**, Measurement of blood volume using red cells labelled with radioactive carbonmonoxyde, *J. Nucl. Med.*, 9, 571, 1968.

50. **Clark, J. C. and Buckingham, P. D.**, *Short Lived Radioactive Gases for Clinical Use*, Butterworths, London, 1975.

51. **Phelps, M. E., Hoffman, E. J., Coleman, R. E., Welch, M. J., Raichle, M. E., Weiss, E. S., Sobel, B. E., and Ter-Pogossian, M. M.**, Tomographic images of blood pool and perfusion in brain and heart, *J. Nucl. Med.*, 17, 603, 1976.

52. **Phelps, M. E., Huang, S. C., Hoffman, E. J., and Kuhl, D. E.**, Validation of tomographic measurement of cerebral blood volume with C-11-labelled carboxyhemoglobin, *J. Nucl. Med.*, 20, 328, 1979.

53. **Jones, T., Chesler, D. A., and Ter-Pogossian, M. M.**, The continuous inhalation of oxygen-15 for assessing regional oxygen extraction in the brain of man, *Br. J. Radiol.*, 49, 339, 1976.

54. **Subramanyam, R., Alpert, N. M., Hoop, B., Brownell, G. L., and Taveras, J. M.**, A model for regional cerebral oxygen distribution during continuous inhalation of $^{15}O_2$, $C^{15}O$, and $C^{15}O_2$, *J. Nucl. Med.*, 9, 48, 1978.

55. **Subramanyam, R., Bucelewicz, W. M., Hoop, B., and Jones, S. C.**, A system for Oxygen-15 labeled blood for medical applications, *Int. J. Appl. Radiat. Isot.*, 28, 21, 1977.

56. **Meyer, G. J. and Hundeshagen, H.**, An effective method for the preparation of O-15 labeled carboxy-hemoglobin, in *Nuklearmedizin*, Schmidt, H. A. E. and Rösler, H., Eds., K. F. Schattauer Verlag, Stuttgart, 1982, 340.

57. **Weinreich, R., Ritzl, F., Feinendegen, L. E., Schnippering, H. G., and Stöcklin, G.**, Fixation, retention and exhalation of carrier-free 11-C-labeled carbon monoxide by man, *Radiat. Environ. Biophys.*, 12, 271, 1975.

58. **Turton, D. R., Brady, F., Pike, V. W., Selwyn, A. P., Shea, M. J., Wilson, R. A., and De Landsheere, C. M.**, Preparation of human serum[methyl-^{11}C-]methylalbumin microspheres and human serum[methyl-^{11}C-]methylalbumin for clinical use, *Int. J. Appl. Radiat. Isot.*, 35, 337, 1984.

59. **Hnatowitch, D. J., Kulprathipanja, S., Evans, G., and Elmaleh, D. A.**, Comparison of positron emitting blood pool imaging agents, *Int. J. Appl. Radiat. Isot.*, 30, 355, 1979.

60. **Welch, M. J., Thakur, M. L., Coleman, R. E., Patel, M., Siegel, B. A., and Ter-Pogossian, M. M.**, Gallium-68 labeled red cells and platelets, new agents for positron tomography, *J. Nucl. Med.*, 18, 558, 1977.

61. **Lammertsma, A. A., Brooks, D. J., Beaney, R. P., Turton, D. R., Kensett, M. J., Heather, J. D., Marshall, J., and Jones, T.**, *In vivo* measurements of regional cerebral haematocrit using positron emission tomography, *J. Cereb. Blood Flow Metab.*, 4, 317, 1984.

62. **Oldendorf, W. H., Kitano, M., Shimizu, S., and Oldendorf, S. Z.**, Haematocrit of the human cranial blood pool, *Circ. Res.*, 17, 532, 1965.

63. **Larsen, O. A. and Lassen, N. A.**, Cerebral haematocrit in normal man, *J. Appl. Physiol.*, 19, 571, 1964.

64. **Meyer, G. J., Schober, O., and Hundeshagen, H.**, Gradient effects in extravascular water determination using 15-O-labelled water under steady state conditions: theory and error sensitivity, *Eur. J. Nucl. Med.*, 10, 77, 1985.

65. **Ter-Pogossian, M. M., Eichling, J. O., Davis, D. O., Welch, M. J., and Metzger, J. M.**, The determination of regional cerebral blood flow by means of water labeled with radioactive oxygen-15, *Radiology*, 93, 31, 1969.

66. **Huang, S. C. and Phelps, M. E.,** Principles of tracer kinetic modelling in positron emission tomography and autoradiography, in *Positron Emission Tomography and Autoradiography,* Phelps, M. E., Mazziotta, J. C., and Schelbert, H. R., Eds., Raven Press, New York, 1986, 287.

67. **Lammertsma, A. A., Frackowiak, R. S., Hoffman, J. M., Huang, S. C., Weinberg, I. N., Dahlbom, M., MacDonald, N. S., Hoffman, E. J., Mazziotta, J. C., Heather, J. D., Forse, G. R., Phelps, M. E., and Jones, T.,** The $C^{15}O_2$ build-up technique to measure regional cerebral blood flow and volume of distribution of water, *J. Cereb. Blood Flow Metab.,* 9, 461, 1989.

68. **Lammertsma, A. A.,** Positron emission tomography and *in vivo* measurements of tumor perfusion and oxygen utilization, *Cancer Metastasis Rev.,* 6, 521, 1987.

69. **Raichle, M. E.,** Measurement of local brain blood flow and oxygen utilization using oxygen-15 radiopharmaceuticals, in *Positron Emission Tomography,* Reivich, M. and Alavi, A., Eds., Alan R. Liss, New York, 1985, 241.

70. **Ter-Pogossian, M. M. and Herscovitch, P.,** Radioactive oxygen-15 in the study of cerebral blood flow, blood volume, and oxygen metabolism, *Sem. Nucl. Med.,* 15, 377, 1985.

71. **Mazziotta, J. C. and Phelps, M. E.,** Positron emission tomography studies of the brain, in *Positron Emission Tomography and Autoradiography,* Phelps, M. E., Mazziotta, J. C., and Schelbert, H. R., Eds., Raven Press, New York, 1986, 493.

72. **Lammertsma, A. A., Wise, R. J. S., and Jones, T.,** Regional cerebral blood flow and oxygen utilization in edema associated with cerebral tumors, in *Recent Progress in the Study and Therapy of Brain Edema,* Go, K. G. and Baethman, A., Eds., Plenum Press, New York, 1984, 331.

73. **Beaney, R. P. and Lammertsma, A. A.,** in *Positron Emission Tomography,* Reivich, M. and Alavi, A., Eds., Alan R. Liss, New York, 1985, 425.

74. **Herscowitch, P. and Raichle, M. E.,** What is the correct value for the brain-blood partition coefficient for water?, *J. Cereb. Blood Flow Metab.,* 5, 65, 1985.

75. **Ackerman, R. H., Davis, S. M., Correia, J. A., Alpert, N. M., Buonanno, F., Finkelstein, S., and Brownell, G. L.,** Positron imaging of CBF and metabolism in patients with cerebral neoplasms, *J. Cereb. Blood Flow Metab.,* 1, 575, 1981.

76. **Beaney, R. P., Brooks, D. J., Leenders, K. L., Thomas, D. G., and Jones, T.,** Blood flow and oxygen utilisation in the contralateral cerebral cortex of patients with untreated intracranial tumors as studied by positron emission tomography, with observations on the effect of decompressive surgery, *J. Neurol. Neurosurg. Psychiatry,* 48, 310, 1985.

77. **Ito, M., Lammertsma, A. A., Wise, R. J. S., Bernardi, S., Frackowiak, R. S. J., Heather, J. D., McKenzie, C. G., Thomas, D. G. T., and Jones, T.,** Measurement of regional cerebral blood flow and oxygen utilization in patients with cerebral tumors using ^{15}O and positron emission tomography: analytical techniques and preliminary results, *Neuroradiology,* 23, 62, 1982.

78. **Lammertsma, A. A., Itoh, M., McKenzi, C. G., Jones, T., and Frackowiak, R. S. J.,** Quantitative tomographic measurements of regional cerebral blood flow and oxygen utilization in patients with brain tumors using oxygen-15 and positron emission tomography, *J. Cereb. Blood Flow Metab.,* 1, S567, 1981.

79. **Ter-Pogossian, M. M., Eichling, J. O., Davis, D. O. et al.,** The measurement *in vivo* of regional cerebral oxygen utilization by means of oxyhemoglobin labeled with radioactive oxygen-15, *J. Clin. Invest.,* 49, 385, 1970.

80. **Frackowiak, R. S. J., Jones, T., Lenzi, G. L., and Heather, J. D.,** Regional cerebral oxygen utilisation and blood flow in normal man using oxygen-15 and positron emission tomography, *Acta Neurol. Scand.,* 62, 336, 1980.

81. **Frackowiak, R. S. J., Lenzi, G. L., Jones, T., and Heather, J. D.,** Quantitative measurement of regional cerebral blood flow and oxygen metabolism in man using ^{15}O and positron emission tomography: theory, procedure, and normal values, *J. Comput. Assist. Tomogr.,* 4, 727, 1980.

82. **Jones, T., Frackowiak, R. S. J., Lammertsma, A. A., and Rhodes, C. G.,** Compartmental analysis of the steady-state distribution of 15-O_2 in the body, *J. Nucl. Med.,* 23, 750, 1982.

83. **Biegler, R. E., Kostick, F. A., and Gillespie, J. R.,** Compartment analysis of the steady state distribution of $^{15}O_2$ and $^{15}O-H_2O$ in total body, *J. Nucl. Med.,* 22, 959, 1981.

84. **Biegler, R. E., Gillespie, J. R., and Kostick, F. A.,** Re: compartment analysis of the steady state distribution of $^{15}O_2$ and $H_2^{15}O$ total body, *J. Nucl. Med.,* 23, 750, 1982.

85. **Lammertsma, A. A., Wise, R. J. S., Heather, J. D., Gibbs, J. M., Leenders, K. L., Frackowiak, R. S. J., Rhodes, C. G., and Jones, T.,** Correction for the presence of intravascular oxygen-15 in the steady-state technique for measuring regional oxygen extraction ratio in the brain. II. Results in normal subjects and brain tumor and stroke patients, *J. Cereb. Blood Flow Metab.,* 3, 425, 1983.

86. **Kety, S. S. and Schmidt, C. F.,** The nitrous oxide method for the quantitative determination of cerebral blood flow in man: theory, procedure and normal values, *J. Clin. Invest.,* 27, 476, 1948.

87. **Mintun, M. A., Raichle, M. E.,, Martin, W. R. W., and Herscovitch, P.,** Brain oxygen utilization measured with O-15 radiotracers and positron emission tomography, *J. Nucl. Med.,* 25, 177, 1984.

88. **Raichle, M. E.**, Measurement of local brain blood flow and oxygen utilization using ^{15}O-radiopharmaceuticals: a rapid dynamic imaging approach, in *Positron Emission Tomography*, Reivich, M. and Alavi, A., Eds., Alan R. Liss, New York, 1985, 241.

89. **Ogawa, T., Uemura, K., Shishido, F., Yamaguchi, T., Murakami, M., Inugami, A., Kanno, I., Sasaki, H., Kato, T., Hirata, K., Kowada, M., Mineura, K., and Yasuda, T.**, Changes of cerebral blood flow and oxygen and glucose metabolism following radiochemotherapy of gliomas: a PET study, *J. Comput. Assist. Tomogr.*, 12, 290, 1988.

90. **Sokoloff, L., Reivich, M., Kennedy, C. et al.**, The (C-14)-deoxyglucose method for the measurement of local cerebral glucose utilization: theory, procedure, and normal values in the conscious and anesthetized albino rat, *J. Neurochem.*, 28, 897, 1977.

91. **Sokoloff, L.**, Localization of functional activity in the central nervous system by measurement of glucose utilization with radioactive deoxyglucose, *J. Cereb. Blood Flow Metab.*, 1, 7, 1981.

92. **Ido, T., Wan, C. N., Casella, V., Fowler, J. S., Wolf, A. P., Reivich, M., and Kuhl, D. E.**, Labeled 2-deoxy-D-glucose analogs. ^{18}F-labeled 2-deoxy-2-fluoro-D-glucose, 2-deoxy-2-fluoro-D-mannose, and ^{14}C-2-deoxy-2-fluoro-D-glucose, *J. Labelled Comp. Radiopharm.*, 14, 175, 1978.

93. **Reivich, M., Kuhl, D., Wolf, A. P., Greenberg, J., Phelps, M., Ido, T., Casella, V., Fowler, J., Gallagher, B., Hoffman, E., Alavi, A., and Sokoloff, L.**, The [^{18}F]-Fluorodeoxyglucose method for the measurement of local cerebral glucose utilization in man, *Circ. Res.*, 44, 127, 1979.

94. **Phelps, M. E., Huang, S. C., Hoffman, E. J., Selin, C., Sokoloff, L., and Kuhl, D. E.**, Tomographic measurement of local cerebral glucose metabolic rate in humans with (F-18)2-fluoro-2-deoxy-D-glucose: validation of method, *Ann. Neurol.*, 6, 371, 1979.

95. **Reivich, M.**, Cerebral glucose consumption: methodology and validation, in *Positron Emission Tomography*, Reivich, M. and Alavi, A., Eds., Alan R. Liss, New York, 1985, 131.

96. **Kato, A., Sako, K., Diksic, M., Yamamoto, Y. L., and Feindel, W.**, Regional glucose utilization and blood flow in experimental brain tumors studied by double tracer autoradiography, *J. Neuro. Oncol.*, 3, 271, 1985.

97. **Kato, A., Diksic, M., Yamamoto, Y. L., and Feindel, W.**, Quantification of glucose utilization in an experimental brain tumor model by the deoxyglucose method, *J. Cereb. Blood Flow Metab.*, 5, 108, 1985.

98. **Blasberg, R. G., Groothuis, D., and Molner, P.**, Application of quantitative autoradiographic measurements in experimental brain tumor models, *Semin. Neurol.*, 1, 203, 1981.

99. **Blasberg, R. G., Shinohara, N., Shapiro, W. R., Patlak, C. S., Pettigrew, K. D., and Fenstermacher, J. D.**, Apparent glucose utilization in Walker 256 metastatic brain tumors, *J. Neuro. Oncol.*, 3, 153, 1985.

100. **Hossmann, K. A., Niebuhr, I., and Tamura, M.**, Local cerebral blood flow and glucose comsumption of rats with experimental gliomas, *J. Cereb. Blood Flow Metab.*, 2, 25, 1982.

101. **Graham, M. M., Spence, A. M., Muzi, M., and Abbott, G. L.**, Deoxyglucose kinetics in a rat brain tumor, *J. Cereb. Blood Flow Metab.*, 9, 315, 1989.

102. **Kirikae, M., Diksic, M., and Yamamoto, Y. L.**, Quantitative measurements of regional glucose utilization and rate of valine incorporation into proteins by double tracer autoradiography in the rat brain tumor model, *J. Cereb. Blood Flow Metab.*, 9, 87, 1989.

103. **DiChiro, G., DeLaPaz, R. L., Smith, B. H., Patronas, N. J., Kufta, C. V., Kessler, R. M., Johnston, G. S., Manning, R. G., and Wolf, A. P.**, ^{18}F-2-fluoro-2-deoxy-glucose positron tomography of human cerebral gliomas, *J. Cereb. Blood Flow Metab.*, 1, S11, 1981.

104. **Di Chiro, G., DeLaPaz, R. L., Brooks, R. A., Sokoloff, L., Kornblith, P. L., Smith, B. H., Patronas, N. J., Kufta, C. V., Kessler, R. M., Johnston, G. S., Manning, R. G., and Wolf, A. P.**, Glucose utilization of cerebral gliomas measured by 18-F-fluorodeoxyglucose and positron emission tomography, *Neurology*, 32, 1323, 1982.

105. **DiChiro, B., Brooks, R. A., Sokoloff, L., Patronas, N. J., DeLaPaz, R. L., Smith, B. H., and Kornblith, P. L.**, Glycolytic rate and histologic grade of human cerebral gliomas: a study with ^{18}F-fluorodeoxyglucose and positron emission tomography, in *Positron Emission Tomography of the Brain*, Heiss, W. P. and Phelps, M. E., Eds., Springer Verlag, Heidelberg, 1983, 181.

106. **Di Chiro, G., Oldfield, E., Bairamian, D., Patronas, N. J., Brooks, R. A., Mansi, L., Smith, B. H., Kornblith, P. L., and Margolin, R.**, Metabolic imaging of the brain stem and spinal cord: studies with positron emission tomography using ^{18}F-2-deoxyglucose in normal and pathological cases, *J. Comput. Assist. Tomogr.*, 7, 937, 1983.

107. **Di Chiro, G., Brooks, R. A., Patronas, N. J., Bairamian, D., Kornblith, P. L., Smith, B. H., Mansi, L., and Barker, J.**, Issues in the *in vivo measurement of glucose metabolism of human central nervous system tumors, Ann. Neurol.*, 15, S138, 1984.

108. **Di Chiro, G., Oldfield, E., Bairamian, D., Patronas, N. J., Kornblith, P. L., Smith, B. H., Brooks, R. A., Margolin, R. A., and Mansi, L.**, *In vivo* glucose utilization of CNS tumors of the brain stem and spinal cord in *The Metabolism of the Human Brain Studied with Positron Emission Tomography*, Greitz, T., Ingvar, D. M., and Widen, L., Eds., Raven Press, New York, 1985, 351.

109. **Di Chiro, G., Brooks, R. A., Bairamian, D., Patronas, N. J., Kornblith, P. L., Smith, B. H., and Mansi, L.,** Diagnostic and prognostic value of positron emission tomography using (F-18)-fluorodeoxy-glucose in brain tumors, in *Positron Emission Tomography,* Reivich, M. and Alavi, A., Eds., Alan R. Liss, New York, 1985, 291.

110. **Di Chiro, G., Oldfield, E., Wright, D. C., De Michele, D., Katz, D. A., Patronas, N. J., Doppman, J. L., Larson, S. L., Ito, M., and Kufta, C. V.,** Cerebral necrosis after radiotherapy and/or intraarterial chemotherapy for brain tumors, *Am. J. Roentgenol.,* 150, 198, 1988.

111. **Patronas, N. J., Di Chiro, G., Brooks, R. A., Kornblith, P. L., Smith, B. H., Rizzoli, H. V., Kessler, R. M., Manning, R. G., Channing, M., Wolf, A. P., and O'Connor, C.,** ^{18}F-fluorodeoxyglucose and positron emission tomography in the evaluation of radiation necrosis in the brain, *Radiology,* 144, 885, 1982.

112. **Patronas, N. J., Brooks, R. A., DeLaPaz, R. L., Smith, B. H., Kornblith, P. L., and DiChiro, G.,** Glycolytic rate (PET) and contrast enhancement (CT) in human cerebral gliomas, *Am. J. Neuroradiol.,* 4, 533, 1983.

113. **Patronas, N. J., DiChiro, G., Smith, B. H., DeLaPaz, R. L., Brooks, R. A., Milam, H., and Kornblith, P. L.,** Depressed cerebellar glucose metabolism in supratentorial tumors, *Brain Res.,* 291, 93, 1984.

114. **Patronas, N. J., DiChiro, G., Kafta, C., Bairamian, D., Kornblith, P. L., Simon, R., and Larson, S. M.,** Prediction of survival in glioma patients by means of positron emission tomography, *J. Neurosurg.,* 62, 816, 1985.

115. **DeLaPaz, R. L., Patronas, N. J., Brooks, R. A., Smith, B. H., Kornblith, P. L., Milam, H., and Di Chiro, G.,** Positron emission tomography study of suppression of gray matter glucose utilization by brain tumors, *Am. J. Nucl. Resonance,* 4, 826, 1983.

116. **Brooks, R. A., Shatzman, B., Tran, A., Di Chiro, G., and Weiss, G. H.,** Dynamic rate constants for F-18 deoxyglucose uptake in normal and tumoral brain tissue, *J. Nucl. Med.,* 25, P8, 1984.

117. **Rhodes, C. G., Wise, R. J. S., Gibbs, J. M., Frackowiak, R. S. J., Hatazawa, J., Palmer, A. J., Thomas, D. G. T., and Jones, T.,** In vivo disturbance of the oxydative metabolism of glucose in human cerebral gliomas, *Ann. Neurol.,* 14, 614, 1983.

118. **Brooks, D. J., Beaney, R. P., and Thomas, D. G.,** The role of positron emission tomography in the study of cerebral tumors, *Semin. Oncol.,* 13, 83, 1986.

119. **Ericson, K., Lilja, A., Bergström, M., Collins, V. P., Eriksson, L., Ehrin, E., von Holst, H., Lundqvist, H., Langström, B., and Mosskin, M.,** Positron emission tomography with 11-C-methyl-L-methionine, 11-C-D-glucose, and 68-Ga-EDTA in supratentorial tumors, *J. Comput. Assist. Tomogr.,* 9, 683, 1985.

120. **Tyler, J. L., Diksic, M., Villemure, J. G., Evans, A. C., Meyer, E., Yamamoto, Y. L., and Feindel, W.,** Metabolic and hemodynamic evaluation of gliomas using positron emission tomography, *J. Nucl. Med.,* 28, 1123, 1987.

121. **Mineura, K., Yasuda, T., Kowada, M., Shishido, F., Ogawa, T., and Uemura, K.,** Positron emission tomographic evaluation of histological malignancy in gliomas using oxygen-15 and fluorine-18-fluoro-deoxyglucose, *Neurol. Res.,* 8, 164, 1986.

122. **Herholz, K., Ziffling, P., Staffen, W., Pawlik, G., Wagner, R., Wienhard, K., and Heiss, W. D.,** Uncoupling of hexose transport and phosphorylation in human gliomas demonstrated by PET, *Eur. J. Cancer Clin. Oncol.,* 24, 1139, 1988.

123. **Meyer, G. J., Schober, O., Gaab, M. R., Dietz, H., and Hundeshagen, H.,** Multiparameter studies in brain tumors, in *Positron Emission Tomography in Clinical Research and Clinical Diagnosis,* Beckers, C., Goffinet, A., and Bol, A., Eds., Kluwer Academy, Dordrecht, 1989, 229.

124. **Doyle, W. K., Budinger, T. F., Valk, P. E., Levin, V. A., and Gutin, P. H.,** Differentiation of radiation necrosis from tumor recurrence by [^{18}F]-FDG and [82]Rb positron emission tomography, *J. Computer Assist. Tomogr.,* 11, 563, 1987.

125. **Warburg, O. H.,** *Über den Stoffwechsel der Tumoren,* Springer Verlag, Berlin, 1926.

126. **Weber, G.,** Enzymology of cancer cells. I, *N. Engl. J. Med.,* 296, 486, 1977.

127. **Weber, G.,** Enzymology of cancer cells. II, *N. Engl. J. Med.,* 296, 541, 1977.

128. **Kuwabara, Y., Ichiya, Y., Otsuka, M., Miyake, Y., Gunareskera, R., Hasuo, K., Masuda, K., Takeshita, I., and Fukui, H.,** High [^{18}F]FDG uptake in primary cerebral lymphoma: a PET study, *J. Comput. Assist. Tomogr.,* 12, 47, 1988.

129. **Minn, H., Paul, R., and Ahonen, A.,** Evaulation of treatment response to radiotherapy in head and neck cancer with fluorine-18 fluorodeoxyglucose, *J. Nucl. Med.,* 29, 1521, 1988.

130. **Di Chiro, G. and Brooks, R. A.,** PET-FDG of untreated and treated cerebral gliomas, *J. Nucl. Med.,* 29, 421, 1988.

131. **Di Chiro, G. and Brooks, R. A.,** PET quantitation: blessing and curse, *J. Nucl. Med.,* 29, 1603, 1988.

132. **Koeppe, R. A., Junck, L., Chen, Y., Betley, A. T., Hutchins, G. D., Rothley, J. M., and Hichwa, R. D.,** Differentiation between glucose transport and phosphorylation process in brain tumors by dynamic PET and FDG, *J. Cereb. Blood Flow Metab.,* 7, S482, 1987.

133. **Brooks, D. J., Beaney, R. P., Lammertsma, A. A., Herold, S., Turton, D. R., Luthra, S. K., Frackowiak, R. S. J., Thomas, D. G. T., Marshall, J., and Jones, T.,** Glucose transport across the blood-brain barrier in normal human subjects and patients with cerebral tumors studied using [^{11}C]-3-O-methyl-D-glucose and positron emission tomography, *J. Cereb. Blood Flow Metab.,* 6, 230, 1986.

134. **Dunlop, D. S.,** Measuring protein synthesis and degradation rates in CNS tissue, in *Research Methods in Neurochemistry,* Vol. 4, Marks, N. and Rodnight, R., Eds., Plenum Press, New York, 1978, 91.

135. **Pardridge, W. M. and Oldendorf, W. H.,** Kinetic analysis of blood brain barrier transport of amino acids, *Biochim. Biophys. Acta,* 401, 128, 1975.

136. **Steinwall, O.,** Transport inhibition phenomena in unilateral chemical injury of blood brain barrier, in *Brain Barrier Systems,* Lajtha, A. and Ford, D., Eds., Elsevier, Amsterdam, 1968, 357.

137. **Oldendorf, W. H. and Szabo, J.,** Brain uptake of radiolabeled amino acids, amines and hexoses after arterial injection, *Am. J. Physiol.,* 221, 1629, 1971.

138. **Oldendorf, W. H. and Szabo, J.,** Amino acid assignment to one of three blood brain barrier amino acid carriers, *Am. J. Physiol.,* 230, 94, 1976.

139. **Vahvelainen, M.-L. and Oja, S. S.,** Kinetic analysis of phenylalanine-induced inhibition in the saturable influx of tyrosine, tryptophan, leucine, and histidine into brain cortex slices from adult and 7-day-old rats, *J. Neurochem.,* 24, 885, 1975.

140. **Pardridge, W. M.,** Kinetics of competitive inhibition of neutral amino acid transport across the blood-brain barrier, *J. Neurochem.,* 28, 103, 1977.

141. **Lajtha, A. and Toth, J.,** The brain barrier system-V: stereospecificity of amino acid uptake, exchange and efflux, *J. Neurochem.,* 10, 909, 1963.

142. **Oldendorf, W. H.,** Stereo-specificity of blood-brain barrier permeability to amino acids, *Am. J. Physiol.,* 224, 967, 1972.

143. **Smith, C. B., Davidsen, L., Deibler, G., Patlak, C., Pettigrew, K., and Sokoloff, L.,** A method for the determination of local protein synthesis in the brain, *Trans. Am. Soc. Neurochem.,* 11, 1980.

144. **Smith, C. B., Deibler, G., Eng, N., Schmidt, K., and Sokoloff, L.,** Measurement of local cerebral protein synthesis: influence of recycling of amino acids in the precursor pool derived from protein degradation, *Proc. Natl. Acad. Sci. U.S.A.,* 85, 9341, 1988; *J. Cereb. Blood Flow Metab.,* 9, S201, 1989.

145. **Smith, Q. R., Takasato, Y., Sweeney, D. J., and Rapoport, S. I.,** Regional cerebrovascular transport of leucine as measured by the *in situ* brain perfusion technique, *J. Cereb. Blood Flow Metab.,* 5, 300, 1985.

146. **Smith, Q. R. and Takasato, Y.,** Kinetics of amino acid transport at the blood-brain barrier studied using an in situ brain perfusion technique, *Ann. N.Y. Acad. Sci.,* 481, 186, 1986.

147. **Smith, Q. R., Momma, S., Aoyagi, M., and Rapoport, S. I.,** Kinetics of neutral amino acid transport across the blood-brain barrier, *J. Neurochem.,* 49, 1651, 1987.

148. **Smith, Q. R., Takasato, Y., and Rapoport, S. I.,** Kinetic analysis of L-leucine transport across the blood brain barrier, *Brain Res.,* 311, 167, 1984.

149. **Phelps, M. E., Barrio, J. R., Huang, S.-C., Keen, R. E., Chugani, H., and Mazziotta, J. C.,** Criteria for the tracer kinetic measurement of cerebral protein synthesis in humans with positron emission tomography, *Ann. Neurol.,* Suppl. 15, S192, 1984.

150. **Keen, R. E., Barrio, J. R., Huang, S. C., Hawkins, R. A., and Phelps, M. E.,** *In vivo* cerebral protein synthesis rates with leucyl-transfer RNA used as precursor pool: determination of biochemical parameters to structure tracer kinetic models for positron emission tomography, *J. Cereb. Blood Flow Metab.,* 9, 429, 1989.

151. **Hawkins, R. A., Huang, S.-C., Barrio, J. R., Keen, R. E., Feng, D., Mazziotta, J. C., and Phelps, M. E.,** Estimation of local cerebral protein synthesis rates with L-[1-^{11}C]leucine and PET: methods, model, and results in animals and humans, *J. Cereb. Blood Flow Metab.,* 9, 446, 1989.

152. **Phelps, M. E., Barrio, J. R., Huang, S. C., Keen, R. E., Chugani, H., and Maziotta, J. C.,** Measurement of cerebral protein synthesis in man with positron computerized tomography: model, assumption, and preliminary results, in *The Metabolism of the Human Brain Studied with Positron Emission Tomography,* Greitz, T., Ed., Raven Press, New York, 1985, 215.

153. **Bustany, P., Sargent, T., Saudubray, J. H., Henry, J. F., and Comar, D.,** Regional human brain uptake and protein incorporation of ^{11}C-L-methionine studied *in vivo* with PET, *J. Cereb. Blood Flow Metab.,* 1, 17, 1981.

154. **Bustany, P., Chatel, M., Derlon, J. M., Darcel, F., Sgouropoulos, P., Soussaline, F., and Syrota, A.,** Brain tumor protein synthesis and histological grades: a study by positron emission tomography (PET) with ^{11}C-L-methionine, *J. Neurooncol.,* 3, 397, 1986.

155. **Lestage, P., Gonon, M., Lepetit, P., Vitte, P. A., Debilly, G., Rosatto, C., Lecestre, D., and Bobillier, P.,** An *in vivo* kinetic model with L-[35S]-methionine for the determination of local cerebral rates for methionine incorporation into protein in the rat, *J. Neurochem.,* 48, 352, 1987.

156. **Mosskin, M.,** Diagnostic Imaging in Glioma, thesis, Karolinska Hospital, Stockholm, Caslon Press, Stockholm, 1987.

157. **Mosskin, M., Ericson, K., Hindmarsh, T., von Holst, H., Collins, V. P., Bergström, M., Eriksson, L., and Johnström, P.,** Positron emission tomography compared with MRI and CT in supratentorial gliomas using multiple stereotactic biopsies as reference, *Acta Radiol.,* 30, 225, 1989.

158. **Ishiwata, K., Ido, T., Abe, Y., Matsuzawa, T., and Iwata, R.,** Tumor uptake studies of S-adenosyl-L-[methyl-¹¹C]-methionine and L-[methyl-¹¹C]-methionine, *Nucl. Med. Biol.,* 15, 123, 1988.

159. **Hatazawa, J., Ishiwata, K., Itoh, M., Kameyama, M., Kubota, K., Ido, T., Matsuzawa, T., Yoshimoto, T., Watanuki, S., and Seo, S.,** Quantitative evaluation of [methyl-¹¹C]-methionine uptake in tumor using positron emission tomography, *J. Nucl. Med.,* 30, 1809, 1989.

160. **Kameyama, M., Shirane, R., Itoh, J., Sato, K., Katakura, R., Yoshimoto, T., Hatazawa, J., Itoh, M., Seo, S., and Ido, T.,** The Accumulation of ¹¹C-Methionine and Histological Grade in Cerebral Glioma Studied with PET, CYRIC Annual Report 1988, Cyclotron and Radioisotope Center, Tohoku University, Sendai, 1989, 228.

161. **O'Tuama, L. A., Guilarte, T. R., Douglass, K. H., Wagner, H. N., Jr., Wong, D. F., Dannals, R. F., Ravert, H. T., Wilson, A. A., La France, N. D., Bice, A. N., and Links, J. M.,** Assessment of [¹¹C]-Methionine transport into the human brain, *J. Cereb. Blood Flow Metab.,* 9, 341, 1988.

162. **Ishiwata, K., Vaalburg, W., Elsinga, P. H., Paans, A. M. J., and Woldring, M. G.,** Comparison of L-[1-¹¹C]-methionine and L-methyl-[¹¹C]-Methionine for measuring *in vivo* protein synthesis rates with PET, *J. Nucl. Med.,* 29, 1419, 1988.

163. **Kirikae, M., Diksic, M., and Yamamoto, Y. L.,** Transfer coefficients for L-valine and the rate of incorporation of L-[1-¹⁴C]-valine into proteins in normal adult rat brain, *J. Cereb. Blood Flow Metab.,* 8, 598, 1988.

164. **Vahvelainen, M.-L. and Oja, S. S.,** Kinetics of influx of phenylalanine, tyrosine, tryptophan, histidine and leucine into slices of brain cortex from adult and 7-day-old rats, *Brain Res.,* 40, 477, 1972.

165. **Pollay, M.,** Regional transport of phenylalanine across the blood-brain barrier, *J. Neurosci. Res.,* 2, 11, 1976.

166. **Casey, D. L., Digenis, G. A., Wesner, D. A., Washburn, L. C., Chaney, J. E., Hayes, R. L., and Callahan, A. P.,** Preparation and preliminary tissue studies of optically active D- and L-[¹¹C]-phenylalanine, *Int. J. Appl. Radiat. Isot.,* 32, 325, 1981.

167. **Momma, S., Aoyagi, M., Rapoport, S. I., and Smith, Q. R.,** Phenylalanine transport across the blood-brain barrier as studied with the in situ brain perfusion technique, *J. Neurochem.,* 48, 1291, 1987.

168. **Choi, B. and Pardridge, W. M.,** Phenylalanine transport at the human blood-brain barrier, *J. Biol. Chem.,* 261, 6536, 1986.

169. **Halldin, C. and Langstöm, B.,** Synthesis of [3-¹¹C]phenylpyruvic acid and its use in a transamination to [3-¹¹C]phenylalanin, *J. Labelled Comp. Radiopharm.,* 23, 715, 1986.

170. **Bolster, J. M., Vaalburg, W., Paans, A. M. J., vanDijk, T. H., Elsinga, P. H., Zijlstra, J. B., Piers, D. A., Mulder, N. H., Woldring, M. G., and Wynberg, H.,** Carbon-11 labelled tyrosine to study tumor metabolism by positron emission tomography, *Eur. J. Nucl. Med.,* 12, 321, 1986.

171. **Johnström, P., Stone-Elander, S., Ericson, K., Mosskin, M., and Bergström, M.,** ¹¹C-labelled glycine: synthesis and preliminary report on its use in the investigation of intracranial tumors using positron emission tomography, *Appl. Radiat. Isot.,* 38, 729, 1987.

172. **Tsukiyama, T., Hara, T., Iio, M., Kido, G., and Tsubokawa, T.,** Preferential accumulation of ¹¹C in human brain tumors after intravenous injection of ¹¹C-1-pyruvate, *Eur. J. Nucl. Med.,* 12, 244, 1986.

173. **Takeda, A., Goto, R., Tamemasa, O., Chaney, J., and Digenis, G.,** Biological evaluation of radio-labelled D-methionine as a parent compound in potential nuclear imaging, *Radioisotopes,* 33, 213, 1984.

174. **Tamemasa, O., Goto, R., and Suzuki, T.,** Preferential incorporation of some ¹⁴C-labelled D-amino acids into tumor-bearing animals, *Gann,* 69, 517, 1978.

175. **Tamemasa, O., Goto, R., Takeda, A., and Maruo, U.,** High uptake of ¹⁴C-labelled D-amino acids by various brain tumors, *Gann,* 73, 147, 1982.

176. **Lauenstein, L., Meyer, G.-J., Sewing, K.-F., Schober, O., and Hundeshagen, H.,** Uptake kinetics of ¹⁴C L-leucine and ¹⁴C L- and ¹⁴C D-methionine in rat brain and incorporation into protein, *Neurosurg. Rev.,* 10, 147, 1987.

177. **Lauenstein, L.,** Aufnahme und Proteininkorporation von L-leucin und L- und D-Methionine ins Gehirn der Ratte und in Hirntumoren, thesis, Medizinische Hochschule, Hanover, 1988.

178. **Schober, O., Dunden, C., Meyer, G.-J., Müller, J. A., and Hundeshagen, H.,** Nonselective transport of [¹¹C-methyl]-L-, and -D-methionine into a malignant glioma, *Eur. J. Nucl. Med.,* 13, 103, 1987.

179. **Meyer, G.-J., Schober, O., and Hundeshagen, H.,** Uptake of ¹¹C-D- and L-methionine in brain tumors, *Eur. J. Nucl. Med.,* 10, 373, 1985.

180. **Bergström, M., Lundqvist, A., Ericson, K., Lilja, A., Johnström, P., Langström, B., von Holst, H., Eriksson, L., and Blomqvist, G.,** Comparison of the accumulation kinetics of L-[methyl-¹¹C]-methionine and D-[methyl-¹¹C]-methionine in brain tumors studied with positron emission tomography, *Acta Radiol.,* 28, 225, 1987.

181. **Bodsch, W., Coenen, H. H., Stöcklin, G., Takahashi, K., and Hossmann, K.-A.,** Biochemical and autoradiographic study of cerebral protein synthesis with [^{11}F]-and [^{14}C]fluorophenylalanine, *J. Neurochem.,* 50, 979, 1988.

182. **Mineura, K., Kowada, M., and Shishido, F.,** Brain tumor imaging with synthesized ^{18}F-fluorophenylalanine and positron emission tomography, *Surg. Neurol.,* 31, 468, 1989.

183. **Murakami, M., Takahashi, K., Kondo, Y. et al.,** 2-^{18}F-phenylalanine and 3^{18}F-tyrosine: synthesis and preliminary data of tracer kinetics, *J. Labelled Comp. Radiopharm.,* 25, 773, 1988.

184. **Murakami, M., Takahashi, K., Kondo, Y., Mizusawa, S., Nakamichi, H., Sasaki, H., Hagami, E., Iida, H., Kanno, I., Miura, S., Itoh, I., and Uemura, K.,** The slow metabolism of L-[2-^{18}F]-fluorophenylalanine, *J. Labelled Comp. Radiopharm.,* 27, 245, 1989.

185. **Coenen, H. H., Kling, P., and Stöcklin, G.,** Cerebral metabolism of L-[2-^{18}F]fluorotyrosine, a new tracer of protein synthesis, *J. Nucl. Med.,* 30, 1367, 1989.

186. **Herhotz, K.,** private communication.

187. **Orth, F.,** Untersuchungen zur Aufnahme und zum Protein-Einbau von Phenylalanin und p-Jod-phenylalanin in das Gehirn der Ratte und in Hirntumoren der Ratte, Dissertation, Medizinische Hochschule, Hannover, 1990.

188. **Meyer, G.-J., Orth, F., Coenen, H. H., Stöcklin, G., and Hundeshagen, H.,** Uptake and protein incorporation of some iodinated amino acids in brain tumors of rats, *Eur. J. Nucl. Med.,* 8, 427, 1989.

189. **Biersack, H. J., Coenen, H. H., Stöcklin, G., Kashab, M., Reichmann, K., and Bockisch, A.,** SPECT of brain tumors with L-3-[123I]iodo-alpha-methyl-tyrosine (IMT), *J. Nucl. Med.,* 29, 911, 1988.

190. **Biersack, H. J., Coenen, H. H., Stöcklin, G., Reichmann, K., Bockisch, A., Oehr, P., Kashab, M., and Rollmann, O.,** Imaging of brain tumors with L-3-[123I]iodo-alpha-methyl-tyrosine and SPECT, *J. Nucl. Med.,* 30, 110, 1989.

191. **Kawai, K., Fujibayashi, Y., Saji, H., Konishi, J., and Yokoyama, A.,** New radioiodinated radiopharmaceutical for cerebral amino acid transport studies: 3-iodo-alpha methyl-L-tyrosine, *J. Nucl. Med.,* 29, 778, 1988.

192. **Bustany, P., Henry, J. F., Sargent, T., Zarifian, E., Cabanis, E., Collard, P., and Comar, D.,** Local brain protein metabolism in dementia and schizophrenia: *in vivo* studies with ^{11}C-L-methionine and positron emission tomography, in *Positron Emission Tomography of the Brain,* Heiss, W. D. and Phelps, M. E., Eds., Springer-Verlag, Berlin, 1983, 208.

193. **Bustany, P., Henry, J. F., and deRotrou, J.,** Local cerebral metabolic rate of ^{11}C-L-methionine in early stages of dementia, schizophrenia and Parkinson's disease, *J. Cereb. Blood Flow Metab.,* 3, 492, 1983.

194. **Bustany, P. and Comar, D.,** Protein synthesis evaluation in brain and other organs in humans by PET, in *Positron Emission Tomography,* Reivich, M. and Alavi, A., Eds., Alan R. Liss, New York, 1985, 183.

195. **Dienel, G. A., Pulsinelli, W. A., and Duffy, T. E.,** Regional protein synthesis in rat brain following acute hemispheric ischemia, *J. Neurochem.,* 35, 1216, 1980.

196. **Dweyer, B. E., Donatoni, P., and Wasterlain, C. G.,** A quantitative autoradiographic method for the measurement of local rates of brain protein synthesis, *Neurochem. Res.,* 7, 563, 1982.

197. **Mies, G., Bodsch, W., Paschen, W., and Hossmann, K. A.,** Experimental application of triple labeled quantitative autoradiography for measurement of cerebral blood flow, glucose metabolism and protein biosynthesis, in *Positron Emission Tomography of the Brain,* Heiss, W. D. and Phelps, M. E., Eds., Springer-Verlag, Berlin, 1983, 19.

198. **Anders, B.,** Anwendung von kinetischen Modellen zur quantitativen Berechnung der Proteinsynthese im Gehirn und Optimierung einer Meßanordnung zur Aufnahme von Plasma-Aktivatätskurven, Diplomarbeit, Medizinische Hochschule, Hannover, and Technische Hochschule, Hannover, 1988.

199. **Ishiwata, K., Vaalburg, W., Elsinga, P. H., Paans, A. M. J., and Woldring, M. G.,** Metabolic studies with L-[1-^{14}C]-tyrosine for the investigation of a kinetic model to measure protein synthesis rates with PET, *J. Nucl. Med.,* 29, 524, 1988.

200. **Diamondstone, T. I.,** Amino acid metabolism. II. Metabolism of the individual amino acids, in *Textbook of Biochemistry with Clinical Correlations,* Derlin, T. M., Ed., Plenum Press, New York, 1982, 563.

201. **Young, S. N.,** The significance of tryptophan, phenylalanine, tyrosine, and their metabolites in the nervous system, in *Handbook of Neurochemistry,* Vol. 3, Lajtha, A., Ed., Plenum Press, New York, 1982, 559.

202. **Ishiwata, K., Hatazawa, J., Kubota, K., Kameyama, M., Itoh, M., Matsuzawa, T., Takahashi, T., Iwata, R., and Ido, T.,** Metabolic fate of L-[methyl-^{11}C]methionine in human plasma, *Eur. J. Nucl. Med.,* 15, 665, 1989.

203. **Hatazawa, J., Ishiwata, K., Itoh, M., Kameyama, M., Kubota, K., Ido, T., Matsuzawa, T., Yoshimoto, T., Watanuki, S., and Seo, S.,** Quantitative evaluation of L-[methyl-C-11]methionine uptake in tumor using positron emission tomography, *J. Nucl. Med.,* 30, 1809, 1989.

204. **Ericson, K., Blomqvist, G., Bergström, M., Eriksson, L., and Stone-Elander, S.,** Application of a kinetic model on the methionine accumulation in intracranial tumors studied with positron emission tomography, *Acta Radiol.,* 28, 505, 1987.

205. **Patlak, C. S., Blasberg, R. G., and Fenstermacher, J. D.,** Graphical evaluation of blood to brain transfer constants from multiple time uptake data, *J. Cereb. Blood Flow Metab.,* 3, 1, 1983.

206. **Gjedde, A.,** A high and low affinity transport of D-glucose from blood to brain, *J. Neurochem.,* 36, 1463, 1981.

207. **Lassen, N. A. and Gjedde, A.,** Kinetic analysis of the uptake of glucose and some of its analogs using the single capillary model. Comments on some points of controversy, in *Tracer Kinetics and Physiologic Modelling,* Lambrecht, R. M. and Rescigno, A., Eds., Springer-Verlag, Berlin, 1983, 348.

208. **Meyer, G.-J., Harre, R., Orth, F., Gaab, M. R., Dietz, H., and Hundeshagen, H.,** *In vivo* protein synthesis in human brain tumors measured with ¹¹C-L-methionine, *Eur. J. Nucl. Med.,* 15, 506, 1989.

209. **Bergström, M., Ericson, K., Hagenfeldt, L., Mosskin, M., von Holst, H., Noren, G., Eriksson, L., Ehrin, E., and Johnström, P.,** PET study of methionine accumulation in glioma and normal brain tissue: competition with branched amino acids, *J. Comput. Assist. Tomogr.,* 11, 208, 1987.

210. **O'Tuama, L. A., La France, N. D., Dannals, R. F., Douglass, K. H., Links, J. M., Bice, A. N., Williams, J. A., Villemagne, V., and Wagner, H. N., Jr.,** Quantitative imaging of neutral amino acid transport by human brain tumors, *J. Cereb. Blood Flow Metab.,* 7, S517, 1987.

211. **Lilja, A., Bergström, K., Hartvig, P., Spännare, B., Halldin, C., Lundquist, H., and Langström, B.,** Dynamic study of supratentorial gliomas with L-methyl-11-C-methionine and positron emission tomography, *Am. J. Neurol. Radiol.,* 6, 505, 1985.

212. **Lilja, A., Lundqvist, H., Olsson, Y., Spännare, B., Gullberg, P., and Langström, B.,** Positron emission tomography and computed tomography in differential diagnosis between recurrent or residual glioma and treatment induced brain lesions, *Acta Radiol.,* 30, 121, 1989.

213. **Mosskin, M., von Holst, H., Bergström, M., Collins, V. P., Eriksson, L., Johnström, P., and Noren, G.,** Positron emission tomography with ¹¹C-methionine and X-ray computed tomography of intracranial tumors compared with histopathologic examination of multiple biopsies, *Acta Radiol.,* 28, 673, 1987.

214. **Mosskin, M., Ericson, K., Hindmarsh, T., von Holst, H., Collins, V. P., Bergström, M., Eriksson, L., and Johnström, P.,** Positron emission tomography compared with magnetic resonance imaging and computed tomography in supratentorial gliomas using multiple stereotactic biopsies as reference, *Acta Radiol.,* 30, 225, 1989.

215. **Schober, O., Creutzig, H., Meyer, G.-J., Becker, H., Schwarzrock, R., Dietz, H., and Hundeshagen, H.,** ¹¹C-methionin-PET, IMP-SPECT, CT und MRI bei Hirntumoren, *Fortschr. Rontgenstr.,* 143, 133, 1985.

216. **Schober, O., Meyer, G.-J., Stolke, D., and Hundeshagen, H.,** Brain tumor imaging using ¹¹C-labelled L-methionine and D-methionine, *J. Nucl. Med.,* 26, 98, 1985.

217. **Schober, O., Meyer, G.-J., Gaab, M. R., Müller, J. A., and Hundeshagen, H.,** Grading of brain tumors by L-[¹¹C]methionine PET, *J. Nucl. Med.,* 27, 890, 1986.

218. **Schober, O., Meyer, G.-J., Duden, C., Lauenstein, L., Niggemann, J., Müller, J.-A., Gaab, M. R., Becker, H., Dietz, H., and Hundeshagen, H.,** Die Aufnahme von Aminosäuren in Hirntumoren mit der Positronen-Emissionstomographie als Indikator für die Beurteilung von Stoffwechselaktivität und Malignität, *Fortschr. Rontgenstr.,* 147, 503, 1987.

219. **Schober, O., Meyer, G.-J., Gaab, M. R., Dietz, H., and Hundeshagen, H.,** Multi-parameter studies in brain tumors by PET, *J. Nucl. Med.,* 29, 853, 1988.

220. **Mineura, K., Sasajima, T., Yoshitaka, S., Kowada, M., Shishido, F., and Uemura, K.,** Early and accurate detection of primary cerebral glioma with interfibrillary growth using ¹¹C-L-methionine positron emission tomography, *J. Med. Imaging,* 3, 192, 1989.

221. **Fujiwara, T., Matsazuwa, T., Kubota, K., Abe, Y., Itoh, M., Fukuda, H., Hatazawa, J., Yoshioka, S., Yamaguchi, K., Ito, K., Watanuki, S., Takahashi, T., Ishiwata, K., Iwata, R., and Ido, T.,** Relationship between histologic type of primary lung cancer and carbon-11-L-methionine uptake with positron emission tomography, *J. Nucl. Med.,* 30, 33, 1989.

222. **Eriksson, L.,** personal communcations: Proc. Eur. Symp. on Positron Emission Tomography in Cancer, Hammersmith Hospital, London, July 5 to 6, 1989.

223. **Bergström, M. and Muhr, C.,** personal communications: Proc. Eur. Symp. on Positron Emission Tomography in Cancer, Hammersmith Hospital, London, July 5 to 6, 1989.

224. **Welch, M. J., Coleman, R. E., Straatman, M. G. et al.,** Carbon-11-labeled methylated polyamine analogs: uptake in prostate and tumor in animal models, *J. Nucl. Med.,* 18, 74, 1977.

225. **Miller, T. R., Siegel, B. A., Fair, W. R., Smith, E. K., and Welch, M. J.,** Imaging of canine tumors with ¹¹C-methylputrescine, *Radiology,* 129, 221, 1978.

226. **McPherson, D. W., Fowler, J. S., Wolf, A. P., Arnett, C. D., Brodie, J. D., and Volkow, N.,** Synthesis and biodistribution of no carrier added [1-¹¹C]-putrescine, *J. Nucl. Med.,* 26, 1186, 1985.

227. **Bolster, J. M., Vaalburg, W., van Dijk, T. H., Zijlstra, J. B., Paans, A. M. J., Wijnberg, H., and Woldring, M. G.,** Synthesis of carbon-11 labelled ornithine and lysine: preliminary accumulation in rats with Walker 256 carcinosarcoma, *Int. J. Appl. Radiat. Isot.,* 36 263, 1985.

228. **Ding, Y.-S., Antoni, G., Fowler, J. S., Wolf, A. P., and Langström, B.,** Synthesis of L-[5-^{11}C]-ornithine, *J. Labelled Comp. Radiopharm.,* 27, 1079, 1989.

229. **Hwang, D.-R., Lang, L., Mathias, C. J., Kadmon, D., and Welch, M. J.,** N-3-[^{18}F]-fluoropropylputrescine as potential PET imaging agent for prostate and prostate derived tumors, *J. Nucl. Med.,* 30, 1205, 1989.

230. **Jänne, J., Pösö, H., and Raina, A.,** Polyamines in rapid growth and cancer, *Biochim. Biophys. Acta,* 473, 241, 1978.

231. **Tabor, C. W. and Tabor, H.,** Polyamines, *Annu. Rev. Biochem.,* 53, 749, 1984.

232. **McCann, P. P., Peg, A. E., and Sjoerdsma, A., Eds.,** *Inhibition of Polyamine Metabolism: Biologic Significance and Basis for New Therapies,* Academic Press, Orlando, 1987.

233. **Hiesiger, E., Fowler, J. S., Wolf, A. P., Logan, J., Brodie, J. D., McPherson, D., MacGregor, R. R., Christman, D. R., Volkow, N. D., and Flamm, E.,** Serial PET studies of human cerebral malignancy with [^{11}C]-putrescine and [1-^{11}C]-deoxy-D-glucose, *J. Nucl. Med.,* 28, 1251, 1987.

234. **Adelstein, S. J., Bloomer, W. D., Kassis, A. I., and Sastry, K. S. R.,** The potential use of alpha and Auger electron emitting radionuclides for therapy, in *Nuclear Medicine in Clinical Oncology,* Winkler, C., Ed., Springer-Verlag, Berlin, 1986, 408.

235. **Christman, D., Crawford, E. J., Friedkin, M., and Wolf, A. P.,** Detection of DNA synthesis in intact organisms with positron emitting [methyl-^{11}C]thymidine, *Proc. Natl. Acad. Sci. U.S.A.,* 69, 988, 1972.

236. **Sundoro-Wu, B. M., Schmall, B., Conti, P. S., Dahl, J. R., Drumm, P., and Jacobsen, J. K.,** Selective alkylation of pyrimidyl-dianions: synthesis and purification of ^{11}C labeled thymidine for tumor visualization using positron emission tomography, *Int. J. Appl. Radiat. Isot.,* 35, 705, 1984.

237. **Poupeye, E., Leeheer, E., de Slegers, G., Goethals, P., and Counsell, R. E.,** Synthesis of ^{11}C-labelled thymidine for tumor visualization using positron emission tomography, *Appl. Radiat. Isot.,* 40, 57, 1989.

238. **Martiat, P., Ferrant, A., Labar, D., Cogneau, M., Bol, A., Michel, C., Michaux, J. L., and Sokal, G.,** *In vivo* measurement of carbon-11 thymidine uptake in non-Hodgkin's lymphoma using positron emission tomography, *J. Nucl. Med.,* 29, 1633, 1988.

239. **Graham, M. M., Ferrant, A., de Schryver, A.,** personal communications; **Graham, M. M. et al.,** *Proc. Eur. Symp. on Positron Emission Tomography in Cancer,* Hammersmith Hospital, London, July 1989.

240. **Shiue, C. Y., Wolf, A. P., and Friedkin, M.,** Synthesis of 5'-deoxy-5-[^{18}F]fluorouridine and related compounds as probes for measuring tissue proliferation *in vivo, J. Labelled Comp. Radiopharm.,* 21, 865, 1984.

241. **Ishiwata, K., Monma, M., Iwata, R., and Ido, T.,** Automated synthesis of 5-[^{18}F]fluoro-2'-deoxyuridine, *Appl. Radiat. Isot.,* 38, 467, 1987.

242. **Diksic, M., Farrokhzad, S., Yamamoto, Y. L., and Feindel, W.,** A simple synthesis of ^{18}F-labelled 5-fluorouracil using acetohypofluorite, *Int. J. Nucl. Med. Biol.,* 11, 141, 1984.

243. **Shani, J., Wolf, W., Schlesinger, T., Atkins, H. L., Bradey-Moore, P. R., Casella, V., Fowler, J. S., Greenberg, D., Ido, T., Lambrecht, R. M., MacGregor, R., Mantescu, C., Neirinckx, R., Som, P., Wolf, A. P., Wodinski, I., and Meaney, K.,** Distribution of ^{18}F-5-fluorouracil in tumor-bearing mice and rats, *Int. J. Nucl. Med. Biol.,* 5, 19, 1978.

244. **Wiley, A. L., Nickels, J., Wessels, B. W., and Lieberman, L. M.,** Preliminary clinical imaging studies with ^{18}F-5-fluorouracil, in *Nuklearmedizin: Stand und Zukunft,* Schmidt, H. A. E. and Woldring, M. G., Eds., F. K. Schattauer Verlag, Stuttgart, 1978, 207.

245. **Young, D., Vine, E., Ghanbarpour, A., Shani, J., Siemsen, J. K., and Wolf, W.,** Metabolic and distribution studies with radiolabeled 5-fluorouracil, *Nucl. Med.,* 21, 1, 1982.

246. **Visser, G. W. M., Gorree, G. C. M., Braakhuis, B. J. M., and Herschied, J. D. M.,** An optimized synthesis of ^{18}F-labelled 5-fluorouracil and a reevaluation of its use as a prognostic agent, *Eur. J. Nucl. Med.,* 15, 225, 1989.

247. **Abe, Y., Fukuda, H., Ishiwata, K., Yoshioka, S., Yamada, K., Endo, S., Kubota, K., Sato, T., Matsuzawa, T., Takahashi, T., and Ido, T.,** Studies on ^{18}F-labeled pyrimidines. Tumor uptakes of ^{18}F-5-fluorouracil, ^{18}F-5-fluorouridine, and ^{18}F-5-fluorodeoxyuridine in animals, *Eur. J. Nucl. Med.,* 8, 258, 1983.

248. **Ishiwata, K., Ido, T., Kawashima, K., Murakami, M., and Takahashi, T.,** Studies on ^{18}F-labeled pyrimidines. II. Metabolic investigations of ^{18}F-5-fluorouracil, ^{18}F-5-fluoro-2'-deoxyuridine and ^{18}F-5-fluorouridine in rats, *Eur. J. Nucl. Med.,* 9, 185, 1984.

249. **Ishiwata, K., Ido, T., Abe, Y., Matsuzawa, T., and Murakami, M.,** Studies on ^{18}F-labeled pyrimidies. III. Biochemical investigation of ^{18}F-labeled pyrimidines and comparison with ^{3}H-deoxythymidine in tumor-bearing rats and mice, *Eur. J. Nucl. Med.,* 10, 39, 1985.

250. **Kameyama, M., Tsurumi, Y., Shirane, R., Itoh, J., Katkura, R., Yoshimoto, T., Hatazawa, J., Itoh, M., Ishiwata, K., and Ido, T.,** Nucleic acid metabolism in gliomas studied with ^{18}Fdurd and PET, *J. Cereb. Blood Flow Metab.,* 9, S221, 1989.

251. **Kameyama, M., Tsurumi, Y., Itoh, J., Sato, K., Katakura, R., Yoshimoto, T., Hatazawa, J., Itoh, M., Watanuki, S., Seo, S., Ishiwata, K., and Ido, T.,** Biological Malignancy and [18]FdUrd Uptake in Glioma Patients—PET Study of Nucleic Acid Metabolism, CYRIC Annual Report 1988, Cyclotron and Radioisotope Center, Tohoku University, Sendai, 1989, 215.

252. **Freed, B. R., McQuinn, R. L., Tilbury, R. S., and Digenis, G. A.,** Distribution of ^{13}N in rat tissue following intravenous administration of nitroso-labeled BCNU, *Cancer Chemother. Pharmacol.*, 10, 16, 1982.

253. **Diksic, M., Farrokhzad, S., Yamamoto, Y. L., and Feindel, W.,** [11]C-, and ^{13}N-labelled BCNU and its *in vivo* pharmacokinetic study with PET, *J. Labelled Comp. Radiopharm.*, 19, 1394, 1982.

254. **Farrokhzad, S., Diksic, M., Yamamoto, Y. L., and Feindel, W.,** Synthesis of ^{18}F-labelled-2-fluoroethyl-nitrosoureas, *Can. J. Chem.*, 62, 2107, 1984.

255. **Diksic, M., Sako, K., Feindel, W. et al.,** Pharmacokinetics of positron labelled 1,3-bis-2-chloroethyl-nitrosourea in human brain tumors using positron emission tomography, *Cancer Res.*, 44, 3120, 1984.

256. **Tyler, J. L., Yamamoto, Y. L., Diksic, M. et al.,** Pharmacokinetics of superselective intra-arterial and intravenous ^{11}C-BCNU evaluated by PET, *J. Nucl. Med.*, 27, 775, 1986.

257. **Conway, T. and Diksic, M.,** Synthesis of ''no carrier added'' carbon-11-SarCNU: the sarcosinamide analog of the chemotherapeutic agent BCNU, *J. Nucl. Med.*, 29, 1957, 1988.

258. **Abe, Y., Matsuzawa, T., Itoh, M., Ishiwata, K., Fujiwara, T., Sato, T., Yamaguchi, K., and Ido, T.,** Regional coupling of blood flow and methionine uptake in an experimental tumor assessed with autoradiography, *Eur. J. Nucl. Med.*, 14, 388, 1988.

259. **Kameyama, M., Tsurumi, Y., Shirane, R., Katakura, R., Suzuki, J., Itoh, M., Fukuda, H., Matsuzawa, T., Watanuki, S., and Ido, T.,** Multiparametric analysis of brain tumor with PET, *J. Cereb. Blood Flow Metab.*, 7, S466, 1987.

Chapter 17

KINETIC ANALYSIS OF RADIOLIGAND BINDING IN BRAIN *IN VIVO*

Albert Gjedde

TABLE OF CONTENTS

I. *IN VIVO* IMAGING OF NEURORECEPTORS

The neuroreceptor concept is based on the observation that exogenous drugs can influence brain function by a specific association with membrane or cytosolic proteins designed to receive endogenous neurotransmitters or neuromodulators. The endogenous agents are not always known. Specific receptors in and on cells were postulated more than a century ago by Langley[24] (1878) as "receptive" substances and by Ehrlich[10] (1909) as "side-chains" of molecules in cells. Chemicals that associate with receptors are known as "ligands" (Latin for "subject to or undergoing binding").

The kinetic description of receptor-ligand interaction followed the discovery of the association between oxygen and myoglobin. When the binding of oxygen to myoglobin was first observed to be a saturable function of the oxygen tension of the solution (Bohr,[4] 1885), the relationship between the quantities of bound and free oxygen corresponded exactly to the relationship between the reaction velocity and substrate concentration of the enzyme invertase, later derived by Michaelis and Menten[27] (1913), and to the relationship between the adsorption capacity of an adsorbing surface and the quantity of adsorbed material, derived by Langmuir[25] (1916). The interaction between a ligand and its receptor is a special case of adsorption by means of a chemical reaction or intermolecular forces. The relationship between the oxygen tension and the quantity of oxygen bound to hemoglobin was considerably more complicated. Bohr[5] (1904) derived an equation which antedated the formulation suggested by Hill[21] (1910), neither description being thermodynamically correct, although Hill's equation has remained in use.

It is possible to study neuroreceptors in tissue samples or sections *in vitro*, using conventional tracer techniques or autoradiography. The advent of autoradiography in the study of receptors led to the synthesis of labeled ligands for most of the known neurotransmitter systems and receptor subtypes in the brain. As a result of these studies, receptor changes have been implicated in numerous diseases.

In vitro, the receptors bind the ligand after the death of the organism and in autoradiography, dehydration, or fixation of the tissue. Hence, neuroreceptor findings obtained *in vitro* do not directly reflect the physiological or pathophysiological state of the subject prior to death. To circumvent this difficulty, it is necessary to study the receptor-ligand interaction *in* or *ex vivo* by administration of the tracer to the living organism. In studies *ex vivo*, the tracer circulates for some time before the quantity accumulated in the target organ is determined post-mortem by tissue sampling or autoradiography. *In vivo*, the tracer circulates while the quantity in the target organ is recorded by external detection of emitted radioactivity.

Although the tomographic images of radioligand binding *in vivo* sometimes match the regional distribution of neuroreceptors obtained *in vitro* by autoradiography quite well, the interpretation of these images often creates problems soluble only by compartmental analysis. For example, unlike the *in vitro* autoradiograms, the *in vivo* tomograms must be reconstructed as functions of the circulation of the radioligand. At early times after the administration, the tomographic images reflect the delivery of the radioligand to the target organ by the circulation, and at later times the images may reflect the metabolism of the radioligand to labeled compounds that bind differently or not at all to the receptors or the more or less complete clearance of the ligand and its metabolites from the target organ. These processes, and the problems of interpretation that they create, are of no concern to autoradiography *in vitro*.

In vitro, the analysis of the neuroreceptor-ligand interaction can be made with high specific activity and excess radioligand. *In vivo*, the free radioligand obscures the bound radioligand when the specific activity is low, or the receptors deplete the tissue of free radioligand when the specific activity is high and the mass of the radioligand low, yielding tomographic images that continue to reflect the capacity of the circulation to deliver the radioligand to the tissue rather than the capacity of the tissue to bind the radioligand specifically.

"Residue detection" is the method of measuring the quantity of tracer left in the target organ at a given moment during circulation, in contrast to the previous methods of "inflow-outflow" detection in which arterial and venous concentrations were measured. Positron emission tomography (PET) is but one of the applications of this method. Other techniques exist, including single photon emission computed tomography (SPECT) and planar imaging. In SPECT, the γ-emission from selected isotopes, mostly ^{99m}Tc or ^{123}I, is also recorded tomographically. Hence, the principles presented below apply to SPECT as well as to PET. Several reviews of the experimental analysis of neuroreceptor-radioligand interaction have been published recently, including Gjedde et al.,[16] Gjedde and Wong,[17] and Gjedde.[18]

II. TRANSIENT ANALYSIS OF NEURORECEPTOR-RADIOLIGAND INTERACTION

A. DIFFERENTIAL EQUATIONS

Radioligands sequester within the tissue with the result that they escape the build-up in the extracellular space that leads to backflux. The "sequestration" imposes an additional state on the tracer. Kinetically, the process generates at least one additional compartment, reflecting transport (e.g., into an intracellular space), binding (e.g., to receptor sites), or metabolism (e.g., by chemical reactions that sequester the label for a shorter or longer time).

In the compartmental analysis, the process is represented by a sequestration coefficient (k_3) describing entry into the additional compartment and an escape coefficient (k_4) describing the exit from this compartment. Entry into the third compartment competes with the likelihood of return of the label to the vascular compartment. Two differential equations govern the competition:

$$\frac{dM_e}{dt} = K_1 C_a - (k_2 + k_3)M_e + k_4 M_b \tag{1}$$

$$\frac{dM_b}{dt} = k_3 M_e - k_4 M_b \tag{2}$$

where M_e and M_b represent the quantities of tracer in the exchangeable extravascular compartment and in the sequestered compartment. The coefficient k_4 is not always a constant, and this may complicate the solution considerably.

It is the purpose of the three-compartment analysis to determine the net clearance, K, into the sequestered compartment and the steady-state volume of distribution of the free tracer in the tissue. The steady-state volume of distribution depends on the reversibility of the binding. For irreversible binding, the volume, V_f, is $K_1/(k_2 + k_3)$. For reversible binding, the volume, V_e, is K_1/k_2 (equal to the partition volume when sequestration is negligible and k_3 hence zero). The net clearance by association is $k_3 V_f$ or $k_3 V_e$, i.e., $K_1 k_3/(k_2 + k_3)$ or $K_1 k_3/k_2$. The ratio between K and V_f or V_e is an estimate of k_3.

The definition and derivation of the blood-brain transfer coefficients K_1 and k_2 have been discussed in Chapter 8 on the kinetic analysis of glucose and glucose tracer uptake and metabolism.

B. SOLUTIONS

The integrated equation for M, the total content of labeled material in the tissue, including the original tracer and its derived forms, has been extended to three compartments (k_4 = 0) by Blomqvist et al.[3] and further extended to include k_4 by Evans,[11]

$$
\begin{aligned}
M = V_a C_a &+ (K_1 + [k_2 + k_3 + k_4]V_a) \int_0^T C_a \, dt \\
&+ (K_1[k_3 + k_4] + k_2 k_4 V_a) \int_0^T \left[\int_0^t C_a \, du \right] dt \\
&- (k_2 + k_3 + k_4) \int_0^T M \, dt - k_2 k_4 \int_0^T \left[\int_0^t M \, du \right] dt
\end{aligned}
\tag{3}
$$

where M is the sum of M_a, M_e, and M_b. This equation yields the coefficients by multiple linear regression of the five independent variables (i.e., C_a, the integrals, and the integrals of integrals) and the one dependent variable (i.e., M). The multiple linear regression is of the form

$$
Y = p_1 X_1 + p_2 X_2 + p_3 X_3 + p_4 X_4 + p_5 X_5
\tag{4}
$$

where p_i denotes the coefficient of independent variable X_i. The normalized solution to Equations 1 and 2 for the volume of distribution of the tracer in brain is

$$
V = \frac{M}{C_a} = K \left(\frac{\displaystyle\int_0^T C_a \, dt - \frac{\displaystyle\int_0^T M_m \, dt}{V_m}}{C_a} \right) + \beta \frac{M_e}{C_a} + V_a
\tag{5}
$$

where K is the unidirectional clearance due to binding, equal to $K_1 k_3/(k_2 + k_3)$, V_m is a sequestration space equal to $K_1 k_3/(k_2 k_4)$, and β is an index of the nonbinding fraction of the tracer, equal to $k_2/(k_2 + k_3)$.

1. Irreversible Binding

The higher the value of the $K_1 k_3$ product relative to the $k_2 k_4$ product, i.e., the larger V_m, the lower the probability of escape of the tracer from brain (irreversible binding). When V_m is sufficiently large (i.e., when k_4 is sufficiently small), Equation 5 reduces to:

$$
V = K\Theta + \beta V_f' + V_a
\tag{6}
$$

where V_f' is the measured (steady- or nonsteady-state) M_e/C_a ratio and Θ is the normalized integral introduced in Equation 5 [$\int_0^T C_a dt/C_a(T)$]. When the ratio between the compound free in brain and the compound free in blood no longer changes rapidly, the ratio M_e/C_a eventually approaches V_f. The graphical representation of this case is a straight line. K is the slope of the line and $\beta V_f' + V_a$ is the ordinate intercept (see "slope-intercept" or "Patlak" plot in chapter on glucose uptake and metabolism).

For values of Θ for which k_4 is still effectively nil, it is an empirical observation that Equation 5 may be simulated by

$$
V - V_a \cong K\Theta + \beta V_f(1 - e^{-K_1\Theta/V_f})
\tag{7}
$$

If divided by the volume of distribution in a region of no binding, the ratio eventually reaches the linear relationship:

$$\frac{V^{(1)} - V_a}{V^{(2)} - V_a} \cong \beta k_3 \Theta + \beta^2 \tag{8}$$

which is the basis for the so-called "ratio" method which derives the value of k_3 from the slope and the intercept of the relationship between the ratio of the measured volumes of distribution in a region of binding ($V^{(1)}$) and a region of no binding ($V^{(2)}$).

2. Reversible Binding

If k_4 is not negligible, the three-compartment tracers must eventually approach a transient equilibrium defined by the four coefficients of Equations 1 and 2. At steady state, the "equilibrium volume" of distribution is

$$V(\infty) \rightarrow \frac{K_1}{k_2} \left(1 + \frac{k_3}{k_4} \right) + V_a = V_e + V_m + V_a \tag{9}$$

where the K_1/k_2 ratio is the volume V_e and the k_3/k_4 ratio is the V_m/V_e ratio. The k_3/k_4 ratio is a convenient measure of the tissue's "binding potential" (p_B), imposed by the presence of binding sites.

C. DEFINITION OF BINDING PARAMETERS

The interaction between a radioligand and a neuroreceptor is a case of sequestration by establishment of bonds between the tracer and specific binding sites in the tissue. In this case, the transfer coefficients k_3 and k_4 acquire definite meaning relevant to the kinetic description of neuroreceptor-radioligand interaction.

1. Michaelis-Menten Equation

The kinetic description of neuroreceptor-radioligand interaction is based on the principles of the Michaelis-Menten analysis of enzymatic reactions. The criteria of simple receptor-ligand interaction are the same as those of enzyme-substrate interaction and facilitated diffusion and include saturability, stereospecificity, and competitive inhibition.

The Michaelis-Menten equation is the equilibrium solution to a differential equation describing the change per unit time of the quantity of ligand bound to receptors. The simplest model of the binding process is in principle a three-compartment micromodel in which the ligand in solution, the receptor-ligand complex, and the internalized ligand or otherwise inactivated receptor-ligand complex represent the three states of the tracer. The change of the quantity of bound ligand equals the difference between ligand associating with the receptor and ligand dissociating from the receptor. The association rate is the probability of each ligand molecule joining the receptor (k_{on}), multiplied by the number of molecules per unit volume and the number of receptor sites. The dissociation rate is the combined likelihood of dissociation by return of the ligand to solution (k_{off}) or internalization through the membrane (k_{in}), multiplied by the number of receptor-ligand complexes,

$$\frac{dB}{dt} = k_{on}C(B_{max} - B) - (k_{off} + k_{in})B \tag{10}$$

where B_{max} is the quantity of sites available for binding, B is the quantity of receptor-ligand complexes ("bound-ligand"), and C is the concentration of ligand molecules in the solution.

The internalization or inactivation rate is $k_{in}B$. At equilibrium, when the number of receptor-ligand complexes is constant in time, i.e., when $dB/dt = 0$, we obtain the Michaelis-Menten equation,

$$B = B_{max} \frac{C}{K_D + C} \qquad (11)$$

where K_D is the Michaelis half-saturation concentration, substituted for $[k_{off} + k_{in}]/k_{on}$. Note that C has units of concentration while B and B_{max} may be expressed in any unit suited to indicate a quantity of bound ligand. Strictly, K_D is a dissociation constant only when k_{in} is negligible compared to k_{off}. Hence, studies of receptor sites (for which k_{in} is low) often indicate a much higher affinity between the receptor and the ligand than observed for facilitated diffusion (for which, of course, k_{in} is substantial).

Since the Michaelis-Menten equation is a hyperbola, nonlinear regression is required to estimate K_D and B_{max} from this equation directly. However, in the past, only linear equations with two parameters were soluble with ease. About 1930, Barnett Woolf (see Haldane[27]) derived the linearized versions of the Michaelis-Menten equation now named after other people. The most famous of Woolf's linearizations is the Scatchard plot, reinvented no less than four times.[9,22,31,32] Woolf showed that Equation 11 can be linearized by rearrangement to:

$$B = -K_D \frac{B}{C} + B_{max} \qquad (12)$$

in which B_{max} is the ordinate intercept and $-K_D$ is the slope. Woolf is usually credited with a plot of his own:

$$\frac{C}{B} = \frac{C}{B_{max}} + \frac{K_D}{B_{max}} \qquad (13)$$

in which the slope is $1/B_{max}$ and the abscissa-intercept is $-K_D$ when C/B is plotted vs. C. The statistical merits of each of the linearized equations were discussed by Cressie and Keightley.[7]

2. Michaelis-Menten Equation *in Vivo*

The definitions of k_3 and k_4 follow from Equation 10 that governs the association and dissociation of the ligand-receptor complex in the presence of ligand and inhibitor(s). However, the aqueous concentration of the ligand is unknown *in vivo* where only the content of exchangeable ligand, M_e, can be estimated, albeit with uncertainty. Equation 10 must be modified accordingly:

$$\frac{dM_b}{dt} = k_{on}[M_{b_{max}} - (M_b + M_{b_l})] \frac{M_e}{V_d} - k_{off} M_b \qquad (14)$$

where V_d is the volume of solution in which the ligand is dissolved. This volume is the ratio between the content of exchangeable ligand and the concentration of exchangeable ligand in aqueous solution. At all times, $M_e(t) = V_d C(t)$. At equilibrium $M_e(\infty) = V_d C(\infty) = V_e C_a(\infty)$. M_b is the quantity of bound ligand, and M_{b_l} is the total quantity of bound inhibitors.

The difference between M_{b_l} is $M'_{b_{max}}$ to indicate the number of receptor sites available for binding of the ligand in question. By simple comparison of Equation 14 with Equations 1 and 2, the transfer coefficients k_3 and k_4 can be identified as follows:

$$k_3 = k_{on}M'_{b_{max}}/V_d \tag{15}$$

$$k_4 = k_{off} + k_{on}\frac{M_e}{V_d} = k_{off}\left(1 + \frac{C}{K_D}\right) \tag{16}$$

where M_e is the quantity (mass) of tracer available for binding (exchangeable ligand).

Since k_3 and k_4 are arbitrary assignations of the different components of the differential equations, k_3 can be defined to be independent of the injected dose of tracer while k_4 is a function of the free tracer concentration and therefore not a constant when M_e is not constant or negligible. This definition of k_4 is not new to receptor kineticists; it is known as the "observed k", employed to calculate k_{on} when k_{off} is known.

According to Equations 15 and 16, the k_3/k_4 ratio is the ratio between $M_{b_{max}}/V_d$ and $K'_D + C$ at steady state, e.g., when C attains a constant level. This ratio is the unitless "binding potential" p_B. Note that the magnitude of p_B depends on the volume of distribution and concentration of the ligand. The higher the concentration of the ligand, the lower the magnitude of p_B. Thus, during the approach to equilibrium, the measured distribution space V, equal to the ratio $[M_a + M_e + M_b]/C_a$, rises to the steady-state value

$$V(\infty) \to V_e\left(1 + \frac{M_{b_{max}}/V_d}{K'_D + C(\infty)}\right) + V_a \tag{17}$$

where K'_D is the product $K_D[1 + (C_I/K_I)]$. In PET, it is often possible to do several studies in the same subject, and these studies may therefore be repeated with different doses of the radioligand. The highest ligand concentrations reduce the binding potential to zero and the equilibrium volume to its minimum value, equal to the partition volume $V_e + V_a$ of the ligand. By nonlinear regression, the equation allows estimates to be made of $V_e + V_a$, $M_{b_{max}}$ and K'_D.

D. COMPETITION

The quantity $M'_{b_{max}}$ represents the number of receptors available for occupation by the labeled ligand. If the receptor is blocked by a competitor, the total quantity of receptor sites in the tissue ($M'_{b_{max}}$) is reduced by a factor that depends on the receptor's affinity for the competitor:

$$M'_{b_{max}} = M_{b_{max}}\frac{K_I}{K'_I + C_I} \tag{18}$$

where $M_{b_{max}}$ is the maximal quantity of bound ligand, K'_I is the competitor's apparent half-inhibition constant ("IC_{50}"), and C_I is its aqueous concentration. The apparent inhibition constant is the affinity of the competitor in the presence of the ligand:[18]

$$K'_I = K_I\left(1 + \frac{C}{K_D}\right) \tag{19}$$

where C is the free ligand concentration and K_D is the dissociation constant of the ligand. Both of these equations follow from simple inhibition kinetics when an inhibitor competes

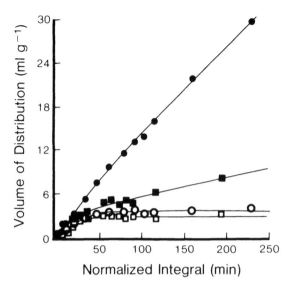

FIGURE 1.   *N*-Methylspiperone (NMSP) binding to neostriatum (closed symbols) and cerebellum (open symbols), plotted as volume of distribution vs. normalized integral in normal human volunteer. Circles: Before haloperidol blockade; squares: after 90% blockade of NMSP binding sites in neostriatum by haloperidol, 2.5 ng ml^{-1} in plasma. (Redrawn from Wong, D. F., Gjedde, A., Wagner, H. N., Jr., Dannals, R. F., Douglass, K. H., Links, J. M., and Kuhar, M. J., *J. Cereb. Blood Flow Metab.*, 6, 147, 1986b.)

with the ligand for binding. Note that $M'_{b_{max}}$ is defined only when C and C_I are constant or negligible. Thus, the quantity of bound inhibitor is also a function of the ligand concentration:

$$M_{b_I} = \frac{M_{b_{max}} C_I}{C_I + K_I \left(1 + \dfrac{C}{K_D}\right)} \tag{20}$$

and K'_I can be a constant only when C is negligible or invariant. This condition also applies to $M'_{b_{max}}$ and k_3.

Blockade involves prior occupancy of the receptor by the blocking agent, unlike saturation which involves significant occupation of the receptor by the labeled ligand during the course of study, and displacement which involves clearance of the labeled ligand from the receptor by an unlabeled competitor or by the unlabeled ligand itself. The effect of haloperidol blockade of N-[^{11}C]methylspiperone trapping in striatum *in vivo* is shown in Figure 1.

The measure of blockade is saturation, defined as the ratio between bound inhibitor and the maximal binding capacity at equilibrium. In a receptor-controlled brain function, occupation of the receptor by an agonist is the factor responsible for the magnitude of the response. Occupancy is a function of the affinity and concentrations of the ligands that compete for occupation. When a single competitor contributes to the occupancy, the degree of saturation by the inhibitor can be calculated from the ratio,

$$s = \frac{C_I}{K'_I + C_I} = 1 - \frac{k_3^{(I)}}{k_3} \tag{21}$$

where $K_3^{(I)}$ is the value of k_3 measured in the presence of inhibition. The degree of saturation is also an index of the degree of occupancy established by any pharmacological agent bound to the receptor.

The apparent volume of distribution of the ligand increases when binding is blocked because k_3 declines as less exchangeable tracer attaches to binding sites.

E. NONSPECIFIC BINDING

Nonspecific binding refers to the sequestration of some fraction of the tracer by attachment to sites that cannot be saturated in the range of concentrations available for study. Depending on the mechanism involved, nonspecific binding can also be manifested as a change of solubility. In this sense, nonspecific binding is all binding to sites of sufficiently low affinity and high capacity to escape saturation at the chosen ligand concentrations.

The nonspecific binding simulates an expansion of distribution space for exchangeable ligand which renders the concentration of exchangeable ligand in tissue water lower than suspected from the sample content. The degree of nonspecific binding is practically impossible to determine under nonsteady-state conditions because the ratio between free and nonspecifically bound tracer changes as a function of time. It also introduces a formal flaw in the solutions of Equations 1 and 2 because k_2 and k_3 vary when the solubilities in blood and tissue are functions of time.

At steady state, nonspecific binding can be simulated by an expansion of the steady-state distribution volume of the form expressed in Equation 9,

$$V_d = V_w(1 + \rho) \tag{22}$$

where ρ is the "binding potential" of nonspecific or low-affinity binding (see Table 2), indicating the k_3/k_4 ratio of this binding, and V_w is the volume of the aqueous solution measured in the absence of the nonspecific binding. The ratio between V_d and V_w is the solubility factor. At steady state, the concentration, C, is M_e divided by V_d.

Although it is not correct, strictly speaking, to measure nonspecific binding at nonsteady state, the error may be small. Mintun et al.[28] used Equations 1 and 2 to estimate the nonspecific binding of a radiolabeled ligand in brain from measurements of the free fraction in plasma *in vitro*. The reciprocal of the volume of distribution in brain was defined as f_2 (although it is not a true fraction):

$$f_2 = \frac{1}{V_d} \tag{23}$$

The same authors defined the free fraction in arterial blood as

$$f_1 = \frac{1}{\alpha} \tag{24}$$

where α is the solubility of the ligand in the sample relative to water. Thus,

$$f_2 = \frac{f_1}{V_e} \tag{25}$$

according to which f_2 can be calculated when f_1 and V_e are known. If the ligand distributes only in a fraction of the total tissue water, e.g., only in extracellular fluid, V_w refers only to the volume of water in the extracellular space. Note that f_1 is a proper fraction while f_2 is the reciprocal of a volume with a unit of $g\ ml^{-1}$ or $ml\ ml^{-1}$. The effect of plasma protein binding on radioligand permeability in blood-brain barrier is listed in Table 1.

TABLE 1
Influence of Plasma Protein Binding on Apparent
Permeability of Radioligands

Tracer	PS (ml g^{-1} min^{-1})	Correction for protein binding	Ref.
Spiperone	2.80	Yes	30[a]
Methylspiperone	0.22	No	33[a]
Spiperone	0.20	No	26[b]
Fluoroethylspiperone	0.09	No	23[b]

[a] Humans.
[b] Baboons.

Values of f_1 and f_2 of [^{18}F]spiperone were measured in monkey and man by Perlmutter et al. The values were 4 to 6% for f_1 and 1 to 2% for f_2 in monkey[29] and 5% for f_1 and 0.7% for f_2 in man.[30] Values of f_1 for other ligands have been reported also to be close to 5%. Hence, a typical value of the partition volume, V_e, is 5 ml g^{-1} for radioligands. Since low-affinity binding sites contribute to the partition volume, it may not be correct to assume that all regions of the brain share the same value of the partition volume. In one study, this was true of the nonspecific binding of N-methylspiperone in striatum and cerebellum.[34]

F. METABOLITES

Metabolism of a radioligand is a form of sequestration which interferes with the interpretation of sequestration as binding. Metabolism contributes to a mixed signal and must be corrected for, either chemically or by mathematical analysis.

Labeled metabolites of the tracer may be generated in the circulation and then pass into the brain. Labeled metabolites may also be generated in brain tissue, pass back into the circulation or bind to the specific sites in question, or both. If the labeled ligand is present at concentrations above tracer concentrations, the metabolites may block the specific sites to a significant extent if they retain an affinity for these sites.

It is not possible to correct all forms of interference by metabolites. However, a partial correction can be made by consideration of the volume distribution (V_e) recorded in a brain region of no known specific binding or a region chosen to indicate baseline binding.

The steady-state volume of distribution, V_e, may decline significantly during the study. Kinetically, the decline is impossible in the absence of (1) continued generation of labeled material in the circulation which fails to achieve steady-state distribution in brain, or (2) significant slowing of the decline of the tracer concentration in the circulation.[5a,33] As discussed by Wong et al.,[33] the effect of metabolites of this nature may be corrected mathematically by adjustment of the tracer concentration in the circulation to the level that yields a nondeclining steady-state volume of distribution in a nonbinding region. The presence of metabolites consistent with this adjustment must be confirmed by HPLC analysis of arterial samples of the tracer.

If the steady-state volume of distribution remains constant, any labeled metabolites in the circulation and in the brain have achieved an apparent steady-state and thus merely contribute to the magnitude of the nonspecific binding estimated by the methods discussed above.

If the steady-state volume of distribution continues to increase with time, the case can be interpreted as (1) the appearance of labeled metabolites in brain that simulate binding or (2) significant acceleration of the decline of the tracer concentration in the circulation. A mathematical adjustment of the radioactivity in brain may be performed to yield a nonincreasing steady-state volume of distribution.

TABLE 2
Distribution and Binding Constants for
N-[¹¹C]Methylspiperone Binding in Human Caudate in
Absence and Presence of Unlabeled Haloperidol *In Vivo*

Variable	Before haloperidol	After haloperidol
K_1 (ml g^{-1} min^{-1})	0.15	0.21
k_2 (min^{-1})	0.051	0.077
k_3 (min^{-1})	0.087	0.010[a]
V_e (cerebellum; ml g^{-1})	3.0	2.7
V_f (ml g^{-1})	1.1	2.4[a]
ρ (relative to cerebellum; ratio)	4.1	0.0[a]

[a]   Significantly different (paired *t*-test) at $p < 0.05$.

Adapted from Wong, D. F., Gjedde, A., and Wagner, H. N., Jr., *J. Cereb. Blood Flow Metab.*, 6, 137, 1986a.

When the labeled metabolites bind to specific sites only, correction may be possible, but the error will chiefly affect the calculation of K_D if the labeled metabolites have a different affinity, to be lumped as "ligand-like" entities if the affinity is the same, in which case no real error is incurred.

III. EXPERIMENTAL ANALYSIS OF BINDING *IN VIVO*

A. TRANSIENT ANALYSIS OF IRREVERSIBLE BINDING

The main distinction observed below is between "reversible" and "irreversible" binding, referring to the presence or absence of noticeable loss of bound tracer from the brain during the study. Irreversibility is a function of the k_3/k_4 ratio (binding potential) and can be predicted from the volume of distribution at equilibrium; the higher the equilibrium M/C_a ratio, the lower the degree of reversibility in a given period of study.

The $V(\infty)/K_1$ value of any potential radioligand indicates the order of magnitude of the half-time (in min) of the approach toward equilibrium. Thus, steady-state M/C_a ratios of the order of 10 ml g^{-1} or more indicate irreversible binding during the first hour. Irreversible trapping is characteristic of the high-affinity binding of certain radioligands, e.g., the substituted butyrophenones spiroperidol and methylspiperone, to dopamine D_2 receptors in the basal ganglia.

The purpose of the analysis of irreversible binding is the estimation of k_3. Regression analysis of the irreversible binding yields estimates of K_1, k_2, k_3, K, and V_f. Since the irreversible case, by definition, fails to reach equilibrium in the period of study, it is not possible to determine the equilibrium constants $M_{b_{max}}$ and K_D' directly by this procedure. However, when k_{on} is a constant, the value of k_3 is an observed index of the receptor density that varies with $M_{b_{max}}'$ and V_d. The value of $M_{b_{max}}'$ may vary because the actual receptor number varies or because some receptors are blocked by endogenous or exogenous competitors. V_d may vary with variations of protein or other low-affinity binding in the tissue.

Equation 21 predicts an inverse proportionality between the concentration of a blocking agent (inhibitor) and the value of k_3. Wong et al.[33,34] determined the values of K_1, k_2, and k_3 for tracer N-methylspiperone (NMSP) in neostriatum and cerebellum in the human brain, using Equation 7. The NMSP binding was blocked with a therapeutic dose of haloperidol. Unblocked, the net binding rate of NMSP (K) in the striatum was 0.094 ml g^{-1} min^{-1}, and blocked it was 0.023 ml g^{-1} min^{-1}. As listed in Table 2, the volume of distribution of unbound NMSP (V_f) in the striatum increased from 1.1 to 2.4 ml g^{-1} and hence approached

TABLE 3
Transfer Coefficients for Radioligands

Radioligand	Species	K_1	k_2	k_3	k_4	Ref.
		(ml g^{-1} min^{-1})	(min^{-1})			
[^{18}F]Spiperone	Baboon	0.15	0.014	0.012	0.002	26
	Baboon	0.11	0.038	0.024	0.012	29
	Human	0.12	0.016	0.055	0.020	30
[^{18}F]Fluoroethylspiperone	Baboon	0.064	0.031	0.040	0.008	2
	Human	0.050	0.019	0.041	0.002	2
[^{11}C]Raclopride	Human	0.15	0.37	0.51	0.14	14

the value of V_e observed in cerebellum, a region with no binding, and the corresponding value of k_3 declined from 0.09 to 0.01 min^{-1} after blockade by haloperidol. Thus, at plasma concentrations of 2 to 3 nM, haloperidol blocked 90% of the NMSP binding sites. A representative example of such an experiment is shown in Figure 1.

Equation 21 can be rearranged to yield the formula for a modified Woolf plot:

$$\frac{1}{k_3^{(I)}} = \frac{1}{k_3} + \left[\frac{V_{I_e}}{k_{on}M_{b_{max}}K_I} \right] C_{I_a}(\infty) \qquad (26)$$

where V_{I_e} is the steady-state partition of haloperidol between brain and plasma and C_{I_a} is the steady-state arterial concentration of the inhibitor. Assuming constant $V_{I_e}/[k_{on}K_I]$ ("D_w") and negligible C, Wong et al.[34] calculated the $M_{b_{max}}$ and K_I' of the D_2 dopamine receptors and haloperidol, respectively, in the human brain from the slope and abscissa intercept of this relationship. A disadvantage of this indirect approach is the need to assume values for constants that cannot be verified in each experiment. Advantages to the use of unlabeled ligands different from the tracer include the safety of giving compounds long in use as medication, and the avoidance of binding to nonspecific but saturable sites.

B. TRANSIENT ANALYSIS OF REVERSIBLE BINDING

Completely irreversible trapping or binding is rare. In most cases, sooner or later, a transient equilibrium (i.e., dC/dt = 0) is approached between the quantities of tracer in the circulation, free tracer in brain, and tracer bound to receptors or converted to metabolites.

The purpose of the kinetic analysis of reversible binding is the estimation of the maximal value of the "binding potential" p_B which equals the k_3/k_4 ratio in the presence of negligible levels of the tracer, i.e., when $C(\infty)$ is negligible compared to K_D in Equations 17 and 28. The values of k_3 and k_4 are estimated by multiple linear regression to Equation 3. The analysis must therefore be extended to include a significant portion of the approach to equilibrium, although the equilibrium cannot be reached completely to estimate the binding potential. Table 3 lists values of the transfer coefficients determined for the radioligand [^{18}F]spiperone in baboons and humans. Either from measurements of the transfer coefficients, or from measurements of volumes of distribution at steady state (see below), it is possible to calculate binding potentials for variety of radioligands. Results of such experimental calculations of the binding potentials are exemplified in Table 4.

The transfer coefficients cannot be determined by Equations 1 and 2 when k_4 is not a constant. The transfer coefficient ceases to be a constant when the ligand concentration C changes significantly relative to K_D (Equation 16). Therefore, under ordinary circumstances, it is not possible to determine the transfer coefficients and hence the k_3/k_4 ratio under

TABLE 4
Binding Potentials at Tracer Doses of Selected Radioligands

Ligand	Receptor	k_3/k_4	Region	Ref.
Methylspiperone	Dopamine D_2	>10	Caudate-putamen	1[a]
Bromospiperone	Dopamine D_2	6.7	Caudate-putamen	6[a]
Spiperone	Dopamine D_2	6.4	Caudate-putamen	26[b]
Raclopride	Dopamine D_2	3.7	Putamen	14[a]
Carfentanil	μ-Opiate	3.4	Thalamus	15[a]
SCH-23 390	Dopamine D_1	2.0	Putamen	13[a]
Carfentanil	μ-Opiate	1.8	Frontal cortex	15[a]
Methylspiperone	Serotonin HT_2	1.5	Frontal cortex	36[a]
Haloperidol	Dopamine D_2	1.1	Caudate-putamen	26[b]

[a] Humans.
[b] Baboons.

nonsteady-state conditions when the ligand concentration is significant, i.e., when the radioligand is not subject to tracer kinetics. However, when the unlabeled ligand can be given so far in advance that its concentration in the circulation and brain tissue reaches approximate steady-state, as required by Equation 16, and when the plasma concentration of the ligand can be determined, it is then possible to determine the k_3/k_4 ratio at different degrees of saturation of the receptor sites and hence to calculate the binding constants. This experiment is essentially a competition with the ligand as its own inhibitor.

Huang et al.[23] solved Equations 1 and 14 directly by a double-injection procedure in which the coefficients V_a, K_1, k_2, $M_{b_{max}}$, and k_{off} were estimated in two steps. A constraint was imposed on the result; the authors fixed the $k_{off}V_d/k_{on}$ ratio at 1 pmol g^{-1}. In principle, the first injection, after which ligand concentrations were negligible, yielded estimates of K_1, k_2, $k_{on}M_{b_{max}}/V_d$, and k_{off}. This step was performed as a standard determination of K_1—k_4 according to Equation 5. The second step, during which all coefficients except K_1 assumed the values estimated for the first step, yielded k_{on}/V_d and $M_{b_{max}}$ separately. The procedure did not require complete equilibrium. In theory, a single low-specific activity injection would have sufficed but the number of parameters was too great to yield accurate estimates.

Farde et al.[14] solved the problem of the significant cross-product term (nonlinear factor) somewhat differently, using the information that $M_e(t)$ in Equation 14 can be separately determined as,

$$M_e(t) = V_e C_a(t) - \frac{V_e}{K_1}\left[\frac{dM(t)}{dt}\right] \tag{27}$$

using values of V_e and K_1 determined in the high specific activity injection experiment.

C. STEADY-STATE ANALYSIS OF REVERSIBLE BINDING

Equation 9 defines an equilibrium volume of distribution. According to Equation 17, the equilibrium volume declines when the receptors are saturated. Thus, when the ligand concentration rises in the distribution space, the equilibrium volume falls, as shown by Equation 17. The actual amounts of ligand held in the compartment can be determined by multiplying the volume of distribution with the arterial concentration of the ligand, $C_a(\infty)$. Multiplication with the arterial concentration yields the total quantity of tracer in the region at equilibrium:

$$M(\infty) = M_e + \frac{M_{b_{max}}C}{K_D + C} + M_a \tag{28}$$

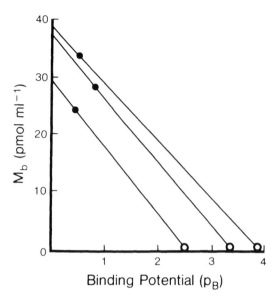

FIGURE 2.   Eadie-Hofstee-Scatchard plot according to Equation 29. Ordinate: binding potential p_B ("bound/free" ratio); abscissa: quantity of bound ligand in putamen of normal human volunteer injected with [^{11}C]raclopride at different specific activities. (Redrawn from Farde, L., Eriksson, L., Blomquist, G., and Halldin, C., *J. Cereb. Blood Flow Metab.*, 9, 696, 1989.)

Equation 28 can be rearranged to yield the equation underlying the linearized plot attributed to Rosenthal, Scatchard, Eadie, and Hofstee:

$$M_b = -p_B[V_d K_D] + M_{b_{max}} \tag{29}$$

where p_B is the binding potential equal to the steady-state M_b/M_e ratio. The ordinate intercept is $M_{b_{max}}$ and the slope is $-V_d K_D$. By the definition underlying Equation 14, at equilibrium, M_e and M_b can be calculated from the volumes of distribution V_e and V_m:

$$M_e(\infty) = V_d C(\infty) = V_e C_a(\infty) \tag{30}$$

and

$$M_b = V_m C_a \tag{31}$$

where $V_e = K_1/k_2$ and $V_m = p_B V_e$. An example of the use of Equation 29 is given in Figure 2.

Since k_3 and k_4 cannot be estimated easily under non-steady-state circumstances when the ligand concentration is significantly above tracer level, it is usually not a simple task to calculate M_b and M_e by regression. Fortunately, M_e and M_b can be determined in several other ways. If a particular region of the brain is known not to contain any receptor sites for the ligand under investigation, at steady-state, $M_e + M_a$ is by definition the tracer content in that region. Alternatively, if no such region exists, then Equation 17 must be used to determine the volume of distribution of the free ligand (V_e) as the volume of distribution of the ligand at saturating levels of $C(\infty)$. This can be accomplished by nonlinear regression of Equation 17 to simultaneously determined values of $V(\infty)$ and $C_a(\infty)$. The third method

TABLE 5
$M_{b_{max}}$ of Dopamine D_2 Receptor Sites in Neostriatum of Male Human Volunteers and Baboons

Subjects	$M_{b_{max}}$ (pmol g^{-1})	Type of binding	Type of analysis	Ref.
Humans	9	Irreversible	Transient	34
	14	Reversible	Steady state	12
	17	Irreversible	Transient	35
	25	Reversible	Steady state	13
	34	Reversible	Steady state	14
	35	Reversible	Transient	14
Baboons	30	Reversible	Steady state	26
	26	Reversible	Transient	23

of estimating M_e uses the volume of distribution of an enantiomer of the tracer which does not bind to the receptor sites under study but to all other sites in exactly the same manner and with the same physical properties as the tracer itself.

Similarly, the bound quantity M_b can be determined as the difference between the quantity of tracer in a binding region and the quantity of tracer in a nonbinding region. If a nonbinding region does not exist, the bound quantity must be determined from the volumes $V_e + V_a$, recorded at complete saturation,

$$M_b = M(\infty) - (M_a + M_e) = M(\infty) - C_a(V_e + V_a) \tag{32}$$

according to which

$$V_m = \frac{M}{C_a} - (V_e + V_a) \tag{33}$$

The Woolf plot of Equation 28 is

$$\frac{1}{p_B} = \frac{M_e}{M_{b_{max}}} + \frac{V_d K_D}{M_{b_{max}}} \tag{34}$$

which is identical to Equation 26.

Calculation of the binding potential p_B at (at least) two different ligand concentrations allowed Farde et al.[12,13] to determine the neostriatal $M_{b_{max}}$ and K_D values of the dopamine D_2 radioligand raclopride in this manner. The concentration of free ligand was estimated as the quantity of tracer in cerebellum at each dose of raclopride injected. The same procedure allowed Logan et al.[26] to calculate the neostriatal D_2 receptor $M_{b_{max}}$ value in baboons. The published estimates of neostriatal $M_{b_{max}}$ values in normal subjects are summarized in Table 5.

Recently, Farde et al.[14] compared the transient and steady-state analyses of the reversible binding of radiolabeled raclopride in striatum of normal humans. In this comparison, the authors reported no difference between the results of the two approaches but the comparison revealed an unexplained increase of the values of $M_{b_{max}}$ reported by the same authors, using the same equipment and analysis, from 1986 to 1989, as shown in Table 5.

IV. PROSPECTS FOR THE FUTURE

The advent of *in vivo* imaging of neuroreceptors and the quantitative analysis of neuroreceptor-radioligand interaction has significantly changed the perspectives of the neurochemistry of the human brain. It is now possible to assay the changes of receptor occupancy and enzyme activity that accompany the performance of specific functions of the human brain. These assays have the potential of establishing a link between the higher integration functions of cognition and behavior and fundamental neurochemical events which may significantly improve our understanding of normal brain function and our ability to treat disorders of cognition and behavior in diseases of the brain.

V. GLOSSARY

α — Solubility of ligand in sample of blood or plasma (source fluid) relative to water (ratio)

β — "Backflux" fraction of tracer subject to sequestration in tissue; the closer to unity β is, the higher the fraction of tracer that escapes sequestration; equal to $k_2/(k_2 + k_3)$ (ratio)

B — Quantity of bound ligand (pmol)

B_{max} — Maximal quantity of bound ligand (pmol)

C — Ligand concentration in water (nM)

C_a — Ligand concentration in arterial sample of source fluid (plasma or blood) (nM)

C_I — Inhibitor concentration in water (nM)

C_{I_a} — Inhibitor concentration in arterial sample (nM)

f_1 — Fraction of "free" ligand in an arterial sample; reciprocal of α (ml ml^{-1})

f_2 — "Fraction" of "free" ligand in a sample of brain tissue; reciprocal of V_d (g ml^{-1} or ml ml^{-1})

K_D — Michaelis half-saturation concentration of ligand (nM)

K_I — Inhibitor constant (nM)

K'_I — Concentration-dependent inhibitor constant; IC$_{50}$ of inhibitor (nM)

k_{in} — Frequency of internalization, transport, or inactivation of ligand-receptor complex (min^{-1})

k_{on} — Transfer coefficient for association of ligand to receptor sites (min^{-1} nM^{-1})

k_{off} — Frequency of dissociation of ligand from receptor site (min^{-1})

M — Total quantity of ligand and all labeled materials derived from ligand, in a sample (pmol g^{-1} or ml^{-1})

M_a — Total quantity of ligand and all labeled materials derived from ligand, in the vascular compartment of a sample of brain tissue (pmol g^{-1} or ml^{-1})

M_b — Quantity of specifically bound ligand in brain (pmol g^{-1} or ml^{-1})

M_{b_I} — Quantity of specifically bound inhibitor(s) in brain (pmol g^{-1} or ml^{-1})

$M_{b_{max}}$ — Maximal binding capacity of tissue receptor sites (pmol g^{-1}, or ml^{-1})

$M'_{b_{max}}$ — Maximal binding capacity of tissue receptor sites for ligand after blockade by inhibitors (pmol g^{-1} or ml^{-1})

M_e — Quantity of exchangeable (i.e., nonspecifically bound) ligand in brain (pmol g^{-1} or ml^{-1})

M_i — Total quantity of ligand and all labeled materials derived from ligand, held in compartment i (pmol g^{-1} or ml^{-1})

M_m — Total quantity of ligand and all labeled materials derived from ligand, specifically bound in brain, equivalent to M_b (pmol g^{-1} or ml^{-1})

p_i — Coefficient of independent variable X_i

p_B — "Binding potential" of ligand in brain, equivalent to k_3/k_4 (ratio)

ρ — Nonspecific and low-affinity "binding potential" of ligand, equal to $([V_d/V_w] - 1)$ (ratio)

t — Time (min)

T	Time of termination of experiment (min)
Θ	Normalized time-integral of C_a (min)
u	Dummy time variable (min)
V	Measured volume of distribution; ratio between M and C_a (ml g^{-1} or ml^{-1})
V_a	Actual vascular volume of distribution of ligand in sample of brain; ratio between M_a and C_a (ml g^{-1} or ml^{-1})
V_d	Volume of distribution of ligand in sample of brain relative to water; ratio between M_e and C (ml g^{-1} or ml^{-1})
V_e	Partition volume of exchangeable ligand in sample of brain relative to source fluid (plasma or blood); ratio between $M_e(\infty)$ and $C_a(\infty)$ (ml g^{-1} or ml^{-1})
V_f	Steady-state non-equilibrium volume of distribution of precursor of binding in sample of brain; equal to βV_e (ml g^{-1} or ml^{-1})
V_f'	Presteady-state non-equilibrium volume of distribution of precursor of binding in sample of brain; equal to M_e/C_a ratio (ml g^{-1} or ml^{-1})
V^i	Measured volume of distribution; ratio between M and C_a of region i (ml g^{-1} or ml^{-1})
V_{l_e}	Equilibrium volume of distribution of exchangeable inhibitor in sample of brain relative to source fluid (plasma or blood) (ml g^{-1} or ml^{-1})
V_m	Measured volume of distribution of specifically bound ligand in brain; equivalent to $p_B V_e$ and M_b/C_a (ml g^{-1} or ml^{-1})
V_w	Actual volume of water in sample of brain (ml g^{-1} or ml^{-1})
X_i	Independent variable i
Y	Dependent variable

REFERENCES

1. **Arnett, C. D., Wolf, A. P., Shiue, C. Y., Fowler, J. S., MacGregor, R. R, Christman, D. R., and Smith, M. R.,** Improved delineation of human dopamine receptors using [^{18}F]-N-methylspiroperidol and PET, *J. Nucl. Med.*, 27, 1878, 1986.
2. **Bahn, M. M., Huang, S.-C., Hawkins, R. A., Satyamurthy, N., Hoffman, J. M., Barrio, J. R., Mazziotta, J. C., and Phelps, M. E.,** Models for *in vivo* kinetic interactions of dopamine D$_2$-neuroreceptors and 3-(2'-[^{18}F]fluoroethyl)spiperone examined with positron emission tomography, *J. Cereb. Blood Flow Metab.*, 9, 840, 1989.
3. **Blomqvist, G.,** On the construction of functional maps in positron emission tomography, *J. Cereb. Blood Flow Metab.*, 4, 629, 1984.
4. **Bohr, C.,** *Experimentelle Untersuchungen über die Sauerstoffaufnahme des Blutfarbstoffes*, Copenhagen, 1885.
5. **Bohr, C.,** Theoretische Behandlung de quantitativen Verhältnisse bei der Sauerstoffaufnahme des Hämoglobins, *Zentralbl. Physiol.*, 23, 1, 1904.
5a. **Carson, R. E., Blasberg, R. G., Channing, M. A., Stein, S. D., Simpson, N. R., Cohen, R. M., and Herscovitch, P.,** Tracer infusion for equilibrium measurements: applications to ^{18}F-cyclofoxy opiate receptor with PET, *J. Cereb. Blood Flow Metab.*, 9(Suppl.), S203, 1989.
6. **Crawley, J. C. W., Crow, T. J., Johnstone, E. C., Oldland, S. R. D., Owens, D. G. C., Poulter, M., Pufter, O., Smith, T., Veall, N., and Zanelli, G. D.,** Dopamine D$_2$ receptors in schizophrenia studied *in vivo*, *Lancet*, II, 224, 1986.
7. **Cressie, N. A. C. and Keightley, D. D.,** Analysing data from hormone-receptor assays, *Biometrics*, 37, 235, 1981.
8. **Dixon, M.,** The determination of enzyme inhibitor constants, *Biochem. J.*, 55, 170, 1953.
9. **Eadie, G. S.,** The inhibition of cholinesterase by physostigmine and prostigmine, *J. Biol. Chem.*, 146, 85, 1942.
10. **Ehrlich, P.,** Über den jetzigen Stand der Chemotherapie, *Ber. Deutsch. Chem. Ges.*, 42, 17, 1909.
11. **Evans, A. C.,** A double integral form of the three-compartmental, four-rate-constant model for faster generation of parameter maps, *J. Cereb. Blood Flow Metab.*, 7(Suppl. 1), S453, 1987.

12. **Farde, L, Hall, H., Ehrin, E., and Sedvall, G.**, Quantitative analysis of D_2 dopamine receptor binding in the living human brain by PET, *Science*, 231, 258, 1986.

13. **Farde, L., Halldin, C., Stone-Elander, S., and Sedvall, G.**, PET analysis of human dopamine receptor subtypes, using ^{11}C-SCH 23390 and ^{11}C-raclopride, *Psychopharmacology*, 92, 278, 1987.

14. **Farde, L., Eriksson, L., Blomqvist, G., and Halldin, C.**, Kinetic analysis of central [11]Craclopride binding to D_2-dopamine receptors studied by PET — a comparison to the equilibrium analysis, *J. Cereb. Blood Flow Metab.*, 9, 696, 1989.

15. **Frost, J. J., Douglass, K. H., Mayberg, H. S., Dannals, R. F., Links, J. M., Wilson, A. A., Ravert, H. T., Crozier, W. C., and Wagner, H. N., Jr.**, Multicompartmental analysis of ^{11}C-carfentanil binding to opiate receptors in human measured by positron emission tomography, *J. Cereb. Blood Flow Metab.*, 9, 398, 1989.

16. **Gjedde, A., Wong, D. F., and Wagner, H. N., Jr.**, Transient analysis of irreversible and reversible tracer binding in human brain *in vivo*, in *PET and NMR: New Perspectives in Neuroimaging and in Clinical Neurochemistry*, Battistin, L. and Gerstenbran, F., Eds., Alan R. Liss, New York, 1986, 223.

17. **Gjedde, A. and Wong, D. F.**, Modeling neuroreceptor binding of radioligands *in vivo*, in *Quantitative Imaging: Neuroreceptors, Neurotransmitters, and Enzymes*, Frost, J. J. and Wagner, H. N., Jr., Eds., Raven Press, New York, 1990, 51.

18. **Gjedde, A.**, Tracer studies of neuro-receptor kinetics in vivo, in *Functional Imaging in Movement Disorders*, Martin, W. R. W., Ed., CRC Press, Boca Raton, FL, 1990.

19. **Gjedde, A. and Wong, D. F.**, Positron tomographic quantitation of neuroreceptors in human brain *in vivo*-with special reference to the D_2 dopamine receptors in caudate nucleus, *Neurosurg. Rev.*, 10, 9, 1987.

20. **Haldane, J. B. S.**, Graphical methods in enzyme chemistry, *Nature*, 179, 832, 1957.

21. **Hill, A. V.**, The possible effects to aggregation of the molecules of haemoglobin on its dissociation curves, *J. Physiol.*, 40, 4, 1910.

22. **Hofstee, B. H. J.**, Graphical analysis of single enzyme systems, *Enzymologia*, 17, 273, 1954—1956.

23. **Huang, S.-C., Bahn, M. M., Barrio, J. R., Hoffman, J. M., Satyamurthy, N., Hawkins, R. A., Mazziotta, J. C., and Phelps, M. E.**, A double-injection technique for *in vivo* measurement of dopamine D_2-receptor density in monkeys with 3-(2′-^{18}F] fluoroethyl)spiperone and dynamic positron emission tomography, *J. Cereb. Blood Flow Metab.*, 9, 850, 1989.

24. **Langley, J. N.**, On the physiology of the salivary secretion. Part II. On the mutual antagonism of atropin and pilocarpin, having especial reference to their relation in the submaxillary gland of the cat, *J. Physiol.*, 1, 339, 1878.

25. **Langmuir, I.**, The constitution and fundamental properties of solids and liquids, *J. Am. Chem. Soc.*, 37, 2221, 1916.

26. **Logan, J., Wolf, A. P., Shiue, C. Y., and Fowler, J. S.**, Kinetic modeling of receptor-ligand binding applied to positron emission tomographic studies with neuroleptic tracers, *J. Neurochem.*, 48, 73, 1987.

27. **Michaelis, L. and Menten, M. L.**, Die Kinetik der Invertinwirkung, *Biochem. Z.*, 49, 333, 1913.

28. **Mintun, M. A., Raichle, M. E., Kilbourn, M. R., Wooten, G. F., and Welch, M. J.**, A quantitative model for the *in vivo* assessment of drug binding sites with positron emission tomography, *Ann. Neurol.*, 15, 217, 1984.

29. **Perlmutter, J. S., Larson, K. B., Raichle, M. E., Markham, J., Mintun, M. A., Kilbourn, M. R., and Welch, M. J.**, Strategies for *in vivo* measurement of receptor binding using positron emission tomography, *J. Cerebr. Blood Flow Metab.*, 6, 154, 1986.

30. **Perlmutter, J. S., Kilbourn, M. R., Raichle, M. E., and Welch, M. J.**, MPTP-induced up-regulation of *in vivo* dopaminergic radioligand-receptor binding in humans, *Neurology*, 37, 1575, 1987.

31. **Rosenthal, H.**, A graphic model for the determination and presentation of binding parameters in a complex system, *Anal. Biochem.*, 20, 525, 1967.

32. **Scatchard, G.**, The attractions of proteins for small molecules and ions, *Ann. N.Y. Acad. Sci.*, 51, 600, 1949.

33. **Wong, D. F., Gjedde, A., and Wagner, H. N., Jr.**, Quantification of neuroreceptors in the living human brain. I. Irreversible binding of ligands, *J. Cereb. Blood Flow Metab.*, 6, 137, 1986a.

34. **Wong, D. F., Gjedde, A., Wagner, H. N., Jr., Dannals, R. F., Douglass, K. H., Links, J. M., and Kuhar, M. J.**, Quantification of neuroreceptors in the living human brain. II. Inhibition studies of receptor density and affinity, *J. Cereb. Blood Flow Metab.*, 6, 147, 1986b.

35. **Wong, D. F., Wagner, H. N., Jr., Tune, L. E., Dannals, R. F., Pearlson, G. D., Links, J. M., Tamminga, C. A., Broussole, E. P., Ravert, H. T., Wilson, A. A., Toung, J. K. T., Malat, J., Williams, J. A., O'Tuama, L. A., Synder, S. H., Kuhar, M. J., and Gjedde, A.**, Positron emission tomography reveals elevated D2 dopamine receptors in drug-naive schizophrenics, *Science*, 234, 1558, 1986c.

36. **Wong, D. F., Gjedde, A., Dannals, R. F., Lever, J. R., Hartig, P., Ravert, H., Wilson, A., Links, J., Villemagne, V., Braestrup, C., Harris, J., and Wagner, H. N., Jr.,** Quantification of neuroreceptors in living human brain: equilibrium binding kinetic studies with dopamine and serotonin receptors, *J. Cereb. Blood Flow Metab.,* 7(Suppl. 1), S356, 1987.
37. **Young, S. W. and Kuhar, M. J.,** Serotonin receptor localization in rat brain by light microscopic autoradiography, *Eur. J. Pharmacol.,* 62, 237, 1980.

Chapter 18

IN VIVO STUDIES OF THE CENTRAL BENZODIAZEPINE RECEPTORS IN THE HUMAN BRAIN WITH POSITRON EMISSION TOMOGRAPHY

Pascale Abadie and J. C. Baron

TABLE OF CONTENTS

I. INTRODUCTION

Since their discovery in 1977 by Squires and Braestrup[1] and Möhler and Okada,[2] the central-type benzodiazepine receptor (cBZR) has attracted considerable interest because it has been shown to mediate the sedative, anticonvulsant, and anxiolytic properties of the benzodiazepines[3] and hence may be implicated in the pathophysiology of epilepsy and anxiety. It has been clearly demonstrated that the cBZR is part of the macromolecular $GABA_A$ receptor, on which it acts by modulating the interaction between GABA and its receptor and, in turn, the frequency of opening of the Cl^- channel.[3] The cBZR therefore modulates the activity of the main inhibitory neuronal system in the CNS; it is activated by agonists such as the benzodiazepines and related compounds (e.g., zolpidem), the action of which is antagonized by flumazenil, while inverse-agonists such as the β-carbolines induce convulsions and anxiety.[3] In the mammalian brain, the density of the cBZR is high in the cerebral cortex, lesser in the cerebellar cortex, low in the basal ganglia and thalamus, and negligible in white matter.[3] A number of putative endogenous ligands have been investigated, including the so-called diazepam binding inhibitor, but none has been fully validated as yet.[3] For review on the general properties of the cBZR, the reader is referred to the excellent book by Müller.[3]

The cBZR must be distinguished from the "peripheral-type benzodiazepine receptor" which binds with variable affinity the benzodiazepines, but with high affinity nonbenzodiazepine compounds such as PK-1195 and RO-5-4864; flumazenil has no affinity for this site. This receptor is present in low density in the normal brain, but is very densely present in lung, kidneys, and heart; in the brain, it is mainly localized on glial cells and is found in high density on glial and macrophagic cells in tumoral and inflammatory conditions. There is as yet no established functional role assigned to this binding site.[3]

In what follows we will describe the state of the art in the investigation of the cBZR in the human brain *in vivo* by means of PET.

II. RADIOLIGANDS

A. ¹¹C-FLUNITRAZEPAM

Flunitrazepam (FNZ) is an agonist of the benzodiazepine family, with affinity in the nanomolar range for the cBZR and no affinity for the peripheral site. It has been and is still

TABLE 1
In Vitro Binding Studies of Human Brain for Central Benzodiazepine Receptor

Brain regions	Number of samples	Binding (temp)	Ligand	B_{max} (fmol/mg protein)	K_d (nM)	Ref.
Frontal cortex	3	Homogenates	^{3}H-Diazepam	860 ± 130[a]	3.5 ± 0.3	14
Temporal cortex		(4°C)		670 ± 150[a]	4.9 ± 1.3	
Occipital cortex				840 ± 110[a]	4.8 ± 0.8	
Hippocampus				670 ± 120[a]	4.2 ± 0.6	
Cerebellar cortex				580 ± 40[a]	4.2 ± 0.5	
Pons				160 ± 20[a]	5	
Cerebral cortex	6	Homogenates	^{3}H-RO-15-1788	2300 ± 100[b]	1.2 ± 0.1	15
		(0°C)	^{3}H-Flunitrazepam	2200 ± 100[b]	2.7 ± 0.5	
Hippocampi	5	Homogenates	^{3}H-Flunitrazepam	111 ± 8.5[ac]	1.94 ± 0.3	16
Frontal cortex	20	Homogenates	^{3}H-Flunitrazepam	1728 ± 55[a]	1.86 ± 0.06	17
Temporal cortex	21	(0°C)		1385 ± 96[a]	1.63 ± 0.06	
Frontal cortex	4	Homogenates	^{3}H-RO-15-1788	1289 ± 34[a]	0.69 ± 0.03	18
Temporal cortex	4	(0°C)		851 ± 160[a]	0.61 ± 0.03	
Occipital cortex	5			789 ± 118[a]	0.66 ± 0.04	
Gray matter	6	Homogenates	^{3}H-RO-15-1788	500 ± 70[b]	3.7	19
		(0°C)		30 ± 5[b]	3.5	
White matter		(samples obtained at biopsy during brain surgery)				

[a] mean ± SEM.
[b] mean ± SD.
[c] fmol/mg tissue.

used as a reference ligand for *in vitro* studies of the cBZR in the human brain (Table 1). For PET use, it has been labeled in high specific radioactivity (SRA = 500 to 1000 mCi/μmol) with the 20.3-min half-life radionuclide ^{11}C in the methyl position by methylation of nor-FNZ with methyl-iodide.[4,5]

1. Validation Studies in Baboons

Comar et al.[4,5] have reported pharmacokinetic and validation studies following intravenous bolus injection of ^{11}C-FNZ in baboons using the single slice ECAT II device. In whole blood, these authors showed an essentially stable radioactive concentration at about 5% of injected dose per liter (DI/L), lower than the value obtained in brain. In the latter organ taken as a whole because of the poor resolution of the camera used, ^{11}C-FNZ penetration was very rapid with a peak at 2 min of about 15% DI/L, followed by a rapid washout for about 30 min at which time a lesser clearance is seen (with values of about 10% DI/L). Due to the lower concentration in blood, the contribution of the vascular compartment to the PET measured radioactivity in brain was deemed negligible; extracerebral radioactivity was also found much lower than brain.[4,6]

Because of the labeling position, demethylation of ^{11}C-FNZ was possibly involved in the PET findings reported above. This was evaluated by measurement of expired [11]CO_2 over 1 h,[7] which was found to account for about 10% of the total radioactivity injected. The brain images published[8] suggest a preferential distribution of ^{11}C-FNZ in cerebral cortex and cerebellum, within the first 5 min after tracer administration. However, 20 to 30 min later, this initial distribution to gray matter structures is now changed to a more homogenous pattern, indicating a faster clearance of ^{11}C-FNZ from gray matter despite its known enrichment in cBZR.

The contribution of specific binding of the radioligand to the total brain radioactivity registered in high SRA studies was demonstrated by Comar et al.[4,5] in pioneering experiments using cold loads of lorazepam as the displacer. When administered intravenously in therapeutic amounts 30 min after ^{11}C-FNZ, this high affinity competitor induced a rapid displacement of the radiotracer from brain, which can be evaluated at about 35% of total radioactivity. This indicates a large nonspecific binding of ^{11}C-FNZ *in vivo* at t = 30 min; presumably, a large part of the specific binding is rapidly washed out from brain in the first few minutes following injection, a fact already noted in rats by Duka et al.,[9] as a result of reduced affinity at physiological temperature (the K_D of ^{3}H-FNZ, which is low at 0°C, see Table 1, is fivefold larger at 35°C, with a twofold increase in k_{off}.

Stereoselectivity of *in vivo* binding of ^{11}C-FNZ was assessed by using two benzodiazepinic stereoisomers, RO-11-6896 and RO-11-6893.[8] Both isomers, when injected in loading doses at 30 min post-^{11}C-FNZ, induced a large increase in brain radioactivity as well as blood radioactivity. This unexpected result was interpreted as due to displacement of specific binding of ^{11}C-FNZ to peripheral benzodiazepine receptors in peripheral organs by the two stereoisomers, both having affinity for these sites; the newly circulating ^{11}C-FNZ would then have access to brain tissue, hence masking any underlying displacement. Similar, although less striking results, were also obtained when chlordiazepoxide was used as the displacer.

2. Human Studies

Only preliminary results were published,[8] in the form of PET images obtained 5 to 10 min following administration of ^{11}C-FNZ at tracer doses. These images show a perfusion-like distribution with similarly high uptake in all gray matter structures and low uptake in white matter. Problems were apparently encountered with respect to the product's solubility (for intravenous injection the alcohol medium was painful) and unpracticality of displacement studies due to the pharmacological effects of the drugs.

3. Conclusion

Several problems in the use of the agonist ^{11}C-FNZ for the study of the cBZR with PET were apparent:

1. Large nonspecific binding
2. Rapid initial wash-out of tracer from brain tissue
3. *In vivo* demethylation of ^{11}C-FNZ
4. Unpractical displacement studies with benzodiazepines because of pharmacological effects such as displacement of peripherally-bound radioligand, allosteric changes in the cBZR, and sedation
5. An intrinsically variable and relatively low affinity of ^{11}C-FNZ for the cBZR due to sensitivity to both GABA and temperature[10]

These problems did not qualify ^{11}C-FNZ as a useful ligand for PET studies. Recently, ^{18}F-labeled-fluorodiazepam has been tested in rats with similarly disappointing *in vivo* features.[11]

B. ^{11}C-SURICLONE

Suriclone is a nonbenzodiazepine agonist for the cBZR of the cyclopyrrolone family. It was selected by Frost et al.[12] because it would give complementary information on the cBZR *in vivo*. Also, this radioligand was shown *in vitro* to increase its affinity for the cBZR by a factor of 10 from 0 to 37°C.[13] In the baboon, following intravenous injection of 20 mCi at a SRA of 1200 mCi/μmol, the PET images taken at 55 to 65 min indicated a high uptake in cerebral cortex and cerebellum, with negligible binding in the caudate nucleus. Pretreatment with 1 mg/kg of the antagonist flumazenil induced a fourfold reduction in brain uptake of ^{11}C-Suriclone at 55 to 65 min without affecting uptake in extracerebral tissues. These encouraging results led the authors[12] to conduct PET studies in three human beings (20 mCi, 6 to 12 μg). The summed images 30 to 60 min postinjection indicate a high uptake in cerebral cortex and cerebellum, intermediate uptake in thalamus, and low uptake in caudate, with stable kinetics in brain between 3 and 72 min. However, no quantitative data were provided, and displacement studies were not performed. In addition, the PET images published[12] indicate a distribution of tracer among gray matter structures in human brain far too uniform compared to the known heterogeneity of the cBZR and the distribution of ^{11}C-flumazenil (see below). The lack of further publication with this radioligand since the initial work of Frost et al.[12] indeed suggests it is not a very useful method for PET studies of the cBZR.

C. ^{11}C-FLUMAZENIL (^{11}C-FLU)

This radioligand has become the reference for PET studies of the cBZR. Flumazenil (RO-15-1788) is an imidazo-benzodiazepine which lacks the aromatic cycle in position 5 of the C cycle that is typical of the benzodiazepines. It is a highly specific, selective antagonist of the cBZR which has been in clinical use for several years in toxicology and anesthesia. It can be labeled with ^{11}C at high SRA (0.5 to 1.8 Ci/μmol) by methylation of its nor-precursor using ^{11}C-methyl-iodide derived from ^{11}C-CO_2.[20,21]

A number of investigations have shown flumazenil to be a better radioligand than flunitrazepam for the study of the cBZR in vivo. First, its dissociation from the receptor is much slower in vitro, and its binding to the cBZR is not GABA dependent.[10,22] Second, it has no pharmacological effect and displacing doses can be used without risk of inconvenience.[23] Third, in vivo studies in rodents with ^{3}H-flumazenil have shown a highly detectable specific binding, a regional tracer distribution similar to that obtained in in vitro binding studies, no binding to the peripherical BZR, and negligible nonspecific binding[24,25] (^{3}H-flunitrazepam, in contrast, showed very low specific binding in vivo), while brain radio-activity was found as almost exclusively unchanged flumazenil.[26] Finally, plasma protein binding of ^{3}H-flumazenil in mice was favorably low (around 45% of total).[27]

1. Validation Studies in Baboons with PET

Extensive PET investigations have been performed in baboons by the group of Mazière at Orsay,[7,28] using ^{11}C-FLU injected intravenously as a bolus.

a. Blood Kinetics of ^{11}C-FLU

Following the initial peak, a very slow wash-out of whole blood ^{11}C occurred during the PET studies (80 min). Metabolism of ^{11}C-FLU in the baboon has not been studied, but total demethylation, as assessed by expired [11]CO_2, was found to be negligible (less than 1% in 1 h).[7,28]

b. Brain Distribution and Kinetics of Tracer Amounts of ^{11}C-FLU

There is a considerable initial penetration of the radiotracer in brain, indicating easy transfer across the blood-brain barrier (BBB). This is followed by a slow rise, a maximum being reached in grey matter structures at around t + 10 min; thereafter, a progressive wash-out of ^{11}C is observed in all regions. The brain images show first a tracer distribution roughly according to perfusion, soon with relative redistribution in favor of regions known to be rich in cBZR such as the cerebral cortex and the cerebellum. In the latter structure, the maximum tracer uptake is around 5% DI/L.

c. Displacement and Inhibition of Specific Binding[7,28,29]

The intravenous injection of a pharmacological amount of cold flumazenil, 20 min following that of tracer doses of the ^{11}C-labeled tracer, induces an immediate, abrupt decrease of ^{11}C-radioactivity in gray matter structures to reach almost, as a step function, a lower level, the value of which is strictly dose dependent. However, beyond 1600 nmol/kg of flumazenil, no additional displacement is seen indicating full saturation of the cBZR. At this dose, 85 to 90% of the radioactive concentration of control experiments has been displaced, indicating that the total control radioactivity is composed of 85 to 90% specific binding and only 10 to 15% of nondisplaceable binding accounting for both free ligand and nonspecific-binding (if any). Following partial displacement with nonsaturating doses, the residual brain ^{11}C-kinetics are grossly similar and parallel to control kinetics.

Inhibition experiments with dilution of ^{11}C-FLU with increasing amounts of cold flumazenil have produced results similar to displacement experiments. There was a dose-dependent reduction in maximal uptake and pharmacokinetics resembling control values but at a lower range.[28]

Displacement has also been demonstrated using a benzodiazepine, lorazepam. Using this cBZR agonist as the displacer, the results have been identical to those obtained with the parent drug, although maximal displacement (88% of control) was obtained for amounts twice as large with lorazepam as with flumazenil (3200 nmol/kg vs. 1600 nmol/kg), indicating a more effective interaction with cBZR for the latter compound.[28]

d. Stereospecificity

Clear-cut displacement of ^{11}C-FLU was obtained with the active D-enantiomer of the agonist RO-11-6896, while a marginal displacement was seen following administration of identical amounts of the inactive enantiomer RO-11-6893, indicating stereoselectivity of the specific binding of ^{11}C-FLU in the baboon's brain.[28]

e. Pharmacological Investigations

Hantraye et al.[28] showed that the dose of pentylentretetrazol (PTZ) necessary to induce a generalized seizure was two to three times higher in PET studies which used ^{11}C-FLU at drug doses ranging 64 to 320 nmol/kg than in those at <2 nmol/kg. Since PTZ acts on the GABA$_A$ complex at the picrotoxin site, the pharmacological interaction observed indicated that higher receptor *in vivo* occupancy with ^{11}C-FLU occurred when the dose of cold drug was increased. In another series of experiments, the same authors[29] actually estimated the occupancy of cBZR in the temporo-occipital cortex by increasing doses of cold flumazenil added 20 min after tracer amounts of ^{11}C-FLU. To that end, they used the following formula: % occupancy = % ^{11}C-FLU displaced/specific binding, the latter being defined in full saturation studies at about 90% of total ^{11}C-FLU. The results showed that flumazenil induced a dose-dependent saturation, such as occupation of receptors which reached its maximum (98%) for doses beyond 0.33 µmol/kg. The monophasic relationship obtained suggested that a single population of receptors sites is labeled *in vivo* with ^{11}C-FLU, in agreement with *in vitro* studies (two subtypes of the cBZR have, however, been identified with the non-benzodiazepine compounds CL 218872 and zolpidem[3]). The Orsay team has subsequently published a number of PET studies in baboons on the *in vivo* interaction between ^{11}C-FLU and various ligands of the GABA$_A$ complex (partial or inverse agonists and antagonists) which have confirmed the pharmacological reliability of the method.[30-34]

2. Validation Studies in Humans

The validation studies in baboons just described established ^{11}C-FLU as an ideal radioligand for PET studies of the cBZR. Application to human beings was all the more possible in that flumazenil had been marked and used in several countries without any serious adverse effects reported following intravenous administration of very large amounts (100 mg).[23,35] Only very mild, inconsistent effects of the agonist or the inverse agonist-type have been occasionally reported.[36,38] This allowed displacement studies in humans with flumazenil without the fear of adverse, unethical effects.

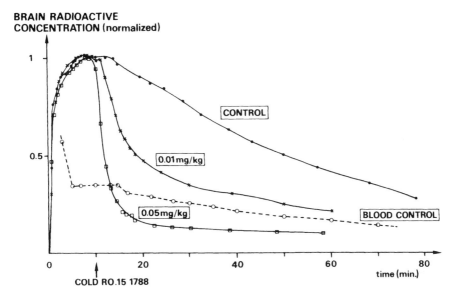

BRAIN RADIOACTIVE
CONCENTRATION (normalized)

FIGURE 1. [¹¹C]-flumazenil brain kinetics in occipital cortex: one control study and two displacement studies are shown. Curves are expressed in arbitrary units, the individual ¹¹C concentration values being normalized to that obtained at t = 9.5 min (i.e., immediately before the intravenous injection of unlabeled flumazenil). Also shown is the time course of ¹¹C radioactive concentration in venous blood during the control study, relative to the maximum concentration of brain radioactivity. (From Samson, Y., Hantraye, P., Baron, J. C., Soussaline, F., Comar, D., and Mazière, M., *Eur. J. Pharmacol.*, 110, 247, 1985. With permission.)

a. Blood Kinetics of ¹¹C-FLU (Figure 1)

Following an initial peak, a progressive decline in whole blood radioactivity is observed.[29,39,41] However, a slight or moderate rebound of radioactivity is generally present around t + 15 min, indicating the appearance of labeled metabolites in blood. These labeled metabolites have been investigated and identified (see below).

b. Regional Brain Pharmacokinetics of ¹¹C-FLU (Intravenous Bolus Injection of Trace Amounts)

A recent study in humans has demonstrated that brain penetration of ¹¹C-RO-15-3890, the almost exclusive labeled metabolite of ¹¹C-FLU, is almost nonexistent[42] indicating that the observed ¹¹C kinetics after ¹¹C-FLU injection represent unchanged ligand (see below). In all brain regions a rapid uptake of ¹¹C-FLU is observed, reflecting easy transfer across the blood-brain-barrier. A slower rise is seen in grey-matter regions until about t + 10 min, followed by a quasi-plateau of about 5 min and then a slow decline (see Figure 1). In white matter, the maximum uptake occurs earlier (2 to 7 min) and at a value lower than in gray matter regions. While the initial uptake is grossly proportional to perfusion, with all gray matter structure having a somewhat identical uptake, there is rapid redistribution in favor of the cerebral cortex, followed in order by the cerebellar cortex and the basal ganglia-thalamus[29,39,43] (see Plate 1* and Figure 2) in agreement with known differences in density of the cBZR as determined for the human brain *in vitro* (see Table 1).

c. Displacement and Inhibition Studies

Samson et al.[39] and Persson et al.[29] demonstrated that, as in baboons,[28,29] intravenous administration of cold loads of flumazenil 10 to 20 min following trace amounts of ¹¹C-

---

* Plate 1 follows page 198.

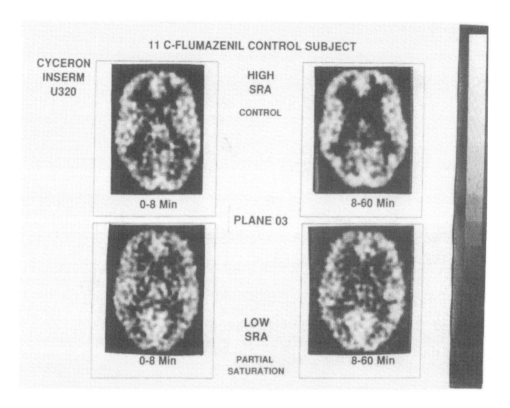

FIGURE 2.   ^{11}C-flumazenil brain images obtained in the same subject as in Plate 1, showing only plane 03 (basal ganglia-thalamus level). The subject was studied twice, first in the control (high SRA; top row image) and then in the partial (low SRA, bottom row images by co-injection of about 1.5 mg of unlabeled flumazenil) condition, but receiving 10 mCi on both occasions. Initial images (accumulated from 0 to 8 min postinjection) are shown on the left side, and late images (8 to 60 min) are shown on the right. The early images are identical in both conditions, showing a perfusion-like distribution of ^{11}C-FLU. In the control conditions, the late image shows redistribution of the radiotracer with high concentration in the cerebral cortex and low radioactivity in the thalamus and basal ganglia (see Plate 1). In the partial saturation study, lower radioactive concentrations in brain tissue are obtained relative to extracerebral tissues, but the distribution is similar to that seen in the control state.

FLU induced an immediate, abrupt, dose-dependent displacement of radioligand from gray matter structures (around 90% for a dose of 20 mg) with smaller effects in white matter (see Figure 1). Inhibition of specific binding of ^{11}C-FLU was achieved either by pretreatment with cold flumazenil (given orally[40] or intravenously[43]) or by coinjection with the radioligand.[41,43] As in baboons,[28] these studies demonstrated a dose-dependent inhibition of ^{11}C-FLU uptake in gray matter structures, the maximum of which was in the order of 90% of control for doses of 10 mg or 0.1 mg/kg[41-43] (Figures 2 and 3). These studies concurred in showing a very large specific binding of ^{11}C-FLU in the human brain *in vivo* in humans. The occurrence of partial displacement or inhibition of ^{11}C-FLU uptake in white matter and pons, of the order of 30 to 40% of total uptake in control state,[39,41,43] was unexpected because of the negligible density of cBZR in these areas (see Table 1); it was interpreted as an artifact of imaging due to the partial volume effect from highly radioactive neighboring gray matter structures.

d. Pharmacological Correlations

Persson et al.[29] showed in two patients treated with large dose of benzodiazepines a moderately faster dose-dependent wash-out of ^{11}C-FLU from cerebral cortex (despite a similar

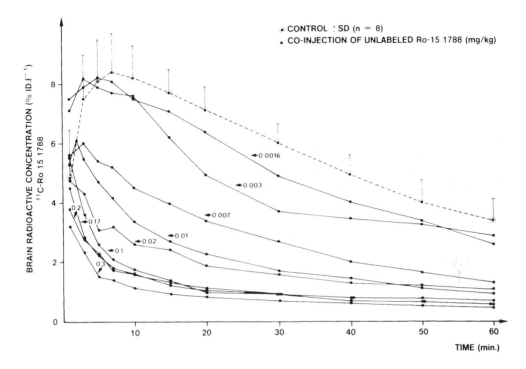

FIGURE 3. Kinetics of ^{11}C-flumazenil obtained in the occipital cortex in nine co-injection studies with increasing amounts of unlabeled flumazenil. For comparison, the control kinetics are also shown. Inhibition of ^{11}C-flumazenil uptake is apparent, with a clear dose dependence. Beyond 0.1 mg/kg of flumazenil, there is no significant further inhibition of ^{11}C-flumazenil brain uptake at times > 20 min. (From Pappata, S., Samson, Y., Chavoix, C., Prenant, C., Mazière, M., and Baron, J. C., *J. Cereb. Blood Flow Metab.*, 8, 304, 1988. With permission.)

initial uptake), indicating pharmacological competition with inhibition of ^{11}C-FLU binding reflecting cBZR occupancy. Further studies on occupancy of cBZR by orally given benzodiazepines as measured by ^{11}C-FLU and PET have also appeared;[44,45] they will be described later.

e. In Conclusion

[11]C-FLU was established as the ligand of choice for the investigation of the cBZR in the humans brain with PET. It has most of the characteristics that define ideal radioligands for PET studies:[46] selectivity for the cBZR high penetration through the BBB, negligible entry of the labeled metabolites in brain, negligible nonspecific binding, lack of pharmacological untoward effects, rapid rate of association and adequate rate of dissociation, and easily displaceable specific binding.

III. QUANTITATION

In what follows, methods for quantitation of regional specific binding of ^{11}C-FLU to cBZR and, in turn, of density/affinity of the cBZR will be described. Two main approaches have been used: "equilibrium" methods and "dynamic" methods.

A. "EQUILIBRIUM" METHODS

They are all based on the basic pharmacological equilibrium between concentrations of the specifically bound and the free radioligand which, if nonspecific binding is assumed negligible, can be written as: $B = T - F$ (1), where T is the total radioligand concentration in the region of interest and B and F the specifically bound and the free radioligand concentrations in the same region, respectively. If B can be determined in this way, then either the density (B_{max}) and affinity (K_d) or the ratio B_{max}/K_d (the so-called "binding potential") can be determined using appropriate methods. However, there are a number of underlying assumptions to this approach which will be discussed now.

1. Assumptions and Validation

a. Brain Radioactivity Consists Only of the Parent Radioligand and not of Labeled Metabolites

Although this has not been proven for ^{11}C-FLU in the human brain, studies in rats 30 min after an *in vivo* administration of ^{3}H-RO-15-1788 demonstrated that only pure radioligand was recovered in brain.[26] The main metabolite of flumazenil is the acid derivative, RO-15-3890 (see below). Labeling with ^{11}C of this metabolite has been carried out,[47] and this compound has been administered intravenously to man,[48] the results convincingly demonstrated that this metabolite has no access to brain tissue, with a brain/blood ratio of radioactivity between 12 to 36 min of about 0.042 (which is exactly what would be expected for a purely intravascular tracer), compared to a corresponding ratio of 6.40 for ^{11}C-flumazenil (corrected for circulating ^{11}C-metabolites). Hence, although it remains possible that minute amounts of the acid metabolite are formed within the brain, it can be said that brain radioactivity measured by PET following ^{11}C-FLU injection is wholly or quasi-exclusively in the parent form.

It is appropriate to review now the analysis of labeled metabolites of ^{11}C-FLU in blood. Two main methods for separation have been used: (1) thin layer chromatography (TLC) on silica gel with chloride-methanol (9:1) and hexane-benzene-dioxane-ammonium hydroxide (70:50:45:5) as solvent[25,40] and (2) extraction of flumazenil by the organic solvent methylene chloride.[40] Because the former method is complex and slow relative to the decay of ^{11}C, the latter was prefered by Shinotoh et al.[40] for its simplicity. As a validation, they just indicated that TLC analysis "showed that >99% of the radioactivity in the extracted fraction of the blood was that of unmetabolized ^{11}C-RO-15-1788." They demonstrated in two volunteers that there was a rapid metabolism to water-soluble compounds which accounted for the secondary rise in whole blood ^{11}C concentration around the tenth minute (see above). In an extensive validation study, Barré et al. showed in rabbits that the extraction procedure (with chloroform and buffer pH 11) was efficient (yield for ^{11}C-FLU recovered in organic fraction = 98%; the acid metabolite wholly remained in the aqueous fraction) and reasonably accurate as compared to TLC, the latter showing only one metabolite peak (with Rf = 0 and corresponding to the acid); the percentage of circulating metabolites was similar to those reported by Potier et al.[25] in rats injected with ^{3}H-flumazenil (i.e., 48% at 3 min and 65% at 20 min). In two baboons, Barré et al.[59] were able to confirm that the extraction method was as accurate as TLC, with percentages of metabolites around 25, 40, and 44% of total activities at 5, 20, and 60 min following ^{11}C-FLU injection respectively. Similar prelimary findings have been obtained in humans.

b. Nonspecific Binding (NS) in Brain is Negligible

In agreement with *in vivo* binding and autoradiographic studies,[22] Goeders and Kuhar[24] found that NS was much less than 10% of total binding in brain membranes of rat following intravenous injection of ^{3}H-flumazenil and could be neglected for quantitative and autoradiographic studies (it was around 5% for Potier et al.[25]). *In vivo* PET studies with ^{11}C-FLU have demonstrated in both baboons and humans that the displaceable fraction in cortical areas 10 to 20 min following injection was about 90% of total ^{11}C-FLU (see above). The remaining 10% hence must represent the sum of free and NS, indicating that the latter can safely be assumed negligible.

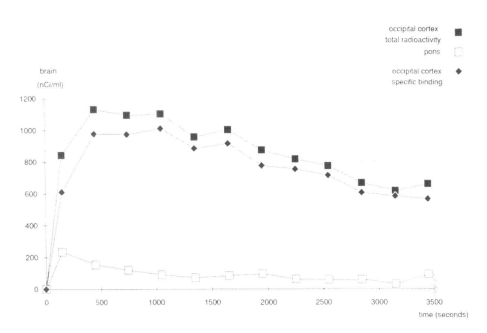

FIGURE 4. Kinetics of ^{11}C-flumazenil in the occipital cortex and in the pons measured in a young healthy untreated volunteer in a control (high SRA) study. The data were obtained as described in Plate 1. Subtraction of the pons kinetics, which reflect essentially the free radioligand, from the occipital kinetics (bound + free) radioligand provides the kinetics of the specifically bound ligand in the region.

c. Determination of Free and Bound Ligand in Brain Regions

Because, as just discussed, nondisplaceable ^{11}C-FLU in brain is mainly or exclusively in the form of free ligand, concentration of the latter in various brain regions can be estimated in situations or areas where specific binding does not occur. Hence, when full saturation of the cBZR by cold competitor (especially with cold flumazenil) is achieved, the concentration of ^{11}C-FLU in brain regions can be taken to represent free radioligand[40,41] (see Figure 3). Alternatively, radioactivity in pons (which, based on *in vitro* data, contains virtually no cBZR) can be used to estimate nondisplaceable ^{11}C-FLU (Figure 4).

Pappata et al.[41] study — In their extensive studies, Pappata et al.[41] reported in four volunteers that brain uptake of ^{11}C-FLU (in % DI/L) at times >20 min did not decrease further for amounts of coinjected cold flumazenil beyond 0.1 mg/kg (see Figure 4). For example, the mean percent DI/L for this nondisplaceable fraction of ^{11}C-FLU was 0.98, 0.85, 0.71, and 0.66 (average SD ≃ 0.11, N = 4) for the occipital cortex data measured at 20, 30, 40, and 50 min, respectively; this can be compared to the corresponding average uptakes (in percent DI/L) of 7.1, 6, 4.8, and 4 found in eight control (high SRA) studies, respectively.[41] Hence by this approach, these authors were able to determine the time course of free ^{11}C-FLU uptake in various brain regions; for example, in the occipital cortex, there was an initial peak of ^{11}C-FLU at ~2 min followed by a rapid wash-out until ~10 min and a very slow clearance thereafter; both the peak and the rest of the brain curve were well below "control" uptakes, and the kinetics of brain tracer were very different from "control" brain curves (see Figure 3). Similiar observations were made by Persson et al.[43] and Savic et al.[49]

The time course of free ^{11}C-FLU differed, however, among cerebral cortex, white matter, and cerebellum.[41] Hence, at t = 30 min, it represented on average 15, 14, and 14% of the "control" (high SRA) studies in the occipital, frontal, and parieto-occipital cortex, respectively, compared to 24 and 41% for the cerebellum and corona radiata, respectively.[41] While the difference between cerebral cortex and cerebellum seemed to indicate a true nonuniformity in free ^{11}C-FLU uptake, the result in white matter was unexpected as it suggested that a 59% fraction of total ^{11}C-FLU actually bound to receptor sites.

The interpretation made by Pappata et al.[41] was that due to the rather low spatial resolution of the camera used, there occurred contamination of counts from surrounding gray matter (partial volume effect). This important issue will be further discussed later.

Having determined the average time course of the free (F) ^{11}C-FLU in various cortical regions and cerebellum, Pappata et al.[41] went on to estimate the specifically bound (B) ^{11}C-FLU (B) in these regions by simply subtracting this value from the total uptake measured in the corresponding regions in both controls (N = 8, high SRA) and partial saturation studies (N = 5, dose of cold flumazenil ranging from 0.0016 to 0.02 mg/kg), expressed in percent DI/L. In controls they obtained time courses of B showing a slow rise over 10 min, a smooth plateau for about 5 min, and a slow washout thereafter; the rank of B was occipital > temporo-parietal > frontal > inferior temporal > cerebellum, in agreement with known densities of cBZR *in vitro* (see Table 1). The determination of B in the five partial saturation studies further supported the validity of the approach used, as typical inhibition (sigmoid) curves were obtained for all regions and all times, when expressing percentage B as a function of either the total amount of flumazenil administered (in nanomoles per kilogram) or the corresponding F in the same region (in nanomoles). When B was expressed in molar amounts as a function of F, and all 13 studies were combined, typical saturation curves were obtained in all regions and all times, allowing calculation by Scatchard analysis regional B_{max} and K_d values that appeared satisfactory relative to *in vitro* data (see later), although the exact values clearly depended on time of determination. Overall, Pappata et al.[41] concluded that their method of estimation of F was validated.

Shinotoh et al. study[40]—These authors performed saturation experiments in four subjects using orally administered flumazenil in doses of 20, 30, 50, and 150 mg, respectively, given 30 min before intravenous injection of ^{11}C-FLU. In the frontal cortex, for example, the remaining uptake of ^{11}C-FLU at t = 20 min was 48, 22, 31, and 23% of control (high SRA) studies, respectively. The effect was similar in other cortical structures and in cerebellum, but was highly variable in brain stem and white matter (residual ^{11}C-FLU ranging from ~30 to 100% in the latter structure). In the subject with the largest effect of cold flumazenil, these authors showed subtraction curves of ^{11}C-FLU uptake (in percent DI/L) for the frontal cortex and cerebellum between the high SRA and the saturation studies; these curves, taken to represent specific binding, showed features similar to those just described, including frontal cortex > cerebellum. However, the variability and lack of dose dependence of effects seen indicate that the oral administration of flumazenil is not adequate if consistent saturation of the cBZR is desired.

Persson et al. study[43]—These authors designed a wholly different approach than the previous two groups to determine F and B in brain regions. They chose to use a reference structure in brain, namely, white matter, to estimate F in all neocortical areas because white matter is known from *in vitro* studies to have negligible density of cBZR (see Table 1). Hence, since the contribution of intravascular unmetabolized ^{11}C-FLU to regional radioactivity (including white matter) is negligible at times >10 min, white matter can be taken to represent F in all cortical regions.

To validate this hypothesis, these authors[43] reported that the time course of ^{11}C-FLU uptake in the pons (at t > 20 min) in high SRA experiments was similar to that obtained in neocortex of five subjects under full saturation (conditions pretreatment with >10 mg of cold flumazenil). It is unfortunate that in their paper[43] only one of these five experiments is shown on graph, and no uptake data are actually presented either for pons or for corona radiata. Despite the use of a high-resolution PET camera, these authors also found that "the pons region was affected by the protective dose, but to a much smaller extent" (than the cerebral cortex), and that "this was also the case" for the corona radiata, presumably as a result of the partial volume effect. In their Figure 4, the effect of pretreatment on the pons ^{11}C-FLU time course is not so small, being about 10% of that occurring in the neocortex. Also, the human pons region is known from *in vitro* studies to contain cBZR, probably present in the nuclei of the pons, although in very low concentration (see Table 1). Finally, there is no attempt in their paper[43] to quantitate B in brain structures other than the neocortex so that it is not known whether the use of white matter to estimate F would be valid in these regions.

Validation of this approach was provided by determining B in the neocortex of five young untreated volunteers who each underwent four to five ^{11}C-FLU PET studies, including one at control state (high SRA) and the remaining at increasing amounts of flumazenil (whole dose range, 4 μg to 2.5 mg). Using the pons and the corona radiata as reference structures to determine F, neocortical B (mean of four frontal and four occipital regions) was calculated for each PET study, choosing the time of 25 min (see below); in each volunteer, a typical saturation curve of B (nM) was obtained either vs. the dose of flumazenil administered (mg) or F (nM), allowing to measure B_{max} and K_d values (either by Scatchard plots or by nonlinear fitting) that compared well to *in vitro* data; Hill plots showed linear data in each case. The comparison of the B_{max} and K_d values obtained with the use of pons and that of corona radiata showed to significant difference, indicating that both could be used as reference structure. In addition, when the pons and corona radiata of the full saturation studies (flumazenil, >10 mg) were used, there was no significant effect on the B_{max} and K_d values obtained. Finally, the use of the "full-saturation" study performed in each of the volunteers to determine F in the same neocortical regions provided B_{max} and K_d values not significantly different from the above methods, indicating the validity of using pons and corona radiata data obtained from the same PET study as that for neocortex avoiding the need of the additional "full-saturation" PET study.

d. Pharmacologic Pseudoequilibrium

Perfect equilibrium is reached *in vitro* when the concentration of specifically bound radioligand is stable, indicating that there is no net flux in the association/dissociation reaction. *In vivo*, however, this state cannot be reached because the free ligand concentration in brain depends on that in plasma which itself is necessarily subject to biological removal. However, it has been assumed that if the concentrations of F and that of B are in equilibrium to one another, i.e., if the ratio B/F is stable, the steady-state equilibrium equations can apply with acceptable uncertainty: this state has been termed "pseudo-equilibrium". Because it is possible to estimate F in ^{11}C-FLU experiments with PET (see above), the time course of the B/F ratio has been evaluated for pseudo-equilibrium.

Pappata et al.[41] plotted the mean B/F ratio calculated in the control (high SRA) studies for various brain regions and observed that following a slow rise until 20 min, it then reached an essentially stable value in all regions until 60 min ranging from 4 to 6 in cortical structures and around 3.5 in the cerebellum, with highest values in the occipital cortex. In most regions, however, they noted a slight decline in this value after 40 min (Figure 5).

Persson et al.,[43] using pons/corona radiata to determine F, also found that the B/F ratio for the neocortex rose until 25 to 30 min and was essentially stable thereafter; the mean B/F was around 7, but with a very large variability, particularly after 40 min. These authors used the time 25 min to determine B in their subsequent analysis for three reasons: (1) B/F was stable at this time; (2) B was close to its maximum (peak around 10 to 15 min); and (3) the pons/plasma ^{11}C radioactivity ratio became stable. However, because the ^{11}C-radioactivity in blood was not corrected for metabolites, a true equilibrium cannot be assumed in these studies. In a further work, however, Persson et al.[48] indicates that in a single subject, the neocortex/venous plasma ^{11}C-FLU (corrected for labeled metabolites) was stable between 12 and 36 min; obviously this would need to be confirmed for the arterial plasma in several

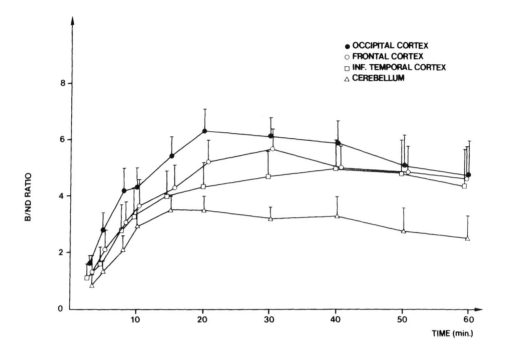

FIGURE 5.   Regional brain kinetics of the specifically bound/nondisplaceable (B/ND) ^{11}C-flumazenil. Mean ± SD of the B/ND ratio was emitted individually as a function of time in eight control studies. Nondisplaceable fraction was estimated in four independent subjects under saturation condition and specifically bound by subtraction from total radioactivity. The B/ND ratio is seen to remain essentially constant at times > 20 min, although a slight decline is apparent. (From Pappata, S., Samson, Y., Chavoix, C., Prenant, C., Mazière, M., and Baron, J. C., *J. Cereb. Blood Flow Metab.*, 8, 304, 1988. With permission.)

subjects. In their studies, Savic et al.[49] also showed stability of the B/F ratio for the parietal cortex between 18 and 54 min and used a pseudo-equilibrium interval of 15 to 40 min, consistent with both Persson's and Pappata's works. Reasons for the lack of complete stability of the B/F ratio were speculated by Pappata et al.:[41] in addition to uncertain estimation of F and, in turn, of B, a lack of true equilibrium, or the slow appearance of labeled or unlabeled metabolites in brain tissue, were proposed.

2. Determination of B_{max} and K_d of the cBZR in Brain Regions

As a result of the just described procedures, Pappata et al.[41] and Persson et al.[43] were able to calculate regional B_{max} and K_d values. In the former study, these varibles were obtained in 4 different regions from 13 healthy volunteers, each subject contributing one value (Figure 6). In the latter, they were obtained in a single large area of neocortex (both frontal and occipital cortex lumped together), but measured in five healthy volunteers each subjected to four to five PET ^{11}C-FLU studies. Hence, the results obtained can barely be compared (see Table 2). In both cases, however, linearity of the Scatchard plots was demonstrated,[41-43] with Hill coefficients >0.98 in four fifths of Persson et al.'s subjects,[43] indicating that flumazenil apparently binds *in vivo* to a single class of receptors, as it also does *in vitro*.

Pappata et al.[41] pointed out that the B_{max} and K_d values obtained showed a dependence on time of determination, with measured B_{max} declining by 45 to 60% and K_d declining by 28 to 65% between 20 and 50 min, while the ratio B_{max}/K_d remained essentially constant. This effect appeared to reflect the similar slight fall in the B/F ratio over time alluded to above. Persson et al.[43] reported a similar, although less conspicuous, fall in both B_{max} and

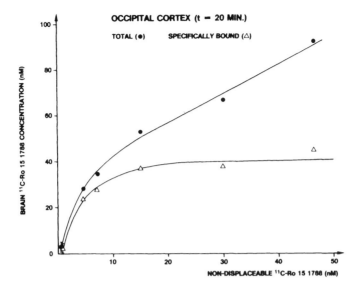

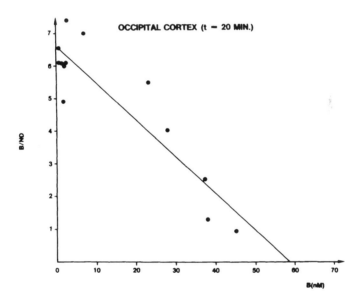

FIGURE 6. Total and specifically bound ^{11}C-flumazenil as a function of the molar concentration of nondisplaceable ligand and corresponding Scatchard plot measured in the occipital cortex at 20 min postinjection. The values found in eight control (high SRA) studies are shown as mean ± SD. The remaining data were extracted from five partial saturation co-injection studies. The values obtained from these data were B_{max} = 58.6 nM and K_d = 8.8. nM. (From Pappata, S., Samson, Y., Chavoix, C., Prenant, C., Mazière, M., and Baron, J. C., *J. Cereb. Blood Flow Metab.*, 8, 304, 1988. With permission.)

K_d of neocortex with later determination times, in spite of a different method used to determine F. The reasons for this phenomenon remain unclear.

A simplification of the method of Persson et al.[43] has been designed by Savic et al.,[49,50] in which each subject undergoes two ^{11}C-FLU PET studies at widely different SRA in order to calculate B_{max} and K_d based on two data pairs only. The free ^{11}C-FLU is measured from the corona radiata, and all measurements are averaged over the 15- to 40-min "pseudo-

TABLE 2

B_{max} and K_d Values in Healthy Volunteers Measured *In Vivo* with PET and
^{11}C-Flumazenil

N	Time (min)	Regions	B_{max} nM (mean ± SD)	K_d nM (mean ± SD)	Ref.
13	20—40	Occipital cortex	44.6 ± 13.0	6.9 ± 1.8	41
		Frontal cortex	36.6 ± 9.1	6.6 ± 1.5	
		Inferior temporal cortex	25.2 ± 8.8	5.3 ± 2.3	
		Cerebellum	23.7 ± 6.5	7.0 ± 1.5	
5	25	Neocortex			43
		(reference: pons)	87 ± 17	8.6 ± 4	
		(reference: corona radiata)	80 ± 13	11 ± 5.6	
5	15—36	Temporal-medial cortex (right)	71 ± 35	24 ± 15	49, 50
		Sensori motor cortex (right)	107 ± 47	26 ± 15	
		Occipital cortex	141 ± 69	27 ± 17	
		Lateral temporal cortex (right)	102 ± 46	26 ± 16	
		Frontal cortex (right)	107 ± 47	26 ± 15	
		Parietal cortex (right)	98 ± 22	24 ± 11	

equilibrium'' time interval in order to improve the statistics of acquisition. Also, the approach is made acceptable by a strict choice of the SRA in the two studies, which should be very high in the first study and sufficiently low in the second (flumazenil doses in the order of a few micrograms and 1.5 mg, respectively) in order to obtain the two extremes in this simplified Scatchard plot. The values by Savic et al.[49,50] obtained in control subjects are also shown in Table 2.

Compared to *in vitro* human data on the cBZR (Table 1), the values measured *in vivo* appear roughly similar. They show the same pattern of regional B_{max}, with highest values in the occipital cortex, intermediate values in other cortical areas, and lowest values in the cerebellum. This pattern is also quite consistent with *in vitro* specific binding with ^{11}C-FLU on human brain sections.[51] The B_{max} values are similar to *in vitro* data for Persson et al.[41] and Savic et al.,[49,50] but lower for Pappata et al.[41] The K_d values are higher than *in vitro* data when the latter are measured at low temperature, but similar to those measured at physiological temperature; Savic et al.,[49,50] however, reported inordinately high K_d values. A striking feature remains the large variance of the values obtained *in vivo*, indicating a variability in the measurement that reflects not only methodological errors (e.g., uncertainly in the SRA value, poor counting statistics — particularly in reference structures — due to high resolution PET and restricted dosimetry, partial volume effects, errors in patient's head repositioning at repeated scans, variability in region placement), but also oversimplification of pharmacological assumptions (especially, uniformity of free ligand concentration, negligible nonspecific binding, and equilibrium conditions). It is clear that individual variations will be difficult to detect by these approaches, mainly if the Savic method, which appears to be the only practical one, is applied. However, studies of large patient samples, or using the patient as his own control (e.g., right-left asymmetries), should be quite acceptable. An alternative was proposed by Pappata et al.[41] who noticed that because of very low receptor occupation (≤ 5%) and low F relative to K_d (<5%) at high SRA studies, the B/F ratio reflected closely the B_{max}/K_d ratio and could be used to estimate the "binding potential" regionally in a practical single scan approach. A wholly different alternative is to investigate the use of a "dynamic" approach for determination of B_{max} and K_d.

B. DYNAMIC METHODS

Dynamic methods to measure binding parameters are attractive because they use the whole time course of brain radioactivity following administration of the tracer and because the assumption of pharmacologic equilibrium is not required.[46] Under specific assumptions and experimental conditions, a single PET study can be used to provide the "binding potential" B_{max}/K_d (also referred to as $k_{on} \times B_{max}$). If B_{max} and K_d need to be determined separately, however, two or more administrations of the radioligand, including low SRA and/or displacement studies are necessary.[46] Major issues in these "dynamic" studies are their complexity in term of computer science, the difficulty in validating the models used, the statistical assessment of the parameters obtained, and the limited regional capability of the measurement due to inadequate count rate.

With respect to ^{11}C-FLU, there are two reports on dynamic methods to estimate binding parameters. Yamasaki et al.[52] briefly reported results of the use of a three-compartment model (plasma, receptor-bound, and nonspecifically bound) to fit the ^{11}C-FLU brain dynamics obtained in a single (high SRA) PET study. The input function was obtained from venous blood samples from which unchanged ^{11}C-FLU was extracted. Using this method, these authors obtained pixel-by-pixel maps of the plasma-receptor-bound forward and reverse transfer constants k_1 and k_2 and of the ratio k_1/k_2 which reflects the "binding potential". In the frontal cortex, this variable was found to range from 4.23 to 10.5 in seven healthy volunteers (age range of 23 to 70 years). The validity of these findings appears uncertain, as arterial blood sampling and independent measurement of cerebral blood flow and volume were lacking, while the justification for not including a cerebral free ligand compartment is unclear.

Blomquist et al.[53] have developed a kinetic method based on two PET studies with different amounts of unlabeled flumazenil to measure, on a pixel-by-pixel basis, the parameters B_{max}, k_{on}, and k_{off} and the ratio k_{off}/k_{on} ($= K_d$). They used a standard, three-compartment model (protein unbound ^{11}C-FLU in plasma, tissue free and tissue specifically bound ^{11}C-FLU), neglecting nonspecific binding and nonlinear equations accounting for receptor occupation. Because the equations cannot be solved for so many parameters, the time course of the free radioligand was determined first, using a reference region (the pons) and assuming a similar time course of F in the reference region and in the region of interest (see previous discussions of this assumption). One advantage of this approach is that it eliminates the need to measure the arterial blood input function. Two PET studies of ^{11}C-FLU are performed, one at high SRA where the ratio specific-binding/free approaches B_{max}/K_d, and another at low SRA where the specifically bound ligand approaches B_{max}. Using data between 2 and 55 min postinjection, they report preliminary findings in two subjects, in whom mean B_{max} and K_d (brain region not indicated) were 77 and 13.4 nM, respectively, compared to 100 and 12.9 nM, obtained in the same subjects with the "equilibrium" method of Savic et al.,[49] using the data between 22 and 28 min. They also display quantitative brain maps of B_{max} and K_d obtained in one subject, showing confinement of positive parameter values to the grey matter. However, to obtain these maps, "the algorithm was applied only to those pixels with positive time integral of (specific binding, and) only positive parameter values are displayed." Obviously, these results are only very preliminary and the method used remains to be validated. In addition, its complexity and the fact that two PET studies are required does not make it very attractive to apply on a routine basis, especially if the results do not differ from those obtained with the more simple "equilibrium" method, and if similar assumptions (e.g., the use of a reference region to determine F) are required.

IV. CLINICAL APPLICATIONS

A. PHARMACOLOGICAL STUDIES

Pioneer studies on the determination by ^{11}C-FLU-PET of cBZR occupancy *in vivo* in humans treated with benzodiazepines have appeared recently.[44,45] Pauli et al.[44] briefly indicate (in abstract form) that they estimated the percentage occupancy by calculating the change in the B_{max}/K_d ratio that occurred during treatment. Using the pons as a reference to estimate F, they determined B in cortical regions and employed a linear least squares method to calculate B_{max}/K_d. Following (2 h) oral uptake of diazepam (30 mg), the reduction in B_{max}/K_d from the untreated state measured in four subjects was 20 ± 6% (SEM) for the frontal cortex. Using a more simple approach, the ratio (total ^{11}C-uptake in one PET slice)/(total venous plasma uptake) measured over 60 min after injection provided a value of 16 ± 3% (SEM), comparable to that obtained with the B_{max}/K_d approach. Overall, the results indicate a limited occupancy of the cBZR despite the use of relatively large doses of diazepam. This is consistent with the lower affinity of the agonists for the cBZR, with the rapid elimination of ^{11}C-flunitrazepam from brain observed *in vivo* with PET[4-8] and with the rapid *in vivo* dissociation of the agonist-cBZR binding reported in rodents by Kuhar and Goeders[24] and Potier et al.[25]

Shinotoh et al.[45] studied cBZR occupation by orally given clonazepam (30 μg/kg in six subjects and 50 μg/kg in one); ^{11}C-FLU-PET was performed before and again 1.5 h following clonazepam. Receptor occupancy was estimated by simply calculating the percentage reduction of the ^{11}C-FLU uptake (in units of percent DI/L) measured in the cerebral cortex 30 min after tracer injection, assuming negligible NS binding and equilibrium conditions. In the treated conditions, the distribution and kinetics of ^{11}C-FLU in brain are similar to the control state, except that the peak of tracer kinetics occurs earlier (7 to 8 min compared to 10 to 12 min) and that it is reduced from 15.3 to 23.5%, depending on the cortical region studied in the lower-dose study (N = 6) and around 30% in the high-dosage study (N = 1). Clinically, these effects were accompanied by drowsiness, ataxia, and significant delay of the P 300 (auditory evoked potential). Although these results clearly demonstrate that clinical effects of benzodiazepines do not require large occupancy of the cBZR, the method used actually underestimates the latter by about 5 to 10% since nondisplaceable ^{11}C-FLU was not taken into account in the calculations.

B. PATHOPHYSIOLOGICAL STUDIES
1. Epilepsy
Savic et al.[49] have reported a reduction in cBZR density in the epileptogenic focus of ten patients with poorly controlled lateralized partial simple or complex seizures. They included only patients treated with carbamazepine or phenytoin (no interaction with GABA) and without focal changes on CT and MRI scans. The epileptogenic focus was defined based on clinical and EEG findings, drawn on MRI scans and computer copied on ^{11}C-FLU-PET scans performed interictally. The focus was located in the medial-temporal, the sensorimotor, or the occipital cortex. In order to measure regionally both B_{max} and K_d, two PET scans were performed, one at high and the other at low SRA, as described earlier: in the high SRA study, the uptake of ^{11}C-FLU measured 8 to 54 min after injection was reduced 15 to 35% in the focus relative to the homologous contralateral area, compared to a maximum of 4% in nonepileptogenic control areas. For the measurement of B_{max} and K_d, the reference structure used to estimate F was the white matter of the corona radiata. In five patients, an additional PET study performed at presumably full saturation was reported to show that the kinetics of ^{11}C-FLU in cortical areas was similar to that obtained in white matter in the high SRA studies, indicating that the latter adequately approximates the free radioligand; they did not demonstrate, however, whether "full saturation" had any effect on white matter ^{11}C-FLU kinetics. In addition, the ^{11}C-FLU kinetics at full saturation were similar in the epileptogenic focus and homologous areas, indicating no alteration in free ligand space and nonspecific binding in the epileptogenic focus. Overall, the results convincingly showed a significant reduction in B_{max} of 29 ± 13% in the epileptogenic focus, without significant change in K_d; there were no changes in nonepileptogenic areas. Compared to ^{11}C-deoxy-glucose scans performed in one patient, the cBZR deficit appeared better correlated to the

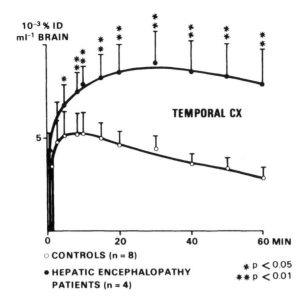

FIGURE 7. Kinetics of [¹¹C]flumazenil in the temporal cortex (CX) of four patients with a history of hepatic encephalopathy. The cerebral ¹¹C uptake was dramatically increased in patients as compared with control. Similar changes were found in other cortical areas and in the cerebellum. Values for uptake (means ± SD) are expressed as percentage of the injected dose (ID) per milliliter of brain. Significant differences were detected with Student t tests corrected for unequal variances. (From Samson, Y., Bernuau, J., Pappata, S., Chavoix, C., Baron, J. C., and Mazière, M., *N. Engl. J. Med.*, 316, 414, 1987. With permission.)

clinical assessment of focus topography. The authors concluded that the PET study appeared useful in the clinical evaluation of partial complex seizures and that the apparent loss of cBZR (or increased concentration of competing endogenous ligand) strengthened the hypothesis that inhibitory mechanisms are disturbed in the epileptogenic focus.

The same authors further reported[50] results of similar studies performed in ten patients with Grand Mal epilepsy with normal MRI. Although the B_{max} values appeared lower in all cortical regions in patients compared to both healthy controls and focal epileptic subjects, the difference did not reach significance.

2. Hepatic Encephalopathy

Samson et al.[54,55] reported in abstract form their findings using high SRA ¹¹C-FLU-PET in four patients with history of recurrent episodes of hepatic encephalopathy due to liver cirrhosis with hepatic failure. The PET studies were performed within 1 month of the last episode in benzodiazepine untreated patients. Compared to healthy subjects, there was a two- to threefold increase in ¹¹C brain uptake in all cortical regions which was highly significant at all times > 8 min (Figure 7), despite roughly similar venous blood ¹¹C kinetics. Wash-out from brain was apparently slower in patients, indicating increased ¹¹C-FLU retention. In two patients with liver cirrhosis, but no history of hepatic encephalopathy, the ¹¹C brain kinetics were reported as unremarkable. The author's interpretation of these findings was that of an increased brain specific binding of ¹¹C-FLU in patients with history of hepatic encephalopathy, as a result of either an increased accessibility of ¹¹C-FLU to the cBZR (lesser peripherical metabolism of the drug or leaky BBB or both) or an elevated density of the cBZR. Regardless of the mechanism, the results provided an explanation for the hypersensitivity to benzodiazepines of patients with hepatic failure, with the attendant risk for acute encephalopathy; the GABA related hypothesis for hepatic encephalopathy and the beneficial effects of flumazenil are also well established.[56]

3. Alzheimer's Disease

Preliminary findings with ^{11}C-FLU-PET in two patients with probable Alzheimer's disease have been reported by Yamasaki et al.[52] They used the ratio of ^{11}C-FLU uptake brain/venous blood (corrected for labeled metabolites) to estimate roughly the "binding potential", which was found to be decreased in all cortical areas, except the primary visual cortex, in the most severely affected patient and in the parieto-temporal cortex bilaterally in the moderately affected case. In both patients, the binding potential appeared preserved in the thalamus. The results suggested a loss of cBZR in cortical areas in Alzheimer's disease.

4. Normal Aging

Yamasaki et al.[52] also reported preliminary findings in 11 healthy male volunteers in whom the binding potential for ^{11}C-FLU, estimated as indicated above for the whole cerebral cortex at 20 min postinjection, was found to decrease linearly with age in an apparently significant way, from around 8 at 20 years of age down to 5 at 60 years. Using a compartmental model to calculate dynamically a "binding potential" (see previous paragraph), these authors reported data showing no significant age-related decline. This important issue will require elucidation in further studies.

5. Spinocerebellar Degeneration

Shinotoh et al.[57] have reported (in Japanese) results of a ^{11}C-FLU-PET spinocerebellar degeneration. Using the brain/blood ratio method of Yamasaki et al.,[52] they observed a significantly increased ratio in the cerebellar cortex in seven patients with olivo-ponto-cerebellar atrophy compared to six control subjects of similar age. In four patients with degenerative cerebellar atrophy, the cerebellar ratio was lower than controls, but the difference did not reach statistical significance. The interpretation made by the authors implicated a compensatory up-regulation of the cBZR as a result of loss of the cerebellar Purkinje cells in some variants of spinocerebellar degeneration.

6. Anxiety

In two patients with severe anxiety, Persson et al.[29] reported an accelerated wash-out of ^{11}C-FLU from frontal cortex, following an unchanged initial uptake, as compared to control subject. However, both patients were treated with large doses of benzodiazepines at time of the PET study, so that the accelerated washout presumably reflected cBZR occupancy and was not necessarily related to the psychiatric condition.

Note: Shinotoh et al.[58] calculated estimated radiation doses due to ^{11}C-FLU in two young healthy volunteers. The target structures were the small intestine and the bladder (78 and 51 mrad/mCi, respectively), while all other organs were below 12 mrad/mCi.

REFERENCES

1. **Squires, R. F. and Braetstrup, C.**, Benzodiazepine receptor in rat brain, *Nature*, 266, 732, 1977.
2. **Möhler, H. and Okada, T.**, Benzodiazepine receptor: demonstration in the central nervous system, *Science*, 198, 849, 1977.
3. **Müller, W. E.**, *The Benzodiazepine Receptor*, Shepherd, M., Häfner, H., McHugh, P., and Sartorius, Eds., Cambridge University Press, New York, 1987.

4. **Comar, D., Maziére, M., Godot, J. M., Berger, G., Soussaline, F., Menini, C., Arfel, G., and Naquet, R.,** Visualisation of ^{11}C-flunitrazepam displacement in the brain of the live baboon, *Nature,* 280, 329, 1979.

5. **Comar, D., Mazière, M., Cepeda, C., Godot, J. M., Menini, C., and Naquet, R.,** The kinetics and displacement of ^{11}C-flunitrazepam in the brain of the living baboon, *Eur. J. Pharmacol.,* 75, 21, 1981.

6. **Mazière, M., Godot, J. M., Berger, G., Baron, J. C., Comar, D., Cepeda, C., Menini, C., and Naquet, R.,** Positron emission tomography. A new method for *in vivo* brain studies of benzodiazepine in animal and in man, in *GABA and Benzodiazepine Receptors,* Costa, E., Chiara, G., and Gessa, G. L., Eds., Raven Press, New York, 1981, 273.

7. **Mazière, M., Prenant, C., Sastre, J., Crouzel, M., Comard, D., Hantraye, P., Kaijima, M., Guibert, B., and Naquet, R.,** Etude "*in vivo*" des récepteurs aux benzodiazepines par tomographie par émission de positons, *Encéphale,* 9, 151B, 1983.

8. **Mazière, M., Prenant, C., Sastre, J., Crouzel, M., Comard, D., Hantraye, P., Kaijima, M., Guibert, B., and Naquet, R.,** C11-RO 15 1788 et C11-flunitrazépam, deux coordinats pour l'étude par tomographie par émission de positons des sites de liaison des benzodiazépines, *C. R. Acad. Sci. Paris,* 293, 871, 1983.

9. **Duka, Th., Höllt, V., and Hertz, A.,** *In vivo* receptor occupation by benzodiazepines and correlation with the pharmacological effect, *Brain. Res.,* 179, 147, 1979.

10. **Möhler, H. and Richards, J. G.,** Agonist and antagonist benzodiazepine receptor interaction *in vitro, Nature,* 294, 763, 1981.

11. **Grafton, S. T., Ikezaki, K., Luxen, A., Chugani, D., Black, K. L., Maziotta, J. C., Barrio, J. R., and Phelps, M. E.,** Labeling of rat C-6 glioma implants with 3-18F-fluorodiazepam and 3H diazepam: *in vitro* tissue homogenate and *in vivo* autoradiographic results, *J. Cereb. Blood Flow Metab.,* 9, S18, 1989.

12. **Frost, J. J., Wagner, H. N., Dannals, R. F., Ravert, H. T., Wilson, A. A., Links, J. M., Rosenbaum, A. E., Trifiletti, R. R., and Snyder, S.,** Imaging benzodiazepine receptors in man with 11C-suriclone by positron emission tomography, *Eur. J. Pharmacol.,* 122, 381, 1986.

13. **Trifiletti, R. R. and Snyder, S. H.,** Anxiolytic cyclopyrrolones zopiclone and suriclone bind to a novel site linked allosterically to benzodiazepine receptors, *Mol. Pharmacol.,* 26, 458, 1984.

14. **Breastrup, C., Albrechtsen, R., and Squires, R. F.,** High densities of benzodiazepine receptors in human cortical areas, *Nature,* 269, 702, 1977.

15. **Sieghart, W., Eichinger, A., Riederer, P., and Jellinger, K.,** Comparison of benzodiazepine receptor binding in membranes from human or rat brain, *Neuropharmacology,* 8, 751, 1985.

16. **Manchon, M., Kopp, N., Rouzioux, J. M., and Miachon, S.,** Etude des recepteurs aux benzodiazepines dans l'hippocampe des suicidés, *C. R. Acad. Sci. Paris,* 4, 131, 1986.

17. **Cheetham, S. C., Crompton, M. R., Katopa, C. L. E., Parker, S. J., and Horton, R. W.,** GABA A/benzodiazepine binding sites and glutamic acid decarboxylase activity in depressed suicide victims, *Brain Res.,* 460, 114, 1988.

18. **Trifiletti, R. R., Snowman, A. M., Whitehouse, P. J., Marcus, K. A., and Snyder, S. H.,** Huntington's disease: increased number and altered regulation of benzodiazepine receptor complexes in frontal cerebral cortex, *Neurology,* 37, 916, 1987.

19. **Ferrarese, C., Appolonio, I., Frigo, M., Gaini, S. M., Piolti, R., and Frattola, L.,** Benzodiazepine receptors and diazepam binding inhibitor in human cerebral tumors, *Ann. Neurol.,* 26, 564, 1989.

20. **Mazière, M., Hantraye, P., Prenant, C., Sastre, J., and Comar, D.,** Synthesis of RO 15 1788 and ^{11}C: a specific radioligand for the *in vivo* study of central benzodiazepine receptors by positron emission tomography, *Int. J. Appl. Radiat. Isot.,* 35, 973, 1984.

21. **Ehrin, E., Johnstrom, P., Stone-Elander, S., and Nilsson, J. L. G.,** Preparation and preliminary positron emission tomography studies of 11C-RO 15 1788, a selective benzodiazepine receptor antagonist, *Acta Pharm. Suec.,* 21, 183, 1984.

22. **Brown, C. L. and Martin, I. L.,** Kinetics of 3H-RO 15 1788 binding to membrane-bound rat brain benzodiazepine receptors, *J. Neurochem.,* 42, 918, 1984.

23. **Darragh, A., Lambe, R., Kenny, M., and Brick, I.,** Tolerance of healthy volunteers to intravenous administration of the benzodiazepine antagonist RO 15 1788, *Eur. J. Clin.,* 24, 569, 1983.

24. **Goeders, N. E. and Kuhar, M. J.,** Benzodiazepine receptor *in vivo* with 3H-RO 15 1788, *Life Sci.,* 37, 345, 1985.

25. **Potier, M. C., Prado de Carvalho, L., Dodd, R. H., Brown, C. L., and Rossier, J.,** *In vivo* binding of 3H-RO 15 1788 in mice: comparison with the *in vivo* binding of 3H-flunitrazepam, *Life Sci.,* 43, 1287, 1988.

26. **Inoue, O., Akimoto, Y., Hashimoto, K., and Yamasaki, T.,** Alterations in biodistribution of ^{3}H-RO 15 1788 in mice by acute stress: possible changes in *in vivo* binding availability of brain benzodiazepine receptor, *Int. J. Nucl. Med. Biol.,* 12, 369, 1985.

27. **Hashimoto, K., Inoue, O., Itoh, T., Goromaru, T., and Yamashaki, B.,** Study on measurement of free ligand concentration in blood and quantitative analysis of brain benzodiazepine receptors, *Kaku Igaku,* 11, 1235, 1988 (in Japanese, English abstract).

28. **Hantraye, P., Kaijima, M., Prenant, C., Guibert, B., Sastre, J., Crouzel, M., Naquet, R., Comar, D., and Mazière, M.,** Central type benzodiazepine binding sites: a positron emission tomography study in the baboon's brain, *Neurosci. Lett.,* 48, 115, 1984.

29. **Persson, A., Ehrin, E., Eriksson, L., Farde, L., Hedstrom, C. G., Litton, J. E., Mindus, P., and Sedvall, G.,** Imaging of 11C-labelled RO 15 1788 binding to benzodiazepine receptors in the human brain by positron emission tomography, *J. Psychiat. Res.,* 4, 609, 1985.

30. **Hantraye, P., Chavoix, C., Guibert, B., Fukuda, H., Brouillet, E., Dodd, R. H., Prenant, C., Crouzel, M., Naquet, R., and Mazière, M.,** Benzodiazepine receptors studied in living primates by positron emission tomography: inverse agonist interactions, *Eur. J. Pharmacol.,* 138, 239, 1987.

31. **Mazière, M., Hantraye, P., Kaijima, M., Dodd, R., Guibert, B., Prenant, C., Sastre, J., Crouzel, M., Comar, D., and Naquet, R.,** Visualisation by positron emission tomography of the apparent regional heterogeneity of the central type benzodiazepine receptors in the brain of living baboons, *Life Sci.,* 36, 1609, 1985.

32. **Hantraye, P., Brouillet, E., Fukuda, H., Chavoix, C., Guibert, B., Dodd, R. H., Prenant, C., Crouzel, M., Naquet, R., and Mazière, M.,** Benzodiazepine receptors studied in living primates by positron emission tomography, *Eur. J. Pharmacol.,* 153, 25, 1988.

33. **Chavoix, C., Hantraye, P., Brouillet, E., Guibert, B., Fukuda, H., de la Sayette, V., Fournier, D., Naquet, R., and Mazière, M.,** *Status epilepticus* induced by pentylenetetrazole modulates *in vivo* [11]C RO 15 1788 binding to benzodiazepine receptors. Effects of ligands acting at the supramolecular receptor complex, *Eur. J. Pharmacol.,* 146, 207, 1988.

34. **Brouillet, E., Chavoix, C., de la Sayette, V., Hantraye, P., Kunimoto, M., Khalili-Varasteh, M., Guibert, B., Fournier, D., Dodd, R. H., Naquet, R., and Mazière, M.,** Anticonvulsant activity of the diaryltriazine LY 81067: studies using electroencephalographic recording and positron emission tomography, *Neuropharmacology,* 4, 351, 1989.

35. **Darragh, A., Lambe, R., O'Boyle, C., Kenney, M., and Brick, I.,** Absence of central effects in man of the benzodiazepine antagonist RO 15 1788, *Psychopharmacology,* 80, 192, 1983.

36. **Klotz, U. and Kanto, J.,** Pharmacokinetics and clinical use of flumazenil (RO 15 1788), *Clin. Pharmacokinet.,* 14, 1, 1988.

37. **File, S. and Pellow, S.,** Intrinsic actions of the benzodiazepine receptor antagonist RO 15 1788, *Psychopharmacology,* 88, 1, 1986.

38. **Higgit, A., Lader, M., and Fonagy, P.,** The effects of benzodiazepine antagonist RO 15 1788 on psychological performance and subjective measures in normal subjects, *Psychopharmacology,* 89, 395, 1986.

39. **Samson, Y., Hantraye, P., Baron, J. C., Soussaline, F., Comar, D., and Mazière, M.,** Kinetics and displacement of 11C-RO 15 1788, a benzodiazepine antagonist studied in human brain *in vivo* by positron emission tomography, *Eur. J. Pharmacol.,* 110, 247, 1985.

40. **Shinotoh, H., Yamasaki, T., Inoue, O., Itoh, T., Suzuki, K., Hashimoto, K., Tateno, Y., and Ikehira, H.,** Visualisation of specific binding sites of benzodiazepine in human brain, *J. Nucl. Med.,* 27, 1593, 1986.

41. **Pappata, S., Samson, Y., Chavoix, C., Prenant, C., Mazière, M., and Baron, J. C.,** Regional specific binding of 11C-RO 15 1788 to central type benzodiazepine receptors in human brain: quantitative evaluation by PET, *J. Cereb. Blood Flow Metab.,* 8, 304, 1988.

42. **Persson, A., Pauli, S., Carl-Gunnar, S., Halldin, C., and Sedvall, G.,** Cerebral uptake of 11C-RO 15 1788 and its acid metabolite 11C-RO 3890, a PET-study in healthy volunteers, *Hum. Psychopharmacol.,* 1989.

43. **Persson, A., Pauli, S., Halldin, C., Stone-Ellander, S., Farde, L., Sjögren, I., and Sedvall, G.,** Saturation analysis of specific 11C-RO 15 1788 binding to the human neocortex using positron emission tomography, *Hum. Psychopharmacol.,* 4, 21, 1989.

44. **Pauli, S., Blomqvist, G., Persson, A., Farde, L., Halldin, C., and Sedvall, G.,** Benzodiazepine receptor occupancy in the living human brain determined by positron emission tomography, *J. Cereb. Blood Flow Metab.,* 9, S131, 1989.

45. **Shinotoh, H., Iyo, M., Yamada, T., Inoue, O., Suzuki, K., Itoh, T., Fukuda, H., Yamasaki, T., Tateno, Y., and Hirayama, K.,** Detection of benzodiazepine receptor occupancy in the human brain by positron emission tomography, *Psychopharmacology,* 99, 202, 1989.

46. **Baron, J. C.,** PET studies of brain receptors in neurological disorders, in *Neuroimaging,* Aichner, F., Gerstenbrand, F., and Grcevic, N., Eds., Gustav Fischer, Stuttgart, 1989, 3.

47. **Halldin, C., Stone-Elander, S., Thorell, J. O., Persson, A., and Sedvall, G.,** 11C-labelling of RO 15 1788 in two different positions and also 11C labelling of its main metabolite RO 15 3890, for PET studies of benzodiazepine receptors, *J. Appl. Radiat. Isot.,* 39, 993, 1988.

48. **Persson, A., Pauli, S., Swahn, C. G., Halldin, C., and Sedvall, G.,** Cerebral uptake of 11C-RO 15 3890: a PET study in healthy volunteers, *J. Cereb. Blood Flow Metab.,* 9, S122, 1989.

49. **Savic, I., Roland, P., Sedvall, G., Persson, A., Pauli, S., and Widén, L.,** *In vivo* demonstration of reduced benzodiazepine receptor binding in human epileptic foci, *Lancet,* II, 863, 1988.

50. **Savic, I., Roland, P., Sedvall, G., Persson, A., Pauli, S., and Widén, L.,** Positron emission tomography studies of benzodiazepine receptor binding in patients with partial and generalized epilepsy, *J. Cereb. Blood Flow. Metab.,* 9, S230, 1989.

51. **D'Argy, R., Gilberg, P. G., Stalnacke, C. G., Persson, A., Bergström, M., Langström, B., Schoeps, K. O., and Aquilonius, S. M.,** *In vivo* and *in vitro* receptor autoradiography of the human brain using an ^{11}C-labelled benzodiazepine analogue, *Neurosci. Lett.,* 85, 304, 1988.

52. **Yamasaki, T., Inoue, O., Shinoto, O., Itoh, T., Iyo, M., Tateno, Y., Suzuki, K., Kashida, Y., Yashimoto, K., and Tadokoro, H.,** Benzodiazepine receptor study in the elderly using PET and clinical application of a new tracer ^{11}C a methyl-N-methyl-benzyl amine, in *Liver and Aging, 1986, Liver and Brain,* Kitani, K., Ed., Elsevier, New York, 1986, 265.

53. **Blomqvist, G., Pauli, S., Farde, L., Eriksson, L., Persson, A., and Halldin, C.,** Dynamic model for reversible ligand binding, in *Emission Tomography in Clinical research and Clinical Diagnosis: Tracer Modeling and Radioreceptors,* Beckers et al., Eds., ECSC, EEC, EAEC, Brussels, 1989, 35.

54. **Samson, Y., Bernuau, J., Chavoix, C., Pappata, S., Baron, J. C., Crouzel, M., Bert, C., Prenant, C., Benhamou, J. P., and Mazière, M.,** Central type benzodiazepine receptors (BZR) in chronic hepatic encephalopathy: a PET study of humans, *J. Cereb. Blood Flow. Metab.,* 7, S345, 1987.

55. **Samson, Y., Bernuau, J., Pappata, S., Chavoix, C., Baron, J. C., and Mazière, M.,** Cerebral uptake of benzodiazepine measured by positron emission tomography in hepatic encephalopathy, *N. Engl. J. Med.,* 316, 414, 1987.

56. **Mullen, K. D., Mendelson, W. B., Martin, J. V., Bassett, M. L., and Jones, E. A.,** Could an endogenous benzodiazepine ligand contribute to hepatic encephalopathy?, *Lancet,* 1, 457, 1988.

57. **Shinotoh, H., Tateno, Y., and Hirayama, K.,** Benzodiazepine receptors in spino-cerebellar degeneration studied with positron emission tomography, *Clin. Neurol. (Tokyo),* 28, 437, 1988.

58. **Shinotoh, H., Yamasaki, T., Inoue, O., Itoh, T., Hashimoto, K., Tateno, Y., Ikehira, H., Suzuki, K., and Kashida, Y.,** A study of benzodiazepine receptor in human brain using ^{11}C-RO 15 1788 and positron emission tomography, *Clin. Neurol. (Tokyo),* 1789, 1985.

59. **Barré, L., Debruyne, D., Abadie, P., Moulin, M., and Baron, J. C.,** Methods for [^{11}C] Ro 15 1788 radioactive metabolite assay in rabbit, baboon and human blood, *Int. J. Appl. Radiat. Isot. A,* submitted.

Chapter 19

POSITRON EMISSION TOMOGRAPHY (PET) IMAGING OF NEURORECEPTORS IN MENTAL ILLNESS

Henry N. Wagner, Jr.

TABLE OF CONTENTS

I. INTRODUCTION

Regional neuronal activity stimulated in human beings or experimental animals by sensory or motor activity is reflected in increased regional blood flow or increased glucose or oxygen metabolism, which can be used to identify anatomical regions of the brain involved in such mental processes.[1-16] Thus, we can visualize the spatial and temporal patterns of neuronal activation. Together with electrical measurements, we can examine the hierarchical "circuitry" of the brain by relating blood flow and metabolism to the performance of various mental tasks and during different states, such as sleep or sleep deprivation.

For example, we can begin to ask whether diseases involve primarily cortical or subcortical regions. Hughlings Jackson, the father of modern neurology, postulated that mental illness results from dissolution of the highest cerebral centers.[17,18] His hypothesis extended Herbert Spencer's idea that consciousness evolved in human beings because of its survival value for the human species. Jackson conceived that "positive" symptoms of mental illness were caused by overactivity of lower brain structures released from cortical control. According to him, mental illness begins with dissolution of the cortical substrata of consciousness, reversing the process of evolution, and reducing complex and specialized brain functions to simpler and more general functions, i.e., overactivity of the remaining, unaffected lower levels of brain function.

Eugen Bleuler postulated that in mental illnesses, such as schizophrenia, emotions controlled by lower centers gained dominance over disturbed intellectual functions controlled by higher cortical centers. Overactivity of the spared lower centers is a release phenomenon. These hypotheses can now be examined by positron emission tomography (PET).

Regional cerebral blood flow and glucose utilization had been examined in patients with schizophrenia, both during "resting" conditions and during the performance of various mental tasks.[19-22] In the early 1970s, Ingvar and Franzen reported that blood flow to the prefrontal regions was less in patients with schizophrenia than in normal persons during the performance of mental tasks.[23,24] While these findings have not been confirmed by some investigators, others have found changes, including overactivity of the left cerebral hemisphere during spatio-temporal tasks that in normal persons activate the right cerebral hemisphere.[25,26] The frontal lobes and limbic system remain of special interest as a possible site of disease in some patients with schizophrenia.

Mental illness can be the result of structural changes in the brain. A brain tumor can produce almost every type of mental dysfunction. Most, however, are the result of functional or biochemical rather than structural abnormalities of the brain. For example, in Parkinson's disease, dopamine concentrations in the caudate nuclei and putamen are very low. In other diseases, neuroreceptors are involved.

At the turn of the century, Paul Ehrlich proposed that the pharmacological response to drugs might be due to the drug's binding to a specific chemical group on a cell.[27] It took until 1973 to prove the existence of opiate receptors in the mammalian brain.[28] Radioactive naloxone, a drug that blocks the pharmacological action of opiates, made it possible to prove the existence of opiate receptors. Autoradiography of the binding of radiolabeled naloxone to brain tissue extracts from experimental animals and from human brain specimens obtained at autopsy indicated that opiate receptors were located at sites where opiates would be expected to act — in the respiratory center (opiates suppress respiration) and the centers involved in perception of pain (e.g., the medial thalamus). Such findings provided important evidence that opiate receptors were important links along the chain of neurotransmission.

The finding of opiate receptors in the brain intensified the search for endogenous brain substances with opiate-like properties. Two such substances were found — endorphin and enkephalin. Since then, over 100 different types or subtypes of neuroreceptors have been discovered.

Most drugs act by stimulating or blocking receptors. For example, drugs that block dopamine receptors of the D2 type diminish delusions and hallucinations in psychotic patients and improve their cognitive function. Other drugs that block receptors are cimetidine, which blocks histamine receptors; propranolol, which blocks beta adrenergic receptors; and haloperidol, which blocks D2 dopamine receptors. The widely used drug Valium, with sales in hundreds of millions of dollars per year, stimulates benzodiazepine receptors, producing an inhibitory effect on neurotransmission, which accounts for its tranquilizing effects. Neuroreceptors not only can vary in number in different parts of the brain and in different persons, but can alter their molecular structure, switching from an excitatory to an inhibitory functional state.

To extend the autoradiographic studies of neuroreceptors in animals to living human beings, N-methyl spiperone (NMSP), which has a high affinity for binding to D2 dopamine receptors, was labeled with ^{11}C and used to obtain the first successful imaging of neuroreceptors in the brain of a living human being.[29] Most of the dopamine receptors of the D2 type were found in the caudate nucleus and putamen, parts of the brain concerned with movement and emotion.

In studies of normal persons, due to the great sensitivity with which radioactivity can be measured, receptor density could be measured in various brain regions in units of picomoles per gram, an exceedingly small concentration.[30] Thus, PET was able to measure regional concentrations of important molecules in the living body with a sensitivity and specificity comparable to that obtained by radioimmunoassay in the study of body fluids, such as blood.[31,32] For example, it has been shown that D2 dopamine receptors decrease dramatically between the ages of 19 and 73 years, with most of the decrease occurring before the age of 40.[30]

One year after the imaging of dopamine receptors, opiate receptors were imaged in living human beings with ^{11}C carfentanil, a drug that stimulates opiate receptors.[33] Soon thereafter, a simple procedure was devised that could be used to measure directly the blocking effect of drugs, such as naloxone and naltrexone, that does not require a PET imaging device, but rather is based on the use of a dual radiation detector system that provides measurements of the time course of the radioactive drug in the part of the brain toward which the detector is directed.[34]

Naloxone and naltrexone are drugs used to treat patients suffering from narcotic abuse by blocking the opiate receptors. They belong to the class of drugs called ''antagonists''. Such drugs bind to receptors without inducing the biochemical and behavioral effects normally produced by opiate narcotics, but they prevent subsequent binding to the receptor of the stimulating ''agonist'' drugs.

With a simple dual-probe radiation detector, it was found that the administration of a single 50-mg oral dose of naltrexone blocks opiate receptors in the human brain for almost a week, at a time when blood levels of the drug have fallen to about one twentieth of the maximum levels.[35] Animal experiments and plasma measurements in human beings had led the manufacturer of the drug to recommend a dose 50 times greater than that required to produce nearly complete blockade of the receptors. Administration of smaller doses could decrease side effects and lessen patients' reluctance to take the drug, which has been a problem. Since the effects of the drugs on the brain often do not require a PET scanner, it may be possible to monitor drug effects with positron-emitting tracers and simple detectors in individual patients.

Just as in the treatment of drug addicts with naltrexone, side effects of the drug treatment of schizophrenic patients often present problems. Use of the simple dual detector probe system makes it possible to measure the dose of neuroleptic drugs that block dopamine receptors to a desired level of occupancy. Measurement of the degree of occupancy of the receptors by direct monitoring of the blocking effects of drugs, such as haloperidol, could result in an increase in effectiveness and decrease the side effects.

II. RECEPTORS AND MENTAL ILLNESS

For decades, it has been suspected that abnormalities of the dopaminergic system might be involved in patients with schizophrenia. Drugs that help schizophrenic patients block D2 dopamine receptors; amphetamines, which elevate synaptic dopamine concentrations, exacerbate symptoms in schizophrenic patients and can produce psychotic symptoms in otherwise normal persons; post-mortem studies of the brains of schizophrenic patients reveal increased numbers of D2 dopamine receptors in the caudate nucleus and putamen. One cannot always be certain that the increases are not the result of the patient's having received drugs that block dopamine receptors, resulting in a compensatory increase in their number or affinity.

The increased concentrations of D2 dopamine receptors in the caudate and putamen of patients with schizophrenia could be the result of deficiencies of presynaptic neurons that could result in a decreased release of neurotransmitters into synapses in the caudate and putamen. The postsynaptic neurons of subcortical structures would be "released" in the sense described by Jackson.

Using ^{11}C NMSP as a radioligand, normal persons were found to have a receptor availability index in the caudate nuclei averaging 17 ± 2.5 pmol/g. Post-mortem data indicate that D2 dopamine receptor densities range from about 10 to 20 pmol/g in the human striatum in adults, the higher values being in young persons. In patients with schizophrenia, whether or not they had been treated previously with neuroleptic drugs, D2 dopamine receptor densities were significantly higher.[36]

Using a different radioligand, ^{11}C raclopride, values in the putamen of normal persons averaged 25 ± 1.6 pmol/g. Studies with this radioligand, employing a different analytical approach, failed to find an increase in D2 dopamine receptors in schizophrenia.[37]

It remains to be determined whether increased D2 dopamine receptors in some schizophrenic patients precede or follow the development of symptoms of schizophrenia. Increased numbers of dopamine receptors do not cause symptoms of schizophrenia per se because patients who receive antipsychotic drugs do not improve their thought disorders for at least a week, although a calming effect may be observed immediately after starting treatment.

Further studies are needed in regions other than the caudate/putamen. The important role of the limbic system, especially the hippocampus, in emotional behavior makes it a region of special interest for PET studies of patients with schizophrenia.

III. DEPRESSION

Dopamine receptors of the D2 subtype are elevated in other types of mental illness than schizophrenia.[38] Some patients with psychotic depression, who have delusions, hallucinations, and thought disorder, but are classifiable as suffering from bipolar illness rather than schizophrenia using standardized criteria, have been found to have elevated D2 dopamine receptor densities. In 15 patients with bipolar manic/depressive illness and not on medication, receptor levels in nonpsychotic patients did not differ from age-matched normal persons, but psychotic patients, whether in a manic or depressed phase of illness at the time of PET imaging, had dopamine receptor density values that were two or three times higher than age-matched control subjects.

Patients suffering from mood disorders following a stroke may also have abnormalities in certain neuroreceptor levels in some parts of the brain.[39] Between 30 and 50% of patients who have had a stroke suffer from persistent depression not related to the degree of physical impairment caused by the stroke, but related to involvement of the left cerebral hemisphere. The presence of mood disorders is related to the region damaged by the stroke. Although there are only a few reported cases of mania occurring after a stroke, the majority of such patients have lesions of the right cerebral hemisphere. Mayberg and colleagues observed in patients with strokes damaging the right cerebral hemisphere that there was evidence of increased availability of serotonin receptors of the S2 type, due to an increased number of receptors, increased affinity of the receptors, or decreased occupancy by endogenous serotonin.[40] These preliminary findings support the concept that there may be "seats" of emotion in the brain. Nearly 100 years ago, William James commented on the question of localization of emotions in brain centers: " . . . concerning the emotions . . . either separate and special centers, affected to them alone, are their brain-seat, or else they correspond to processes occurring in the motor and sensory centers already assigned, or in others like them, not yet known." His question remains unanswered today, but is now being addressed by PET.

IV. PET AND PSYCHIATRY

Despite the enormous contributions of persons such as Jackson and Bleuler, the study of mental illness has remained largely subjective, because of the paucity of biological markers. The history of psychiatry reflects a continuing search for objectivity. At times, one finds specific and unmistakable anatomic lesions, as in Pick's and Wilson's disease, in which the clinical and pathological findings were described simultaneously. Even when structural abnormalities are found, it is difficult to know whether the structural changes are a cause or an effect of the disease. The age-old problem is the relationship between brain structures and mental functions.

In many diseases, functional or biochemical abnormalities are often detectable before structural changes can be found. Today, it is possible to study high risk populations to see if regional biochemical abnormalities occur before symptoms or anatomical changes take place. An example is the reduction of glucose metabolism in the basal ganglia of asymptomatic persons at risk of Huntington's disease.

PET now provides a new way to obtain objective biological markers if they are present in mental disorders. Just as it is difficult to conceive of diagnosing anemia without blood counts, or hypertension without a blood pressure cuff, it may some day be impossible to think of mental diseases without biochemical markers.

ACKNOWLEDGMENT

This work was supported by U.S. Public Health Service Grant NS 15090. I wish to acknowledge the contributions of many investigators who took part in this work, including J. J. Frost, D. F. Wong, R. F. Dannals, J. M. Links, H. S. Mayberg, U. S. Scheffel, E. D. London, A. A. Wilson, and J. R. Lever.

REFERENCES

1. **Reivich, M., Kuhl, D., Wolf, A., Greenberg, J., Phelps, M., Ido, T., Casella, V., Fowler, J., Hoffman, E., Alavi, A., Som, P., and Sokoloff, L.,** The [18F]fluorodeoxyglucose method for the measurement of local, cerebral glucose utilization in man, *Circ. Res.,* 44(1), 127, 1979.
2. **Reivich, M. and Alavi, A., Eds.,** *Positron Emission Tomography,* Alan R. Liss, New York, 1982.
3. **Phelps, M. E., Huang, S. C., Hoffman, E. J., Selin, C., Sokoloff, L., and Kuhl, D. E.,** Tomographic measurement of local cerebral glucose metabolic rate in humans with (F-18) 2-fluoro-2-deoxy-D-glucose: validation of method, *Ann. Neurol.,* 6(5), 371, 1979.
4. **Phelps, M. E. and Mazziotta, J. C.,** PET: human brain function and biochemistry, *Science,* 228, 799, 1985.

5. **Phelps, M. E., Mazziotta, J. C., and Schelbert, H. R.,** *Positron Emission Tomography and Autoradiography: Application in the Brain and Heart,* Raven Press, New York, 1986.

6. **Sokoloff, L., Reivich, M., Kennedy, C., Des Rosiers, M. H., Patlak, C. S., Pettigrew, K. D., Sakurada, O., and Shinohara, M.,** The [^{14}C]deoxyglucose method for the measurement of local cerebral glucose utilization: theory, procedure, and normal values in the conscious and anesthetized albino rat, *J. Neurochem.,* 28(5), 897, 1977.

7. **Sokoloff, L.,** The radioactive deoxyglucose method, theory, procedure and applications for the measurement of local glucose utilization in the central nervous system, in *Advances in Neurochemistry,* Vol. 4, Agranoff, B. W. and Aprison, M. H., Eds., Plenum Press, New York, 1982, 1.

8. **Ter-Pogossian, M. M., Phelps, M. E., Hoffman, E. J., and Mullani, N. A.,** A positron emission transaxial tomograph for nuclear imaging (PETT), *Radiology,* 114(1), 89, 1975.

9. **Ter-Pogossian, M. M., Mullani, N. A., Hood, J., Higgins, C. S., and Currie, C. M.,** A multislice positron emission computed tomograph (PETT IV) yielding transverse and longitudinal images, *Radiology,* 18(2), 477, 1978.

10. **Ido, T., Wan, C. N., Casella, V., Fowler, J. S., and Wolf, A. P.,** Labeled 2-deoxy-D-glucose analogs. ^{18}F-labeled 2-deoxy-2-fluoro-D-gluocse, 2-deoxy-2-fluoro-D-mannose and ^{14}C-2-deoxy-2-fluoro-D-glucose, *J. Labelled Compd. Radiopharm.,* 14, 175, 1978.

11. **Tewson, T. J., Welch, M. J., and Raichle, M. E.,** [^{18}F] labeled 3-deoxy-3-fluoro-D-glucose: synthesis and preliminary biodistribution data, *J. Nucl. Med.,* 19(12), 1339, 1978.

12. **Raichle, M. E.,** Measurement of local brain blood flow and oxygen utilization using oxygen-15 radiopharmaceuticals: a rapid dynamic imaging approach, in *Positron Emission Tomography,* Reivich, M. and Alavi, A., Eds., Alan R. Liss, New York, 1985, 241.

13. **Raichle, M. E., Martin, W. R. W., Herscovitch, P., Mintun, M. A., and Markham, J.,** Brain blood flow measured with intravenous H$_2$^{15}O. II. Implementation and validation, *J. Nucl. Med.,* 24(9), 790, 1983.

14. **Mintun, M. A., Raichle, M. E., Martin, W. R., and Herscovitch, P.,** Brain oxygen utilization measured with O-15 radiotracers and positron emission tomography, *J. Nucl. Med.,* 25(2), 177, 1984.

15. **Ter-Pogossian, M. M. and Hersovitch, P.,** Radioactive oxygen-15 in the study of cerebral blood flow, blood volume and oxygen metabolism, *Semin. Nucl. Med.,* 15(4), 377, 1985.

16. **Frackowiak, R. S. J. and Lammertsma, A. A.,** Clinical measurement of cerebral blood flow and oxygen consumption, in *Positron Emission Tomography,* Reivich, M. and Alavi, A., Eds., Alan R. Liss, New York, 1985, 153.

17. **Roback, A. A.,** *History of Psychology and Psychiatry,* Philosophical Library, New York, 1961.

18. **Gladston, I.,** *Historic Derivations of Modern Psychiatry,* McGraw-Hill, New York, 1967.

19. **Buchsbaum, M. S., Holcomb, H., Kessler, R., Johnson, J., King, A. C., Cappelletti, J., Bisserbe, J. C., Van Kammen, D. P., Selub, S., Wu, J., Manning, R. G., and Channing, M.,** Lateralized asymmetries in glucose uptake assessed by positron emission tomography in patients with schizophrenia and normal controls, in *Laterality and Psychopathology,* Flor-Henry, P. and Gruzelier, J., Eds., Elsevier, Amsterdam, 1983b, 559.

20. **DeLisi, L. E., Holcomb, H. H., Cohen, R. M., Pickar, D., Carpenter, W., Morihisa, J. M., King, A. C., Kessler, R., and Buchsbaum, M. S.,** Positron emission tomography in schizophrenic patients with and without neuroleptic medication, *J. Cereb. Blood Flow Metab.,* 5, 201, 1985.

21. **DeLisi, L. E., Buchsbaum, M. S., Holcomb, H. H., Dowling-Zimmerman, S., Pickar, D., Boronow, J., Morihisa, J. M., van Kammen, D. P., Carpenter, W., Kessler, R. et al.,** Clinical correlates of decreased anteroposterior metabolic gradients in positron emission tomography (PET) of schizophrenic patients, *Am. J. Psychiatry,* 142(1), 78, 1985.

22. **Farkas, T., Wolf, A. P., Jaeger, J., Brodie, J. D., Christman, D. R., and Fowler, J. S.,** Regional brain glucose metabolism in chronic schizophrenia. A positron emission transaxial tomographic study, *Arch. Gen. Psychiatry,* 41(3), 293, 1984.

23. **Franzen, G. and Ingvar, D. H.,** Absence of activation in frontal structures during psychological testing of chronic schizophrenics, *J. Neurol. Neurosurg. Psychiatry,* 38, 1027, 1975.

24. **Ingvar, D. H. and Franzen, G.,** Abnormalities of cerebral blood flow distribution in patients with chronic schizophrenia, *Acta Psychiat. Scand.,* 50, 425, 1974.

25. **Ingvar, D. H., Rosen, I., Eriksson, M., and Elmqvist, D.,** Activation patterns induced in the dominant hemisphere by skin stimulation, in *Sensory Functions of the Skin in Primates,* Zotterman, Y., Ed., Pergamon Press, New York, 1976, P549.

26. **Berman, K. F., Zec, R. F., and Weinberger, D. R.,** Physiologic dysfunction of dorsolateral prefrontal cortex in schizophrenia. II. Role of neuroleptic treatment, attention, and mental effort, *Arch. Gen. Psychiatry,* 43, 126, 1986.

27. **Ehrlich, P.,** Address in pathology on chemotheraspeutics: scientific principles, methods and results, *Lancet,* 2, 445, 1913.

28. **Pert, C. B. and Snyder, S. H.,** Opiate receptors: demonstration in nervous tissue, *Science,* 179, 1011, 1973.

29. Wagner, H. N., Jr., Burns, H. D., Dannals, R. F., Wong, D. F., Langstrom, B., Duelfer, T., Frost, J. J., Ravert, H. T., Links, J. M., Rosenbloom, S. B., Lukas, S. E., Kramer, A. V., and Kuhar, M. J., Imaging dopamine receptors in the human brain by positron tomography, *Science,* 221(4617), 1264, 1983.

30. Wong, D. F., Wagner, H. N., Jr., Dannals, R. F., Links, J. M., Frost, J. J., Ravert, H. T., Wilson, A. A., Rosenbaum, A. E., Gjedde, A., Douglass, K. H. et al., Effects of age on dopamine and serotonin receptors measured by positron tomography in the living human brain, *Science,* 226, 1393, 1984.

31. Wong, D. F., Gjedde, A., and Wagner, H. N., Jr., Quantification of neuroreceptors in the living human brain. I. Irreversible binding of ligands, *J. Cereb. Blood Flow Metab.,* 6(2), 137, 1986b.

32. Wong, D. F., Gjedde, A., Wagner, H. N., Jr., Dannals, R. F., Douglass, K. H., Links, J. M., and Kuhar, M. J., Quantification of neuroreceptors in the living human brain. II. Inhibition studies of receptor density and affinity, *J. Cereb. Blood Flow Metab.,* 6(2), 147, 1986a.

33. Frost, J. J., Wagner, H. N., Dannals, R. F., Ravert, H. T., Links, J. M., Wilson, A. A., Burns, H. D., Wong, D. F., McPherson, R. W., Rosenbaum, A. E., Kuhar, M. J., and Snyder, S., Imaging opiate receptors in the human brain by positron tomography, *J. Comput. Assist. Tomogr.,* 9, 231, 1985.

34. Bice, A. N., Wagner, H. N., Frost, J. K. J., Natarajan, T. K., Lee, M. C., Wong, D. F., Dannals, R. F., Ravert, H. T., Wilson, A. A., and Links, J. M., Simplified detection system for neuroreceptor studies in the human brain, *J. Nucl. Med.,* 27, 184, 1986.

35. Lee, M. C., Wagner, H. N., Tanada, S., Frost, J. J., Bice, A. N., and Dannals, R. F., Duration of occupancy of opiate receptors by naltrexone, *J. Nucl. Med.,* 29, 1207, 1988.

36. Wong, D. F., Wagner, H. N., Tune, L. E., Dannals, R. F., Pearlson, G. D., Links, J. M., Tamminga, C. A., Broussolle, E. P., Ravert, H. T., Wilson, A. A., Toung, J. K. T., Malat, J., Williams, J. A., O'Tuama, L. A., Snyder, S. H., Kuhar, M. J., and Gjedde, A., Positron emission tomography reveals elevated D2 dopamine receptors in drug-naive schizophrenics, *Science,* 234, 1558, 1986.

37. Sedvall, G., Farde, L., Persson, A., and Wiesel, F. A., Imaging of neurotransmitter receptors in the living human brain, *Arch. Gen. Psychiatry,* 43, 995, 1986.

38. Wong, D. F., Pearlson, G., Tune, L. E. et al., In vivo measurement of D2 dopamine receptor abnormalities in drug-naive and treated manic depressive patients, *J. Nucl. Med.,* 28, 611, 1987.

39. Robinson, R. G. H., Kubos, K. L., Starr, L. B., Rao, K., and Price, T. R., Mood disorders in stroke patients: importance of lesion location, *Brain,* 107, 81, 1984.

40. Mayberg, H. S., Robinson, R. G., Wong, D. F., Parikk, R. M., Bodluc, P. L., Starkstein, S. E., Price, T. P., Dannals, R. F., Links, J. M., Wilson, A. A., Ravert, H. T., and Wagner, H. N., PET imaging of cortical S2 serotonin receptors following stroke: lateralized changes and relationship to depression, *Am. J. Psychiatry,* 145, 937, 1988.

Chapter 20

THE STUDY OF SCHIZOPHRENIA WITH SPECT

Jörg J. Pahl, Karim Rezai, Daniel S. O'Leary, and Nancy C. Andreasen

TABLE OF CONTENTS

I. INTRODUCTION

Schizophrenia is a complex, psychiatric disorder of unknown etiology. The predominant age of onset is in the late teens, with males and females affected equally. One percent (1%) of the general population is afflicted with schizophrenia at any given time.[1] The phenotypic expression of schizophrenia appears to involve multiple brain systems. This is reflected in the symptomatology which is accordingly diverse and includes not only auditory, visual, and tactile hallucinations, but also disorganized speech and intellectual impoverishment.

A number of factors have hampered progress in our understanding of the etiology, pathophysiology, and management of schizophrenia. Firstly, phenotypic expression is so extremely varied in this illness that there is considerable overlap between it and other psychiatric conditions. Eminent psychiatrists including Kraepelin and Bleuler have argued at length about the nosology of the major psychoses.[2,3] The demarcation lines separating schizophrenia from bipolar disease and delusional disorder remain unclear. Various sets of criteria (Feighner, RDC, DSM-III, and DSM-III-R) have been advocated as a basis with which to differentiate schizophrenia from other psychotic conditions, but none is definitive.[4] Secondly, psychiatrists have repeatedly debated the question of whether schizophrenia represents a single illness or a spectrum of disorders. The latter position is now more commonly accepted.

Genetic studies have not supported the presence of homogeneity in schizophrenia.[5] The disorder thus seems to be both phenomenologically and etiologically heterogeneous. Neuroscientists anticipate that neuroimaging will help solve some of these problems.

II. BRAIN IMAGING IN SCHIZOPHRENIA

Brain imaging techniques have traditionally been divided into structural and functional modalities. The former supply considerable anatomic but little physiological detail. Examples include X-ray CT and magnetic resonance imaging (MRI). Of the two techniques, MRI offers considerable advantages over X-ray CT. It allows us to visualize small limbic structures such as the amygdala in all three orthogonal planes without the use of ionizing radiation. The MRI scan acquisition parameters (e.g., pulse sequences with different echo and repetition times) can be varied to obtain images offering optimal anatomic detail as seen in Figure 1. Shown in the figure are coronal sections of the brain depicting structures of particular interest to schizophrenia. Visualized are the lateral ventricles and the basal ganglia (caudate nucleus, putamen and globus pallidus), as well as small limbic structures (amygdala and hippocampus). Individual structures have been outlined as regions of interest (ROI). Special computer software has been used to reconstruct these ROIs into a three-dimensional MRI composite (see Plate 1*).

A number of post-mortem X-ray CT and MRI studies have demonstrated focal changes in overall size and gray matter content of medial temporal lobe regions including the hippocampus.[6,7] Some of these investigations have also found a corresponding ipsilateral dilatation of the inferior temporal horn.[8]

Most schizophrenic patients, however, show minimal or no structural brain alterations. Despite the paucity of histological changes, considerable pharmacological evidence points to the probability that brain neurochemistry is disturbed in schizophrenia.[9] Regional chemical composition of the brain can be studied at post-mortem with autoradiography or *in vivo* with functional imaging techniques. Single photon emission computed tomography (SPECT) and/or positron emission tomography (PET) are currently being used for this purpose in living human beings.[10] Physiological and chemical brain parameters that are being measured

* Plate 1 follows page 198.

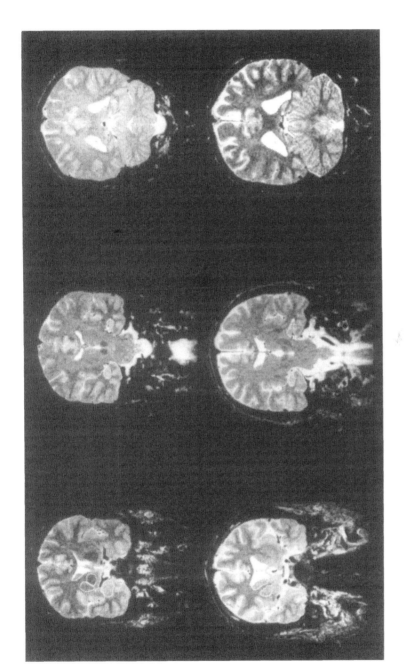

FIGURE 1. Magnetic resonance imaging (MRI) in schizophrenia. The figure depicts three corresponding coronal planes for a normal individual (top panel) and a schizophrenic patient (bottom panel). The first set of patient-control images (from left to right) was taken at the level of the optic chiasm; the next was taken at the level of the internal auditory meatus; and the last was taken at the level of the trigonum of the lateral ventricle. Pulse sequences used (echo time, TE = 90 msec; and relaxation time, TR = 3000 msec) resulted in the T-2 weighted images which are ideal for determining ventricular size. Lateral and third ventricular dilatation is present in the schizophrenic patient.

in schizophrenia include regional glucose metabolism, cerebral blood flow (CBF), and neuroreceptor concentrations. Information gained from studies during life rather than from post-mortem brain tissue is expected to be critical to our understanding of the pathophysiology of schizophrenia.

A. SINGLE PHOTON EMISSION COMPUTED TOMOGRAPHY (SPECT)

SPECT is the technique of generating tomographic images from distribution of an internally administered radionuclide. The radiopharmaceuticals and instrumentation used for SPECT and PET imaging are discussed in detail in Chapters 1 to 6. Briefly, the term "computed tomography" implies, SPECT imaging requires acquisition of data from a number of angular positions around the target area. These are subsequently backprojected with the aid of a computer to reconstruct tomographic images in orthogonal or oblique planes. A correction is ordinarily applied to compensate for the attenuation of photons within the patient's body. SPECT images are generally superior to planar scans with respect to image contrast and their ability to localize structures in three-dimensional space. For these reasons, practically all newly developed brain radiopharmaceuticals are imaged with SPECT and the classic ^{133}Xe methodology for rCBF measurements has also been adapted for imaging in SPECT mode.

B. EXPERIMENTAL DESIGN OF COGNITIVE EVALUATIONS

SPECT is well suited for testing schizophrenia-related hypotheses.[11] It is cost-effective, uses commercially available radiotracers, and is technically less demanding than PET. Hypotheses that are readily testable with SPECT include the frontal lobe dysfunction and hemispheric laterality hypotheses.[12-14] Several such studies exist and will be reviewed in this chapter.

Apart from determining the functional response of a single lobe (i.e., frontal, temporal, etc.), one can also activate individual cerebral systems of interest (CSOI) by having the subject perform a standardized task during SPECT data acquisition. Such systems subserve specific cognitive processes involved in audition, language, attention, and memory and are potentially abnormal in schizophrenic patients. Thus, there is evidence that tasks requiring conscious, serial, effortful attentional processes are impaired in schizophrenia, while more automatic parallel processes are normal.[15] An integrated approach which involves a collation of state-of-the-art anatomical (MRI), neurophysiological (SPECT), and cognitive techniques is well suited for exploring the anatomical and functional substrates of such cognitive processes.

Functional imaging is routinely conducted under baseline (i.e., controlled) conditions and during activation procedures. Image subtraction techniques are used to determine the areas of regional activation and/or depression in paired baseline/activation studies. Sequential increases in stimulus content that can vary from total or partial sensory deprivation through increasing complexity of the sensory stimulation pattern can be used to induce a progressive rise in cerebral blood flow and metabolism above a reference state.

The resting unstimulated state is usually characterized by minimal external sensory stimulation in an awake, relaxed subject who is not engaged in mental or motor activity. Ambient conditions during such studies consist of diffuse dim lighting and low noise levels emanating mainly from SPECT camera and computer operations. Ingvar has described a "hyperfrontal" resting cerebral blood flow pattern (prefrontal 30% greater than hemispheric mean; frontal flow higher than parietal lobe flow; temporal lobe flow the lowest) in such resting subjects using the intracarotid ^{133}Xe method.[16] The significance of the hyperfrontal pattern remains uncertain. Anxiety, emotional factors, and intent to act may all underlie this phenomenon. The resting state is also characterized by cortical and subcortical left-right

TABLE 1
Criteria To Be Met for Cognitive Activation Tasks

1. The task chosen should activate a specific functional cognitive system (e.g., memory, attention, language) that has been shown to be abnormal in schizophrenia.

2. Performance of the task by the subject should be assessable behaviorally, independently of the imaging technique being used.

3. The task should be one that is well-researched behaviorally and that has a firm (and testable), theoretical basis.

4. The task should be as simple as possible in sensory and motor requirements and should allow differentiation of simpler from more complex processes through a hierarchical or subtractive approach.

5. The task should be as pure a measure of the functional system as possible, in the sense that minimal participation of other functional systems should be required to perform the task. For example, a task designed to assess an aspect of attention should, ideally, not require memory and/or language. Since this is rarely possible in practice, the task should allow componential assessment of the chosen functional system through a hierarchical or subtractive technique.

6. The chosen functional cognitive system should map onto potentially "localizable brain systems", i.e., there should be reason to believe that the task chosen will activate specific brain systems and that the activation can be seen with the chosen imaging technique.

7. The task must be well suited to the technical constraints of the imaging technique to be used. The temporal parameters of the neuroimaging technique are of primary importance, followed by its temporal resolution.

metabolic symmetry in normal subjects in the eyes and ears open condition and during states of partial sensory deprivation (eyes patched or ears plugged).[17]

Controlled baseline states have been devised by some investigators to reduce the effects of anxiety and thinking on neuronal activity. Reivich et al. have, for instance, shown with PET that anxiety increases the glucose metabolic rate in the posterior-fronto-orbital and middle frontal regions.[18] Uncontrolled thinking represents brain work and therefore also affects blood flow and metabolism. A possible disadvantage of controlled baseline states is that they may activate regions of interest (ROI) that have been specifically targeted by the investigation. Thus, there is no easy solution to the problem of choosing the baseline state.

Ideally, a controlled baseline task should activate the same regions as the experimental activation task, apart from the ones specific to the experimental task. This goal is, however, difficult to achieve in reality. Table 1 lists the criteria to be met for cognitive activation tasks in schizophrenia. Current baseline tasks, including the simple number matching task used by Berman et al., do potentially result in activation of frontal lobe structures in individual patients while not showing a significant response when averaged over the entire group.[19] As pointed out in criterion 5 of Table 1, cognitive stimulation tasks inherently activate multiple functional systems. One therefore needs to devise a hierarchical stimulation paradigm which divides tasks into their component parts. Thus, in a mathematical task (e.g., push button if $x + y > z$) requiring both motor activity and cognition, one would use two stimulation steps and subtraction image analysis techniques to determine the regional functional responses to either motor or cognitive component of the task.

III. SPECT RADIOPHARMACEUTICALS AND INSTRUMENTATION FOR SINGLE PHOTON BRAIN IMAGING

Nearly all radiopharmaceuticals currently available for SPECT brain imaging depict regional cerebral blood flow. These can be broadly divided into two categories: those that freely diffuse in and out of the brain by virtue of their lipophylicity, as exemplified by ^{133}Xe and other noble gases, and those that are efficiently extracted by the brain upon their first transit through the brain circulation, as exemplified by ^{123}I-IMP and ^{99m}Tc-HMPAO. Information about the brain blood flow can be derived with either class of pharmaceuticals by externally recording their photon emissions and applying mathematical models that derive physiologic data from radiotracer distribution. For the diffusible tracer group, rCBF is derived principally from the slope of the clearance of the radiotracer in the brain. Thus, data have to be recorded dynamically, i.e., in multiple discrete samples as a function of time. The units of measurement are in ml/100 g/min. For the first pass tracers, however, the brain content of the radiotracer is stable for relatively extended periods of time and therefore a set of only static images would suffice. Images so obtained depict relative distribution of the blood flow in the brain and do not quantify cerebral perfusion in absolute units. To do so would require arterial blood sampling coincident with the tracer injection, which is invasive and rarely practiced.

[133]Xe methodology is by far responsible for most of the investigative work done to date on CBF, including those in the field of psychiatry. This methodology is based on the pioneering work of Kety and Schmidt in 1948 with diffusible tracers and has undergone several evolutionary phases in the last decades. Its initial implementation involved intra-carotid injection of the tracer and external monitoring of its photon emission by an array of probe-type detectors placed around the convexity of the head. Obrist et al. later modified the technique to administer xenon by inhalation.[20] This necessitated dynamic recording of the xenon input to the brain which is easily accomplished by placement of an additional probe over the lungs or the carotid arteries in the neck.

Multiprobe devices have a number of inherent limitations. Their sensitivity is maximal only for the surface events which on the one hand limits their usefulness for rCBF measurement primarily from the cortex of the brain and on the other hand makes them vulnerable to signal contamination from external carotid circulation in the scalp and the skull. Furthermore, their output does not constitute a true rCBF ''image'' of the brain and should be considered only as spot measurements at selected loci over the brain surface. In recent years, Lassen et al. have developed a SPECT device for rCBF measurements with [133]Xe (Tomomatic-64).[21] True, transaxial images of the brain perfusion are obtained with this device based on a set of four 1-min images acquired during and after inhalation of xenon for 1 min.

The ^{133}Xe approach has two unique advantages. First, it provides absolute measurement of rCBF in milliliters per 100 g per minute which is not easily feasible with either of the first-pass tracers. Second, xenon clears the body rapidly via exhalation and therefore a second study can be obtained in the same subject in less than 1 h. This is especially useful for assessment of brain response to physiologic or psychologic interventions. The main disadvantages of ^{133}Xe relate to its physical attributes. Only a small fraction of the inhaled xenon is dissolved in blood and subsequently reaches the brain. Furthermore, the photon emissions of ^{133}Xe are of relatively low energy. This results in significant image degradation due to attenuation losses and inefficient scatter rejection.

The problem of low photon flux in xenon work is also compounded by the need to acquire data dynamically, i.e., in short time intervals. The detector systems thus have to be maximized for sensitivity, usually at the expense of compromising spatial resolution. The devices currently available have a nominal resolution in the range of 2 cm or better, but for reliable rCBF sampling, one needs to select even larger-sized ROIs. Rezai et al. have shown that rCBF estimates are valid only when drawn from brain regions of 3 cm or larger and composed primarily of gray matter.[22] Smaller regions are subject to partial voluming errors and blood flow in low perfusion areas of the brain such as the white matter is overestimated due to scattering of photons from adjacent high flow gray matter areas.

Multiprobe devices have given us a wealth of information on brain blood flow, yet, they are subject to the constraints of xenon methodology and also suffer from their own limitations described earlier. While it is a tribute to human ingenuity that such a body of knowledge could be amassed from such imperfect devices, it is imperative that we reexamine our understanding of brain perfusion with the more refined tools that have become available to us by means of SPECT imaging and the new radiopharmaceuticals.

The new radiopharmaceuticals for brain imaging provide improved image resolution over the xenon methodology. Both IMP and HMPAO achieve stable concentrations in the brain shortly after intravenous injection and permit extended imaging times to obtain sufficient information density with high resolution collimators. The photon emission of ^{123}I and ^{99m}Tc used with IMP and HMPAO, respectively, are in the optimal range for standard gamma cameras used for SPECT imaging in nuclear medicine installations. However, the spatial resolution of these devices (1.2 to 1.5 cm, FWHM) is still inadequate for showing fine anatomical detail, especially in deeper brain regions. Figure 2 shows a series of sagittal ^{99m}Tc-HMPAO SPECT images that have been acquired with a Siemens gamma camera (Orbiter). Efforts are currently underway in several centers to design SPECT devices capable of image resolution under 1 cm. New camera designs can also benefit from enhanced detector sensitivity. The count rates available in clinical SPECT imaging are limited by the practical aspects of time spent per study and the maximum permissible doses of the radiopharmaceutical. A significant gain in image quality is yet possible by maximizing the information content of the SPECT images. For example, Figure 3 shows an idealized performance of the same camera used in Figure 2. These high count images depict sections of a special brain phantom devised by Chang et al.[23]

The drawback of the first pass radiotracers is their inability to measure absolute rCBF without an arterial reference sample. A common practice is therefore to express the brain perfusion for any given region as a ratio referenced to the cerebellum or the total brain. Furthermore, the first pass tracers do not clear the brain rapidly and as such, rest and activation studies cannot be performed back to back. In practice, such studies are performed 24 h apart. A corollary to this, however, is that the activation and scanning procedures need not be coincident. The subjects can be injected under the well-controlled conditions of a psychophysiological laboratory and imaged 15 to 30 min later for a considerable gain in operational facility. The resultant images reflect a brain blood flow pattern present during a narrow time window of about 5 to 10 min following the injection of either ^{99m}Tc-HMPAO or ^{123}I-IMP.

IV. SPECT IMAGE ANALYSIS

The images acquired with SPECT devices are useful for qualitative assessment of gross, but not subtle findings in the cortical and subcortical areas. Thus, one can demonstrate decreased CBF in the primary cortical visual area during visual sensory deprivation. However, demonstration of subtle changes, and in general those related to cognitive activation procedures, requires a more rigorous analysis. This is typically carried out in two steps. In the first step, anatomical localization schemes are used to rigidly define the regions of interest, and in the second step, mathematical and statistical procedures are applied to extract physiologic data from those regions. Needless to say, these procedures should also take into account the pharmacokinetic model of the radiotracer under investigation. For the first pass tracers, the procedure is rather simple. The static images obtained are considered directly representative of the relative distribution of the blood flow in the brain. For ^{133}Xe, however, rCBF is derived by multiplying the observed clearance rate of the radiotracer by a fixed value representing the blood-brain partition coefficient for aggregate white and gray matter. The xenon raw data are deconvolved with the tracer input curve recorded via a probe over the lungs or the carotid arteries prior to clearance measurements. There is also potential for SPECT to map the density of various receptors in the brain provided appropriate radiopharmaceuticals become available. Such quantitative analyses, however, require application of far more elaborate modeling schemes. Typically, multiple differential equations need to be

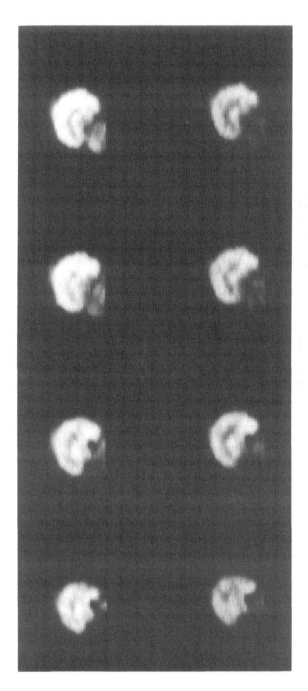

FIGURE 2.   SPECT images of a schizophrenic patient during cognitive activation. ^{99m}Tc HMPAO sagittal images are shown of a schizophrenic patient under resting conditions (top row) and during activation with the Wisconsin Card Sorting Test (WCST). The frontal lobes fail to show regional perfusion changes. A number of patient-related and technical factors could account for this finding. These include true nonactivation of the frontal lobes, choice of radiotracer (^{99m}Tc HMPAO rather than ^{133}Xe), and poor spatial resolution characteristics of the SPECT camera.

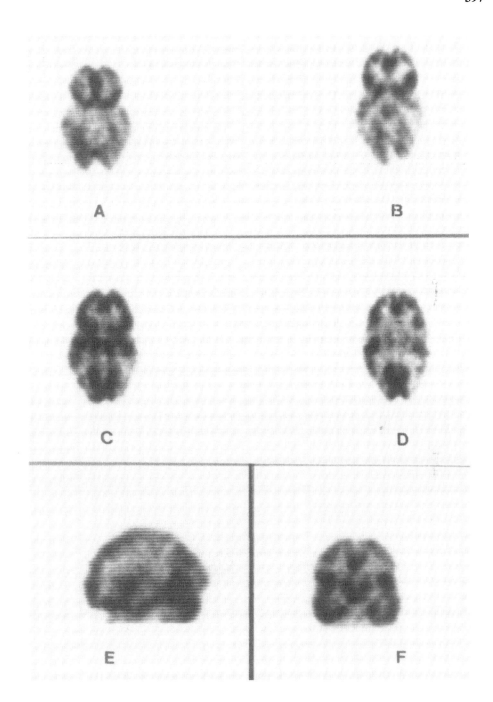

FIGURE 3. High-resolution SPECT. Transaxial (A to D), sagittal (E), and coronal (F) SPECT images generated under ideal conditions with a rotating SPECT camera system (similar to that used in Figure 2) from a 3-D anthropormorphic brain phantom. This set of images shows the excellent anatomical detail. The caudate nucleus (CN), the thalamus (TH), and the primary visual cortex (PVC) are, as seen, well defined in these images. (From Chang, W., Madsen, M. T., Lanfen, W., and Kirchner, P. T., *Nucl. Med. Commun.*, 11, 11, 1990. With permission.)

solved simultaneously in order to define the rate constants governing the interaction of the natural ligands and the analogues with the receptor in question.

V. ANATOMICAL LOCALIZATION SCHEMES

Anatomical resolution achieved with MRI is superior to that seen with either ^{133}Xe or ^{99m}Tc-HMPAO SPECT imaging by an order of magnitude. Thus, various MRI based anatomical localization schemes have been devised to improve the accuracy with which ROIs are defined in functional images. Superimposition of SPECT on MRI scans represents one such approach. Others include using stereotactic localization schemes. A report from a recent NIMH workshop provides a comprehensive overview of these methodologies.[24]

The following sections review the findings that various authors have reported with SPECT brain imaging in normal controls and subjects suffering from schizophrenia. Findings are not always directly comparable since studies differ in patient variables, radiotracer choice, and SPECT camera design as well as image analysis technique applied.

VI. FUNCTIONAL BRAIN IMAGING IN NORMAL SUBJECTS

The investigation of brain function in normal control groups represents an integral component of SPECT studies in schizophrenia. SPECT imaging is performed in such individuals under defined conditions (''resting wakefulness'' or baseline task) during sensory deprivation or cognitive activation and during pharmacological challenge.

A. BASELINE STATE

Considerable controversy surrounds the definition of the resting state cerebral blood flow landscape in normal subjects. It is most frequently described as hyperfrontal with gray matter rCBF normally 10 to 15% above the mean hemispheric flow level in frontal regions, at mean level centrally, and 5 to 10% below the mean postcentrally.[25] A number of authors including Reivich have, however, questioned the topography of the resting rCBF pattern.[18] Numerous factors, including anxiety, could account for increased frontal rather than parietotemporal blood flow in both normal and sick individuals.

The baseline conditions are either resting or cognitively controlled states. ''Resting wakefulness'' is usually defined as an eyes and ears open paradigm in which the patient is exposed to minimal sensory stimulation while being scanned in a quiet and dimly lit room. Such individuals are often very anxious as pointed out above. This can potentially lead to considerable activation of the prefrontal cortical region. Neuropsychologists have therefore developed a number of baseline states which engage the subject in tasks that are designed not to activate the target region and, at the same time, to lessen the effects of anxiety on CBF in this region. An example of such a baseline state is the number matching task used by Berman et al.[19] It is paired with the Wisconsin Card Sorting Test (WCST) when performing activation of the frontal lobes. The images representing the activated state can then be subtracted from the baseline state during image analysis with the ''difference'' image representing the regions that are either activated or depressed.

B. COGNITIVE ACTIVATION

In designing a specific cognitive activation protocol, it is essential that the radiotracer selected and the SPECT camera used be highly suited to the proposed research paradigm. Furthermore, it is important that the task be both simple and readily understood by normal control subjects as well as patients who often suffer from severe cognitive defects or are distracted by, for instance, auditory hallucinations. Ideally, cognitive activation tasks used in SPECT imaging should meet the criteria listed in Table 1. These criteria are only partly met by activating procedures in current use. Such tasks include the WCST, the Tower of London Test (TL), the Continuous Performance Task (CPT), and the Porteus Maze (PM). These tasks have been selected for their purported ability to activate cerebral regions that are hypothesized to be abnormal in schizophrenia.

The WCST is the most commonly used functional activating procedure in the schizophrenia SPECT literature. It is considered a frontal lobe task since it requires changes in cognitive set for its performance. Schizophrenic patients perform poorly as a group on the WCST. It is appropriate to note in passing that a significant number of normal subjects (>40%) fail to show activation of the frontal lobes with the WCST when imaged with ^{133}Xe.[11] Devous has even shown that activation with this task sometimes leads to *depression* of CBF in the frontal lobes in normals![26] Thus, there is a need for a frontal lobe task that will reliably activate the targeted region in normal controls. This may not be possible, however, because of the complexity of prefrontal cortical connections and individual variations in cognitive processing strategies.

VII. FUNCTIONAL BRAIN IMAGING IN SCHIZOPHRENIC SUBJECTS

Brain imaging in schizophrenia has profited over the years by important innovations that have been made in nuclear medicine instrumentation. Paradigm design has also been marked by increasing levels of sophistication. This has allowed investigators to replace global estimations of, for instance, cerebral blood flow and metabolism with regional determinations of these parameters. Studies in the resting state have been complemented by cerebral activation procedures that are based on specific theories of cognitive dysfunction in schizophrenia. Thus, brain imaging has been used to activate selectively the frontal lobes of patients, based on the hypothesis that negative symptom schizophrenics would demonstrate a frontal lobe syndrome and that dorsolateral prefrontal cortex (DLPFC) blood flow would correlate significantly with such symptoms.[12]

Functional brain imaging in schizophrenia began with the pioneering work of Kety and Schmidt (1948).[27] It has progressed to present day state-of-the-art investigations of cognitive dysfunction in schizophrenia. Initial examinations were performed with cortical probes. SPECT scanning with three-dimensional imaging capability represents a fairly recent addition to the single photon functional imaging field.

A. CORTICAL PROBE MEASUREMENTS: RESTING STATE

Physiological brain imaging is based on the observations of Roy and Sherrington (1898), who were the first to demonstrate tight coupling of CBF and metabolism.[28] They also showed that these parameters were directly related to brain function in healthy individuals. This fundamental concept shaped the Kety and Schmidt nitrous oxide determinations of global CBF and metabolism in schizophrenia.[29] Their initial report of normal global CBF and oxygen values in the illness has since been replicated in numerous additional studies. They encouraged the search for focal CBF abnormalities by airing the possibility that regional rather than global CBF changes would be found to underlie the schizophrenic syndrome.[27]

Advances in instrumentation led to the development of the ^{133}Xe clearance technique that allowed investigators to measure regional cerebral blood flow (rCBF) in addition to global CBF using an array of sodium iodide probes positioned about the head. This method generates two-dimensional functional landscapes of rCBF reflecting underlying brain activity. Ingvar used the ^{133}Xe clearance technique to study 11 "normal" patients and demonstrated that rCBF was significantly higher (20 to 40%) in the frontal region than in the temporal, parietal, or occipital regions. The subjects were examined in a state of "resting wakefulness". The eyes were kept occluded with a sandbag. Ingvar coined the term "hyperfrontality" to describe this "resting" rCBF pattern.[25]

In 1974 Ingvar and Franzen published a CBF study of 20 chronic schizophrenics of whom the younger group (mean age 25 years; disease duration 5 years) showed a normal hyperfrontal rCBF pattern while the older group (mean age 61 years; disease duration 40 years) had a relatively low frontal CBF pattern and, in most cases, high flow rates in the occipito-temporal region. CBF in the occipito-temporal region correlated positively with the degree of cognitive disturbance in the schizophrenic population. The appearance of relatively low frontal CBF in the patients was termed "hypofrontality" as opposed to the normal "hyperfrontality" pattern.[30] The hypofrontality finding in schizophrenia remains controversial having been replicated by some but not all investigators.[31]

Further evidence for a positive relationship between cerebral dysfunction and symptomatology came from the studies of Hoyer and Osterreich.[32] They performed a global study of CBF and oxygen utilization on schizophrenic patients subtyped according to clinical symptoms. They found highest flows in patients with productive schizophrenia and other acute psychoses and reduced CBF and oxygen utilization in patients with nonproductive schizophrenia.

The initial resting CBF studies of David Ingvar and Hoyer and Osterreich were followed by a number of replication studies that only partly corroborated their findings. Thus, Mathew et al. showed diffuse rCBF reductions which were statistically significant only in the right frontal area while Ariel et al. found a diminished anterior-posterior rCBF gradient.[33,34] In a later report, Mathew et al. described an inverse correlation between postcentral rCBF and hallucinatory behavior.[35] This result directly contrasts that of Ingvar.[30] The discrepancy may be due to significant differences in subject age and disease duration that were present in Ingvar's patient population as compared with that of Mathew's. In addition, the very young ages of the patients studied by Matthew et al. in the 1981 and 1982 studies might explain the lack of clear-cut frontal hypoperfusion in this age group.

B. CORTICAL PROBE MEASUREMENTS: COGNITIVE ACTIVATION

The absence of abnormal "resting state" rCBF values in most acute and many chronic schizophrenic patients led investigators to use cognitive activation procedures during CBF measurements. The rationale for using such cognitive probes was that these tests could be used to investigate functional reserve in schizophrenia. It was furthermore felt that, by so doing, patients with cognitive defects could be more readily differentiated from those without and from normal controls on the basis of CBF patterns.

Gur et al., in a test of the laterality hypothesis of schizophrenia, studied 15 medicated patients and 25 matched controls with the xenon inhalation technique during a resting baseline and during the performance of verbal and spatial tasks.[36] Controls demonstrated an expected increase in left hemisphere CBF during the verbal task and right hemisphere increases during the spatial task. In contrast, patients showed no flow asymmetry for the verbal task and greater left hemisphere increase for the spatial task. The authors concluded that their finding was consistent with the hypothesis that schizophrenia is associated with left hemisphere overactivation for spatial tasks.

In a follow-up study, Gur et al. examined 19 unmedicated schizophrenics and 19 matched controls.[37] Resting flows were higher in the left hemisphere in schizophrenics, again supporting the hypothesis of left hemisphere overaction. Furthermore, CBF was found to be correlated with the severity of the psychosis. The pattern of rCBF changes during activation with the verbal and spatial tasks was also different in schizophrenics. Schizophrenics demonstrated greater increases in flow for the verbal than the spatial tasks. This effect was found to be more pronounced in the severely disturbed patients. Comparison with the earlier sample of medicated schizophrenics suggested that neuroleptics restored symmetry of resting flows before they produced symptomatic relief. Medication did not appear to affect the rCBF abnormalities during cognitive activation.

TABLE 2

SPECT/Cortical Probe (CP) Findings in Schizophrenia

Year	S = schizophrenics C = controls Neuroleptic (NL)	Test[a] conditions status parameter	Tracer CP/SPECT[b]	Results[c]	Ref.
1948	S = 22 C = 35 All medicated	Resting	N₂O CP Global CBF	Normal whole brain CBF	27
1974	S = 20 C = 11 All medicated	Resting	¹³³Xe CP rCBF	Frontal rCBF reduced ("hypofrontality"); occipito-temporal rCBF increased	30
1979	S = 31 C = 11 All medicated	Resting	¹³³Xe CP rCBF	Low frontal rCBF correlates with negative symptoms; high postcentral rCBF with positive symptoms	16
1982	S = 23 C = 18 13/23 medicated	Resting	¹³³Xe CP rCBF	Low postcerebral flow seen with hallucinatory behavior	35
1983	S = 29 C = 22 Most medicated	Resting	¹³³Xe CP rCBF	Decreased anterior-posterior rCBF gradient; Mean CBF decreased	34
1985	S = 19 C = 19 Off NL > 1 week	1. Resting 2. Verbal task 3. Spatial task	¹³³Xe CP rCBF	1. Increased (L) hemisphere CBF pattern 2. Higher than normal CBF increase 3. Lower than normal CBF increase	37
1985	S = 34 C = 32 Off NL > 1 week	1. Resting 2. WCST	¹³³Xe SPECT rCBF	1. Hypofrontal CBF pattern 2. Reduced frontal (L>R) CBF	26
1986	S = 10 C = 10 Never medicated	Resting	¹³³Xe CP rCBF	Hypofrontal CBF pattern	42

TABLE 2 (continued)
SPECT/Cortical Probe (CP) Findings in Schizophrenia

Year	S = schizophrenics C = controls Neuroleptic (NL)	Test[a] conditions status parameter	Tracer CP/SPECT[b]	Results[c]	Ref.
1986	S = 20 C = 20 Off NL > 4 weeks	1. Resting 2. WCST 3. NM	^{133}Xe CP rCBF	1. Relative DLPFC rCBF reduced 2. DLPFC not activated 3. Normal rCBF pattern	38
1986	S = 24 C = 25 All medicated	1. Resting 2. WCST 3. NM	^{133}Xe CP rCBF	1. DLPFC rCBF values normal 2. DLPFC not activated 3. DLPFC not activated	39
1987	S = 51 C = 36 Off NL > 2 weeks	1. Resting 2. Piribel	^{133}Xe CP rCBF	Hypofrontal CBF pattern reversible: 1. During exacerbation of illness 2. After dopaminergic stimulation	43
1988	S = 16 C = 25 Off NL > 4 weeks	1. Resting 2. NM 3. WCST	^{133}Xe CP rCBF	1. Normal resting pattern 2. DLPFC not activated 3. DLPFC not activated	40
1988	S = 24 C = 25 Off NL > 4 weeks	1. Resting 2. SMT 3. RPM	^{133}Xe CP rCBF	1. Normal resting pattern 2. DLPFC not activated 3. DLPFC not activated	41

a   WCST, Wisconsin Card Sorting Test; NM, Number Matching Test; RPM, Ravens Progressive Matrices; SMT, Symbols Matching Task; Piribel: dopaminergic stimulation.

b   N$_2$O, nitrous oxide gas; CP cortical probe.

c   DLPFC, dorsolateral prefrontal cortex.

Weinberger and Berman tested the hypofrontality hypothesis with paired baseline/activation task paradigms in a series of four ^{133}Xe inhalation studies.[38] The DLPFC was the primary region of interest in all investigations. In the first study they combined the WCST with a number-matching baseline task. (As noted, the WCST challenges the subject's ability to think abstractly in that patients are required to alter their cognitive set in response to changes in visual instructions and cues.) The simple number-matching task failed to activate the DLPFC in both subject groups and did not otherwise differentiate controls from patients. Subjects in the study consisted of 20 medication-free chronic schizophrenic patients and 25 normal controls. During the resting state the authors recorded a reduction of relative, but not absolute, rCBF values. Controls showed a clear increase in DLPFC that was regionally specific involving only the DLPFC during the WCST. Unlike normals, nonmedicated schizophrenics failed to show activation of the DLPFC on the WCST. The authors' conclusion was that their findings were consistent with the general hypothesis of frontal lobe dysfunction in schizophrenia.

Further ^{133}Xe investigations at NIMH by Berman and Weinberger measured rCBF in neuroleptically medicated schizophrenics while at rest, during the number-matching baseline task and the WCST.[39] The results were qualitatively identical to those reported previously for medication-free patients by this group in 1986. In a second study rCBF was determined while medication free patients (N = 18) performed two versions of a visual CPT that measures attention.[39] The investigators found no differences in DLPFC blood flow between subject populations in either condition. The authors concluded that DLPFC dysfunction in schizophrenia was independent of medication status and not determined simply by state factors such as attention, mental effort, or severity of psychotic symptoms. Dysfunction of the DLPFC appeared to be cognitively linked to a physiological trait in the illness.

In the third study in the series, Wienberger et al. examined a new patient cohort consisting of 25 medication-free chronic schizophrenics as part of a replication study.[40] The chronic schizophrenic group again demonstrated reduced DLPFC rCBF during performance of the WCST, but not during the simple number-matching task. In addition, indices of dopaminergic and serotonergic neurotransmitter function (CSF homovanillic acid and 5-hydroxyindoleacetic acid concentrations) correlated with prefrontal rCBF during the WCST, but not during rest or the number matching task. The authors concluded that behavior-specific hypofunction of DLPFC in schizophrenia is reproducible and may be mediated by monoaminergic mechanisms.

Berman, in the last of the ^{133}Xe NIMH reports, used additional cognitive probes to investigate the role of the DLPFC in abstract reasoning.[41] In the study, she examined 24 medication-free patients and 25 age- and sex-matched normal control subjects while performing two tasks: Raven's Progressive Matrices (RPM) and the number matching baseline task. The RPM is a nonverbal, abstract reasoning problem-solving task. Subjects with postrolandic lesions show impaired abilities on this test, suggesting that posterior areas are critical for the cognitive functions necessary to perform the test; there is no evidence that patients with prefrontal lesions have difficulty with this particular task. Both schizophrenics and controls failed to activate the DLPFC on these tasks. The authors concluded that DLPFC dysfunction in schizophrenia is linked to pathophysiology of a regionally specific neural system rather than to global cortical dysfunction and that this pathophysiology is more apparent under prefrontally specific cognitive demand.

The ^{133}Xe studies of Chabbrol and Geraud offer some insight into the natural history of CBF abnormalities in schizophrenia. Chabbrol et al. performed a resting ^{133}Xe study in ten neuroleptically naive adolescent schizophrenics with hebephrenia.[42] They found no differences in mean hemispheric blood flows between groups. Regional CBF in the prefrontal lobes compared to the rest of the brain showed loss of the hyperfrontal pattern in the schizophrenic patients. The authors interpreted their findings as evidence that abnormal CBF findings are present at the outset of the illness.

Geraud used the ^{133}Xe inhalation method to study the evolution of the hypofrontal CBF pattern in order to determine whether it represented a state or trait marker for schizophrenia.[43] He found a hypofrontal pattern in young chronic schizophrenics whose disease had evolved for more than 2 years and who were in remission. This hypofrontal pattern was reversible, as it disappeared during exacerbation of the illness. Geraud did not find a correlation between the DSM-III clinical subtype and hemodynamic activity of the frontal lobes. He showed that hypofrontality rarely occurred in schizophrenics in whom the disorder had lasted less than 2 years, but found hypofrontality in nearly all chronic schizophrenics in remission. In his study, the frontal pattern of most schizophrenics with exacerbation fell within the normal range. These findings concur with those of the majority of other investigators who have failed to report hypofrontality in subacute schizophrenics (<2 years disease duration) in the active phase of their disorder.[43]

Geraud also demonstrated that the hypofrontal pattern seen in young schizophrenics did not constitute an irreversible situation, as dopaminergic stimulation restored near-normal frontality. Chronic schizophrenics in remission demonstrated significant decreases in rCBF in frontal and prefrontal regions (hypofrontal pattern) during the ^{133}Xe inhalation study. Their hypofrontal pattern was also reversible, as it disappeared during exacerbation of the disease. Furthermore piribedil, a dopaminergic agonist, restored near normal frontality in ten medication-free schizophrenics.[43] The authors concluded that their results might reflect the effects of neuroleptic washout or dopaminergic depletion in chronic schizophrenia.

VIII. SPECT STUDIES IN SCHIZOPHRENIA

The schizophrenia SPECT literature is still rather limited. Devous, in a recent review that included his own investigations, reported a decrease in frontal CBF over resting values in schizophrenics during a special version of the WCST.[11] Low frontal rCBF was observed in 43% of schizophrenics (DSM-III) and 12% of volunteers. Hypofrontality was more common in paranoid (45%) than nonparanoid (21%) schizophrenics. Temporal lobe hypoperfusion was the second most common rCBF abnormality and was more prevalent in nonparanoid (27%) than in paranoid (7%) schizophrenics. Positive symptoms correlated with temporal lobe flow, but not with frontal lobe flow. Negative symptoms were inversely correlated with right temporal lobe flow, but not with left or right frontal flows. Neuroleptic treatment history did not affect rCBF. This interesting finding, which may point to functional reorganization of the brain in schizophrenia, is in partial agreement with the earlier cortical probe studies. The authors concluded that their data supported the hypothesis of frontal lobe dysfunction in schizophrenia and at the same time indicated that this dysfunction is more significant in the left hemisphere than the right.

A number of additional SPECT studies of schizophrenia are underway at various institutions. Many of these studies use rotating gamma SPECT cameras and the new iodinated or ^{99m}Tc labeled tracers.

IX. HYPOTHESES EXAMINED WITH CORTICAL PROBES AND SPECT

A. HYPOFRONTALITY

One theoretical mechanism that could account for hypofrontality in schizophrenia is dysfunction of dopaminergic neural transmission in the prefrontal cortex. As proposed by Ingvar, schizophrenic subjects show normal mean hemispheric flows with a shift in rCBF distribution so that high flows are relatively less common in frontal structures and relatively more common in postcentral structures.[29] Since frontal structures control intentional behavior while postcentral structures are concerned with perceptual processes (gnosis), schizophrenia may be thought of as a hypointentional, hypergnostic state: and this state may be caused by defective function of the nonspecific mediothalamic frontocortical projection system. Decreased tone in the nonspecific mediothalamic frontocortical projection system could result in reduction of resting and activated rCBF in areas of the prefrontal cortex. Present ^{133}Xe studies appear to support the hypothesis.

B. HEMISPHERIC ASYMMETRY

Flor-Henry remains the main proponent of hemispheric dysfunction in schizophrenia.[13,14] The cortical probe investigations of Gur et al. and the SPECT studies of Devous et al. lend support to the hemispheric asymmetry hypothesis by demonstrating left hemisphere over-action in both cases.[11,37] It is worthwhile remembering, however, that these are group effects and that a significant number of patients have rCBF values in the normal range. This in itself does not distract from the finding of hemispheric asymmetry in schizophrenia which, after all, is considered a heterogeneous condition.

X. CONCLUSIONS

Research endeavors in schizophrenia have been aided by advances in *in vitro* and *in vivo* analytical techniques. Brain imaging modalities currently depict brain structure and function in both living human beings and post-mortem brain specimens with surprising accuracy. These studies have shown that cerebral structures can be defined nearly as precisely with MRI as at post-mortem. SPECT is being used to determine rCBF in discreet areas of the brain. PET has become a chemical analytical device. It is capable of detecting chemical species including dopamine receptors in picomolar concentrations. These newly developed tools are opening up new vistas of the brain. They are currently being used to study the pathophysiology of schizophrenia. With them we will be able to investigate the natural history of this condition.

Ultimately, it may also be possible to define the structural and functional substrates of individual symptoms such as persecutory delusions or auditory hallucinations as derangements in specific components of cerebral functional systems in affected individuals.

REFERENCES

1. **Wyatt, R. J., Alexander, R. C., Egan, M. F., and Kirch, D. G.,** Schizophrenia, just the facts. What do we know, how well do we know it?, *Schizophr. Res.,* 1, 3, 1988.
2. **Kraepelin, E.,** *Dementia praecox* and paraphrenia, E. & S. Livingstone, Edinburgh, 1919.
3. **Bleuler, E.,** Physisch und psychisch in der pathologie, *Z. Ges. Neurol. Psychiatr.,* 30, 426, 1916.
4. *DSM-III: Diagnostic and Statistical Manual of Mental Disorders,* 3rd ed., American Psychiatric Association, Washington, D.C., 1980.
5. **Garver, D. L., Reich, T., Isenberg, K. E., and Cloninger, C. R.,** Schizophrenia and the question of genetic heterogeneity, *Schizophr. Bull.,* 15, 421, 1989.
6. **Bogerts, B., Meertz, E., and Schonfeldt-Bausch, R.,** Basal ganglia and limbic system pathology in schizophrenia: a morphometric study of brain volume and shrinkage, *Arch. Gen. Psychiatry,* 42, 784, 1985.
7. **Suddath, R. L., Casanova, M. F., Goldberg, T. E. et al.,** Temporal lobe pathology in schizophrenia: a quantitive magnetic resonance imaging study, *Am. J. Psychiatry,* 146, 464, 1989.
8. **Brown, R., Colter, N., Corsellis, J. A. N. et al.,** Postmortem evidence of structural brain changes in schizophrenia: differences in brain weight, temporal horn area, and parahippocampal gyrus compared with affective disorder, *Arch. Gen. Psychiatry,* 43, 36, 1986.
9. **Reynolds, G. P.,** Beyond the dopamine hypothesis: the neurochemical pathology of schizophrenia, *Br. J. Psychiatry,* 155, 305, 1989.
10. **Phelps, M. E. and Mazziotta, J. C.,** Positron emission tomography: human brain function and biochemistry, *Science,* 228, 799, 1985.

11. **Devous, M. D.,** Imaging brain function by single-photon emission computer tomography, in *Brain Imaging: Applications in Psychiatry,* Andreasen, N. C., Ed., American Psychiatric Association, Washington, D.C., 1988, 147.

12. **Weinberger, D. R.,** Schizophrenia and the frontal lobe, *TINS,* 8, 367, 1988.

13. **Flor-Henry, P.,** Lateralized temporal-lobe dysfunction and psychopathology, *Ann. N.Y. Acad. Sci.,* 280, 777, 1976.

14. **Flor-Henry, P.,** Functional hemispheric asymmetry and psychopathology, *Integrative Psychiatry,* 46, 1983.

15. **Callaway, E. and Naghdi, S.,** An information processing model for schizophrenia, *Arch. Gen. Psychiatry,* 39, 339, 1982.

16. **Ingvar, D. H.,** "Hyperfrontal" distribution of the cerebral grey matter flow in resting wakefulness: on the functional anatomy of the conscious state, *Acta Neurol. Scand.,* 60, 12, 1979.

17. **Phelps, M. E., Mazziotta, J. C., and Sung-Cheng, H.,** Study of cerebral function with positron computed tomography, *J. Cereb. Blood Flow Metab.,* 2, 113, 1982.

18. **Reivich, M., Gur, R., and Alavi, A.,** Positron emission tomographic studies of sensory stimuli, cognitive processes and anxiety, *Hum. Neurobiol.,* 2, 25, 1983.

19. **Berman, K. F. and Weinberger, D. R.,** Cerebral blood flow studies in schizophrenia, in *Handbook of Schizophrenia,* Vol. 1, *The Neurology of Schizophrenia,* Nasrallah, H. A. and Weinberger, D. R., Eds., Elsevier, New York, 1986.

20. **Obrist, W. D.,** Stability and sensitivity of CBF indices in the noninvasive ^{133}Xe method, in *Cerebral Blood Flow and Metabolism Measurement,* Hartmann, A. and Hoyer, S., Eds., Springer-Verlag, Berlin, 1985.

21. **Stokely, E. M., Sveinsdottir, E., Lassen, N. A., and Rommer, P.,** A single photon dynamic computer-assisted tomograph (DCAT) for imaging brain function in multiple cross sections, *J. Comput. Assist. Tomogr.,* 4, 230, 1980.

22. **Rezai, K., Kirchner, P. T., Armstrong, C., Ehrhardt, J. C., and Heistad, D.,** Validation studies for brain blood flow assessment by radioxenon tomography, *J. Nucl. Med.,* 29, 348, 1988.

23. **Chang, W., Madsen, M. T., Lanfen, W., and Kirchner, P. T.,** A 3-D anthropomorphic brain phantom for ECT applications, *Nucl. Med. Commun.,* 11, 11, 1990.

24. **Mazziotta, J. C. and Koslow, S. H.,** Assessment of goals and obstacles in data acquisition and analysis from emission tomography: report of a series of international workshops, *J. Cereb. Blood Flow Metab.,* 7, S1, 1987.

25. **Ingvar, D. H.,** Functional landscapes of the dominant hemisphere, *Brain Res.,* 107, 181, 1976.

26. **Devous, M. D., Raese, J. D., Herman, J. H., Paulman, R. G., Gregory, R. R., Rush, A. J., Chehabi, H. H., and Bonte, F. J.,** IX-12. Regional cerebral blood flow in schizophrenic patients at rest and during Wisconsin card sort tasks, *J. Cereb. Blood Flow Metab.,* 5 (Suppl. 1), S201, 1985.

27. **Kety, S. S., Woodford, R. B., Harmel, M. H., Freyhan, F. A., Appel, K. E., and Schmidt, C. F.,** Cerebral blood flow and metabolism in schizophrenia, *Am. J. Psychiatry,* 104, 765, 1948.

28. **Roy, C. S. and Sherrington, C. S.,** The regulation of the blood supply of the brain, *J. Physiol.,* 11, 85, 1890.

29. **Kety, S. S. and Schmidt, C. F.,** The nitrous oxide method for the quantitative determination of cerebral blood flow in man: theory, procedure and normal values, *J. Clin. Invest.,* 27, 476, 1948.

30. **Ingvar, D. H. and Franzen, G.,** Abnormalities of cerebral blood flow distribution in patients with chronic schizophrenia, *Acta Psychiatr. Scand.,* 50, 425, 1974.

31. **Pahl, J. J., Swayze, V., and Andreasen, N. C.,** Diagnostic advances in anatomical and functional brain imaging in schizophrenia, in *Recent Advances in Schizophrenia,* Kales, A., Stefanis, C. N., and Talbott, J. A., Eds., Springer-Verlag, Berlin, 1989.

32. **Hoyer, S. and Oesterreich, K.,** Blood flow and oxidative metabolism of the brain in the course of acute schizophrenia, in *Cerebral Function, Metabolism and Circulation,* Ingvar, D. H. and Lassen, N. A., Eds., Munksgaard, Copenhagen, 1977.

33. **Mathew, R. J., Meyer, J. S., Francis, D. J. et al.,** Regional cerebral blood flow in schizophrenia: a preliminary report, *Am. J. Psychiatry,* 138, 112, 1981.

34. **Ariel, R. N., Golden, C. J., Berg, R. A. et al.,** Regional cerebral blood flow in schizophrenics: tests using the Xenon-133 inhalation method, *Arch. Gen. Psychiatry,* 40, 258, 1983.

35. **Mathew, R. J., Duncan, G. C., Weinman, M. L., and Barr, D. L.,** Regional cerebral blood flow in schizophrenia, *Arch. Gen. Psychiatry,* 39, 1121, 1982.

36. **Gur, R. E., Skoinick, B. E., Gur, R. C., Caroff, S., Rieger, W., Obrist, W. D., Younkin, D., and Reiv'ch, M.,** Brain function in psychiatric disorders. I. Regional cerebral blood flow in medicated schizophrenics, *Arch. Gen. Psychiatry,* 40, 1250, 1983.

37. **Gur, R. E., Gur, R. C., Skolnick, B. E., Caroff, S., Obrist, W. D., Resnick, S., and Reivich, M.,** Brain function in psychiatric disorders. III. Regional cerebral blood flow in unmedicated schizophrenics, *Arch. Gen. Psychiatry,* 42, 329, 1985.

38. **Weinberger, D. R., Berman, K. F., and Zec, R. F.,** Physiologic dysfunction of dorsolateral prefrontal cortex in schizophrenia. I. Regional cerebral blood flow evidence, *Arch. Gen. Psychiatry,* 43, 114, 1986.

39. **Berman, K. F., Zec, R. F., and Weinberger, D. R.,** Physiologic dysfunction of dorsolateral prefrontal cortex in schizophrenia. II. Role of neuroleptic treatment, attention, and mental effort, *Arch. Gen. Psychiatry,* 43, 126, 1986.
40. **Weinberger, D. R., Berman, K. F., and Illowsky, B. P.,** Physiologic dysfunction of dorsolateral prefrontal cortex in schizophrenia. III. A new cohort and evidence for a monoaminergic mechanism, *Arch. Gen. Psychiatry,* 45, 609, 1988.
41. **Berman, K. F., Illowsky, B. P., and Weinberger, D. R.,** Physiologic dysfunction of dorsolateral prefrontal cortex in schizophrenia, IV. Further evidence for regional and behavorial specificity, *Arch. Gen. Psychiatry,* 45, 616, 1988.
42. **Chabrol, H., Guell, A., Bes, A., and Moron, P.,** Cerebral blood flow in schizophrenic adolescents, *Am. J. Psychiatry,* 143, 130, 1986.
43. **Geraud, G., Arne-Bes, M. C., Guell, A., and Bes, A.,** Reversibility of hemodynamic hypofrontality in schizophrenia, *J. Cereb. Blood Flow Metab.,* 7, 9, 1987.

Chapter 21

PET STUDIES OF MOVEMENT DISORDERS

Mark Guttman and Gabriel Léger

TABLE OF CONTENTS

I. INTRODUCTION

Movement disorders is a growing subspecialty in neurology dealing with conditions associated with involuntary movements. Examples of these involuntary movements include tremor, chorea, dystonia, tics, or myoclonus. These clinical signs are often not specific to a single disease entity. One of the problems in making a specific diagnosis is the lack of sensitive and specific diagnostic testing that will confirm or refute the clinical impression. Most of the usual diagnostic radiological tests are normal since these diseases often have normal brain anatomy. Neurologists have been very interested in developing diagnostic tests to help in the differential diagnosis and to answer research questions in subjects with movement disorders.

Neuroscience research has identified important biochemical alterations that are likely responsible for the signs and symptoms in some patients with movement disorders. This has led to great advances in the understanding of the pathophysiology of the disease processes and has resulted in new treatment strategies. Most of this work has been performed on post-mortem samples of brain tissue from patients with these conditions. There are many problems in correlating post-mortem observations with clinical findings. Investigators are unable to perform studies that prospectively evaluate the biochemical changes that are related to the different stages of the disease, the response to medication, and alterations due to side effects of the medication. Positron emission tomography (PET) studies have the potential to evaluate these biochemical alterations during life. It is hoped that the use of this exciting new technology will allow us to understand the clinical problems, treatment, and differential diagnosis in the field of movement disorders. This chapter will review the current information on the use of PET and movement disorders. This subject has been dealt with in previous publications.[1,2] We must emphasize that this subject is in its infancy and that investigators have barely scratched the surface in this field. We look forward to the future when further studies will provide more information of the biochemical alterations in these diseases.

II. PARKINSON'S DISEASE

Parkinson's disease is a degenerative neurological disorder characterized by tremor, rigidity, bradykinesia, and postural instability.[3] It is associated with cell loss and dopamine depletion in the nigrostriatal pathway.[3] The usual age of onset is in the fifth or six decade. The etiology of this disease is currently unknown, with viral, genetic, and environmental factors being questioned. Patients with toxic exposure to manganese and carbon disulfide also have symptoms of parkinsonism which suggests that a neurotoxin may be involved in the etiology of Parkinson's disease.[4] In support of this hypothesis, a major breakthrough occurred with the serendipitous finding of parkinsonism after administration of a street drug in a group of addicts.[5] 1-Methyl-4-phenyl-1,2,3,6-tetrahydropyridine (MPTP) has since been shown to cause histological and biochemical lesions that are similar to Parkinson's disease in animal models.

Therapy is aimed at replacing the neurochemical defect in this condition. L-dopa, the precursor of dopamine, has been used successfully in the treatment of patients with parkinsonism. Dopamine does not cross the blood brain barrier and cannot be used as a therapeutic agent. In clinical practice, L-dopa is combined with a peripheral dopa decarboxylase inhibitor to increase its bioavailability and to reduce the systemic adverse effects. Despite clinical improvement with L-dopa and direct acting dopaminomimetic agents, these drugs do not alter the natural history of the disease. Clinical deterioration proceeds over time and the therapeutic response often declines and is complicated by numerous adverse effects.

Parkinsonism may be present in conditions other than Parkinson's disease. Disease processes that affect the substantia nigra or caudate nucleus and putamen may present with similar symptoms. In this chapter we will discuss PET studies of Parkinson's disease separately from studies of patients with other types of parkinsonism. These include MPTP-induced parkinsonism, progressive supranuclear palsy, striatonigral degeneration, olivopontocerebellar atrophy, ALS-PD Complex of Guam, cyanide-induced parkinsonism, and tumor-induced parkinsonism. We will also include a discussion of PET studies in primates with MPTP-induced parkinsonism.

A. PET STUDIES OF PARKINSON'S DISEASE

Patients with Parkinson's disease have been studied with PET with a variety of tracers to examine cerebral blood flow, oxygen metabolism, glucose metabolism, presynaptic dopamine function, and dopamine receptors. Overviews on the subject have been published.[6-8]

B. OXYGEN STUDIES

Oxygen studies with [$^{15}O_2$] have been performed with the steady-state inhalation technique[9,10] and with the water bolus technique with [^{15}O]-labeled water.[11] Hemiparkinsonian patients have been shown to have an increased cerebral blood flow and oxygen utilization in the contralateral basal ganglia.[9] Perlmutter and Raichle have also observed increased blood flow in the contralateral basal ganglia, but have localized this to the globus pallidus employing a stereotactic technique for identifying anatomical structures on PET images.[11] These studies are in agreement with the theory that nigrostriatal lesions will produce a disinhibition of the striato-pallidal outflow which will result in increased pallidal activity. Patients with bilateral parkinsonism were shown to have reduced blood flow globally but did not have changes in oxygen metabolism.[9] Both Wolfson and Perlmutter found that the contralateral frontal cortex had reduced blood flow in the hemiparkinsonian patients studied.[9,11] L-dopa administration to patients did not change global cerebral blood flow, but reversed the pallidal changes in one study,[11] but increased blood flow diffusely in another study without affecting oxygen metabolism.[10] These differences are thought to be related to the different doses of L-dopa employed.

C. GLUCOSE METABOLISM

Cerebral metabolism studies with [^{18}F]fluorodeoxyglucose (FDG) have been performed in patients with unilateral[12,13] and bilateral parkinsonism.[14,15] Martin reported that the basal ganglia contralateral to the symptomatic side had an increased regional metabolic rate for glucose, confirming the blood flow and oxygen metabolism studies discussed above.[12] Further data analysis suggests that there is an impaired relationship between thalamus, caudate, and putamen contralateral to the affected limbs.[13] In bilateral patients, Rougemont and colleagues also found an increased glucose utilization in the basal ganglia and normal metabolism in cortical structures.[14] They also reported that these alterations did not change after treatment with L-dopa. Kuhl, however, observed diffusely reduced cerebral glucose metabolism in bilateral patients that were more severely affected than the previous study.[15] These patients were on L-dopa preparations and there was a high incidence of dementia in this group. The authors comment that the distribution of reduced cortical glucose utilization was similar to that seen in Alzheimer's disease.[15]

D. PRESYNAPTIC DOPAMINE SYSTEM

PET may be employed to study the nigrostriatal dopamine pathway during life. 6-[^{18}F]fluoro-L-dopa (6-FD) is an analogue of L-dopa, the precursor of dopamine that is used in the treatment of Parkinson's disease. 6-FD has been employed with PET to visualize the presynaptic component of the nigrostriatal dopamine system in man.[16] The cerebral radioactivity measured after the administration of 6-FD is a function of uptake of the isotope into the brain and its subsequent metabolism to dopamine. Quantitative analysis of the PET data may be performed and numerous models are being developed to interpret the data. One model estimates the index of 6-FD uptake from blood to striatum.[17] Other methods of analysis have been proposed to estimate the blood brain barrier transfer rate and the dopa decarboxylase rates separately in various regions of the brain.[100]

Garnett and colleagues studied six patients with hemiparkinsonism with PET and 6-FD.[18] They reported reduced accumulation of radioactivity in the contralateral striatum. This group extended these observations in a later paper which reported nine hemiparkinsonian subjects and two subjects with bilateral symptoms.[19] They observed reduced dopaminergic activity in the contralateral putamen, while the caudate nucleus was normal in subjects with normal neuro-psychological assessments. This distribution of altered 6-FD uptake is similar to the distribution of maximal dopamine loss reported by Kish and colleagues in subjects with Parkinson's disease.[20] The post-mortem studies showed a greater dopamine depletion in the putamen compared to the caudate nucleus.[20] The putamenal alterations correlated with poor motor performance, since this structure plays a role in motor function. The caudate nucleus, however, is thought to be involved in cognitive pathways.[19] Other groups have also reported striatal reductions with 6-FD and PET using qualitative[21] and semiquantitative analytic techniques.[22,23] An example of a 6-FD PET studies comparing a control subject with an individual with Parkinson's disease is shown in Plate 1.* Leenders employed a similar model to Martin[17] to analyze the PET data, but unfortunately this is not valid since the authors did not adequately quantify the arterial input function.[22] This report has also been criticized because of the contamination of 6-FD with other fluorodopa compounds. Leenders and colleagues reported further studies of parkinsonian subjects with 6-FD.[23] They analyzed the data by examining decay corrected striatal time-radioactivity curves and found that the capacity of the striatum to retain radioactivity was impaired in patients compared to controls. Furthermore, subjects with "on-off" phenomenon had even greater decrease in storage capacity of 6-FD.

In summary, 6-FD PET scans may be used to study the presynaptic dopaminergic function of the nigrostriatal dopamine system in patients with Parkinson's disease. Many centers have shown that the accumulation of radioactivity after 6-FD accumulation is impaired in these patients. One of the major problems that must be solved is consistency of data analysis. Simple ratio methods comparing dopamine rich areas of brain to nondopaminergic regions do not provide accurate data for quantitative analysis. Graphical analysis may provide a better approach to the interpretation of these studies but is not perfect. The ideal method would include separate estimations of the blood brain barrier transfer and dopa decarboxylase rates. The ultimate goal is to implement a standard analytical technique that would promote comparison of data from different centers and provide a sensitive method of quantifying these studies in the future.

Another method of evaluating the presynaptic component of the nigrostriatal dopamine system with PET is using ligands that bind to the dopamine reuptake sites, located on the presynaptic nerve terminals. Dopamine is cleared from the synapse by binding to these uptake sites and reenters the presynaptic terminal for incorporation into vesicles. There are many specific ligands for the measurement of these sites. [^{11}C]nomifensine has been employed with PET to measure dopamine uptake sites in living subjects.[24] Administration of this ligand results in radioactivity accumulation in the striatum of monkeys and man.[24] The thalamus also shows radioactive uptake most likely from adrenergic sites which are very similar to the dopaminergic sites. Patients with Parkinson's disease were found to have reduced accumulation of radioactivity in the putamen contralateral to the clinical symptoms.[25] Another dopamine reuptake site ligand has been labeled for use with PET, [^{11}C]GBR 13119, which is more specific than nomifensine.[26] It is unclear what the advantages of the dopamine reuptake site ligands are compared to 6-FD. Both will evaluate the presynaptic terminal function in the striatum. Lesions of the nigrostriatal pathway which will reduce the number of uptake sites, vesicles and quantity of dopa decarboxylase should produce parallel results for each of these compounds. It is not surprising that

---

* Plate 1 follows page 198.

studies comparing 6-FD with labeled nomifensine have found only minor differences in patients with Parkinson's disease.[27,28]

E. DOPAMINE RECEPTOR STUDIES

The evaluation of D-1 and D-2 receptors *in vivo* has many advantages compared to post-mortem studies with either autoradiography or homogenate binding techniques. In human studies, post-mortem analysis of striatal D-2 receptors has shown supersensitivity in parkinsonian patients not treated with dopamine agonists.[29] This supersensitivity is reversed if agonists have been administered.[30]

PET studies of dopamine receptors in Parkinson's disease have been very preliminary. The literature is complicated by competing centers using different ligands and nonuniform analytical techniques. Leenders and colleagues reported reduced [^{11}C]*N*-methylspiperone (NMSP) binding in patients with Parkinson's disease.[22] This study is problematic because of the analytical technique employed which has since been replaced by other methods because of inaccuracies. Hemiparkinsonian subjects have been studied with NMSP ligand by two other groups.[31,32] The former reported no change in the binding of NMSP between the ipsilateral and contralateral striata.[31] The latter study used the same problematic analytic technique as Leenders and reported higher values in patients with early Parkinson's disease compared to later stages.[32] Rinne and colleagues have used [^{11}C]raclopride to study D-2 receptor changes in parkinsonian patients.[33] They found an increase in the striatum/cerebellum ratio on the contralateral side in drug free hemiparkinsonian subjects using a ratio method of analysis. Unfortunately, no control subjects were studied.

In summary, results of PET studies of dopamine D-2 receptors have been conflicting and have not reproduced the alterations found on post-mortem analysis. D-1 studies have not been reported in patients with Parkinson's disease. Further work is necessary to draw conclusions concerning the role of these receptors in the pathophysiology of this disorder.

F. PET STUDIES TO EVALUATE DOPAMINERGIC IMPLANTS IN PARKINSON'S DISEASE

Over the last decade, neuroscience researchers have been actively investigating the potential of implanting dopaminergic cells into the brain of subjects with nigrostriatal dopamine depletion. The ultimate goal is to treat patients with Parkinson's disease with an implant that would reverse the neurochemical deficit. A number of patient studies have been carried out to assess the outcome of implantation procedures and PET has been used to provide an objective assessment of graft survival in a small number of studies.

Autologous adrenal implants into the caudate nucleus have been performed in many centers and are no longer thought to be of benefit. 6-FD PET studies were performed in one series before and after surgery.[34] Five patients studied did not show consistent changes in striatal 6-FD accumulation when preoperative studies were compared to scans performed 6 weeks after surgery. Gallium EDTA PET studies identified a defect in the blood brain barrier that complicated the analysis of the 6-FD scans. Out of five patients, only one improved clinically and this was after the 6-week scan. Another study of adrenal implants for the treatment of Parkinson's disease employed [^{11}C]raclopride to evaluate D-2 receptor changes.[35] These patients also did not show clinical improvement and there was no change in the PET results. Human fetal dopamine neurons have been implanted into the striatum of two patients with Parkinson's disease.[36] 6-FD PET studies did not show any improvement five to six months after graft placement and the patients did not improve clinically.

G. PET STUDIES OF MPTP-INDUCED PARKINSONISM IN MAN

1. Presynaptic Studies

In a hallmark study, Calne and colleagues studied subjects exposed to MPTP who were clinically asymptomatic with 6-FD and PET.[37] These subjects had reduced striatal uptake of 6-

FD in scans analyzed using a ratio technique. They were successful in identifying *in vivo* evidence of subclinical alterations of the nigrostriatal dopamine system with PET. Subjects exposed to MPTP who displayed signs of parkinsonism were also studied with 6-FD and PET. Calne found similarities in the distribution of the reduction in the 6-FD related striatal radioactivity compared to idiopathic parkinsonism in two subjects.[101]

2. Receptor Studies

Perlmutter and colleagues have performed PET studies with [^{18}F]spiperone in a clinically symptomatic, untreated subject exposed to MPTP.[38] This ligand binds to D-2 and S-2 serotonin receptors. The subject showed increased association and dissociation rates in the striatum. The dissociation rate was increased to a greater extent than the association rate suggesting an increased number of sites present. A direct estimate of the receptor number is not possible because of limitations of the model being used for this compound. Since the patient had not been treated, this study provides further evidence of the upregulation of D-2 receptors using PET. The authors also suggest that PET may provide insights into rate constants that are not easily measured with *in vitro* techniques. These may be very relevant to clinical pharmacology and therapeutics.

3. Progressive Supranuclear Palsy

Progressive supranuclear palsy (PSP) or Steele, Richardson, Olszewski syndrome is a disorder that may present with parkinsonism as a part of the clinical presentation. This syndrome also includes impairment of extraocular movements, dystonic posturing of the neck, subcortical dementia, pseudobulbar palsy, and pyramidal tract signs. Post-mortem examination reveals neurofibrillary tangles and neuronal loss in specific brainstem nuclei. Patients often do not respond to therapy and progress relatively rapidly compared to individuals with Parkinson's disease. There may be difficulty in making the diagnosis at the onset of clinical symptoms. Patients with PSP have been studied with PET to examine cerebral metabolism. Leenders and colleagues reported that cerebral blood flow was decreased globally in five patients with PSP and that the most marked reductions were found in the frontal cortex.[39] Oxygen utilization was also decreased, but to a lessor extent than blood flow, resulting in a raised oxygen extraction. It is not clear what the significance of these findings are. The impairment in oxygen utilization in the frontal lobe paralleled the duration of the disease. PSP patients have been studied with FDG and PET by a number of centers.[40,41,42] These studies confirm the previous observations that the frontal lobes have decreased regional metabolism with reduced glucose utilization. Since there is little or no cortical pathology in PSP, it is interesting to note that a subcortical dementing process would cause cortical abnormalities. This is in contrast to Alzheimer's disease which has predominant cortical pathology and is associated with parietal cortex hypometabolism. In addition to the frontal changes in glucose utilization, there are reductions in the striatum, cerebellum, and thalamus.[42] Foster and colleagues did not find cerebellar changes in their patients.[41] They speculate from their experience with another parkinsonism syndrome, olivo-pontocerebellar degeneration (OPCA), that cerebellar hypometabolism may be more suggestive of OPCA rather than PSP. Leenders and colleagues also studied these patients with 6-FD scans. They found that there was reduced accumulation of radioactivity in the striatum similar to that found in patients with Parkinson's disease.[39] [^{76}Br]bromospiperone has been used to evaluate D-2 dopamine receptors in patients with PSP.[42] Baron and colleagues observed reduced striatal binding in these patients compared to controls, assessed by a ratio method of analysis. These authors speculate that this may be the reason that these patients do not respond to dopaminergic therapy. Similar results have been found in post-mortem analysis in patients with this condition.

H. PET STUDIES OF OTHER PARKINSONIAN SYNDROMES

There have been a number of studies of rare parkinsonian syndromes with PET. Striatonigral degeneration (SND) is associated with parkinsonism that is not responsive to dopaminergic treatment and has cell loss in the putamen, caudate nuclei, and substantia nigra. CT and MRI may reveal atrophy or iron deposition in the striatum. De Volder and colleagues have studied seven patients with this disorder with FDG and PET.[43] They found that these patients had reduced glucose utilization most marked in the striatum and frontal lobe, but also had decreased values throughout the brain. They suggest that striatal hypometabolism may be an indicator of poor response to L-dopa.

OPCA is another syndrome that involves striatal degeneration and parkinsonism which may be inherited or occur spontaneously. Patients also display prominent cerebellar signs. Gilman and colleagues have studied patients with this condition with FDG and PET and have observed reductions in glucose utilization in the brainstem and cerebellum.[44] These studies are difficult to interpret because of the concomitant atrophy in the affected structures for which the authors have attempted to correct. They have suggested that this pattern of metabolic abnormalities may be specific for this disorder.

On the island of Guam in the South Pacific, there is an unusual condition that has components of parkinsonism, amyotrophic lateral sclerosis, and Alzheimer's-like dementia. The ALS-PD complex of Guam does not have a clear etiology and is associated with neurofibrillary tangles throughout the brain. Patients with this disorder have been studied with 6-FD and PET. Individuals with parkinsonism have reduced striatal 6-FD uptake compared to controls.[6] Subjects with ALS without parkinsonian signs also have been found to have reduced striatal uptake, suggesting subclinical dysfunction of the nigrostriatal dopamine pathway in a similar manner to the asymptomatic MPTP subjects.[45]

Corticodentatonigral degeneration or corticobasil degeneration is a newly described condition of asymmetrical parkinsonism associated with mild dementia, apraxia and cortical parietal lobe signs. Eidelberg and colleagues performed FDG PET scans on six patients with this disorder and found asymmetries in thalamic and parietal glucose utilization which were not present in subjects with typical hemiparkinsonism.[46] Sawle and colleagues also studied patients with this disorder and found reduced oxygen metabolism in the medial frontal, parietal, and temporal cortices.[47] In addition, there was a reduced influx constant for 6-FD in the striatum and median frontal cortex. The authors propose to employ the combination of these two studies to differentiate corticobasal degeneration from other causes of parkinsonism.[47]

6-FD PET studies have been used to study a patient with cyanide induced parkinsonism.[48] This subject had reduced striatal uptake of 6-FD in a pattern similar to that seen in Parkinson's disease. Another individual with tumor induced parkinsonism has been studied with PET.[49] This patient had a meningioma with cerebral edema and was found to have reduced oxygen utilization in the basal ganglia. This returned to normal after the successful surgical treatment of the lesion.

I. PET STUDIES OF MPTP-INDUCED PARKINSONISM IN PRIMATES

Primates are the only animals that develop a substantial and enduring syndrome similar to that seen in Parkinson's disease.[4] When MPTP is administered to monkeys, a parkinsonian state develops characterized by bradykinesia, rigidity, and, in some cases, tremor. These animals respond to L-dopa treatment and have been shown to develop medication-related dyskinesias after long-term treatment.[50] Pathologically, there is a reduction in the striatal dopamine content by more than 70%.[51] The locus ceruleus is also affected by this neurotoxin.[52] Recently, intraneuronal, eosinophilic inclusions have been observed and appear to be very similar to Lewy bodies.[53]

Recently investigators at the NIH developed a model for unilateral parkinsonism in cynomolgus monkeys.[54] The MPTP-induced hemiparkinsonian primate model may be used for the study of parkinsonism with PET. If the contralateral striatum remains intact, there is an internal control for neurochemical and receptor studies. This internal control may allow the study of important biological questions while models are being designed to enable quantitative analysis of the scans. Validation of this model has recently been published.[55]

1. Presynaptic Studies

Monkeys which have had systemic MPTP administration have been studied with PET using 6-FD.[56,57] The McMaster group studied a monkey with 6-FD scans at the time of MPTP administration, at 2 h, 3 d, and 10 d postadministration. The results showed that the 2 h scan was unchanged; at 3 d the specific striatal radioactivity was increased to 160% and at 10 d the activity was reduced to 45% of the initial value. Clinical correlation was not given. The scan data were compared to post-mortem striatal biochemical analysis on other animals that were given MPTP and sacrificed at similar times; good correlation was found. Another aspect of the McMaster study examined three animals with different stages of parkinsonism. One animal with subclinical damage had 60% reduction of striatal dopamine measured by HPLC and only 20% reduction of the 6-FD related specific striatal activity. The mildly affected monkey had 80% reduction of striatal dopamine and 55% reduction of 6-FD related activity. The severely affected animal had good correlation of the dopamine content (97% reduction) to 6-FD related specific activity (100% reduction). One possibility for the poor correlation of the PET data and the HPLC findings in the subclinical and mildy affected animals is that the data from the 6-FD scans were compared to similar data from other animals. Hemiparkinsonian primates unilaterally lesioned with MPTP have been studied with 6-FD PET (Plate 2*).

The MPTP hemiparkinsonian primate model has been used to assess *in vivo* subclinical dysfunction of the nigrostriatal dopamine system.[58] Since clinical symptoms are thought to occur after 80% of this pathway is lesioned, less dysfunction should be able to be assessed with 6-FD and PET. Three monkeys were given repeated injections of MPTP with the end point of clinical hemiparkinsonism. The first monkey had three separate injections in the left internal carotid artery approximately 2 weeks apart and remained asymptomatic. 6-FD PET scans were performed approximately 10 d after each infusion. There was a stepwise reduction of 6-FD related striatal uptake on the lesioned side of 10, 29, and 60% when compared to the contralateral side. The other two monkeys had single injections. A 6-FD PET study was performed 4 weeks after the MPTP injection in the second animal and showed 40% reduction of 6-FD related striatal activity when clinical signs were absent. The third monkey had a 6-FD PET scan 11 d after the infusion. This showed 34% reduction of striatal activity. The distribution of loss of radioactivity in the 6-FD PET scans is also similar to the distribution of dopamine depletion found in Parkinsonian patients.[20] This may correspond to earlier and more severe putamenal damage with MPTP which is further evidence that this toxin provides an appropriate model for the study of Parkinson's disease. These studies suggest that PET employing 6-FD can provide data on subclinical damage to the nigrostriatal system *in vivo*.

6-FD PET studies have been used to evaluate neural implant viability in parkinsonian monkeys. Miletich and colleagues have shown that after fetal midbrain implantation 6-FD scans identified the tissue in an animal who displayed clinical improvement.[59] Stereotactically implanted sympathetic ganglia and adrenal medulla tissue into the caudate and putamen of parkinsonian monkeys has also been evaluated by PET.[60] 6-FD PET scans were employed to evaluate the viability of the implants. Unfortunately all four animals did not show improvement in their clinical status or their PET studies.

Leenders and colleagues have performed PET scans with [^{11}C]nomifensine to study the presynaptic dopamine system.[61] They have assessed a single cynomolgus monkey who was unilaterally lesioned by the intracarotid administration of MPTP. They were able to show that the MPTP-induced lesion caused significant reduction in the uptake of [^{11}C]nomifensine.

2. Receptor Studies

Alterations of striatal dopamine receptors secondary to denervation have been studied with PET in MPTP treated primates. Increased D-2 receptor binding has been confirmed in MPTP

---

*   Plate 2 follows page 198.

treated primates by autoradiographic techniques[62] and homogenate binding studies.[63] Hantraye and colleagues administered MPTP to a baboon and performed sequential PET studies with [^{76}Br]bromospiperone to investigate the alterations of striatal D-2 receptors *in vivo*.[64] The baboon had three series of MPTP injections which resulted in transient parkinsonism. PET studies showed reduction in D-2 binding using a semiquantitative analysis technique. The finding of decreased D-2 sites is difficult to interpret in light of the histological alterations found in monkeys that were used as *in vitro* controls for the PET studies. These histological studies were performed on another species of monkey that displayed different clinical features compared to the baboon studied with PET. *In vitro* receptor studies on the post-mortem tissue also showed reduction in D-2 sites. This was most likely secondary to the pathological findings in the striatum. Parkinson's disease is not associated with ultrastructural changes in these regions and other investigators have not found evidence of striatal damage in nonhuman primates.[65]

Dopamine receptors have been evaluated in unilaterally lesioned monkeys with PET. Leenders and colleagues used [^{11}C]raclopride to assess a rhesus monkey lesioned with intracarotid administration of MPTP.[61] They performed sequential studies and report an increase in uptake compared to the contralateral side. We have confirmed this finding in a cynomolgus monkey lesioned in a similar manner (Plate 3*). D-1 receptors have also been assessed in unilaterally lesioned animals.[66] A reduction in the striatal uptake of SCH 23390, a specific D-1 receptor antagonist, has been demonstrated (Plate 4**). The loss of D-1 receptors in the striatum after nigrostriatal denervation is not a consistent finding using *in vitro* analytic techniques. Rat studies have shown increased numbers of receptors,[67] no change[68] or reduced binding[69] by various methods of analysis with SCH 23390. If the PET studies in MPTP monkeys are indicative of reduced D-1 receptors in patients with parkinsonism, this may be relevant to the therapeutics of this condition.

III. HUNTINGTON'S DISEASE

Huntington's disease (HD) is an autosomal dominant disorder characterized by choreiform movements, dementia, and psychiatric disturbances.[70] The onset of this neurodegenerative disease is variable, but most individuals commence their symptoms in the third or fourth decades of life. Recent advances in molecular biology have shown a linkage with the HD gene to chromosome four and predictive testing is available to identify subjects at risk of developing this disorder. Unfortunately, neuroscientists do not know the neurochemical defect specific to HD, although a reduced concentration of GABA has been found in the striatum. Replacement therapy has not been successful and there is currently no therapeutic strategy that is effective. Choreiform movements have been described in other neurological disorders that must be considered in the differential diagnosis before the diagnosis of HD is made.

A. PET STUDIES OF PATIENTS WITH HD

Subjects with symptomatic HD have been studied with FDG and PET. Kuhl and colleagues were the first to report caudate nucleus hypometabolism in subjects with various stages of HD.[71] This landmark study found reduced striatal glucose utilization even before caudate atrophy was present on CT. These observations have been criticized because of the high incidence of neuroleptic use in the patient population. In this and a subsequent report, these authors have found that cortical glucose metabolism is normal, in contrast to Alzheimer's disease which is associated with abnormal cortical glucose utilization.[71,72] Others have confirmed the observation of caudate hypometabolism in patients with early symptoms of HD without caudate atrophy[73] and with psychiatric manifestations of the disease only.[74]

---

* Plate 3 follows page 198.

** Plate 4 follows page 198.

Childhood onset of the disease is usually associated with a paternal inheritance and may manifest with parkinsonian symptoms. De Volder and colleagues reported two patients with childhood onset HD and found similar metabolic defects.[75] They concluded that the metabolic defects are similar to those found in adult patients. Investigators at the University of Michigan have correlated FDG scans with neurological decline[76] and with cognitive function.[77] Cerebral glucose metabolism was normal in patients with various stages of HD, but the caudate metabolism declined proportionally with the findings of bradykinesia, rigidity, and overall functional capacity.[76] The putamen glucose utilization correlated with motor defects, chorea, eye movement abnormalities, and timed motor coordination. Indices of thalamic metabolism correlated with dystonia.[76] Measures of verbal learning and memory correlated with caudate metabolism.[77] Leenders and colleagues have studied patients with HD with a variety of tracers.[78] Labeled oxygen studies were reported to show reduced cerebral blood flow in the striatum and frontal cortex which paralleled the reduced glucose metabolism found in a single patient with HD. Furthermore, a 6-FD scan was normal and an NMSP study showed reduced striatal binding.[78]

B. PET STUDIES OF SUBJECTS AT RISK FOR HD

Since the early report by Kuhl and colleagues of the presymptomatic identification of subjects at risk of developing HD,[71] investigators have hoped that this finding would make a significant impact on the management of families with HD. The UCLA group followed their initial report with a study of 58 at risk individuals.[79] They confirmed the reduced caudate to cortex ratio of glucose metabolism and suggested that this method may provide a direct means of monitoring experimental therapy in subjects at risk for HD. Young and colleagues, however, failed to show reduced caudate metabolism in their at risk population.[80] The UBC group combined FDG PET studies with DNA polymorphisms for the preclinical detection of HD and confirmed Mazziotta's results.[81] This controversy lead to a number of criticisms, which have included differences in data analysis, clinical identification of soft neurological signs and the use of different PET tomographs.[82] It is unclear if FDG PET is sensitive to detect preclinical HD. With the availability of DNA testing, it is more cost effective to utilize this type of test since there is a high degree of sensitivity. This is possible only when the family structure is adequate. For those individuals who do not have an adequate family structure, FDG PET may reveal predictive results when the diagnosis is not in question.

C. PET STUDIES IN OTHER CHOREIFORM DISORDERS

There are other neurological disorders that may manifest choreiform movements. Benign hereditary chorea is another autosomal dominant disorder manifesting chorea. Suchowersky and colleagues reported three subjects with this condition studied with FDG and PET.[83] They found relative reduction in caudate nucleus metabolism in these individuals and conclude that caudate hypometabolism is not specific for HD. FDG and PET have been employed to study chorea associated with systemic lupus erythematosus.[84] Striatal glucose metabolism was within the normal range in these patients and the ratio of striatal to cortical metabolism was increased, a markedly different result compared to HD patients. The authors conclude that caudate glucose hypometabolism is not the PET correlate of chorea. Hosokawa and colleagues offer a differing opinion with the report of reduced caudate metabolism in individuals with choreacanthocytosis, sporadic progressive chorea and dementia, and dentato-rubro-pallido-luysian atrophy.[85] These studies indicate that other pathological processes that may manifest chorea may be associated with caudate nucleus glucose hypometabolism. FDG PET studies should not be used to confirm the diagnosis of HD in individuals without family histories compatible with this disorder. We conclude that FDG PET does not provide a specific index to permit the diagnosis of HD and it may not be sensitive in identifying individuals at risk for the disorder before symptoms appear. Clinical studies should not be performed until these controversies are resolved.

D. PET STUDIES IN DYSTONIA

Dystonia is a hyperkinetic movement disorder involving sustained twisting involuntary movements.[86] The most useful classification for dystonia is according to etiology. In this scheme, possible causes are divided into those with an underlying neurologic disorder, termed *secondary* or *symptomatic dystonia,* and those without an identifiable origin, called *primary* or *idiopathic dystonia.* Patients with idiopathic dystonia fail to reveal any significant findings on CT or MRI. With the exception of a limited number of reports, post-mortem studies give little insight into anatomical or biochemical alterations which underly this disease process. PET remains one of the only *in vivo* techniques available which permits the investigation of the pathophysiology of this disorder. Patients with secondary dystonia often have basal ganglia lesions as the substrate for their disorder. This has caused investigative efforts to focus onto the extrapyramidal pathways as the site of abnormality in primary dystonia.

1. Cerebral Blood Flow and Oxygen Metabolism

Perlmutter and Raichle were the first to report the *in vivo* demonstration of a localized functional abnormality in a patient with hemidystonia.[87] PET scans showed significant changes in the contralateral basal ganglia. They demonstrated decreased cerebral metabolic rate for oxygen ($CMRO_2$) and oxygen extraction, but increased cerebral blood flow (CBF) and volume (CBV). Resolution limits at that time did not permit further demarcation of the area involved.

In an application of somatosensory stimulation studies developed by Fox and Raichle, Perlmutter and Raichle studied seven dystonic patients. In baseline studies, mean hemispheric and global CBF values were normal for all patients.[88] Left-right ratios for five specific regions were not different from controls. Two of the four patients with unilateral symptoms, however, showed significant left-right putaminal differences. Subtraction of baseline CBF images from images acquired during stimulation produces "activation" images in which only regions showing task or stimulus specific increases in CBF can be identified. Activation of the somatosensory cortex is well visualized by this manipulation. In one patient suffering from left hemidystonia, cortical activation was significantly reduced on the right side only. The authors suggest that this patient may suffer from aberrant connections to or dysfunctional response of the contralateral sensory motor cortex. Because this subject demonstrated increased dystonic posturing upon stimulation of the affected hand, they speculate that dystonic responses may be actively inhibited by the somatosensory cortex. In opposition to this hypothesis is the fact that this increased dystonic posturing could be elicited in a patient showing normal activation.

2. Cerebral Metabolic Rate for Glucose (CMRG)

Torticollis is a form of dystonia in which the muscles of the neck undergo sustained contraction resulting in a rotation of the head to one side. In a novel approach to the analysis of PET data, the UBC group showed that in patients with this condition, disruptions of pallidothalamic projections could be demonstrated by means of correlational methods.[89,90] Sixteen torticollis patients studied for CMRG showed no consistent abnormalities in glucose metabolism. Statistical correlation of the activity in a given region with that of all other selected regions revealed an "uncoupling" of the thalamus and basal ganglia. Such uncoupling may represent an abnormality of the projections to and from the basal ganglia rather than pathology within the structure itself.

Gilman performed FDG studies in five patients with predominantly unilateral idiopathic dystonia.[91] All showed normal glucose metabolism; two patients demonstrated asymmetries.

One was observed in the caudate nucleus, with the contralateral side showing increased metabolism. The other demonstrated a cerebellar asymmetry with increased activity ipsilateral to the symptoms. Both patients also showed contralateral parietocortical asymmetries (though the significance of this finding was questioned).

Lang and colleagues studied seven dystonic patients with FDG.[92] Patients with symptomatic hemidistonia showed lesion specific putaminal decreases in glucose metabolism. None of the idiopathic subjects showed significant glucose utilization abnormalities.

Chase and co-workers reported preliminary results of FDG studies on six patients with idiopathic torsion dystonia.[93] By way of a computerized image processing system which scales intensities independent of region size or position, this group was able to reveal substantial left-right regional activity differences in the lenticular nucleus and inferior caudate in three patients. Lenticular changes showed marked asymmetry with hypermetabolism contralateral to the side of predominant symptomatic involvement. Though there was a significant correlation between the degree of metabolic asymmetry within other basal structures with the disease severity, the association of these changes with the clinical presentation was less consistent. Image resolution was not sufficient to specifically implicate or exclude a putaminal or pallidal contribution.

3. The Dopaminergic System

The nigrostriatal dopamine system has been implicated in the pathophysiology of dystonia. The UBC group reported 6-FD studies in two patients, one with hemidystonia and the other with generalized dystonia.[94] Both had demonstrable lesions involving the putamen and reduced 6-FD uptake. Leenders et al. studied six patients with 6-FD.[95] Four suffered from unilateral limb dystonia (three of which had associated hemiparkinsonism) and two suffered from spasmotic torticollis. All four hemidystonics (including the one with no parkinsonian signs) showed decreased tracer uptake in the striatal regions contralateral to the side of symptomatic involvement. Both patients with torticollis showed a bilateral decrease in striatal uptake. They concluded that for selected dystonic patients, the striatal presynaptic dopaminergic pathway may be abnormal. Further investigation of the hemidystonic patients with NMSP showed increased D-2 binding in one subject.

The Hammersmith group also reported severely diminished 6-FD uptake in the contralateral striatum of a patient with combined hemisydtonia/hemiparkinsonism and ipsilateral blepharospasm.[96] A contralateral calcified rostral brainstem-thalamic lesion was seen on MRI. The investigators attributed the dystonic symptoms to the rostral thalamic calcification, stating that such a lesion is capable of "disconnecting" the striopallidal imput to the thalamus or the thalamic output to motor cortex. D-2 receptor binding evaluated with NMSP was normal.

Patients suffering from *diurnal dystonia* have shown symptomatic relief with the administration of levodopa. In such patients, studies with 6-FD are likely to reveal abnormalities. Lang studied seven dystonic patients with 6-FD.[92] Those with unilateral involvement always showed a decrease in 6-FD uptake localized to the contralateral putamen. Parkinsonian signs were not reported for any of the subjects. The magnitude of the decrease in their patient with diurnal variation led them to state that their data present the strongest *in vivo* evidence to date implicating a dysfunction of the nigrostriatal dopaminergic system in this disorder.

In summary, dystonia is a heterogeneous group of disorders, probably without a single underlying etiology. PET has the potential to aid in specifically identifying the underlying disorders involved and thus further define the pathway by which a common symptomatology, the dystonic syndrome, can arise.

E. PET STUDIES IN OTHER MOVEMENT DISORDERS

Wilson's disease is an autosomal recessive degenerative disorder associated with altered copper metabolism. Patients may present with liver failure or with neurological and psychiatric disturbances. Involuntary movements are common and include chorea, dystonia, and tremor. FDG PET studies have been performed in subjects with this disorder[97] and have been used to monitor its therapy.[98] Patients with Wilson's disease have been reported to have reduced putamen glucose utilization.[97] In a single subject that was followed through a treatment regime for this disorder, there was improvement in caudate nucleus and putamen glucose metabolism that coincided with clinical improvement.[98]

Palatal myoclonus is a disorder characterized by rhythmic movements of the soft palate larynx, and pharyngeal muscles. Pathological examination has shown abnormalities in the inferior olives. FDG PET scans have been performed in six patients with this disorder. These individuals were found to have increased metabolism in the inferior olivary nucleus when compared to cortex.[99]

F. SUMMARY AND FUTURE PROSPECTS

PET has been employed to study subjects with movement disorders with a variety of tracers. In Parkinson's disease and other forms of parkinsonism, the dopamine system has been the main focus. Subjects have reduced accumulation of radioactivity in the striatum after 6-FD administration. Preclinical subjects exposed to MPTP have also shown reduced uptake compared to controls. 6-FD and PET may provide a presymptomatic method of diagnosing patients at risk of developing parkinsonism in the future. The study of dopamine receptors has been slow to develop, but many centers are now beginning to examine D-1 and D-2 receptors with PET.

Patients with Huntington's disease have been shown to have reduced glucose utilization in the caudate nucleus with FDG and PET. Preclinical subjects have been identified in some centers with this technique. It is hoped that new therapeutic strategies may be evaluated with serial FDG studies.

One of the important issues that must be clarified in this field is the analytic methods chosen to evaluate the data. The research centers that are involved in these studies must develop techniques that will permit the comparison of data from one center to another. This need is readily apparent in the data analysis of scans of the dopamine system. 6-FD and receptor ligands are currently being analyzed by many different methods, making systematic analysis of the literature difficult. With new mathematical models for data analysis becoming available, it is hoped that a better understanding of the pathophysiology of movement disorders will be attainable from *in vivo* studies with PET.

REFERENCES

1. **Brooks, D. J. and Frackowiak, R. S. J.,** PET and movement disorders, *J. Neurol. Neurosurg. Psychiatry,* 52, 68, 1989.
2. **Leenders, K. L., Gibbs, J. M., Frackowiak, R. S. J., Lammertsma, A. A., and Jones, T.,** Positron emission tomography of the brain: new possibilities for the investigation of human cerebral pathophysiology, *Prog. Neurobiol.,* 23, 1, 1984.
3. **Hornykiewicz, O.,** Dopamine (3-hydroxytyramine) and brain function, *Pharmacol. Rev.,* 18, 925, 1966.
4. **Langston, J. W. and Irwin, I.,** MPTP: current concepts and controversies, *Clin. Neuropharmacol.,* 9, 485, 1986.
5. **Langston, J. W., Ballard, P., Tetrud, J. W., and Irwin, I.,** Chronic parkinsonism in humans due to a product of meperidine-analog synthesis, *Science,* 219, 979, 1983.
6. **Guttman, M. and Calne, D. B.,** *In vivo* characterization of cerebral dopamine systems in human parkinsonism, in *Parkinson's Disease and Movement Disorders,* Jankovic, J. and Tolosa, E., Eds., Urban & Schwarzenberg, Baltimore, 1988, 49.
7. **Leenders, K. L.,** Parkinson's disease and PET tracer studies, *J. Neurol. Transm.,* 27, 219, 1988.
8. **Martin, W. R. W.,** Imaging techniques in Parkinson's disease, *Movement Disorders,* 4(1), S63, 1989.

9. **Wolfson, L. I., Leenders, K. L., Brown, L. L., and Jones, T.,** Alterations of regional cerebral blood flow and oxygen metabolism in Parkinson's disease, *Neurology,* 35, 1399, 1985.

10. **Leenders, K. L., Wolfson, L., Gibbs, J. M., Wise, R. J. S., Causon, R., Jones, T., and Legg, N. J.,** The effects of L-dopa on regional cerebral blood flow and oxygen metabolism in patients with Parkinson's disease, *Brain,* 108, 171, 1985.

11. **Perlmutter, J. S. and Raichle, M. E.,** Regional blood flow in hemiparkinsonism, *Neurology,* 35, 1127, 1985.

12. **Martin, W. R. W., Beckman, J. H., Calne, D. B., Adam, M. J., Harrop, R., Rogers, J. G., Ruth, T. J., Sayre, C. I., and Pate, B. D.,** Cerebral glucose metabolism in Parkinson's disease, *Can. J. Neurol. Sci.,* 11, 169, 1984.

13. **Martin, W. R. W., Stoessl, A. J., Adam, M. J., Ammann, W., Bergstrom, M., Harrop, R., Laihinen, A., Rogers, J. G., Ruth, T. J., Sayre, C. I., Pate, B. D., and Calne, D. B.,** Positron emission tomography in Parkinson's disease: glucose and dopa metabolism, *Adv. Neurol.,* 45, 95, 1986.

14. **Rougemont, D., Baron, J. C., Collard, P., Bustany, P., Comar, D., and Agid, Y.,** Local cerbral glucose utilisation in treated and untreated patients with Parkinson's disease, *J. Neurol. Neurosurg. Psychiatry,* 47, 824, 1984.

15. **Kuhl, D. E., Metter, E. J., and Riege, W. H.,** Patterns of local cerebral glucose utilization determined in Parkinson's disease by the [^{18}F]fluorodeoxyglucose method, *Ann. Neurol.,* 15, 419, 1984.

16. **Garnett, E. S., Firnau, G., and Nahmias, C.,** Dopamine visualized in the basal ganglia of living man, *Nature,* 305, 137, 1983.

17. **Martin, W. R. W., Palmer, M. R., Patlak, C. S., and Calne, D. B.,** Nigrostriatal function in humans studied with positron emission tomography, *Ann. Neurol.,* 26, 535, 1989.

18. **Garnett, E. S., Nahmias, C., and Firnau, G.,** Central dopaminergic pathways in hemiparkinsonism examined by positron emission tomography, *Can. J. Neurol. Sci.,* 11, 174, 1984.

19. **Nahmias, C., Garnett, E. S., Firnau, G., and Lang, A.,** Striatal dopamine distribution in parkinsonian patients during life, *J. Neurol. Sci.,* 69, 223, 1985.

20. **Kish, S. J., Shannak, K., and Hornykiewicz, O.,** Uneven pattern of dopamine loss in the striatum of patients with idiopathic parkinson's disease, *N. Engl. J. Med.,* 318(14), 876, 1988.

21. **Martin, W. R. W., Adam, M. J., Bergstrom, M., Ammann, W., Harrop, R., Laihinen, A., Rogers, J., Ruth, J., Sayre, C., Stoessl, J., Pate, B. D., and Calne, D. B.,** *In vivo* study of DOPA metabolism in Parkinson's disease, in *Recent Development in Parkinson's Disease,* Fahn, S., Marsden, C. D., Jenner, P., and Teychenne, P., Raven Press, New York, 1986, 97.

22. **Leenders, K., Palmer, A., Turton, D., Quinn, N., Firnau, G., Garnett, S., Nahmias, C., Jones, T., and Marsden, C. D.,** DOPA uptake and dopamine receptor binding visualized in the human brain *in vivo,* in *Recent Development in Parkinson's Disease,* Fahn, S., et al., Raven Press, New York, 1986, 103.

23. **Leenders, K. L., Palmer, A. J., Quinn, N., Clark, J. C., Firnau, G., Garnett, E. S., Nahmias, C., Jones, T., and Marsden, C. D.,** Brain dopamine metabolism in patients with Parkinson's disease measured with positron emission tomography, *J. Neurol. Neurosurg. Psychiatry,* 49, 853, 1986.

24. **Aquilonius, S.-M., Bergström, K., Eckernäs, S.-A., Hartvig, P., Leenders, K. L., Lundquist, H., Antoni, G., Gee, A., Rimland, A., Uhlin, J., and Langström, B.,** *In vivo* evaluation of striatal dopamine reuptake sites using ^{11}C-nomifensine and positron emission tomography, *Acta Neurol. Scand.,* 76, 283, 1987.

25. **Tedroff, J., Aquilonius, S.-M., Hartvig, P., Lundqvist, H., Gee, A. G., Uhlin, J., and Langtröm, B.,** Monoamine re-uptake sites in the human brain evaluated in vivo by means of ^{11}C-nomifensine and positron emission tomography: the effects of age and Parkinson's disease, *Acta Neurol. Scand.,* 77, 192, 1988.

26. **Kilbourn, M. R., Ciliax, B. J., Haka, M. S., and Penney, J. B.,** *In vivo* autoradiography of [^{18}F]-GBR 13119 binding in rat brain, *Soc. Neurosci. Abstr.,* 376.11, 1988.

27. **Leenders, K. L, Salmon, E., Tyrrell, P., Perani, D., Brooks, D., Frackowiak, R. S. J., Sagar, H. J., and Marsden, C. D.,** Comparison of cerebral L-(^{18}F)-6-fluorodopa and (^{11}C)-nomifensine uptake in healthy volunteers and patients with Parkinson's disease using PET, *Neurology,* 39 (Suppl. 1), 272, 1989.

28. **Laihinen, A., Aquilonius, S.-M., Haaparanta, M., Hartvig, P., and Langström, B.,** Similar reductions in ^{18}F-fluoro-L-dopa and ^{11}C-(+)-nomifensine kinetics in Parkinson's disease studied with PET, *Neurology,* 39 (Suppl. 1), 272, 1989.

29. **Guttman, M. and Seeman, P.,** L-Dopa reverses the elevated density of D2 dopamine receptors in Parkinson's diseased striatum, *J. Neural Transm.,* 64, 93, 1985.

30. **Guttman, M., Seeman, P., Reynolds, G. P., Riederer, P., Jellinger, K., and Tourtellotte, W. W.,** Dopamine D2 receptor density remains constant in treated Parkinson's disease, *Ann. Neurol.,* 19, 487, 1986.

31. **Rutgers, A. W. F., Lakke, J. P. W. F., Paans, A. M. J., Vaalburg, W., and Korf, J.,** Tracing of dopamine receptors in hemiparkinsonism with positron emission tomography (PET), *J. Neurol. Sci.,* 80, 237, 1987.

32. **Hägglund, J., Aquilonius, S.-M., Eckernäs, S.-A., Hartvig, P., Lundquist, H., Gullberg, P., and Langström, B.,** Dopamine receptor properties in Parkinson's disease and Huntington's chorea evaluated by positron emission tomography using ^{11}C-N-methyl-spiperone, *Acta Neurol. Scand.,* 75, 87, 1987.

33. **Rinne, U. K., Laihinen, A., Rinne, J. O., Nagren, K., Bergman, J., and Ruotsalainen, U.,** Positron emission tomography (PET) demonstrates dopamine receptor supersensitivity in the striatum of patients with early Parkinson's disease, *Neurology,* 39 (Suppl. 1), 273, 1989.

34. **Guttman, M., Burns, R. S., Martin, W. R. W., Peppard, R. F., Adam, M. J., Ruth, T. J., Allen, G., Parker, R. A., Tulipan, N. B., and Calne, D. B.,** PET studies of parkinsonian patients treated with autologous adrenal implants, *Can. J. Neurol. Sci.,* 16, 305, 1989.

35. **Lindvall, O., Backlund, E.-O., Farde, L., Sedvall, G., Freedman, R., Hoffer, B., Nobin, A., Seiger, A., and Olson, L.,** Transplantation in Parkinson's disease: two cases of adrenal medullary grafts to the putamen, *Ann. Neurol.,* 22, 457, 1987.

36. **Lindvall, O., Rehncrona, S., Brundin, P. et al.,** Human fetal dopamine neurons grafted into the striatum in two patients with severe Parkinson's disease, *Arch. Neurol.,* 46, 615, 1989.

37. **Calne, D. B., Langston, J. W., Martin, W. R. W., Stoessl, A. J., Ruth, T. J., Adam, M. J., Pate, B. D., and Schulzer, M.,** Positron emission tomography after MPTP: observations relating to the cause of Parkinson's disease, *Nature,* 317, 246, 1985.

38. **Perlmutter, J. S., Kilbourn, M. R., Raichle, M. E., and Welch, M. J.,** MPTP-induced upregulation of *in vivo* dopaminergic radioligand-receptor binding in humans, *Neurology,* 37, 1575, 1987.

39. **Leenders, K. L., Frackowiak, R. S. J., and Lees, A. J.,** Steele-Richardson-Olszewski syndrome. Brain energy metabolism, blood flow and fluorodopa uptake measured by positrion emission tomography, *Brain,* 111, 615, 1988.

40. **D'Antona, R., Baron, J. C., Samson, Y., Serdaru, M., Viader, F., Agid, Y., and Cambier, J.,** Subcortical dementia. Frontal cortex hypometabolism detected by positron tomography in patients with progressive supranuclear palsy, *Brain,* 108, 785, 1985.

41. **Foster, N. L., Gilman, S., Berent, S., Morin, E. M., Brown, M. B., and Koeppe, R. A.,** Cerebral hypometabolism in progressive supranuclear palsy studied with positron emission tomography, *Ann. Neurol.,* 24, 399, 1988.

42. **Goffinet, A. M., De Volder, A. G., Gillain, C., Rectem, D., Bol, A., Michel, C., Cogneau, M., Labar, D., and Laterre, C.,** Positron tomography demonstrates frontal lobe hypometabolism in progressive supranuclear palsy, *Ann. Neurol.,* 25, 131, 1989.

43. **De Volder, A. G., Francart, J., Laterre, C., Dooms, G., Bol, A., Michel, C., and Goffinet, A. M.,** Decreased glucose utilization in the striatum and frontal lobe in probably striatonigral degeneration, *Ann. Neurol.,* 26, 239, 1989.

44. **Gilman, S., Markel, D. S., Koeppe, R. A., Junck, L., Kluin, K. J., Gebarski, S. S., and Hichwa, R. D.,** Cerebellar and brainstem hypometabolism in olivopontocerebellar atrophy detected with positron emission tomography, *Ann. Neurol.,* 23, 223, 1988.

45. **Guttman, M., Steele, J. C., Stoessl, J., Peppard, R. F., Martin, W. R. W., Walsh, E. M., Ruth, T., Adam, M. J., Pate, B. D., and Tsui, J. K. C.,** 6-(^{18}F)flurodopa PET scanning in the ALS-PD complex of Guam, *Neurology,* 37 (Suppl. 1), 113, 1987.

46. **Eidelberg, D., Moeller, J. R., Sidtis, J. J., Dhawan, V., Strother, S. C., Fahn, S., and Rottenberg, D. A.,** Corticodentatonigral degeneration: metabolic asymmetries studied with ^{18}F-fluorodeoxyglucose and positron emission tomography, *Neurology,* 39 (Suppl. 1), 164, 1989.

47. **Sawle, G. V., Brooks, D. J., Thompson, P. D., Marsden, C. D., and Frackowiak, R. S. J.,** PET studies on the dopaminergic system and regional cortical metabolism in corticobasal degeneration, *Neurology,* 39 (Suppl. 1), 163, 1989.

48. **Rosenberg, N. L., Myers, J. A., and Martin, W. R. W.,** Cyanide-induced parkinsonism: clinical, MRI, and 6-fluorodopa PET studies, *Neurology,* 39, 142, 1989.

49. **Leenders, K. L., Findley, L. J., and Cleeves, L.,** PET before and after surgery for tumor-induced parkinsonism, *Neurology,* 36, 1074, 1986.

50. **Bedard, P. J., DiPaolo, T., Falardeau, P., and Boucher, R.,** Chronic treatment with levodopa, but not bromocriptine induces dyskinesia in MPTP-Parkinsonian monkeys, correlation with [^{3}H]Spiperone binding, *Brain Res.,* 379, 294, 1986.

51. **DiPaolo, T., Bedard, P., Daigle, M., and Boucher, R.,** Long-term effects of MPTP on central and peripheral catecholamine and indolamine concentrations in monkeys, *Brain Res.,* 379, 286, 1986.

52. **Mitchell, I. J., Cross, A. J., Sambrook, M. A., and Crossman, A. R.,** Sites of the neurotoxic action of 1-methyl-4-phenyl-1,2,3,6-tetrahydropyridine in the macaque monkey include the ventral tegmental area and the locus ceruleus, *Neurosci. Lett.,* 61, 195, 1985.

53. **Forno, L. S., Langston, J. W., DeLanney, L. E., Irwin, I., and Ricaurte, G. A.,** Locus ceruleus lesions and eosinophillic inclusions in MPTP-treated monkeys, *Ann. Neurol.,* 20, 449, 1986.

54. **Bankiewicz, K. S., Oldfield, E. H., Chieuh, C. C., Doppman, J. L., Jacobowitz, D. M., and Kopin, I. J.,** Hemiparkinsonism in monkeys after unilateral internal carotid artery infusion of 1-methyl-4-phenyl-1,2,3,6-tetrahydropyridine (MPTP), *Life Sci.,* 39, 7, 1986.

55. **Guttman, M., Fibiger, H. C., Jakubovic, A., and Calne, D. B.,** Intracarotid MPTP administration: biochemical and behavioral observations in a primate model of hemiparkinsonism, *J. Neurochem.,* 54, 1329, 1990.

56. **Chieuh, C. C., Firnau, G., Burns, R. S., Nahmias, C., Chirakal, R., Kopin, I. J., and Garnett, E. S.,** Determination and visualization of damage to striatal dopaminergic terminals in 1-methyl-4-phenyl-1,2,3,6-tetrahydropyridine-induced parkinsonism by [^{18}F]-labeled 6-fluoro-L-dopa and positron emission tomography, *Adv. Neurol.,* 45, 167, 1986.

57. **Chieuh, C. C., Burns, R. S., Kopin, I. J., Kirk, K. L., Firnau, G., Nahmias, C., Chirakal, R., and Garnett, E. S.,** 6-^{18}F-dopa/positron emission tomography visualized degree of damage to brain dopamine in basal ganglia of monkeys with MPTP-induced parkinsonism, in *MPTP: A Neurotoxin Producing a Parkinsonian Syndrome,* Markey, S. P., Castagnoli, N., Jr., Trevor, A. J., and Kopin, I. J., Eds., Academic Press, Orlando, FL, 1986, 327.

58. **Guttman, M., Yong, V. Y. W., Kim, S. U., Calne, D. B., Martin, W. R. W., Adam, M. J., and Ruth, T. J.,** Asymptomatic striatal dopamine depletion: PET scans in unilateral MPTP monkeys, *Synapse,* 2, 469, 1988.

59. **Miletich, R. S., Bankiewicz, K., Plunkett, R., Finn, R., Jacobs, G., Baldwin, P., Adams, R., Kopin, I., DiChiro, G., and Oldfield, E.,** L-[^{18}F]-fluorodopa PET imaging of catecholaminergic tissue implants in hemi-parkinsonian monkeys, *Neurology,* 38 (Suppl. 1), 145, 1988.

60. **Yong, V. W., Guttman, M., Kim, S. U., Calne, D. B., Turnbull, I., Watabe, K., and Tomlinson, R. W. W.,** Transplantation of human sympathetic neurons and adrenal chromaffin cells into parkinsonian monkeys: no reversal of clinical symptoms, *J. Neurol. Sci.,* 94, 51, 1990.

61. **Leenders, K. L., Aquilonius, S.-M., Bergstrom, K., Bjurling, P., Crossman, A. R., Eckernas, S.-A., Gee, A. G., Hartvig, P., Lundqvist, H., Langstrom, B., Rimland, A., and Tedroff, J.,** Unilateral MPTP lesion in a rhesus monkey: effects on the striatal dopaminergic system measured *in vivo* with PET using various novel tracers, *Brain Res.,* 445, 61, 1988.

62. **Joyce, J. N., Marshall, J. F., Bankiewicz, K. S., Kopin, I. J., and Jacobowitz, D. M.,** Hemiparkinsonism in a monkey after unilateral internal carotid artery infusion of 1-methyl-4-phenyl-1,2,3,6-tetrahydropyridine (MPTP) is associated with regional ipsilateral change in striatal dopamine D-2 receptor density, *Brain Res.,* 382, 360, 1985.

63. **Alexander, G. M., Schwartzmann, R. J., and Ferraro, T. N.,** Spiperone binding and dopamine levels in the striatum of MPTP-treated monkeys, *Ann. Neurol.,* 22, 149, 1987.

64. **Hantraye, P., Loc'h, C., Tacke, U., Riche, D., Stulzaft, O., Doudet, D., Guibert, B., Naquet, R., Mazière, B., and Mazière, M.,** *In vivo* visualization by positron emission tomography of the progressive striatal dopamine receptor damage occurring in MPTP-intoxicated nonhuman primates, *Life Sci.,* 39, 1375, 1986.

65. **Gibb, W. R. G., Lees, A. J., Wells, F. R., Barnard, R. O., Jenner, P., and Marsden, C. D.,** Pathology of MPTP in the marmoset, *Adv. Neurol.,* 45, 187, 1986.

66. **Guttman, M., Adam, M., Ruth, T., Calne, D. B., Kebabian, J., and Schoenleber, R.,** SCH-23390 PET scans show reduced D-1 receptor binding in unilaterally MPTP-lesioned monkeys, *Neurology,* 38 (Suppl. 1), 259, 1988.

67. **Porceddu, M. L., Giorgi, O., De Montis, D., Mele, S., Cocco, L., Ongini, E., and Biggio, G.,** 6-Hydroxydopamine-induced degeneration of nigral dopamine neurons: Differential effect on nigral and striatal D-1 dopamine receptors, *Life Sci.,* 41, 697, 1987.

68. **Filloux, F. M., Walmsley, J. K., and Dawson, T. M.,** Presynaptic and postsynaptic D-1 dopamine receptors in the nigrostriatal system of the rat brain: a quantitative autoradiographic study using the selective D-1 antagonist [^{3}H]SCH 23390, *Brain Res.,* 408, 205, 1987.

69. **Marshall, J. F., Navarrete, and Joyce, J. N.,** Decreased [^{3}H]SCH 23390 labeling of striatal D-1 sites after nigrostriatal injury, *Soc. Neurosci. Abstr.,* 13, 1344, 1987.

70. **Martin, J. B. and Gusella, J. F.,** Huntington's Disease. Pathogenesis and management, *Sem. Med.,* 315(20), 1267, 1986.

71. **Kuhl, D. E., Phelps, M. E., Markham, C. H., Metter, E. J., Riege, W. H., and Winter, J.,** Cerebral metabolism and atrophy in Huntington's disease determined by ^{18}FDG and computed tomographic scan, *Ann. Neurol.,* 12, 425, 1982.

72. **Kuhl, D. E., Markham, C. H., Metter, E. J., Riege, W. H., Phelps, M. E., and Mazziotta, J. C.,** Local cerebral glucose utilization in symptomatic and presymptomatic Huntington's disease, in *Brain Imaging and Brain Function,* Sokoloff, L., Ed., Raven Press, New York, 1985, 199.

73. **Hayden, M. R., Martin, W. R. W., Stoessl, A. J., Clark, C., Hollenberg, S., Adam, M. J., Ammann, W., Harrop, R., Rogers, J., Ruth, T., Sayre, C., and Pate, B. D.,** Positron emission tomography in the early diagnosis of Huntington's disease, *Neurology,* 36, 888, 1986.

74. **Garnett, E. S., Firnau, G., Nahmias, C., Carbotte, R., and Bartolucci, G.,** Reduced striatal glucose consumption and prolonged reaction time are early features in Huntington's disease, *J. Neurol. Sci.,* 65, 231, 1984.

75. **De Volder, A., Bol, A., Michel, C., Cogneau, M., Evrard, P., Lyon, G., and Goffinet, A. M.,** Brain glucose utilization in childhood Huntington's disease studied with positron emission tomography (PET), *Brain Dev.,* 10, 47, 1988.

76. **Young, A. B., Penney, J. B., Starosta-Rubinstein, S., Markel, D. S., Berent, S., Giordani, B., Ehrenkaufer, R., Jewett, D., and Hichwa, R.,** PET scan investigations of Huntington's disease: cerebral metabolic correlates of neurological features and functional decline, *Ann. Neurol.,* 20, 296, 1986.

77. **Berent, S., Giordani, B., Lehtinen, S., Markel, D., Penney, J. B., Buchtel, H. A., Starosta-Rubinstein, S., Hichwa, R., and Young, A. B.,** Positron emission tomographic scan investigations of Huntington's disease: cerebral metabolic correlates of cognitive function, *Ann. Neurol.,* 23, 541, 1988.

78. **Leenders, K. L., Frackowiak, R. S. J., Quinn, N., and Marsden, C. D.,** Brain energy metabolism and dopaminergic function in Huntington's disease measured *in vivo* using positron emission tomography, *Movement Disorders,* 1(1), 69, 1986.

79. **Mazziotta, J. C., Phelps, M. E., Pahl, J. J., Huang, S.-C., Baxter, L. R., Riege, W. H., Hoffman, J. M., Kuhl, D. E., Lanto, A. B., Wapenski, J. A., and Markham, C. H.,** Reduced cerebral glucose metabolism in asymptomatic subjects at risk for Huntington's disease, *N. Engl. J. Med.,* 316(7), 357, 1987.

80. **Young, A. B., Penney, J. B., Starosta-Rubinstein, S., Markel, D., Berent, S., Rothley, J., Betley, A., and Hichwa, R.,** Normal caudate glucose metabolism in persons at risk for Huntington's disease, *Arch. Neurol.,* 44, 254, 1987.

81. **Hayden, M. R., Hewitt, J., Stoessl, A. J., Clark, C., Ammann, W., and Martin, W. R. W.,** The combined use of positron emission tomography and DNA polymorphisms for preclinical detection of Huntington's disease, *Neurology,* 37, 1441, 1987.

82. **Hayden, M. R., Hewitt, J., Martin, W. R. W., Clark, C., and Ammann, W.,** Studies in persons at risk for Huntington's disease, Letter to the Editor, *N. Engl. J. Med.,* 382, 1987.

83. **Suchowersky, O., Hayden, M. R., Martin, W. R. W., Stoessl, A. J., Hildebrand, A. M., and Pate, B. D.,** Cerebral metabolism of glucose in benign hereditary chorea, *Movement Disorders,* 1(1), 33, 1986.

84. **Guttman, M., Lang, A. E., Garnett, E. S., Nahmias, C., Firnau, G., Tyndel, F. J., and Gordon, A. S.,** Regional cerebral glucose metabolism in SLE chorea: further evidence that striatal hypometabolism is not a correlate of chorea, *Movement Disorders,* 2(3), 201, 1987.

85. **Hosokawa, S., Ichiya, Y., Kuwabara, Y., Ayabe, Z., Mitsuo, K., Goto, I., and Kato, M.,** Positron emission tomography in cases of chorea with different underlying diseases, *J. Neurol. Neurosurg. Psychiatry,* 50, 1284, 1987.

86. **Jankovic, J. and Fahn, S.,** Dystonic syndromes, in *Parkinson's Disease and Movement Disorders,* Jankovic, J. and Tolosa, E., Eds., Urban & Schwarzenberg, Baltimore, 1988, 283.

87. **Perlmutter, J. S. and Raichle, M. E.,** Pure hemisystonia with basal ganglion abnormalities on positron emission tomography, *Ann. Neurol.,* 15, 228, 1984.

88. **Perlmutter, J. S. and Raichle, M. E.,** Regional cerebral blood flow in dystonia: an exploratory study, in *Advances in Neurology, Dystonia 2,* Fahn, S., Marsden, C. D., and Calne, D. B., Eds., Raven Press, New York, 1988, 50, 255.

89. **Stoessl, A. J., Martin, W. R. W., Clark, C., Ammann, W., Beckman, J. H., Bergstrom, M., Harrop, R., Rogers, J. G., Ruth, T. J., Sayre, C. I., Pate, B. D., and Calne, D. B.,** PET studies of cerebral glucose metabolism in idiopathic torticollis, *Neurology,* 36, 653, 1986.

90. **Martin, W. R. W., Stoessl, A. J., Palmer, M., Adam, M. J., Ruth, T. J., Grierson, J. R., Pate, B. D., and Calne, D. B.,** PET scanning in dystonia, in *Advances in Neurology, Dystonia 2,* Fahn, S., Marsden, C. D., and Calne, D. B., Eds., Raven Press, New York, 1988, 50, 223.

91. **Gilman, S., Junck, L., Young, A. B., Hichwa, R. D., Markel, D. S., Koeppe, R. A., and Ehrenkaufer, R. L. E.,** Cerebral metabolic activity in idiopathic dystonia studied with positron emission tomography, in *Advances in Neurology, Dystonia 2,* Fahn, S., Marsden, C. D., and Calne, D. B., Eds., Raven Press, New York, 1988, 50, 231.

92. **Lang, A. E., Garnett, E. S., Firnau, G., Nahmias, C., and Talalla, A.,** Positron tomography in dystonia, in *Advances in Neurology, Dystonia 2,* Fahn, S., Marsden, C. D., and Calne, D. B., Eds., Raven Press, New York, 1988, 50, 249.

93. **Chase, T. N., Tamminga, C. A., and Burrows, H.,** Positron emission tomographic studies of regional cerebral glucose metabolism in idiopathic dystonia, in *Advances in Neurology, Dystonia 2,* Fahn, S., Marsden, C. D., and Calne, D. B., Eds., Raven Press, New York, 1988, 50, 237.

94. **Fross, R. D., Martin, W. R. W., Li, D., Stoessl, A. J., Adam, M. J., Ruth, T. J., Pate, B. D., Burton, K., and Calne, D. B.,** Lesions of the putamen: their relevance to dystonia, *Neurology,* 37, 1125, 1987.

95. **Leenders, K. L., Quinn, N., Frackowiak, R. S. J., and Marsden, D. C.,** Brain dopaminergic system studies in patients with dystonia using positron emission tomography, in *Advances in Neurology, Dystonia 2,* Fahn, S., Marsden, C. D., and Calne, D. B., Eds., Raven Press, New York, 1988, 50, 243.

96. **Leenders, K. L., Frackowiak, R. S. J., Quinn, N., Brooks, D., Summer, D., and Marsden, C. D.,** Ipsilateral blrepharospam and contralateral hemidystonia and parkinsonism in a patient with a unilateral rostral brainstem-thalamic lesion: structural and functional abnormalities studied with CT, MRI and PET Scanning, *Movement Disorders,* 1, 51, 1986.

97. **Hawkins, R. A., Mazziotta, J. C., and Phelps, M. E.,** Wilson's disease studied with FDG and positron emission tomography, *Neurology,* 37, 1707, 1987.

98. **De Volder, A., Sindic, C. J. M., and Goffinet, A. M.,** Effect of D-penicillamine treatment on brain metabolism in Wilson's disease: a case study, *J. Neurol. Neurosurg. Psychiatry,* 51, 947, 1988.

99. **Dubinsky, R. M. and Hallett, M.,** Palatal myoclonus and facial involvement in other types of myoclonus, in *Advances in Neurology, Facial Dyskinesias,* Jankovic, J. and Tolosa, E., Eds., Raven Press, New York, 49, 1988, 263.

100. **Gjedde, A.,** personal communication.

101. **Calne, D. B.,** unpublished data.

Chapter 22

COGNITIVE AND PHYSIOLOGICAL FACTORS THAT AFFECT REGIONAL CEREBRAL BLOOD FLOW AND OTHER MEASURES OF BRAIN FUNCTION

Karen Faith Berman and Daniel R. Weinberger

TABLE OF CONTENTS

I. INTRODUCTION

In 1896, Charles Roy and Charles Sherrington[1] observed that within seconds of the onset of an epileptic seizure, a swelling of the brain occurred, suggesting an increase in the supply of blood. Some 15 to 20 years later, Joseph Barcroft[2] hypothesized than an enhanced level of function in a tissue is sustained by increasing the rate of oxygen consumption and, therefore, the flow of oxygenated blood to the tissue. Barcroft thus refined the notion that blood flow to a tissue varies with its functional activity and metabolism. Elegantly demonstrating that this concept is applicable to the brain, Schmidt and Hendrix[3] in 1937 recorded a strictly localized increase in blood flow to the visual cortex when a small spot of light was shined on the retina of a cat.

The work of these and other investigators advanced a concept that has now been verified by an extensive body of basic research and clinical studies,[4-6] that is, in the grossly intact brain, "brain work", neuronal metabolism, and blood flow are tightly coupled. An important implication of these observations is that cerebral blood flow can be studied not just to elucidate cerebral hemodynamics per se, but more importantly, it can serve as a marker for brain activity. The theory underlying this important homeostatic relationship was soon applied to investigating the physiology of the human brain by studying the exchange of nonmetabolized (i.e., inert), diffusible molecules, such as nitrous oxide, between capillary and brain tissue. The rate of this exchange would be a function of blood flow and could be measured by the difference between arterial and venous concentrations of the inert tracer.[7] The next phase of *in vivo* neurophysiological measurements involved determining the clearance rate of radioisotopes of inert substances, such as ^{85}Kr and the radioxenons (administered by intracarotid injection or by inhalation), in various brain regions with multiple stationary extracranial radiation detectors.[8] This nontomographic strategy allowed blood flow of individual cortical regions to be measured. Further technological advances in measuring regional cerebral function have culminated in the development of single photon emission computed tomography (SPECT) and positron emission tomography (PET) which provide a three-dimensional picture of cerebral activity.

Since PET and SPECT measure dynamic brain function and since the brain reflects and responds to its surround, it would be expected that the neurophysiological concomitants of a subject's mental state and behavior, sensory input, motor outputs, and cognitive activity can be measured with these methods. A growing body of literature attests to the sensitivity of these methods to such phenomena. Because of this unique capability, functional brain imaging techniques such as PET and SPECT are currently the most powerful in the armamentarium of tools available for investigating the normal human cerebral functional landscape, the physiological responses of the human brain to the challenges of daily activities, and its functional characteristics under pathological conditions.

However, the very facet of these techniques that makes them so potentially valuable — their ability to measure physiological concomitants of behavioral phenomena — makes their application to clinical research quite complex. Cerebral blood flow and metabolism as measured by current *in vivo* methods represent the summation of a set of complex and diverse cerebral physiological phenomena (some of which are related to the behavior and experience of the subject during the procedure, some of which may relate to the physiological concomitants of the health or illness of the subject, and some of which may be extrinsic to both the experience and the illness). It is, therefore, difficult to tease apart which of these physiological phenomena predominate in a given measurement. In order to do so, it is necessary to exert as much control as possible on each of them. For this reason, the importance of controlling testing conditions during functional brain imaging studies is one of the most critical issues in research design. A number of studies of neuropsychiatric patients have been hampered by this methodological problem.

In this chapter, work demonstrating that the physiological effects of mental and other behavioral activities in humans are measurable by functional brain imaging techniques will be reviewed, and some of the information about normal human brain function that has emerged will be summarized. We will also discuss how, on the one hand, failure to control the various physiological phenomena that comprise "brain activity" has presented a major methodological problem in many studies of neuropsychiatric patients, while, on the other hand, strategies can be devised by which these factors can be exploited to provide maximum information about illnesses such as schizophrenia. Peripherally determined physiological factors that affect rCBF, such as the partial pressure of arterial carbon dioxide (pCO_2), and demographic variables, such as age and sex, that must be controlled will also be discussed.

II. STUDIES OF NORMAL BRAIN FUNCTION

A. EFFECTS OF MOTOR ACTIVITY

One of the first indications that techniques such as PET and SPECT are sensitive to functional activation of specific cortical areas was a largely serendipitous observation[9] made with a technique that measures regional cerebral blood flow (rCBF) of a single hemisphere by injecting ^{133}Xe gas dissolved in saline into the carotid artery. When a subject was asked to repeatedly open and close his hand against a blood pressure cuff (in order to increase blood pressure without using drugs), a marked increase in rCBF localized to the contralateral hand motor cortex was seen.[10] This suggested that the measurement of rCBF could be a powerful tool for demonstrating brain regions that become active during specific stimulus conditions. Mapping of cerebral function during a variety of sensory, cognitive, and motor tasks has subsequently been accomplished.

With regard to the motor system, appropriate somatotopic activation has been demonstrated during simple voluntary movement of the hand,[11] arm,[12] foot,[9,13] and eyes.[14] However, primary motor cortex does not seem to be activated during the planning of a motor sequence if it is not executed.[15] The supplementary motor cortex has also been studied with functional brain imaging techniques. While there is clear evidence that this cortical area participates in motor tasks, the literature is not without controversy as to its exact role. Some rCBF studies have suggested that the supplementary motor area participates only in complex motor tasks, serving as a "programming area for motor subroutines".[15] Others maintain that it participates in all motor tasks regardless of the degree of complexity and that it plays a role in "the initiation of movement perhaps through the establishment of motor set or readiness to move".[16] With careful study design, future PET and SPECT experiments have the potential to clarify the different functional neuroanatomical bases of such complex and subtle questions as these.

Voluntary movements guided by sensory information appear to produce increased blood flow in the premotor cortex.[12] Functional brain imaging techniques have also confirmed that movements result in activation of the human basal ganglia and cerebellum, as predicted from lesion and animal experiments. Mazziotta and colleagues (cited in Reference 17) demonstrated differences in ICMRGlu between novel and overlearned motor tasks; novel finger movement produced increases in the contralateral sensorimotor cortex, while an overlearned task (signature writing) produced a bilateral increase in the striata in addition to contralateral sensorimotor cortex.

It is clear that even the smallest voluntary or involuntary movements can be reflected in the cerebral metabolic landscape. On the one hand, this fact implies great potential for functional brain imaging techniques to elucidate the cerebral organization of the motor system; on the other hand, another repercussion of this fact is that experimental design and control are critical if meaningful data are to be gathered. For example, if motor activity per se is being studied, this must be carried out under highly controlled conditions, and movements that are extraneous to those being studied must be avoided. If illnesses involving pathology of the motor system are being investigated, neurophysiological differences between control and experimental subjects may simply be epiphenomena of the movement disorder (e.g., increased activity in those brain regions discussed above in hyperkinetic states). Finally, if behaviors other than motor activity are being studied, movements that are extraneous to the experimental paradigm must be kept to a minimum since the cerebral concomitants of motor activity may contaminate data acquired to study other systems. If movements are necessary to the experimental paradigm (e.g., a finger movement to make a response during a study of cognition), a sensorimotor control task which involves the same movement and sensory phenomena as the cognition being studied, but without the cognitive component, must also be studied.

B. EFFECTS OF SENSORY PHENOMENA

The importance of ambient testing conditions, with regard to amount of sensory input, on the results of PET studies was demonstrated by Mazziotta et al.[18] They examined the effects of various degrees of sensory deprivation on the local cerebral metabolic rate of glucose utilization and found that subjects studied during minimal or no sensory deprivation (i.e., with both eyes and ears open or with eyes or ears selectively occluded) tended to exhibit left-right symmetry of function. However, with progressive sensory deprivation (i.e., both eyes and ears occluded), the functional landscape became more variable and more asymmetries (left greater than right) emerged. Overall glucose metabolism tended to decrease with complete sensory deprivation and the anterior/posterior ratio of activity increased.

The precise meaning of these findings is unclear, but these experiments demonstrate the responsivity of the human brain to changes in ambient sensory conditions and highlighted the need for control of sensory stimuli during functional brain imaging studies. Since all of the conditions under which brain function was measured in these studies could be called ''resting'' states (i.e., no prescribed motor, sensory, cognitive, or other behavioral activities), these data also emphasize the fact that there is probably no such thing as a ''resting state'' in the awake human brain, that data from resting state studies must be interpreted with caution, and that information gleaned from so-called resting state studies that are actually experientially different cannot be directly compared. As discussed below, even the same resting state or ambient conditions may have different experiental effects on patients and normal healthy subjects.

The physiological responses of the human brain to more focussed sensory input have also been studied with PET, SPECT, and nontomographic radioxenon rCBF measurements. Selected studies will be briefly reviewed.

1. Visual Stimuli

Increases in rCBF and ICMRGlu in the visual cortex have been reported for full-field and selective visual stimulation. Half-field stimulation has been shown to produce bilateral increases in visual cortex, with higher values on the side contralateral to the stimulation.[19] Using the nontomographic ^{133}Xe inhalation method, Risberg and Prohovnik demonstrated that even visual aftereffects can produce rCBF changes.[20] They found progressive global blood flow changes when subjects (1) looked at a stationary spiral, (2) looked at a rotating spiral, and (3) performed the spiral-aftereffects test. Local blood flow changes were also observed.

Differential effects of visual stimuli with varying characteristics have also been demonstrated. For example, Fox and Raichle found a linear relationship between visual stimulus frequency and rCBF in the primary visual cortex up to 7.8 Hz, with a decline in rCBF at higher frequencies.[21] Complexity of visual stimuli has also been shown to effect the cerebral functional landscape. Mazziotta and Phelps found successive visual cortex activation from eyes closed to white light, to checkerboard patterns, and to complex visual stimuli such as outdoor scenes.[17] With this paradigm, they found that increases accompanying complex stimuli were larger in higher-order association areas than in primary or striate visual cortex. Their data also suggested that there is equal functional input from each eye to both visual cortices. Retinotopic representations in the visual cortex have also been demonstrated with PET.[22]

Lueck et al.[23] have recently compared rCBF while subjects viewed a black and white display to that while they viewed a multicolored display. They found that color stimuli produced rCBF increases in the lingual and fusiform gyri. On the basis of these data, they proposed a ''color center'' in the human cerebral cortex. Taken together, the studies reviewed above emphasize the need to control visual stimuli in terms of location within the visual field, frequency and magnitude of stimulation, and complexity of visual material during PET and SPECT studies.

2. Auditory Stimuli

Cerebral responses to all types of auditory stimuli studied, both simple and complex, have been demonstrated with functional brain imaging. For example, one study showed a tonotopic rCBF response to tones of varying frequencies in the primary auditory cortex contralateral to the stimulation.[24] In general, more complex stimuli appear to activate larger areas of temporal cortex as well as other cortical and subcortical areas.

The effects of binaural vs. monaural auditory input remain an active area of research in lower animals. Research with PET, SPECT, and other techniques has examined this issue in human subjects. Reivich et al.[25] demonstrated increases in ICMRGlu in the entire right temporal lobe when subjects were monaurally presented factual stories, regardless of which ear was stimulated. This finding held whether the verbal input was presented in the subject's native language or one that he did not know. In contrast, Kushner et al.[26] presented non-English discourse to English speakers monaurally and found activation of the contralateral temporoparietal junction, inferior parietal region and insula, and in the corpus collosum. This nonmeaningful input did not activate frontal speech areas.

Mazziotta and colleagues[17] suggest that the location of the physiological response depends on the content of the auditory stimuli rather than the side stimulated. They found that monaural verbal stimuli resulted in diffuse activation in the left hemisphere and bilateral activation of the transverse and posterior temporal lobes and the thalamus. Nonverbal monaural stimuli (chords) produced bilateral temporoparaietal activation and frontotemporal asymmetries (right greater than left). These authors also proposed that the subject's cognitive strategy plays a role in the area of activation. Subjects who used stereotyped visual imagery or were musically sophisticated tended to show left greater than right asymmetries, while those who used less stereotyped visual imagery or were musically naive showed more right-sided activation. Further work is necessary to clarify this issue, but it is apparent that idiosyncratic processing of sensations can also play a role in the neurophysiological response.

3. Somatosensory Stimuli

Studies of somatosensory stimulation have demonstrated somatotopically represented responses in the postcentral sensory cortex as well as more variable responses in other brain areas. However, there has been relatively little investigation of the effect of stimulus intensity on the magnitude and location of the neurophysiological response. Even anticipation of somatosensory cues appears to alter the cerebral metabolic landscape. Roland[27] demonstrated that when subjects trained their attention to expect a light touch on the tip of the index finger, blood flow in the contralateral somatosensory hand area increased whether or not the stimulus was applied. One interpretation of these interesting data is that when attention is given to a portion of the body in anticipation of a somatosensory event, there is a preparatory tuning of the appropriate somatosensory cortex area.

Other studies have shown concurrent activation of additional cortical areas such as anterior frontal regions during somatosensory stimulation. The interpretation of some of these studies is problematic because in many cases the stimulus condition was compared only with a resting (i.e., no stimulation or task) state. In some studies, different subjects were studied and compared for stimulated vs. resting conditions. For example, in one study,[28] ICMRGlu for a group of subjects receiving mild electrical shocks on the skin of the forearm was compared with ICMRGlu of a separate group of subjects studied at rest. Metabolic asymmetry of postcentral cortex was seen in the stimulated subjects with the greater value on the side contralateral to the side stimulated. Increases anterior to posterior gradient due to relatively greater flow in the inferior frontal gyrus in the stimulated subjects was also reported. In such a paradigm, there is no control for the psychological concomitants of receiving sensory input during a PET or SPECT scan, and the latter finding may reflect emotional and psychological factors rather than the somatosensory stimulation itself.

Somatosensory stimulation studies have also highlighted the issue of tolerance to the physiological effects of activation. Raichle et al.[29] demonstrated that rCBF activations in somatosensory cortex contralateral to vibratory hand stimulation attenuate over time. This finding further illustrates the potential inconsistencies that may arise if time factors are not controlled. This issue becomes particularly important to consider if the scanning time is relatively long, as in ICMRGlu measurements.

A final study that reemphasizes the responsivity of the cerebrum to somatosensory stimuli was carried out by Frieberg et al.[30] using the nontomographic, intracarotid ^{133}Xe technique to measure rCBF during caloric stimulation of the vestibular system. They found a consistent focal activation in the area of the superior temporal gyrus contralateral to the stimulated side.

C. EFFECTS OF COGNITIVE STATE

Perhaps the first empirical observation linking rCBF and mental processes was made by Fulton in 1928.[31] He reported that the bruit over an arteriovenous malformation in a patient's occipital lobe increased when the patient was reading, but not when a light was shined in his eyes. More recently, the investigation of the neurobiological basis of human cognitive function, a line of inquiry for which no adequate animal model exists, has become one of the most challenging and fruitful applications of techniques like PET, SPECT, and nontomographic ^{133}Xe rCBF measurements. The unique capacity of these techniques to measure and localize "brain work" and their relative noninvasiveness make them ideal for such studies. Functional brain imaging studies have yielded a great deal of information about the neurobiology of cognition. A comprehensive review of these data is beyond the scope of this chapter, but a brief overview of the effects of cognition on the cerebral functional landscape will be presented in order to highlight those mental activities that must be controlled as much as possible during functional brain imaging studies. These data also serve to emphasize the need for careful experimental design in studies comparing patients and normal subjects.

That cognitive activities are reflected by measurable regional changes in cerebral blood flow and metabolism has been repeatedly demonstrated since these techniques were first applied to human research over 15 years ago. A variety of cognitive/physiological correlates have been reported. We have recently examined lateralization of function during 11 different cognitive conditions with the nontomographic ^{133}Xe inhalation rCBF method.[32] As can be seen in Figure 1, a wide range of values for relative left vs. right hemisphere activity was seen. On the whole, we found that those tasks that seem to require or to allow for internal verbalization resulted in the greatest activation of the left hemisphere compared to the right. Right hemisphere activation predominated only in the two tasks primarily involving attention and vigilance (the continuous performance tasks). These data demonstrate that the charac-

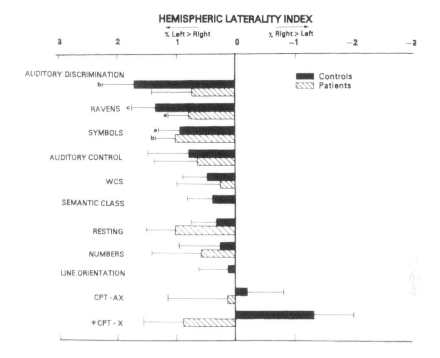

FIGURE 1. Relative right vs. left hemisphere gray cortical blood flow values for normal subjec' r ' patients with schizophrenia during 11 different cognitive conditions. See Reference 32 for s. iple sizes and descriptions of the cognitive conditions. Note that laterality values vary widely with different behavioral states (see text for further discussion). Hemispheric laterality index values were calculated as follows: *(left hemisphere − right hemisphere)* × *100*/whole brain mean CBF. Asterisk (*) = $p \leq 0.05$ for *between*-group comparison of laterality index; a, b, c, d = $p \leq 0.05$, 0.01, 0.005, 0.001, respectively, for *within*-group comparison of left vs. right hemisphere relative CBF.

teristics of the cognitive tasks performed by a subject and the nature of the cognitive operations involved have distinct ramifications for the cerebral functional landscape.

Recent innovations in cognitive psychology applied in conjunction with functional brain imaging techniques have allowed the neural systems subserving various elementary and complex cognitive operations to be explored. Since many radiotracers, such as [15]O water (for PET measurements of rCBF) and the radioxenons (for SPECT and nontomographic rCBF measurements) allow multiple studies to be carried out in the same individual in fairly rapid succession, functional measurements can be made under several different cognitive conditions. Such serial measurements allow each subject to be used as his or her own control to minimized the large interindividual variability in cerebral function that exists and also enables the metabolic concomitants of complex behavioral/cognitive activities to be investigated. One extremely fruitful approach that we have used to study neuropsychiatric patients[33] has been to design pairs of tasks, one of which involves the cognitive activity of interest and the other of which is a "control task" involving the same sensory, motor, and cognitive elements as the first with the exception of the specific cognitive operation of interest itself. Subtracting rCBF or ICMRGlu values for the control task from those for the first-mentioned condition serves to isolate the cerebral functional concomitants of performing the specific cognitive operation being investigated from those that are extraneous.[33] Several cognitive modalities have been explored in detail in this manner.

1. Language

An elegant extension of the approach described above is provided by the experiments of Petersen et al. carried out to investigate lexical processing.[34] They used PET to measure rCBF while subjects (1) visually fixated on a central point, (2) passively looked at single nouns, (3) passively read the noun aloud, and (4) generated a use (i.e., a verb) for every visually presented noun. rCBF for condition 1 was subtracted from condition 2, that for condition 2 was subtracted from condition 3, etc., thereby isolating rCBF changes associated with progressively more complex cognitive operations. A similar experiment was also carried out in which auditory stimuli were presented. They found that each level of the visually presented task resulted in activation of several areas in the occipital cortex, including the calcarine fissure in primary visual cortex bilaterally and in more lateral occipital areas. With auditory stimuli, these occipital activations were not observed, but rCBF increases occurred in primary auditory cortex and a region in the left temporoparietal cortex. In addition to the aforementioned areas, the left anterior frontal lobe was activated during the more complex verb generation condition (with both auditory and visual stimuli). This area was also activated when subjects were asked to monitor a list of visually presented words for a particular semantic category, suggesting that it may be involved in semantic processing.[35]

These investigators also examined the effects of phonological processing (determining whether words pairs rhymed). Activation near the supramarginal gyrus was observed.[35] Taken together, these data highlight the power of functional brain imaging methods to elucidate the neurobiology of complex cognitive processes like language. They also reemphasize the need to control language-related variables during rCBF and metabolic measurements.

2. Attention

While attention is as difficult to measure and to parcel out from other cognitive operations as it is to define, a number of functional brain imaging studies with bearing on this topic have been carried out. Several, including our own work (see Figure 1 as discussed above and in Reference 32), have demonstrated a role for the right hemisphere, particularly the right parietal lobe, in attentional processes. Studies of humans with right hemisphere lesions[36] support this notion. Other areas have also been implicated. LaBerge and Buchsbaum[37] measured ICMRGlu during presentation of an attention-demanding visual identification task to one hemifield and a nonattentional task to the other visual field. Their data suggest that the pulvinar is involved in visual identification demanding attentional filtering.

Still other studies implicate the frontal lobe in attention. However, consensus has not been reached as to the exact locale within the frontal lobe. In a comparison of low-vigilance and high-vigilance conditions of a word-monitoring paradigm, Posner et al.[35] note increased rCBF in supplementary motor cortex when attention is focussed with the intent of action to follow, but find that when the intentional component is not present, the anterior cingulate gyrus is active. These investigators postulate an anterior and a posterior attentional system which interact. Roland[38] found that when subjects were given a particular task and were then presented simultaneously with a potent but irrelevant distracting stimulus, a metabolic increase occurred in the middle portion of the superior prefrontal cortex. He interpreted this observation as evidence for the involvement of this area in attention. An additional interpretation, however, may be that the role of the prefrontal cortex in this situation was to screen out the interference or that it was invoking short-term memory to keep the relevant stimulus "on-line" during the time interval imposed by the distractor.

In another set of studies, Roland[27] demonstrated that when a subject turns his or her attention toward an area of the body with the expectation of receiving sensory input, the metabolism increased in the cortical area that would respond to the somatosensory stimulation. The interpretation of this complex body of data may best be accomplished by a systems analysis approach, i.e., an analysis of the network of brain structures that become active during attentional processes rather that a search for a unitary locus of attention.

Regardless of the definition of attention used or the exact cerebral locale involved, the data reviewed above demonstrate that the attentional state of the subject must also be considered in the design and execution of PET and SPECT studies and in the interpretation of the data. Not only the intensity of the attentional process, but how and where it is focussed can all be reflected in the cerebral metabolic landscape. One approach to minimizing the variability in rCBF or ICMRGlu that may ensue if the subject is allowed to focus attention randomly on the comfort or discomfort of this or that body part, on the various auditory, visual, and tactile sensations attendant upon the scanning procedure, on anxieties about the procedure, on the plan for tomorrow's activities, etc., is to prescribe a well-controlled cognitive, sensory, or motor activity upon which the subject may focus attention during the study.

3. Abstract Reasoning and Problem Solving

Perhaps the most difficult of all human phenomena to study are those cognitive behaviors involved in abstract reasoning and problem solving. Most such behaviors involve a variety of cognitive operations including attention, language, classification, etc., which must be teased apart, and most require varying amounts of visual, auditory, tactile, and motor activity on the part of the subject which must be controlled for. Therefore, the strategy of using multiple measurements in the same individual assumes a position of great importance in the design of such studies. A number of investigations, particularly early ^{133}Xe studies, were constructed such that rCBF during a higher cognitive task was compared with that during a resting state baseline. These paradigms fail to parcel out extraneous sensory, motor, and cognitive factors, and they fail to control for the psychological effects of have rCBF or ICMRGlu measured while performing a cognitive task.

As part of an on-going investigation of cerebral functional reserve in neuropsychiatric patients, we have examined the rCBF response of normal subjects during the performance of two particularly interesting abstract reasoning/problem-solving tasks and compared the results with those obtained while the same subjects performed matched sensorimotor control tasks.[33,39-41] One cognitive task, the Wisconsin Card Sorting Test (WCS), is a neuropsychological test that involves the use of feedback and working memory in the formation of categorical, conceptual sets. It also tests for mental flexibility and perseveration by requiring the subject to switch these conceptual sets when the feedback indicates it is appropriate to do so. Patients with known frontal lobe lesions perform poorly on this test,[42-44] and on the basis of this observation, it has been used as a sensitive indicator of the functional integrity of the frontal lobes in man. Using ^{133}Xe inhalation, we compared rCBF while subjects were performing the WCS to that when they were performing a simple sensorimotor control task, a number-matching task that was identical to the WCS in terms of the visual stimulation and eye scanning involved, the minimal finger movement necessary to indicate a response, and the mode of feedback.[33,38,40] We found a significant rCBF increase localized to the left and right dorsolateral prefrontal cortex (Plate 1A*). These data confirm in healthy individuals the notion, developed on the basis of observations of patients with frontal lobe lesions, that the WCS is a "frontal lobe task".

Another abstract reasoning, problem-solving task that we have studied is Raven's Progressive Matrices (RPM), a nonverbal test that is a well-accepted indicator of general intelligence. Subjects with post-Rolandic lesions are impaired on this test, suggesting that posterior cortical areas are critical for the cognitive functions necessary to perform it. Unlike the WCS, there is no evidence that patients with prefrontal lesions have particular difficulty with it. We found that normal subjects activate posterior cortical areas (parieto-occipital, parietal, and temporal) above levels during a baseline matching, sensorimotor control task

---

* Plate 1A follows page 198.

(Plate 1B*).[41] In contrast to the WCS, no activation of frontal cortical areas above baseline occurred.

These divergent findings in normal subjects during two different abstract reasoning tasks have implications for the role of the prefrontal cortex in normal higher cognitive function; they suggest that not all tests that require problem-solving and complex higher-order cognitive processing are "frontal" tasks. This notion is consistent with the perspective of many current cognitively based theories. Fuster[45] has emphasized the role of dorsolateral prefrontal cortex in the cross-temporal integration of information (or spanning time), while Goldman-Rakic[46] suggests that the dorsolateral prefrontal cortex functions to form "internal representations of situations or stimuli", supporting the "ability to respond to situations on the basis of stored information, rather than on the basis of immediate stimulation". Similarly, Ingvar[47] proposed a primary role for this brain area in "memory of the future", or "the production of serial action plans . . . used as templates with which input is compared." The WCS and RPM differ in several critical aspects that, in light of these cognitive theories, may explain their differential effects on prefrontal cortical physiology.[41] For example, determining the proper category by which to sort a particular WCS stimulus requires that the feedback given on previous trials be considered. This may be conceptualized as the use of working memory to span time and make serial judgments. In contrast to the WCS, no previous experience is necessary to complete each RPM trial, and since all the required information is available to the subject throughout the trial, no working memory or "internal representations" are involved. Carrying out prefrontally specific cognitive operations, such as those mentioned above, may result in physiological activation of prefrontal cortex above the ambient, tonal levels necessary to maintain wakefulness and routine neuronal homeostasis. It appears that successful performance of the prefrontally related WCS requires and results in an augmented prefrontal physiological response, while the RPM does not.

These data underscore the potential power of functional brain imaging to delineate the neural substrate of even very complex human behaviors. Examination of the normal data depicted in Plate 1* also make clear that what and how a subject is thinking can dramatically change the cerebral metabolic landscape. Such changes become most apparent when baseline measures of nonspecific blood flow (i.e., the sensorimotor control task) are subtracted out.

D. EFFECTS OF EMOTIONAL STATE

Several studies have demonstrated that even the mental and emotional state of the subject can have significant effects on rCBF measurements. While most of the literature in this area consists of single, unreplicated reports, several intriguing findings have been reported. Perhaps the first such report was that of Kety,[48] who in 1950 noted a 36% increase in cerebral oxygen consumption in a subject during "grave apprehensiveness". Gur et al.[49] found that during ^{133}Xe inhalation (a relatively lower anxiety situation) subjects with little anxiety exhibited a linear increased in CBF in anxiety, while CBF of highly anxious subjects decreased with increased anxiety; ICMRGlu measured with PET (a relatively more anxiety-provoking situation) manifested a linear decrease in metabolic rate with higher anxiety levels. They postulated a U-shaped relationship between subjects' anxiety levels and cerebral activity. Reiman et al.[50] found that when normal subjects were told that they would be receiving a powerful electric shock during a PET scan, they exhibited increased rCBF in the temporal poles, suggesting a locus for anticipatory anxiety.

Emotional states (as differentiated from clinically significant neuropsychiatric illnesses such as affective disorders) other than anxiety have also been shown to play a role. Mathew et al.[51] reported a relationship between CBF and personality traits. Subjects with high scores

---

*   Plate 1 follows page 198.

on a test of extroversion had lower CBF, suggesting to these authors that extroverted behavior might represent an attempt to compensate for intrinsically low levels of cortical arousal. Using ^{133}Xe inhalation, we found that trained psychodramatists who mentally simulated profound anxiety increased rCBF some 20% over their own resting baseline values. When they simulated a sad, depressed state, their rCBF values decreased.[52] Koyabashi et al.[53] found an inverse relationship between mean CBF and a self-rating scale for "satisfaction with life" (low satisfaction was interpreted as "depression"). However, the literature pertaining to cerebral function in affective disorders per se remains controversial.

While more work needs to be done in this area, one implication of these data is that uncontrolled emotional and psychological factors can add variability to PET or SPECT data or even contaminate a study with false findings that are related to epiphenomena rather than the process or disease being studied. To minimize emotional/psychological variables related to the anxiety of the scanning procedure, subjects should be well acclimatized to the apparatus and the testing conditions. Another approach is to perform multiple studies on each subject so that each can be used as his or her own control as described above. In this manner, the neurophysiological effects on the subject's current emotional state can be subtracted from the study of interest.

E. DEMOGRAPHIC VARIABLES AND PERIPHERAL PHYSIOLOGICAL FACTORS

A number of group variables must be considered in functional brain imaging studies. First, several investigators have reported differences in rCBF and ICMRGlu between men and women,[54,55] suggesting that this variable be taken into account in data analysis and interpretation and necessitating that various experimental populations be matched for sex. Second, although the effect of cerebral dominance on brain functional organization is not clear, differences in rCBF have been reported between left and right handers.[54,56] Finally, aging also appears to affect CBF and metabolism.[57,58] Although some have argued that the observed decreases in various parameters of brain function with age are due to cerebrovascular risk factors rather than to aging per se, the majority of studies confirm the basic finding. Local changes with aging, particularly a diminution of frontal activity,[59] have also been reported.

The most important peripheral physiological determinant of CBF is arterial pCO_2. A solid body of data confirms that CBF varies directly with acute or extreme changes in arterial pCO_2.[60-63] Therefore, changes in CBF may represent an artifactual result of changes in ventilatory function such as hyperventilation accompanying anxiety. The importance of chronic minor changes within the normal range is less clear. In fact, alterations in CBF accompanying acute alterations of pCO_2 have been shown to reverse when the pCO_2 change is chronically maintained.[64] Nonetheless, pCO_2 levels must be monitored to ensure that they remain in the physiological range and to determine whether observed between-group differences in CBF can be accounted for on the basis of ventilatory differences. Several approaches to correcting CBF for changes in pCO_2 have been suggested;[60-63] however, no consensus has been reached as to the proper magnitude of the correction factor. Since pCO_2-related changes in CBF appear to occur on a global rather than regional basis, the use of relative rCBF values normalized to the global mean value represents one method of obviating this problem.

III. STUDIES OF BRAIN FUNCTION IN NEUROPSYCHIATRIC PATIENTS

Those physiological, demographic, cognitive, emotional, and other factors discussed above that can affect measures of brain function become particularly crucial to consider in studies of patient populations. This is essential in order to reduce the "noise" of the physiological measurements in neuropsychiatric disorders in which the pathophysiological findings may be relatively subtle. Failure to control these variables has hampered many attempts to delineate a functional neurobiology of neuropsychiatric disorders. In this section, we will examine studies of rCBF in schizophrenia as an important case in point.

That there is a disorder of the frontal lobes in schizophrenia is a theory that has been in and out of favor virtually since Kraepelin delineated the illness (then called "dementia praecox") as a distinct clinical entity. Some 15 years ago, a series of landmark studies by Ingvar and Franzen[65,66] resulted in a resurgence of interest in the role of the frontal lobes in schizophrenia. Using the intracarotid ^{133}Xe technique to measure rCBF of the lateral aspect of the cortex, they found that while normal subjects exhibited relatively more blood flow to frontal areas of the brain than to other areas visible with this technique (a pattern that they called normal "hyperfrontality"), their patients with schizophrenia failed to show this pattern (i.e., they were relatively "hypofrontal").

This intriguing finding not only revitalized interest in the role of the frontal lobes in schizophrenia, but also heralded the era of functional brain imaging techniques as applied to neuropsychiatry research. However, in the decade following the publication of this work, a number of rCBF and PET studies appeared in the literature, and the results were extremely inconsistent: while some investigators confirmed Ingvar and Franzen's observation of hypofrontality, others did not (for reviews see References 67 and 68). There may be a number of reasons for these inconsistencies, but one of the most important probably relates to a failure on the part of many investigators to control and consider how the brain was engaged when studied. The need for such experimental control to reduce measurement "noise" was not fully appreciated by early investigators in this field. In fact, many of the first rCBF and PET studies of schizophrenia were carried out while subjects were "at rest".

An alternative approach that we have employed is to design "cortical stress tests"[69] in which rCBF is measured while subjects receive controlled sensory input and carry out specified cognitive tasks that reliably activate specific cortical areas in normal subjects and to compare the results to those during a matched sensorimotor control task. This approach can be used to test the ability of patients' brains to physiologically respond to various cognitive or other demands, and it has a number of advantages: measurement "noise" is reduced by controlling inputs and outputs; idiosyncratic responses to the testing procedure and interindividual variability are minimized by using each subject as his or her own control; and, by challenging the brain, subtle abnormalities may be uncovered.

We have measured rCBF using the ^{133}Xe inhalation method during nine different cognitive states in patients with schizophrenia and compared the results with those for age- and sex-matched normal subjects. These conditions included a resting state,[33,39] the two abstract reasoning/problem-solving tasks described above (the WCS and RPM) and their sensorimotor control tasks,[3,39-41] two attentional tasks,[39] and two auditory tasks.[70] We found that while between-group differences in prefrontal blood flow were inconsistent during resting and nonexistent during other tasks that don't activate prefrontal cortex in normals (see Plate 1A*), robust differences could be reliably demonstrated during the condition that in normal subjects activates prefrontal cortex, i.e., while performing the WCS (see Plate 1B*). Even when each subject was used as his or her own control, patients showed a striking failure to activate prefrontal cortex during the WCS.

Had we studied only resting rCBF or rCBF during less relevant behaviors and not included a "prefrontal stress test" in our experimental design, we might have concluded, as other investigators have, that there is no abnormality of prefrontal lobe function in schizophrenia.

---

* Plate 1 follows page 198.

Quite the contrary, we have replicated this finding of a behavior- and region-specific dysfunction of prefrontal cortex in a total of four cohorts of schizophrenic patients, two who were on neuroleptics[39,71] and two who were medication-free.[33,40] A consistent pattern of findings has emerged from these studies and others using PET (see Reference 68 for review). When patients are studied during rest or other nonspecific conditions, they may or may not appear hypofrontal, but virtually all studies carried out during prefrontal stimulation have demonstrated lower prefrontal blood flow or metabolism in schizophrenia. A recent study[71] of monozygotic twins who were discordant for schizophrenia dramatically reconfirmed the importance of testing conditions in the resulting data. We found that hypofrontality could not be reliably demonstrated in the ill twin during resting or simple numbers matching; however, when rCBF was measured during the WCS, the ill twin could be demonstrated to be more hypofrontal than his or her well co-twin in ten out of ten twin pairs.

The preceding discussion demonstrates how the sensitivity of functional brain imaging measurements to behavioral conditions can be utilized to uncover subtle aberrations in neural function. This facet of these techniques can also be exploited to uncover the mechanism of the pathophysiology and the putative site of the underlying primary pathology. Reduced activity of any brain area can arise either from intrinsic pathology of the area that interferes with the function of the intrinsic neural systems or from distant pathology in afferent or efferent circuits that impact on the functional output of the area. By measuring brain function under multiple conditions and considering the findings during each, it can be deduced which of these possibilities operates to produce the pathophysiology. In the face of intrinsic pathology, hypofunction is likely to be observed during all testing conditions. In schizophrenia the observation that prefrontal physiological dysfunction is variable and most apparent when there is need for prefrontally mediated behavior (such as in the WCS) is most consistent with the model of pathology extrinsic to the prefrontal cortex.

The concept of behavior control and stressing the brain during *in vivo* measurements also has important implications for uncovering potentially important correlates of pathophysiology that shed further light on the underlying mechanism. We found that prefrontal rCBF correlated with relevant measures of neurochemical status (cerebrospinal fluid levels of the dopamine metabolite HVA)[40] and of structural pathology (enlargement of the lateral ventricles on CT)[72] when rCBF was determined during the prefrontally specific WCS. These relationships could not be discerned during the nonspecific numbers matching condition, further emphasizing the critical role of behavior in cerebral metabolism and blood flow.

In the future, multiple measurements in the same individual of different functional parameters, such as rCBF, oxidative metabolism, and metabolic rate for glucose, under different behavioral and/or pharmacological conditions, along with *in vivo* local measurements of neurochemistry, such as receptor density and enzyme function, when taken together may help to further dissect the complex set of physiological phenomena that comprise a single rCBF or ICMRGlu measurement. Such detailed information may help to definitively delineate the etiology of subtle illnesses such as schizophrenia and may even suggest novel treatment interventions. For the present, it is clear that vigorous efforts to control the multiplicity of factors that can affect measurements of brain function are necessary in order for meaningful results to emerge.

REFERENCES

1. **Roy, C. S. and Sherrington, C. S.,** On the regulation of the blood supply of the brain, *J. Physiol. (London),* 11, 85, 1896.
2. **Barcroft, J.,** *The Respiratory Function of the Blood,* Cambridge University Press, London, 1914.

3. **Schmidt, C. and Hendrix, J.,** Action of chemical substances on cerebral blood vessels, *Res. Publ. Assoc. Nervous Mental Dis.,* 18, 229, 1938.

4. **Kety, S. S.,** Basic principles for the quantitative estimation of region cerebral blood flow, in *Brain Imaging and Brain Function,* Sokoloff, L., Ed., Raven Press, New York, 1985, 1.

5. **Raichle, M. E., Grubb, R. L., Gado, M. H., Eichling, J. O., and Ter-Pogossian, M. M.,** Correlation between regional cerebral blood flow and oxidative metabolism, *Arch. Neurol.,* 33, 523, 1976.

6. **Siesjo, B. K.,** Cerebral circulation and metabolism, *J. Neurosurg.,* 60, 883, 1984.

7. **Kety, S. S. and Schmidt, C. F.,** The determination of cerebral blood flow in man by use of nitrous oxide in low concentrations, *Am. J. Physiol.,* 143, 53, 1945.

8. **Ingvar, D. H. and Lassen, N. A.,** Quantitative determination of regional cerebral blood flow in man, *Lancet,* 11, 806, 1961.

9. **Lassen, N. A.,** Measurement of regional cerebral blood flow in humans with single-photon-emitting radioisotopes, in *Brain Imaging and Brain Function,* Sokoloff, L., Ed., Raven Press, New York, 1985, 9.

10. **Olesen, J.,** Contralateral focal increase of cerebral blood flow in man during arm work, *Brain,* 94, 635, 1971.

11. **Halsey, J. H., Blaustein, V. W., Wilson, E. M., and Wills, E. H.,** Regional cerebral blood flow comparison of right and left hand movement, *Neurology,* 29, 21, 1979.

12. **Roland, P. E., Skinhoj, E., Lassen, N. A., and Larsen, B.,** Different cortical areas in man in organization of voluntary movements in extrapersonal space, *J. Physiol.,* 43, 137, 1980.

13. **Ingvar, D. H. and Phillipson, L.,** Distribution of cerebral blood flow in the dominant hemisphere during motor ideation and motor performance, *Ann. Neurol.,* 2, 230, 1977.

14. **Fox, P. T., Fox, J. M., Raichle, M. E., and Burde, R. M.,** the role of cerebral cortex in the generation of voluntary saccades: a positron emission tomography study, *J. Neurophysiol.,* 54, 348, 1985.

15. **Roland, P. E., Larsen, B., Lassen, N. A., and Skinhoj, E.,** Supplementary motor area and other cortical areas in organization of voluntary movements in man, *J. Neurophysiol.,* 43, 118, 1980.

16. **Raichle, M. E.,** Circulatory and metabolic correlates of brain function in normal humans, in *Handbook of Physiology: The Nervous System, Higher Function of the Brain,* Mountcastle, V. B. and Plum, F.,Eds., American Physiological Society, Washington, D.C., 1987, 643.

17. **Mazziotta, J. C. and Phelps, M. E.,** Positron emission tomography studies of the brain, in *Positron Emission Tomography and Autoradiography: Principles and Applications,* Phelps, M. E., Mazziotta, J. C., and Schelbert, H., Eds., Raven Press, New York, 1986, 493.

18. **Mazziotta, J. C., Phelps, M. E., Carson, R. E., and Kuhl, D. E.,** Tomographic mapping of human cerebral metabolism: sensory deprivation, *Ann. Neurol.,* 12, 435, 1982.

19. **Reivich, M., Gur, R., and Alavi, A.,** Positron emission tomography studies of sensory stimuli, cognitive processes and anxiety, *Hum. Neurobiol.,* 2, 25, 1983.

20. **Risberg, J. and Prohovnik, I.,** Cortical processing of visual and tactile stimuli studied by non-invasive rCBF measurements, *Hum. Neurobiol.,* 2, 5, 1983.

21. **Fox, P. T. and Raichle, M. E.,** Stimulus rate dependence of regional cerebral blood flow in human striate cortex demonstrated by positron emission tomography, *J. Neurophysiol.,* 51, 1109, 1984.

22. **Fox, P. T., Mintun, M. A., Raichle, M. E., Miezin, F. M., Allman, J. M., and Van Essen, D. C.,** Mapping human visual cortex with positron emission tomography, *Nature,* 323, 806, 1986.

23. **Lueck, C. J., Zeki, S., Friston, K. J., Deiber, M. P., Cope, P., Cunningham, V. J., Lammertsma, A. A., Kennard, C., and Frackowiac, R. S. J.,** The color centre in the cerebral cortex of man, *Nature,* 340, 386, 1989.

24. **Lauter, J. L., Herscovitch, P., Formby, C., and Raichle, M. E.,** Tonotopic organization in human auditory cortex revealed by positron emission tomography, *Hearing Res.,* 20, 199, 1985.

25. **Reivich, M., Greenberg, J., and Alavi, A.,** The use of the [18F]-fluorodeoxyglucose method for mapping functional neural pathways in man, *Acta Neurol. Scand.,* 60(S72), 198, 1979.

26. **Kushner, M. J., Schwartz, R., Alavi, A., Dann, R., Rosen, M., Silver, F., and Reivich, M.,** Cerebral glucose consumption following verbal auditory stimulation, *Brain Res.,* 409, 79, 1987.

27. **Roland, P. E.,** Somatotopic tuning of postcentral gyrus during focal attention in man. A regional cerebral blood flow study, *J. Neurophysiol.,* 46, 744, 1981.

28. **Buchsbaum, M. S., Holcomb, H. H., Johnson, J., King, A. C., and Kessler, R.,** Cerebral metabolic consequences of electrical cutaneous stimulation in normal individuals, *Hum. Neurobiol.,* 3, 35, 1983.

29. **Raichle, M. E., Fox, P. T., and Mintun, M. A.,** Cerebral blood flow and metabolism are uncoupled during somatosensory stimulation in humans, *Soc. Neurosci. Abstr.,* 13, 812, 1987.

30. **Frieberg, L., Olsen, T. S., Roland, P. E., Paulson, O. B., and Lassen, N. A.,** Focal increase of blood flow in the cerebral cirtex of man during vestibular stimulation, *Brain,* 108, 609, 1985.

31. **Fulton, J. F.,** Observations upon the vascularity of the human occipital lobe during visual activity, *Brain,* 51, 310, 1928.

32. **Berman, K. F. and Weinberger, D. R.,** Lateralisation of cortical function during cognitive tasks: regional cerebral blood flow studies of normal individuals and patients with schizophrenia, *J. Neurol. Neurosurg. Psychiatry,* 53, 150, 1990.

33. **Weinberger, D. R., Berman, K. F., and Zec, R. F.,** Physiological dysfunction of dorsolateral prefrontal cortex in schizophrenia. I. Regional cerebral blood flow (rCBF) evidence, *Arch. Gen. Psychiatry,* 43, 114, 1986.

34. **Petersen, S. E., Fox, P. T., Posner, M. I., Mintun, M., and Raichle, M. E.,** Positron emission tomographic studies of the cortical anatomy of single-word processing, *Nature,* 331, 585, 1988.

35. **Posner, M. I., Petersen, S. E., Fox, P. T., and Raichle, M. E.,** Localization of cognitive operations in the human brain, *Science,* 240, 1627, 1988.

36. **Heilman, K. M. and van den Abell, T.,** Right hemisphere dominance for attention: the mechanism underlying hemispheric asymmetries of inattention (neglect), *Neurology,* 30, 327, 1980.

37. **LaBerge, D. and Buchsbaum, M. S.,** Pulvinar activity in humans during a visual attention task as measured by PET images, *Abstr. Soc. Neurosci.,* 15, 481, 1989.

38. **Roland, P. E.,** Cortical organization of voluntary behavior in man, *Hum. Neurobiol.,* 4, 155, 1985.

39. **Berman, K. F., Zec, R. F., and Weinberger, D. R.,** Physiological dysfunction of dorsolateral prefrontal cortex in schizophrenia. II. Role of neuroleptic treatment, attention, and mental effort, *Arch. Gen. Psychiatry,* 43, 126, 1986.

40. **Weinberger, D. R., Berman, K. F., and Illowsky, B. P.,** Physiological dysfunction of dorsolateral prefrontal cortex in schizophrenia. III. A new cohort and evidence for a monoaminergic mechanism, *Arch. Gen. Psychiatry,* 45, 609, 1988.

41. **Berman, K. F., Illowsky, B. P., and Weinberger, D. R.,** Physiological dysfunction of dorsolateral prefrontal cortex in schizophrenia. IV. Further evidence for regional and behavioral specificity, *Arch. Gen. Psychiatry,* 45, 616, 1988.

42. **Milner, B.,** Some effects of frontal lobectomy in man, in *The Frontal Granular Cortex and Behavior,* Warren, J. M. and Akert, K., Eds., McGraw-Hill, New York, 1963, 313.

43. **Milner, B.,** Effects of different brain lesions on card sorting, *Arch. Neurol.,* 9, 100, 1963.

44. **Milner, B.,** Interhemispheric differences in the localization of psychological processes in man, *Br. Med. Bull.,* 27, 272, 1971.

45. **Fuster, J.,** *The Prefrontal Cortex,* Raven Press, New York, 1980.

46. **Goldman-Rakic, P.,** Circuitry of primate prefrontal cortex and regulation of behavior by representational knowledge, in *Handbook of Physiology: The Nervous System, Higher Functions of the Brain,* Plum, F. and Mountcastle, V., Eds., American Physiological Society, Washington, D.C., 1987, 373.

47. **Ingvar, D. H.,** ''Memory of the future'': an essay on the temporal organization of conscious awareness, *Hum. Neurobiol.,* 4, 127, 1985.

48. **Kety, S. S.,** Circulation and metabolism of the human brain in health and disease, *Am. J. Med.,* 8, 205, 1950.

49. **Gur, R. C., Gur, R. E., Resnick, S. M., Skolnick, B. E., Alavi, A., and Reivich, M. E.,** The effect of anxiety on cortical cerebral blood flow and metabolism, *J. Cereb. Blood Flow Metab.,* 7, 173, 1987.

50. **Reiman, E. M., Fusselman, M. J., Fox, P. T., and Raichle, M. E.,** Neuroanatomic correlates of anticipatory anxiety, *Science,* 243, 1071, 1989.

51. **Mathew, R. J., Weinman, M. L., and Barr, D. L.,** Personality and regional cerebral blood flow, *Br. J. Psychiatry,* 144, 529, 1984 (abstract).

52. **Weinberger, D. R., Berman, K. F., Zec, R. F., and Iadarola, M. K.,** Effect of simulated emotional state on cortical blood flow, in *Proc. 39th Annu. Meeting of the Society of Biological Psychiatry,* 1984 (abstract).

53. **Koyabashi, S., Yamaguchi, S., Katsube, T., Arimoto, S., Murata, A., Yamashita, K., and Tsunematsu, T.,** Self-rating depression scales correlated with regional cerebral blood flow in normal volunteers, *Eur. Neurol.,* 26, 199, 1987.

54. **Gur, R. C., Gur, R. E., Obrist, W. K., Hungerbuhler, J. P., Younkin, D., Rosen, A. D., Skolnick, B. E., and Reivich, M.,** Sex and handedness differences in regional cerebral blood flow during rest and cognitive activity, *Science,* 217, 659, 1982.

55. **Shaw, T. and Meyer, J.,** Aging and cerebrovascular disease, in *Diagnosis and Management of Stroke,* Meyer, J. S. and Shaw, T., Eds., Addison-Wesley, Menlo Park, CA, 1982, 1.

56. **Prohovnik, I., Hakansson, K., and Risberg, J.,** Observations on the functional significance of regional cerebral blood flow in resting normal subjects, *Neuropsychologia,* 18, 203, 1980.

57. **Naritomi, H., Meyer, J. S., Sakai, F., Yamaguchi, F., and Shaw, T.,** Effects of advancing age on regional cerebral blood flow, *Arch. Neurol.,* 36, 410, 1979.

58. **Shaw, T. G., Mortel, K. F., Meyer, J. S., Rogers, R. L., Hardenberg, J., and Cutaia, M. M.,** Cerebral blood flow changes in benign aging and cerebrovascular disease, *Neurology,* 34, 855, 1984.

59. **Mamo, H., Meric, P., Luft, A., and Seylaz, A.,** Hyperfrontal pattern of human cerebral circulation, *Arch. Neurol.,* 40, 626, 1983.

60. **Yamamato, M., Meyer, J. S., Sakai, F., and Yamaguchi, F.,** Aging and cerebral vasodilator responses to hypercarbia: responses in normal aging and in persons with risk factors for stroke, *Arch. Neurol.,* 37, 489, 1980.

61. **Maximillian, V. A., Prohovnik, I., and Risberg, J.,** Cerebral hemodynamic responses to mental activation in normo and hypercapnia, *Stroke,* 11, 342, 1980.

62. **Davis, S. M., Ackerman, R. H., Correia, J. A., Alpert, N. M., Chang, J., Buonanno, F., Kelley, R. E., and Rosner, B.,** Cerebral blood flow and cerebrovascular CO_2 reactivity in stroke-age normal controls, *Neurology,* 33, 391, 1983.

63. **Tominaga, S., Strandgaard, S., Uemura, K., Ito, K., and Kutsuzawa, T.,** Cerebrovascular CO_2 reactivity in normotensive and hypertensive man, *Stroke,* 7, 507, 1976.

64. **Evans, M. C. and Cameron, I. R.,** Adaption of rCBF during chronic exposure to hypercapnia and to hypercapnia with hypoxia, *J. Cereb. Blood Flow Metab.,* 1(1), 435, 1981.

65. **Ingvar, D. H. and Franzen, G.,** Abnormalities of cerebral blood flow distribution in patients with chronic schizophrenia, *Acta Psychiatr. Scand.,* 50, 425, 1974.

66. **Ingvar, D. H. and Franzen, G.,** Distribution of cerebral activity in chronic schizophrenia, *Lancet,* ii, 1484, 1974.

67. **Berman, K. F. and Weinberger, D. R.,** Cerebral blood flow studies in schizophrenia, in *The Neurology of Schizophrenia,* Nasrallah, H. and Weinberger, R., Eds., Elsevier/North-Holland, Amsterdam, 1986, 277.

68. **Weinberger, D. R. and Berman, K. F.,** Speculation on the meaning of metabolic hypofrontality in schizophrenia, *Schizophrenia Bull.,* 14, 157, 1988.

69. **Berman, K.,** Cortical "stress tests" in schizophrenia: regional cerebral blood flow studies., *Biol. Psychiatry,* 22, 1304, 1987.

70. **Berman, K. F., Rosenbaum, S. C. W., Brasher, C. A., Goldberg, T. E., and Weinberger, D. R.,** Regional cerebral blood flow during auditory discrimination in schizophrenia, *Abstr. Soc. Neurosci.,* 1987.

71. **Berman, K. F., Torrey, E. F., Daniel, D. G., and Weinberger, D. R.,** Prefrontal cortical blood flow in momozygotic twins concordant and discordant for schizophrenia, *Schizophrenia Res.,* 2, 129, 1989.

72. **Berman, K. F., Weinberger, D. R., Shelton, R. C., and Zec, R. F.,** A relationship between anatomical and physiological brain pathology in schizophrenia: lateral cerebral ventricular size predicts cortical blood flow, *Am. J. Psychiatry,* 144, 1277, 1987.

INDEX

Printed in the United States
by Baker & Taylor Publisher Services